The Chemistry of Antitumour Agents

edited by

Derry E. V. Wilman
Drug Development Section,
Institute of Cancer Research UK

Blackie
Glasgow and London

Published in the USA by
Chapman and Hall
New York

Blackie & Son Limited,
Bishopbriggs, Glasgow G64 2NZ
and
7 Leicester Place, London WC2H 7BP

Published in the USA by
Chapman and Hall
a division of Routledge, Chapman and Hall, Inc.
29 West 35th Street, New York, NY 10001-2291

British Library Cataloguing in Publication Data

The Chemistry of Antitumour Agents.
1. Man. Cancer. Drug therapy
I. Wilman, D.E.V.
616.99'4061

ISBN-13: 978-94-010-6665-5   e-ISBN-13: 978-94-009-0397-5
DOI: 10.1007/978-94-009-0397-5
Library of Congress Cataloging-in-Publication Data

The Chemistry of antitumour agents / D.E.V. Wilman [editor].
    p. cm.
Includes bibliographies and indexes.
ISBN-13: 978-94-010-6665-5
    1. Antineoplastic agents—Structure-activity relationships.
2. Antineoplastic agents—Synthesis.   I. Wilman, D.E.V.
RC271.C5C38   1989
6.16.99'4061—dc20
                                                    89-15754
                                                    CIP

Phototypesetting by Thomson Press (India) Limited New Delhi
Printed in Great Britain by
Thomson Litho Ltd, East Kilbride, Scotland

Foreword

Walter C. J. ROSS
Emeritus Professor, University of London

To paraphrase a statement made by Howard E. Skipper many years ago, 'We cancer chemotherapists have often exploited and overworked our chemist colleagues and they have been conveniently forgotten at award giving times'.

This book is an attempt to rectify this and highlight the contribution of the chemist in modifying the structure of various types of agent to enhance their effectiveness as inhibitors of the growth of neoplastic tissues.

Cancer chemotherapy is a relatively new discipline, coming later than the introduction of sulphonamides and antibiotics. Modern anti-cancer therapy started with the report of the use of a war gas methyl-di-(2-chloroethyl)amine (HN2) in 1946 for the treatment of Hodgkin's disease. The recognition that this compound acted as a bifunctional alkylating agent under physiological conditions led to the synthesis of a wide range of drugs with similar properties. Amongst these were chlorambucil, melphalan, busulphan, and cyclophosphamide which still find use today.

Somewhat later, a range of antibiotics was found to be effective, for example aminopterin (1948) and 6-mercaptopurine (1958) to treat acute leukaemias and 5-fluorouracil and 6-azauracil (1957-8) which were used against a variety of cancers.

Since these early days the net has been cast ever wider and, as well as ingenious modifications of the compounds mentioned above, anticancer drugs now include growing classes of compounds ranging from purely synthetic agents to natural products. Many of these are discussed in the present book. It will be seen that structural modification of all these categories of compound is engaging the attention of chemists.

One of the features of cancer cells that has bedevilled the search for a 'cure' for a given type of tumour has been the lack of specificity—in most instances some normal cell type has features common to the neoplastic cell under attack. As long ago as 1959, Harper wrote 'The development of comparative biochemistry of various tissues and cells has lagged behind the ingenuity of the synthetic chemist, so that although in many cases the gun could be loaded with suitable ammunition we lack the information concerning the target.' Later, in 1969, Haddow remarked 'It is unlikely that the future of cancer chemotherapy rests with these (relatively) small molecules.'

A more recent development promises to overcome these objections. This is the attachment of highly toxic molecules to monoclonal antibodies which are specific in their affinity for antigens on the surface of certain cancer cells. This approach will stretch the ingenuity of the chemist in the selection and conjugation of appropriate 'warheads' and carriers to yield the so-called immunotoxins.

Preface

Cancer as a disease has been known from early times and over the centuries the many attempts to treat it have included herbal remedies and arsenic. Today the cancer patient's treatment involves surgeon, radiotherapist and clinician, with the oncology specialist having an ever increasing part to play. Cancer chemotherapy can be said to have begun some forty years ago, with the first achievement of remissions in childhood acute leukaemias for example by the use of the folic acid antagonists aminopterin and methotrexate. These drugs were soon followed by other antimetabolites such as 6-mecaptopurine and 5-fluorouracil. At the same time, the development of nitrogen mustard alkylating agents was beginning. For many years these two classes of compound were all that was available to the clinician for the drug treatment of cancer, but the field has since diverged in many different directions and today there are a wide range of different types of antitumour drug.

The very diversity of cancer, and the fact that apparently identical tumours in different hosts respond differently to therapy, means that the idea of a 'magic bullet' to cure all tumours has been long discounted. Thus the intervening years have shown a wide range of chemical structures being investigated. These have originated from a variety of sources: extracts from plants, marine life, moulds and fermentation broths; novel products from the chemical laboratory with serendipitous antitumour activity; synthetic mimics of compounds involved in natural metabolic processes. In 1986, the pharmaceutical formulations for 71 compounds in clinical use or under investigation were listed by the National Cancer Institute of the USA alone. These together with many others worldwide have resulted from the initial testing of hundreds of thousands of compounds.

The clinician's current chemotherapeutic arsenal owes much to the synthetic chemists who have been involved in the discovery and development of novel agents. Some scientists have been fortunate in having their name associated with a particular drug, but the contribution by numerous other chemists to our knowledge of the properties of the countless compounds prepared in the search for new agents is often overlooked. This volume might reasonably be regarded as a tribute to them.

To present the whole of antitumour drug chemistry would not have been possible in this book. Nevertheless, the original aim has been achieved: to provide a comprehensive introduction to antitumour drugs, written for chemists by chemists, so that more may apply their talents to the design of new treatments for the group of diseases that are called cancer.

The patient assistance of the publishers and of my colleagues at the Institute of Cancer Research during the preparation of this book is gratefully acknowledged.

D.E.V.W.

Contributors

Michael J. ABRAMS

Johnson Matthey Pharmaceutical Research, West Chester, Pennsylvania 19380, USA.

Alexandru T. BALABAN

Department of Organic Chemistry, Polytechnical Institute, Bucharest, Romania.

Ileana BARACU

Department of Cancer Biochemistry, Institute of Oncology, B-dul 1 Mai nr. 11; sector 1, 78159 Bucharest, Romania.

Marie-Paule COLLARD

Ire-Celltarg, S.A. B-6220 Fleurus, Belgium.

Jean P. C. DEJONGHE

Ire-Celltarg, S.A. B-6220 Fleurus, Belgium.

William A. DENNY

Cancer Reseach Laboratory, University of Auckland, School of Medicine, Private Bag, Auckland, New Zealand.

Richard W. FRANCK

Department of Chemistry, Hunter College, City University of New York, New York City, New York 10021, USA.

Sidney M. HECHT

Departments of Chemistry and Biology, University of Virginia, Charlottesville, Virginia 22901, USA.

Terence C. JENKINS

Cancer Research Campaign Biomolecular Structure Unit, Institute of Cancer Research, Cotswold Road, Sutton, Surrey, SM2 5NG, UK.
Previous address: Medical Research Council Radiobiology Unit, Chilton, Didcot, Oxon, OX11 0RD, UK.

Thomas P. JOHNSON

Southern Research Institute, 2000 Ninth Avenue South, PO Box 55305, Birmingham, Alabama 35255, USA.

Ganesh D. KINI

Nucleic Acid Research Institute, ICN Plaza, 3300 Hyland Avenue, Costa Mesa, California 92626, USA.

Malcolm MacCOSS

Exploratory Chemistry Department, Merck Sharp & Dohme Research Laboratories, Rahway, New Jersey 07065, USA.

Raymond McCAGUE
Drug Development Section, Cancer Research Campaign Laboratory, Institute of Cancer Research, Cotswald Road, Sutton, Surrey, SM2 5NG, UK.

Anthony B. MAUGER
Medlantic Research Foundation, George Hyman Memorial Research Building, 108 Irving Street, N.W., Washington, DC 20010, USA.

John A. MONTGOMERY
Southern Research Institute, 2000 Ninth Avenue South, PO Box 55305, Birmingham, Alabama 35255, USA.

M. G. NAIR
Department of Biochemistry, College of Medicine, University of Alabama, Mobile, Alabama 36688, USA.

Ion NICULESCU-DUVĂZ
Department of Chemical Carcinogenesis, Institute of Oncology, B-dul 1 Mai nr. 11; sector 1, 78159 Bucharest, Romania.

Kandukuri S. P. Bhushana RAO
Omnichem Unit UCL-ICP and Ire-Celltarg, S.A., B-6220 Fleurus, Belgium.

Morris J. ROBINS
Chemistry Department, Brigham Young University, Provo, Utah 84602, USA.

Roland K. ROBINS
Nucleic Acid Research Institute, ICN Plaza, 3300 Hyland Avenue, Costa Mesa, California 92626, USA.

Malcolm SAINSBURY
School of Chemistry, University of Bath, Claverton Down, Bath BA2 7AY, UK.

Antonino SUARATO
Farmitalia Carlo Erba, Viale E. Bezzi 24, 20146 MILANO, Italy.

Michael D. THREADGILL
Medical Research Council Radiology Unit, Chilton, Didcot, Oxon, OX11 0RD, UK

Marie TOMASZ
Department of Chemistry, Hunter College, City University of New York, New York City, New York 10021, USA.

André TROUET
Ire-Celltarg, S.A. B-6220 Fleurus, Belgium.

Keith VAUGHAN
Department of Chemistry, Saint Mary's University, Halifax, Nova Scotia, B3H 3C3, Canada.

Derry E. V. WILMAN
Drug Development Section, Cancer Research Campaign Laboratory, Institute of Cancer Research, Cotswold Road, Sutton, Surrey, SM2 5NG, UK.

Contents

5 Triazenes   159

K. VAUGHAN

6 The chemistry of azolotetrazinones   187

M. D. THREADGILL

7 Chemistry of antifolates   202

M. G. NAIR

15 The chemistry of mitomycins 379

R. W. FRANCK and M. TOMASZ

16 The chemistry of activated bleomycin 395

S. M. HECHT

17 Actinomycins 403

A. B. MAUGER

18 Ellipticines 410

M. SAINSBURY

1 Acridine-based antitumour agents

W. A. DENNY

1.1 Introduction

Chemotherapeutic agents based on the acridine nucleus have a long history, and include the antimalarial compounds mepacrine (**1**) and azacrine (**2**), and the antibacterials proflavine (**3**) and aminacrine (**4**)[1]. More recently, several acridine-based classes of compound have emerged as antitumour drugs. A number of agents from these classes have entered clinical use, and several others are under intensive development by research groups throughout the world. These compounds have quite different ranges of biological activities and modes of action, but all have the common chemistry of acridines; in particular the ability to bind tightly but reversibly to DNA by intercalating between the basepairs of the double helix.

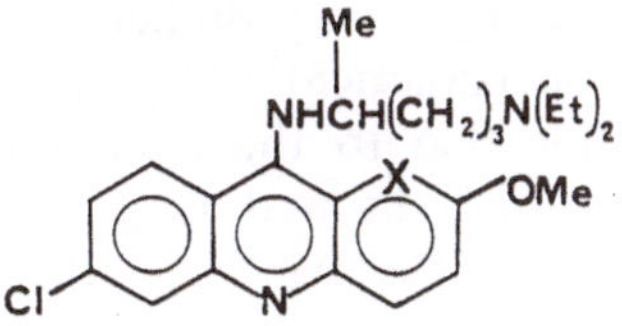

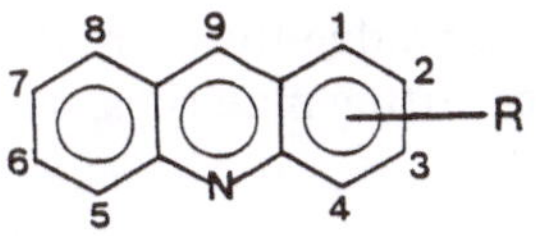

1; $X = -CH=$

2; $X = -N=$

3; $R = 3.6-diNH_2$

4; $R = 9-NH_2$

Acridines have been regarded as the archetypical DNA intercalating agents since proflavine (**3**) was first used to demonstrate this phenomenon[2], and many acridine-based compounds have been used as probes for studying DNA-ligand binding. The main driving forces for intercalative binding are stacking and charge-transfer interactions between the aromatic systems of the ligand and the DNA bases, with polar (H-bond and electrostatic) forces playing a lesser role[3,4]. As a net result of these forces, acridine derivatives possess high association constants for DNA, even under high-salt, physiological conditions. Intracellularly, such compounds are almost totally bound to nucleic acids, and it is this ability which is responsible for the numerous physiological effects of acridine derivatives.

1.2 Nitracrine (NSC 247561) and analogues

The first class of acridine derivatives developed specifically as antitumour drugs were the 9-[(alkylamino)alkyl]-1-nitroacridines. Results from the screening of a large number of acridine derivatives against the sarcoma 180 mouse tumour model showed that activity was associated exclusively with a 1-nitro group on the acridine nucleus[5], and 9-[(dimethylaminopropyl)amino]-1-nitroacridine (**5**; C-283, Ledakrin, nitracrine) was chosen for clinical evaluation[6]. The drug has been quite extensively used in Poland, and the results from several clinical studies have been published[1,7]. Nitracrine was evaluated by the American National Cancer Institute in 1975 (as NSC 247561), but did not proceed to clinical use in the USA[8], although the drug is still used to a limited extent in Poland[9]. More recently, nitracrine has been shown to possess selective toxicity for hypoxic tumour cells[10,11], due to an increased rate of metabolism under such conditions, and to be a hypoxic cell radiosensitising agent[12].

1.2.1 *Synthesis*

Nitracrine is most conveniently prepared in a four-step synthesis from 2-chlorobenzoic acid and 3-nitroaniline[13], beginning with the classical Jourdan–Ullmann condensation[14] (Scheme 1.1). This reaction generally proceeds poorly with weakly basic amines[14], and yields of (**6**) ranging from 35% to 60% have been reported, depending on conditions[15,16]. A higher yield (94%) in this reaction can be achieved by using diphenyliodonium-2-carboxylate (DPIC; **11**) instead of 2-chlorobenzoic acid[17,18]. Although there is no particular advantage in this example due to the availability of the alternative starting materials, the DPIC route is the method of choice in some

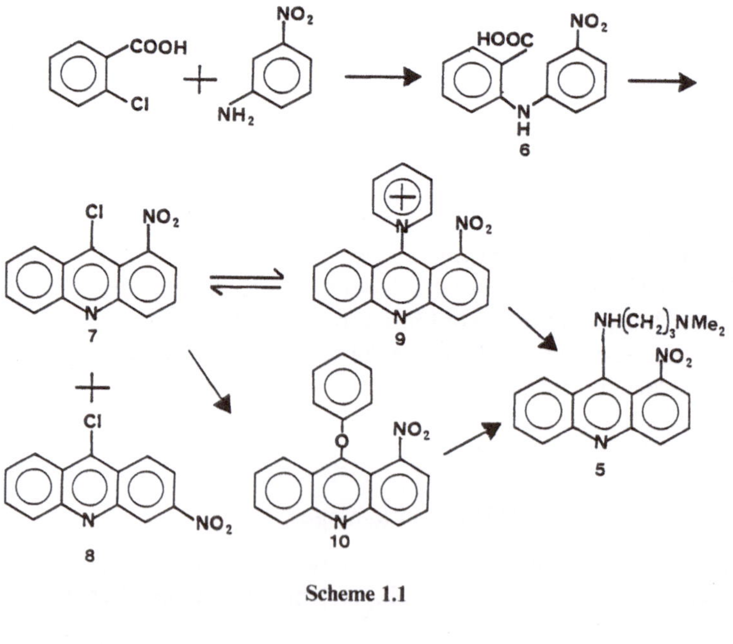

Scheme 1.1

11

12; R=Cl   13

14; R=N(CH₂CH₂OH)₂   15

cases. Thus the very weakly basic 2-chloro-5-nitroaniline (**12**) gives a yield of 5–10% of the condensation product (**13**) using 2-chlorobenzoic acid, but an 81% yield from DPIC[18], while the sensitive amine 2-[*N*, *N*-(bis(2-hydroxyethyl)amino]-5-nitroaniline (**14**) gives no isolatable product with 2-chlorobenzoic acid but a 93% yield with DPIC[19].

Ring closure of (**6**) with POCl₃ proceeds almost quantitatively, but gives a mixture of the 1- and 3-nitro isomers (**7**) and (**8**) in a 65:35 ratio[20]. Small quantities of the desired 9-chloro-1-nitroacridine (**8**) can be readily obtained by flash chromatography[20]. For larger quantities, the mixture of isomers (**7, 8**) is finely powdered and treated briefly with cold pyridine, filtered from the more insoluble 3-isomer, and heated briefly to 60°C to effect conversion to the 9-pyridinium compound (**9**), which crystallises on cooling[16,21]. This may be converted to 9-chloro-1-nitroacridine with POCl₃, or reacted directly with *N*, *N*-dimethylpropylamine to give nitracrine. Reaction of 9-chloro-1-nitroacridine with alkylamines is carried out in excess phenol, and proceeds via the 9-phenoxy derivative (**10**)[22]. The reaction is sensitive to the nature of the alkylamine used[20], and close temperature control is essential. Thus reaction with *N*, *N*-dimethylamino-propylamine to give nitracrine itself proceeds in high yield at 70°C, but at 90°C the yield is greatly reduced[18]. An alternative (and milder) synthesis is formation of the amide from (**6**) with an alkylamine, and ring closure of this (e.g. **16**) with polyphosphate ester to give a 9-alkylamino-1-nitroacridine directly.

16; R = H   5   17

18; R = N(CH₂CH₂OH)₂   19   —

This method is not competitive with those described above for nitracrine itself, due to the necessity of separating the product from the 3-isomer (**17**) (although such a separation by chromatography has been reported[23]). However, it has been used[19] (and is preferred) for the synthesis of 4-*N*, *N*-[bis(2-hydroxyethyl)amino]nitracrine (**19**), where only one isomer is formed

Table 1.1 Physicochemical properties of nitracrine (5).

Alternative names	C-283, Ledakrin, NSC 247561
UV spectrum	
λ_{max} nm	426
log E	3.63
pK_a	
Acridine N	6.21
Dimethylamino N	8.80
Reduction potential $E(1)$ (mV)	-303
DNA association constant (0.01 M) (M^{-1})	2.35×10^5
DNA helix unwinding angle	16°

in the ring closure and the sensitive acridine substituent requires protection and deprotection if the route via the 9-chloro intermediate is used[18].

1.2.2 *Properties*

Nitracrine (5) has the physicochemical properties given in Table 1.1. The solid free base is light-sensitive, while dilute aqueous solutions of the dihydrochloride hydrolyse slowly to 1-nitroacridone (20)[20,24]. Crystallographic studies[25] of the monoiodide salt show the compound exists in the iminoacridan form (21). In contrast, a crystal structure of the dihydrochloride salt[26] shows this to exist as the aminoacridine tautomer (5).

The iminoacridan form of the uncharged acridine species (which will be the predominant species under physiological conditions) has been invoked to explain a number of physicochemical properties of nitracrine which are unique to the 9-alkylamino-1-nitroacridines. These include the pH dependence of hydrolysis (where the rates decrease with increasing pH, as the more hydrolysis-resistant iminoacridan form is adopted)[22], the relatively low DNA binding[1] of nitracrine, and the relatively fast dissociation of nitracrine-DNA complexes compared to isomeric nitroacridines. The crystal structure studies[25,26] of nitracrine and other 1-nitro analogues show the nitro group to be twisted out of the plane of the acridine ring by some 60°, and ^{17}O-NMR studies[27] suggest that a similar situation occurs in solution. This conformation is required to minimise non-bonded interactions with the sidechain in the iminoacridan tautomer, and the relief of these clearly outweighs the concomitant loss of conjugation energy. As expected, the 2-nitro derivative (22) has a coplanar structure[28]. The lack of coplanarity of the nitro group of nitracrine and other 1-nitroacridines contributes to its reactivity, and it is easily replaced by nucleophiles either intramolecularly (e.g. to give (25) from (24))[29], or intermolecularly, particularly by thiols [30-32]. Under certain conditions, thiols give products of nitro reduction as well as displacement, but such products are unstable[31]. The 1-nitro group is also readily reduced electrochemically[33,34] or biologically[12], but in fact the redox potential of nitracrine (-303 mV) is little different to that of the isoelectronic (and presumably coplanar) 3-nitro isomer (23) (-259 mV)[27].

22 : 2- NO$_2$

23 : 3 - NO$_2$

24

25

A recent study[35] of the redox potentials of an homologous series of derivatives shows that the redox potential of the nitro group is dependent on the proximity of the sidechain cationic group. The longer sidechain derivatives (27–29) show even higher redox potentials than nitracrine, with the main reason presumably being the varying pK_a of the acridine chromophore (Table 1.2).

Although nitracrine binds reversibly to DNA by intercalation[10], this is not considered to be the main reason for its unique pattern of biological activity. Rather than acting as a strand-breaking agent, as the majority of intercalating agents appear to do (see later), nitracrine forms DNA adducts[11] and cross-links[36] via bioreductive activation. This explains the absolute requirement for the 1-nitro group, and the fact that it cannot be replaced by groups of similar electronic (e.g. chloro, bromo) or steric (e.g. trifluoromethyl, dimethylamino) properties[37]. Studies of the metabolism of nitracrine[11,33] have tentatively identified the hydroxylamine (30) as a major metabolite. This compound can be synthesized chemically by reduction of nitracrine with $NaBH_4/Pd/C$[11,38], while reduction with $Pd/C/H_2$ gives the 1-amino compound (31). Both reduced products are very reactive: the amine (31) adds rapidly to carbonyl compounds to give tetracyclic derivatives such as (32)[39].

Table 1.2 Redox potential and pK_a values for homologues of nitracrine.

No.	$E(1)$ (mV)	pK_a
(26)	-318	4.92
(5)	-278	6.21
(27)	-235	6.90
(28)	-232	7.40
(29)	-236	7.41

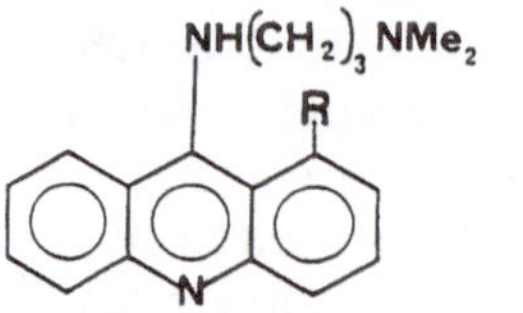

30 ; R = NHOH

31 ; R = NH₂

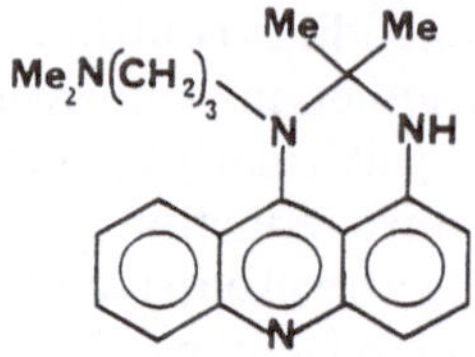

32

The hypoxia-selective toxicity of nitracrine in cell culture studies[10,11] is due to this bioreductive activation occurring more readily under hypoxic than aerobic conditions (where molecular oxygen can reoxidise the initially formed nitro radical anion), resulting in a faster accumulation of nitracrine-induced DNA lesions in the former case. The nature of the ultimate reactive species is unknown, but appears not to be either the hydroxylamine (**30**) or the amine (**31**).

Nitracrine is not active *in vivo* against hypoxic solid tumours, even under favourable conditions where the oxygenated tumour cells are killed by radiation and drug effects are measured precisely by clonogenic assay[11]. Studies of the *in vivo* metabolism of nitracrine show it to be very rapid, with most occurring in the liver. This rapid reductive metabolism prevents distribution of the drug to remote tumour sites, and appears to be the main reason for the lack of *in vivo* activity[11]. Although nitracrine is not used as a conventional antitumour drug, except to a limited extent in Poland[9], it displays a very interesting set of biological activities which are directly traceable to a unique chemistry. This will ensure that nitracrine and its analogues remain a focus of studies in the future.

1.3   Amsacrine (NSC 249992) and analogues

The first acridine derivative to achieve widespread clinical use as an antitumour agent was amsacrine (**35**; *m*-AMSA, NSC 249992), a member of the large family of 9-anilinoacridine compounds[1,40]. The compound was chosen for clinical trials largely on the basis of results against the L1210 leukaemia and B16 melanoma models in mice. Selection was from a number of analogues where the main variations were in the anilino sidechain, and an optimal pattern of a 3′-OCH_3 and a 1′-$NHSO_2CH_3$ group was observed. Early clinical

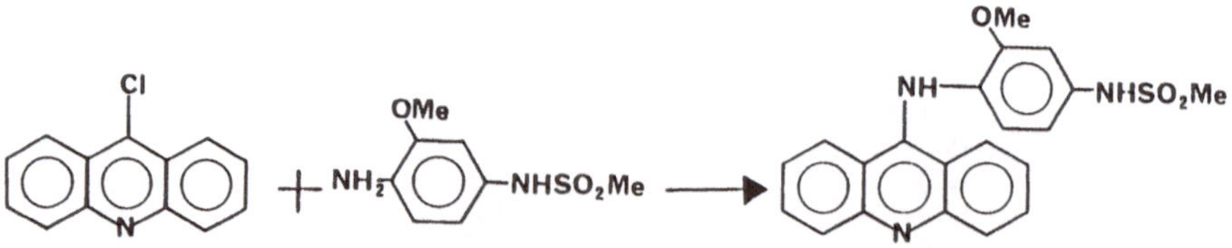

33   34   35

trials showed the usefulness of amsacrine against leukaemias and lymphomas[41,42], with lower levels of response against breast tumours[43]. A potentially useful aspect of the clinical activity of amsacrine was an apparent lack of cross-resistance with adriamycin[44]. The main use of amsacrine is now for leukaemia, where it is used as frontline therapy in combination with antimetabolites[45].

1.3.1 *Synthesis*

Amsacrine is synthesised by mild acid-catalysed coupling of 9-chloro-acridine (**33**) and 4-aminomethanesulfon-*m*-anisidide (**34**) in anhydrous methanol, which gives an excellent yield of the hydrochloride salt[46]. This is converted to the free base, which is dissolved in dimethylacetamide, and provided, together with an aqueous solution of lactic acid, as a two-vial clinical formulation[47]. However, most of the experimental studies have used either the methanesulfonate or isethionate salts, which are more soluble than the initially formed hydrochloride.

Variations in the synthesis of amsacrine are restricted to preparation of the sidechain (**34**), and even more narrowly to variations in the nature of the masking function for the amino group. The simplest preparation is from 3-methoxy-4-nitroaniline (where the nitro group is the precursor), but this compound is not readily available[48]. Large-scale preparation of the sidechain was initially achieved[49] from the readily available isomeric 2-methoxy-4-nitroaniline (**36**) via the butyrylamido synthon (**37**), using hydrolytic demasking (Scheme 1.2). An alternative non-hydrolytic route, suitable also for the preparation of more complex cationic sidechain analogues, uses an

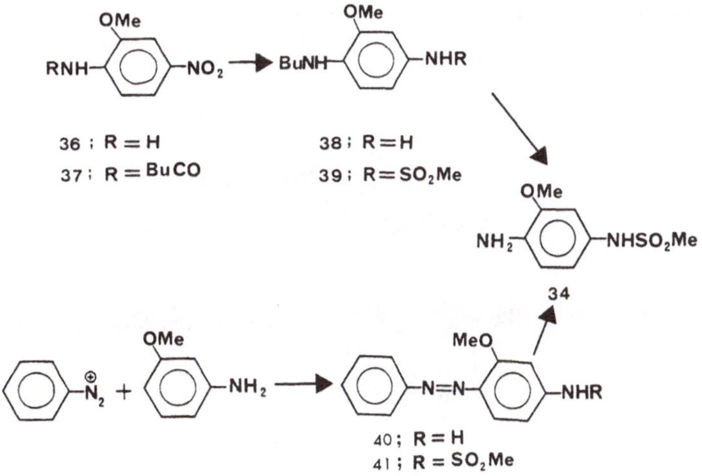
OMe
RNH
NO2
BuNH
NHR
36 ; R=H
37 ; R=BuCO
38 ; R=H
39 ; R=SO2Me
NH2
NHSO2Me
34
N2
NH2
MeO
N=N
40 ; R=H
41 ; R=SO2Me

Scheme 1.2

azobenzene protecting group which provides more crystalline intermediates (**40, 41**) and can be demasked by hydrogenation[18].

1.3.2 Properties

Amsacrine has the physicochemical properties given in Table 1.3. Simple minimum-energy calculations[50] for amsacrine and derivatives predict the anilino ring to lie almost at right angles to the acridine to relieve non-bonded interactions, with the torsion angle having a value of 81° (Figure 1.1). However, crystallographic studies[51,52] show much less rotation, with a value of only 24° for torsion angle, due to the considerable degree of conjugation of the C14–N15 bond. This can be seen directly from the crystal structures, where this bond is significantly shorter (1.35–1.36 Å) than normal (cf. the C19–C22 bond, 1.43 Å). This conjugation is also demonstrated by the high degree of transmission of electronic effects from the anilino to the acridine ring, as summarized by equation (1.1) for the σ effects of anilino ring substituents on

Table 1.3 Physicochemical properties of 9-anilinoacridines (**35**) and (**50**).

Name	Amsacrine (**35**)	CI–921 (**50**)
Alternative names	m-AMSA, Amsidyl, NSC 249992	Amsalog, NSC 343499
UV spectrum		
λ_{max}(nm)	435	448
log E	4.15	4.11
pK_a		
Acridine N	8.02	6.99
Oxidation potential ($E_{1/2}$)(mV)	280	240
DNA association constant (0.01 M) (M^{-1})	1.8×10^5	2.1×10^6
DNA helix unwinding angle	20.5°	18°

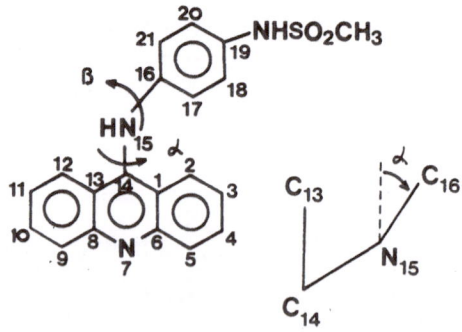

Figure 1.1 Conformation of amsacrine (**35**).

acridine pK_a[53].

$$pK_a = -2.03(\pm 0.08)\sigma + 7.27 \qquad (1.1)$$

$$n = 17, \quad r = 0.984, \quad s = 0.172$$

Amsacrine binds to DNA by reversible intercalation of the acridine chromophore[54], with an association constant of $1.8 \times 10^5 \, M^{-1}$ for calf thymus DNA in 0.01 M salt[55]. By analogy with the crystal structure determined for 9-aminoacridine binding to a dinucleotide[56], amsacrine is postulated[1] to bind with the anilino ring lodged in the minor groove, with the 1'-substituent pointing tangentially away from the helix, in a position to interact with a second macromolecule such as a regulatory protein. This suggestion[1] seems very reasonable in the light of subsequent work on the mode of action of amsacrine (see later). However, it was suggested in a recent study of the rates of dissociation of amsacrine and anilino-substituted analogues from DNA that the anilino ring might bind in the major groove[57].

Amsacrine (**35**) is very difficult to reduce, with a reduction potential of $-803 \, mV$, as determined by pulse radiolysis[58]. However, it does undergo a ready, reversible two-electron oxidation to the quinonediimine AQDI (**43**). This can be achieved electrochemically by cyclic voltammetry, where a redox potential of 280 mV can be measured[59,60] (Figure 1.2), chemically[61] by activated MnO_2 or biologically by liver microsome preparations[62].

For a series of derivatives bearing a 1'-NHR function, two-electron redox potentials were shown[59] to correlate well with substituent electronic properties. The redox process occurs in two distinct steps, and the intermediate one-electron oxidation product of amsacrine (the radical anion **42**) can be detected by pulse radiolysis techniques[63].

The quinonediimine AQDI (**43**) resulting from two-electron oxidation can be isolated, although it is susceptible to hydrolysis to the quinoneimine (**44**) and 9-aminoacridine (**45**)[18,60]. Reaction of the quinonediimine with mild

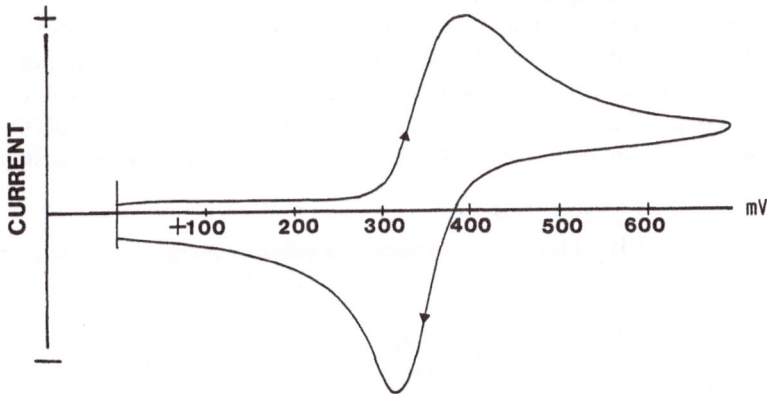

Figure 1.2 Cyclic voltammogram of amsacrine (**35**).

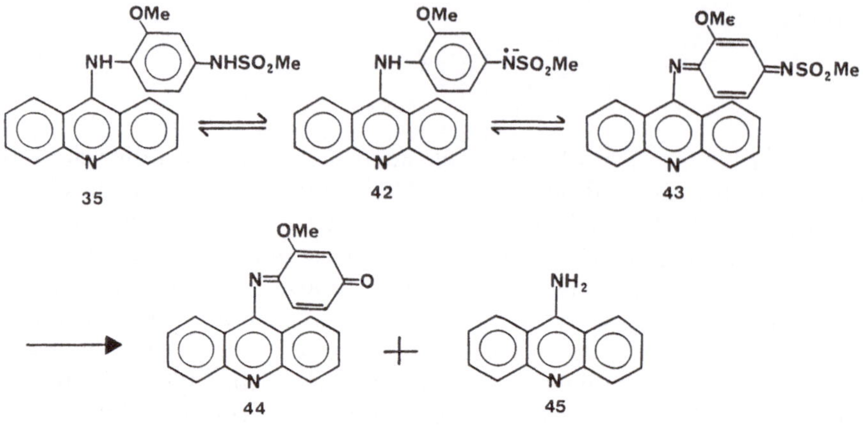

Scheme 1.3

reducing agents such as ascorbic acid results in quantitative reduction to amsacrine[18], while reaction with many nucleophiles such as amines and thiols occurs by 1,4-addition[61,64]. The major product from reaction with thiols is the 6'-adduct (46), although other products are found[61,62]. This chemistry explains the major biliary metabolite observed[61] on treatment of rats with amsacrine, when > 50% of the dose is excreted in the bile as the glutathione conjugate (46; R = glutathione), resulting from attack by GSH on the initially formed quinonediimine (43). The facile oxidation of amsacrine gives it the ability to degrade DNA by a free-radical mechanism in the presence of copper. Amsacrine and Cu(II) species are involved in a rate-determining reaction to form AQDI (43) and Cu(I) species, which are thought to form a complex capable of producing oxygen free radicals[60]. Under suitable conditions, amsacrine efficiently produces single-strand DNA breaks in closed circular supercoiled DNA in the presence of Cu(II) ions[65]. The mechanism is thought to involve singlet oxygen but not hydroxyl radicals[65].

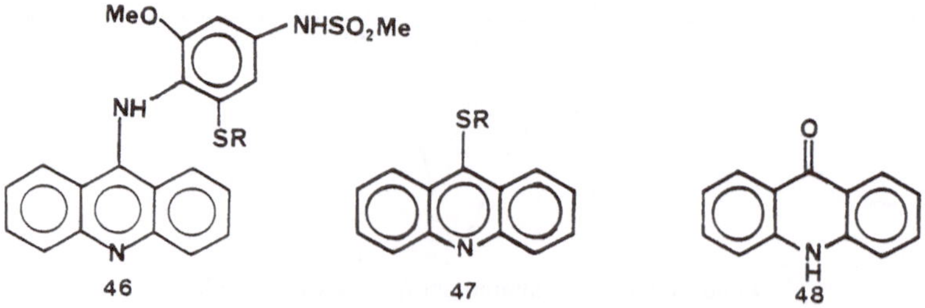

Thiol attack on amsacrine itself results in displacement of the sidechain to give unstable 9-thio species (47), which eventually yield acridone (48). The reaction of 2-mercaptoethanol with amsacrine derivatives has been studied in detail[66]. At pH 7.3, the main substituent effect on the rate of thiolysis was electronic, with electron-donating groups in the anilino ring increasing the rate of thiolysis, by providing a greater proportion of the reactive cationic species. However, even electron-donating groups at the 3′-position greatly inhibited the reaction, presumably by steric hindrance to formation of the tetrahedral transition state[66].

Amsacrine itself reacted rapidly with blood, forming unstable protein adducts which decomposed to acridone[67]. The products of reaction of amsacrine with a number of thiols were isolated[68], and shown to retain the acridine nucleus and the ability to intercalate DNA, but their exact structures have not been determined. A detailed study of the kinetics and mechanism of reaction of amsacrine with thiols[69] showed that the cationic form reacts a thousand-fold faster than the neutral form. The reaction is general-acid-catalysed, with the rate-determining step being the expulsion of the anilino leaving group.

Thus amsacrine is not as chemically inert as might be thought at first glance, and in fact has a rich redox and thiol chemistry. Despite this, neither the thiolysis reaction nor the redox properties of amsacrine appear to have a direct relationship to its biological activity, although the potency of a series of amsacrine derivatives was shown[70] to be dependent on rates of thiolysis. The important biochemical effect of amsacrine is an enzyme-mediated cleavage of DNA involving the enzyme topoisomerase II[71], by a mechanism which appears to be common to many DNA-intercalating agents[72], but which was first described for amsacrine[73].

Amsacrine forms a ternary complex with the enzyme and DNA, altering the position of equilibrium so that a much higher proportion of the fragile covalent complex is formed[73,74]. This mechanism has now been postulated to occur generally[72], and accounts for the protein-associated DNA strand breaks observed with many intercalating antitumour drugs, including amsacrine[75,76].

1.4 CI-921 (NSC 343499) and other amsacrine analogues

Although amsacrine is a successful antileukaemic drug, it lacks broad-spectrum clinical activity[77], and has been difficult to formulate because of low aqueous solubility. Since it was felt that the relatively high aqueous pK_a (8.02 for the acridine nitrogen, implying 87% ionization at physiological pH) played a part in limiting the *in vivo* distribution of amsacrine, analogues were sought which had improved solubility and high DNA binding, but lower pK_a. Early work with the 4-methylcarboxamide (**49**) was encouraging, and a variety of analogues bearing both CONHR groups and additional acridine substitution were evaluated[78,79] before the 4-methyl-5-methylcarboxamide (**50**; CI-921, NSC 343499) was selected for clinical development[80]. As well as showing the sought-after set of physicochemical properties compared to amsacrine (Table 1.3), CI-921 shows superior antileukaemic activity and a broader spectrum of action[80].

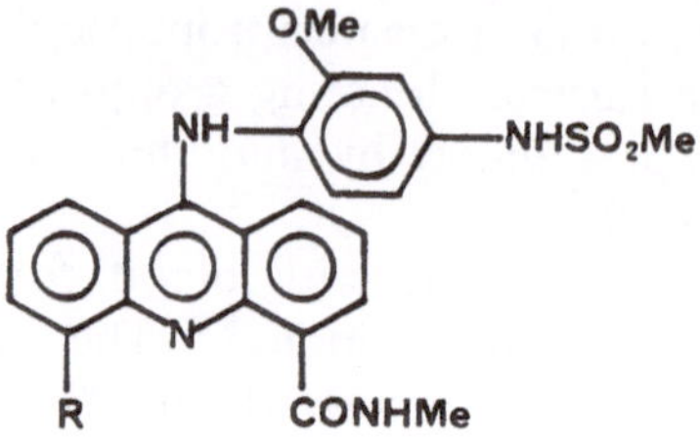

49; R = H

50; R = Me

1.4.1   Synthesis

Three syntheses of the acridine moiety of CI-921 have been evaluated. Reaction of 7-methylisatin (**51**) with 2-chlorobenzoic acid gives the 9-carboxyacridine derivative (**52**), which can be selectively decarboxylated with H_2O_2 to the desired acridone (**53**) (Method A, Scheme 1.4). Although only a

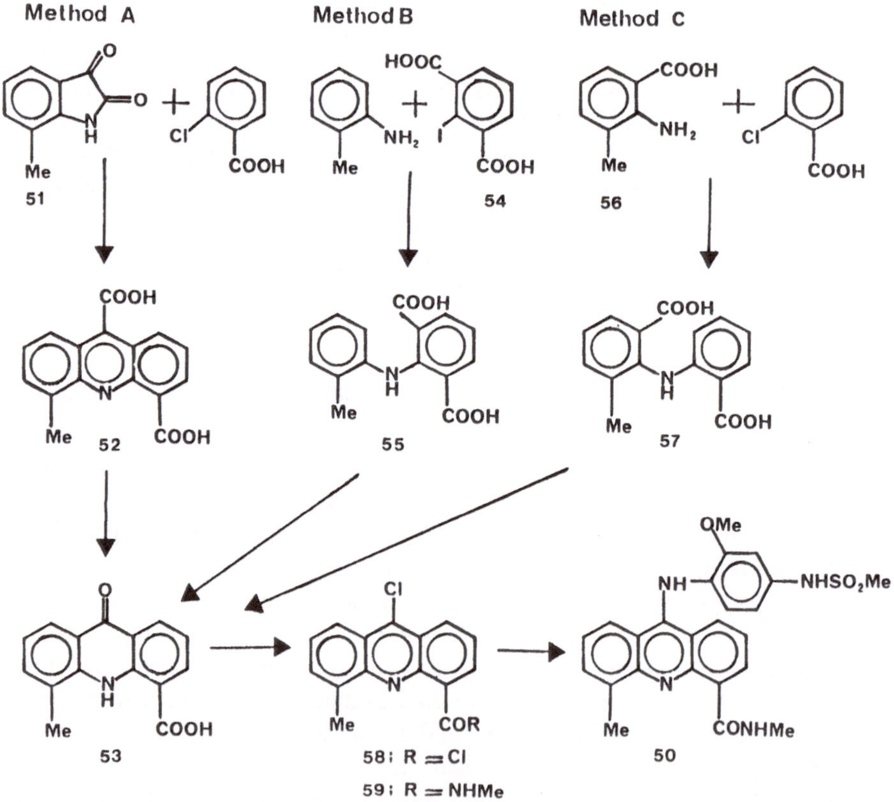

Scheme 1.4

two-step synthesis, the yield was moderate (40% from the isatin, which itself is available in only 60% yield from 2-methylaniline[81]).

A higher yield route (Method B, Scheme 1.4) emerged from efforts to find a general route to substituted acridone-4-carboxylic acids (see later). Reaction of 2-iodoisophthalic acid (54) with 2-methylaniline under anhydrous conditions followed by cyclisation of the intermediate 2-(2-methylphenyl-amino)isophthalic acid (55) with polyphosphoric acid gave the acridone (53) in 65% overall yield. However, the necessity to synthesise 2-iodoisophthalic acid made this route also commercially unattractive, and the most cost-effective synthesis of (53) proved to be the classical Jourdan–Ullmann coupling of 2-chlorobenzoic acid and 3-methylanthranilic acid (56), which are both commercially available (Method C, Scheme 1.4). Electron-deficient anilines usually react poorly in the Jourdan–Ullmann coupling, but anthranilic acids form an exception due to formation of a copper complex which markedly increases reactivity[82]. Thus although 3-methylanthranilic acid is a hindered amine, the reaction gives (57) virtually quantitatively, resulting in an overall 80% yield of (53) by this route. Reaction of (53) with $SOCl_2$ gives 9-chloro-5-methylacridine-4-carbonyl chloride (58), which can be treated selectively with aqueous methylamine to give methyl 9-chloro-5-methylacridine-4-carboxamide (59) in good yield, and this is coupled with 4-aminomethanesulfon-m-anisidide (34) as before to give the hydrochloride salt of CI-921.

1.4.2 Properties

CI-921 is more than four-fold more soluble in water than amsacrine (as the hydrochloride salts, see Table 1.3), and is being used in clinical trials as a single-vial formulation of the isethionate salt[83]. The other significant development in the amsacrine series was prompted by studies which suggested the importance of strongly electron-donating substituents in the anilino ring[39,48]. The success of amsacrine, with a 3'-OCH_3 group ($\sigma - 0.27$) led to the evaluation of compounds bearing other strongly electron-donating 3'-substituents, in particular 3'-$NHCH_3$ and 3'-$N(CH_3)_2$ groups (60, 61). Preparation of the sidechains for these compounds from the known 3-chloro-4-nitroacetamide (62) is straightforward (Scheme 1.5). The most general method[84] involves direct replacement of the halogen by the amine under pressure to give (63), followed by hydrolysis and reaction with methanesulfonyl chloride to give (64). To avoid possible dimesylation in the case of (64a), the N-benzyl derivative (64c) is preferred[85]. Hydrogenation of the nitro compounds (65) in methanol then gives the oxygen-sensitive diamines (66) directly (in the case of (65c) by concomitant removal of the benzyl group).

The dimethylamino derivative (61) has a similar chemistry to amsacrine, being oxidised chemically or electrochemically to the quinonediimine (67)[55], which appears to be more stable than AQDI (4)[3]. In contrast, the 3'-$NHCH_3$

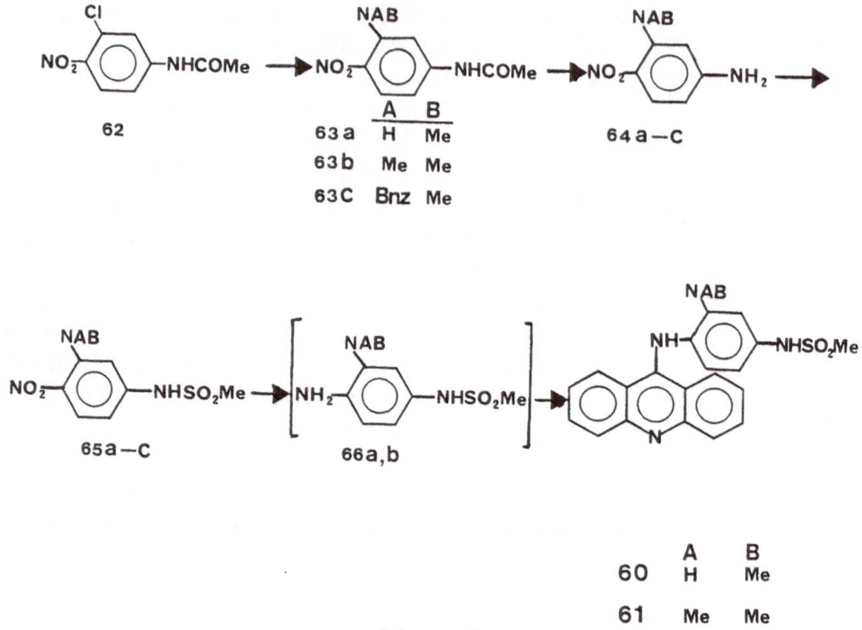

Scheme 1.5

compound (**60**) is irreversibly oxidised to a number of unstable species[55], possibly via the *o*-quinonediimine (**68**). Both compounds (**60, 61**) are more active than amsacrine itself, particularly against the Lewis lung solid tumour, and their derivatives are under evaluation[85,86].

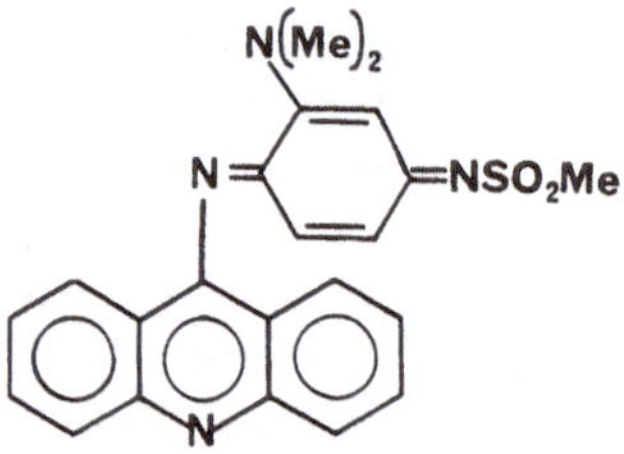

1.5 *N,N'*-Bis-(9-acridinyl)-1,6-hexanediamine (NSC 219733) and related di- and triacridines

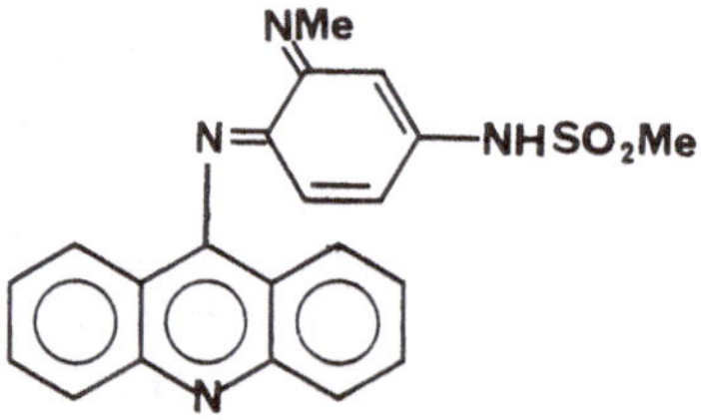

The primary mode of antitumour action of acridines (and other DNA-intercalating ligands) was for a long time considered to be their inhibition of a variety of DNA processing enzymes, possibly by competitive binding to the DNA. This was consistent with observations that antitumour activity among several series of such compounds correlates with high DNA binding affinity[87-89], slow drug/DNA dissociation rates[90-92] and long drug residence times at individual DNA binding sites[93]. Since there are good theoretical reasons why such desirable properties will be enhanced for dimeric molecules containing two intercalating ligands[94,95], there have been many studies of such compounds, and particularly diacridines[1,96,97]. One compound of this class, the hexane-1,6-diamine derivative (69; NSC 219733), shows experimental activity against intercalator-resistant mouse leukaemias[98], and was evaluated for possible clinical trial[99,100].

1.5.1 *Synthesis*

Synthesis of symmetrical diacridines such as (69) is straightforward, involving the coupling of activated acridines (77–79) with the appropriate diamines (80) (Scheme 1.6). 9-Chloroacridines (77) are usually used, in excess phenol, and the reaction proceeds via the 9-phenoxy compound (78) in excellent yields[97,101]. Temperatures above 120–130°C lower the yield considerably[97], but the reaction proceeds even at 60–70°C if dimethyl formamide (DMF) is used as a solvent. If the linker chain component (80) is acid-sensitive, the isolated 9-phenoxy compound (78) can be used to avoid release of HCl, and the 9-methoxy compounds (79) are even more reactive, with the coupling proceeding at 20°C in methanol[102].

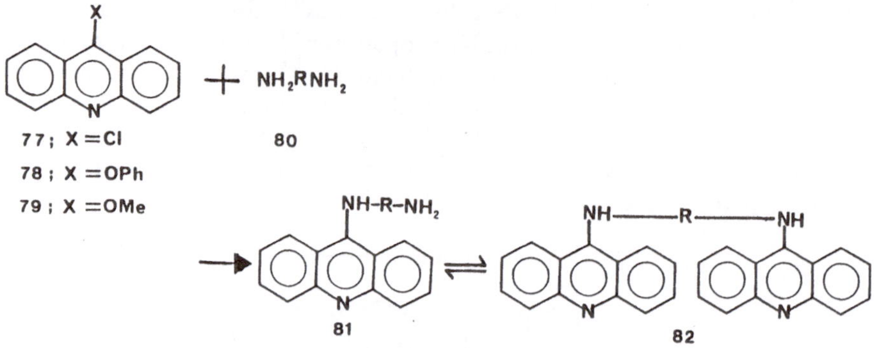

77; X = Cl
78; X = OPh
79; X = OMe

80

Scheme 1.6

Although all the above preparations formally proceed via the monomeric intermediates (**81**) (Scheme 1.6), these compounds cannot be isolated in any significant amounts, even when an excess of the diamine is used[103,104]. They can be prepared by the use of monoprotected diamines[105], and are stable as salts but undergo 'dismutation' to the diacridines (**82**) and diamine under basic conditions[103,104]. The rate of this process is markedly dependent on both the substitution pattern on the acridine and the nature of the linker chain, and it is thought to occur via the iminoacridan tautomer (cf. **21**)[105]. NMR studies suggested that the rate of reaction was related to the ability of the diacridines to undergo intramolecular stacking, which provides the correct orientation for dismutation[105].

1.5.2 *Properties*

Crystallographic studies[106] of the diacridines (**75**) and (**76**) showed them to have an extended conformation, with the acridine chromophores lying approximately parallel and no intramolecular interactions. However, this is probably not the case in solution, where NMR and UV spectroscopy shows[107] that extensive intramolecular stacking occurs, especially above pH 7. Under these conditions, the reaction of diacridines (**75, 76**) with DNA is rate-limited by the necessity for them to unstack before binding to the DNA can take place[95]. Most physicochemical studies with diacridines have been devoted to defining the nature of their interaction with DNA, in particular the requirements for both of the acridines to intercalate, but the question remains controversial. Detailed hydrodynamic studies of the binding of homologous series of polymethylene- and alkylamide-linked diacridines[97,108] show that an abrupt change in binding mode occurs when the linker chain reaches a length of about 8.8 Å, as with the hexanediamine (**69**) and the corresponding amide (**72**). Although electric dichroism studies of (**69**) complexed to DNA show that the plane of the acridine ring is about 10° from complete coplanarity with the DNA basepair plane, both (**69**) and the related amide (**72**) clearly bind under these conditions by intercalation of both chromophores. The linker chain of these compounds is too short to span two basepairs if normal DNA geometry is maintained, and (69) was postulated[108] to bind by bisintercalation at contiguous sites, with one chromophore on either side of a single basepair. This interpretation was controversial, since it violated the 'excluded-site' principle, which states that for intercalators binding to a DNA lattice, ligand binding to a free site then excludes other ligands from binding to the adjacent sites[109]. Furthermore, (69) and (72) were shown[110] by NMR to bind to the self-complementary decadeoxynucleotide d(AT)$_5$·d(AT)$_5$ by monointercalation only, suggesting that the binding mode of these compounds is very condition-dependent.

However, it should be noted that the 'excluded-site' rule is a thermodynamic limitation required in some models for the binding of monointercalators[109], and other interpretations which fit the experimental binding data without requiring this limitation have been proposed[111]. The matter was resolved by the demonstration[112] that the diacridine (83), where the two intercalating chromophores are rigidly held in a coplanar configuration only 7 Å apart by the linker chain, binds to DNA by intercalation of both chromophores. The compound must bind at contiguous sites, showing that interpretation of the hydrodynamic data for the flexibly linked diacridines (69) and (72) in terms of a contiguous binding model is likely to be correct.

83

Studies with derivatives of (69) bearing substituents on the acridine have shown that groups off the long axis of the acridine have more restrictive effects on bisintercalative binding than do groups placed off the short axis[113]. Thus, the 2,6-disubstituted series of diacridines do not show bisintercalative binding until the octanediamine derivative (74), which can form a two basepair sandwich, is reached[113], and similar results have been seen for substituted polyamine-linked diacridines[114]. The nature of the linker chain in 9,9-linked diacridines has also been shown to affect the kinetics of their binding to DNA. The diacridine (71), with a flexible octamethylene chain, dissociates from DNA about 30-fold more slowly than does the monomer 9-aminoacridine (45), as measured by stopped-flow UV spectroscopy[97]. However, the average residence time of each acridine chromophore at a particular DNA site, measured from the jump rate determined by NMR spectroscopy[97,115], is only two-fold slower than that of 9-aminoacridine, implying that the kinetic behaviour of the individual chromophores of (71) is only slightly affected by the presence of a flexible linker chain joining them. Complete dissociation of the molecule from DNA, requiring simultaneous disengagement of both acridines, is a relatively rare event compared to dissociation of a single acridine, with at least fifteen dissociations and reassociations of each chromophore for every complete dissociation of the molecule[97]. This raises the possibility that flexibly linked bisintercalators such as (71) may have considerable mobility when 'bound' to DNA by 'creeping' along the lattice.

However, compounds such as BAPY (73), with more rigid linker chains, have greatly increased residence times for the chromophores, and may act more like 'molecular staples'[1]. For a limited series of diacridines where other parameters such as lipophilicity were taken into account, *in vivo* antitumour activity correlated with long chromophore residence times, indicating the possible utility of rigidly linked bisintercalators as antitumour drugs[97].

A large number of acridine-substituted symmetrical diacridines joined by flexible chains have been synthesised and evaluated for antitumour activity[116], but no compounds markedly superior to NSC 219733 (69) were found. Preclinical studies of (69) indicated only marginal activity in advanced screens, together with pronounced central nervous system (CNS) toxicity, and human trials did not proceed[8]. Although it is probable that (69) was not the best choice, since the heptanediamine analogue (70) has higher biological activity[97,116] and lower adrenergic potency[117], none of the flexibly linked diacridines seem likely to be useful. A smaller number of unsymmetrical diacridines have been prepared[105] using monoprotected diamines in a stepwise synthesis to avoid the possibility of dismutation (see above), but no *in vivo* antitumour activity has been reported.

1.5.3 *Triacridines*

The considerations which led to the study of diacridine derivatives as potential antitumour drugs (see above) have also been behind recent attempts to design and evaluate triacridine species as DNA trisintercalating ligands and antitumour agents. A triacridine (84) joined by simple polymethylene linker chains was reported to trisintercalate DNA on the basis of helix extension data[118], while the amide-linked triacridine (85) was claimed as a trisintercalating ligand from helix unwinding data[119], although this was questioned in a later study[120]. A related amide-linked triacridine (86) was also claimed to trisintercalate on the basis of unwinding data, with the helix extension data being less supportive[121]. A review[96] of these results concluded that, although claims for trisintercalation are more difficult to evaluate than for bisintercalation, some of the molecules probably are true trisintercalating ligands.

Synthesis of these compounds has been by two routes. The linker chain for the simple polymethylene-linked triacridine (84) was constructed[122] from the partially protected triamine (87), by addition of acrylonitrile and reduction to give the triamine (88) (Scheme 1.7). The nature of (88) also allows the synthesis

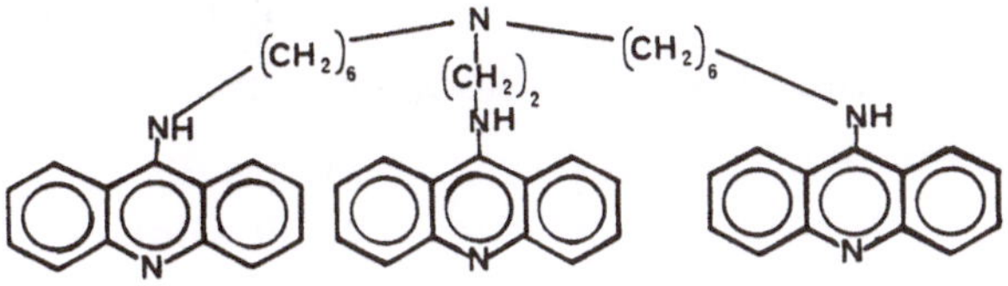

84

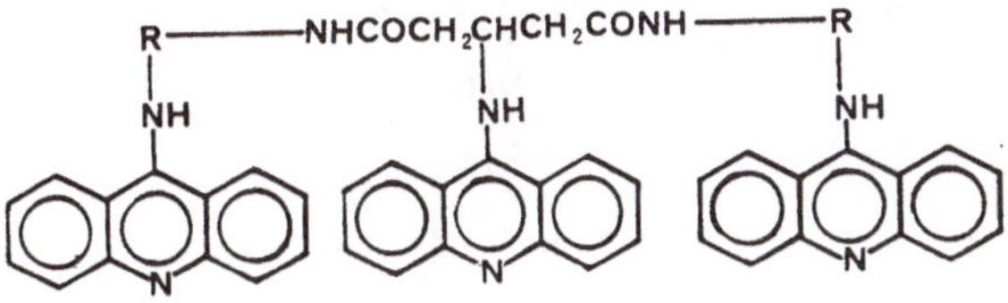

85 ; R = $-(CH_2)_3-$

86 ; R = $-(CH_2)_5NHCO(CH_2)_2-$

of heterogeneous compounds such as (89), with two different chromophores, by stepwise addition of the acridines (Scheme 1.7). The amide linker chains used to prepare (86) and homologues were synthesised from monoprotected

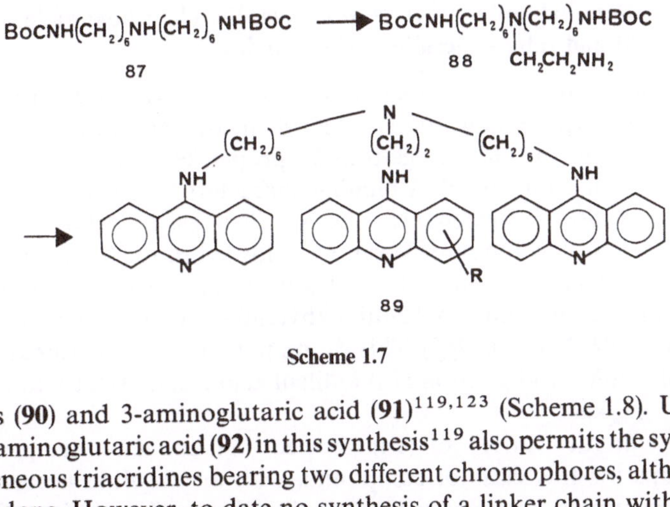

87

88    $CH_2CH_2NH_2$

89

Scheme 1.7

diamines (90) and 3-aminoglutaric acid (91)[119,123] (Scheme 1.8). Use of N-BOC-3-aminoglutaric acid (92) in this synthesis[119] also permits the synthesis of heterogeneous triacridines bearing two different chromophores, although this was not done. However, to date no synthesis of a linker chain with all three amines selectively protected, to permit preparation of a fully heterogeneous molecule with all three acridines different, has been reported.

Although the main aim of the above studies was to determine the mode of DNA binding of the compounds, measurements of strength of binding and also of cytotoxicity showed that there was little improvement over the corresponding diacridine species[96]. A careful study of the biological activity of a series of amide-linked triacridines related to (84) was carried out[123], but the compounds possessed minimal *in vivo* antileukaemic activity.

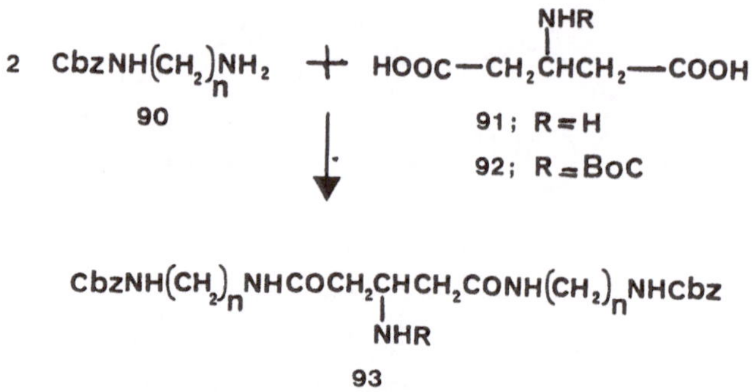

$$2 \quad CbzNH(CH_2)_nNH_2 \quad + \quad HOOC-CH_2CHCH_2-COOH$$

NHR (above)

90   **91; R = H**
                92; R = BoC

$$CbzNH(CH_2)_n NHCOCH_2CHCH_2CONH(CH_2)_n NHCbz$$

NHR

93

Scheme 1.8

Thus the large effort invested in polyacridines has not been rewarded with a useful clinical drug from this class. However, the studies have provided invaluable information about the mode of interaction of polyintercalating agents with DNA, and have suggested clear guidelines for further drug development.

1.6 N-[2-(Dimethylamino)ethyl]-9-aminoacridine-4-carboxamide (NSC 342965) and related acridinecarboxamides

The lead compounds discussed so far are all derivatives of 9-aminoacridine (**45**), where the major modification is a sidechain attached at the 9-position. In an attempt to improve the charge-transfer properties of the 9-aminoacridine chromophore and thus its DNA binding characteristics, a study was carried out[124] where cationic sidechains were attached via a carboxamide group placed at various chromophore positions. This led to the discovery of the 9-aminoacridine-4-carboxamide class of antitumour agents, exemplified by the parent compound N-[2-(dimethylamino)ethyl]-9-aminoacridine-4-carboxamide (**94**; NSC 342965). This shows potent *in vitro* cytotoxicity, with an IC_{50} of 15 nM, and good *in vivo* antileukaemic activity at a low optimal dose of 4.5 mg/kg[124].

Studies with derivatives of (**94**) showed well-defined structure-activity relationships for this class of compound, with an absolute necessity for a sidechain bearing a cationic centre positioned at a fixed orientation with respect to the acridine chromophore. Significant attenuation of the pK_a of the sidechain nitrogen or alteration of its position abolished activity[124]. Thus the 2- and 3-substituted carboxamides (**95, 96**), and even the 4-propylcarboxamide (**97**) and 4-butylcarboxamide (**98**) are inactive. The nature of the group linking the sidechain to the chromophore is also critical, with the sulfonamide (**99**) and N-methylcarboxamide (**100**) compounds being inactive[92]. Substitution of the acridine of (**94**) is also important, with groups in many positions abolishing activity, but those in the 5-position enhancing it[125]. Although the parent

compound (94) shows only antileukaemic activity, some 5-substituted derivatives (e.g. 101) show broader spectrum activity against solid tumours[126].

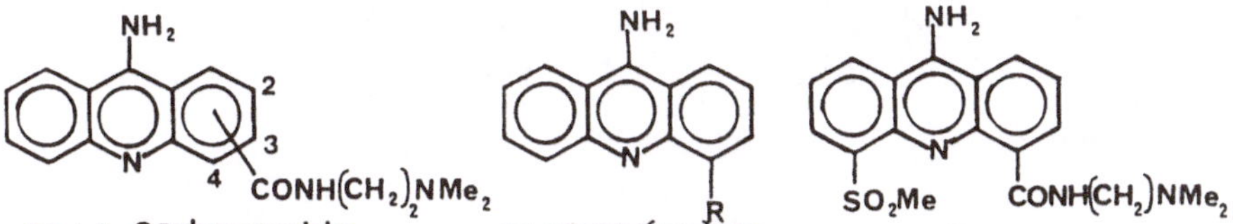

94 ; 4-Carboxamide
95 ; 2-Carboxamide
96 ; 3-Carboxamide

97 ; CONH(CH$_2$)$_3$NMe$_2$
98 ; CONH(CH$_2$)$_4$NMe$_2$
99 ; SO$_2$NH(CH$_2$)$_2$NMe$_2$
100 ; CON(CH$_2$)$_2$NMe$_2$
             |
            Me

101

1.6.1 Synthesis

Synthesis of the parent unsubstituted carboxamide (94) and related sidechain
variants from the known acridone-4-carboxylic acid (102) follows the
preparation of CI-921 (50) outlined in section 1.4.1, involving activation to 9-
chloroacridine-4-carbonyl chloride (103) followed by selective reaction with
N, N-dimethylethylenediamine to form the amide (104) and subsequent phenol-
induced amination (Scheme 1.9)[124].

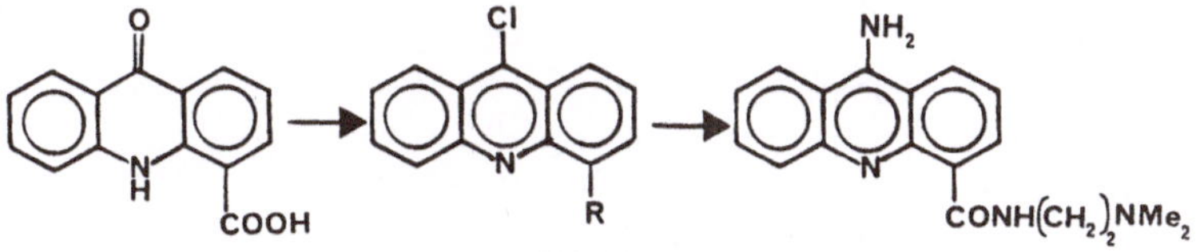

102   103; R=COCl   94
           104; R=CONH(CH$_2$)$_2$NMe$_2$

Scheme 1.9

The substituted acridone-4-carboxylic acids (106) needed for acridine-
substituted derivatives of (94) have been prepared by a variety of methods. The
simple Jourdan–Ullmann condensation of anthranilic and 2-chlorobenzoic
acids to give substituted N-(2-carboxyphenyl)anthranilic acids (105) is quantitative[82], but ring closure of 4-, 5- and 6-substituted derivatives is equivocal
(Scheme 1.10, Method A).

A detailed study[127] of the reaction showed that the direction of ring closure
was dictated by substituent electronic and steric effects and by the cyclising
reagent, but substantially pure single isomers could be obtained in only a few
cases. One alternative method for the synthesis of 5- and 7-substituted
acridone-4-carboxylic acids (106) uses 2-iodoisophthalic acid (107), where
both acid functions are on the same synthon[128] (Scheme 1.10, Method B). A
more general method uses the half esters (108), where one acid function of the

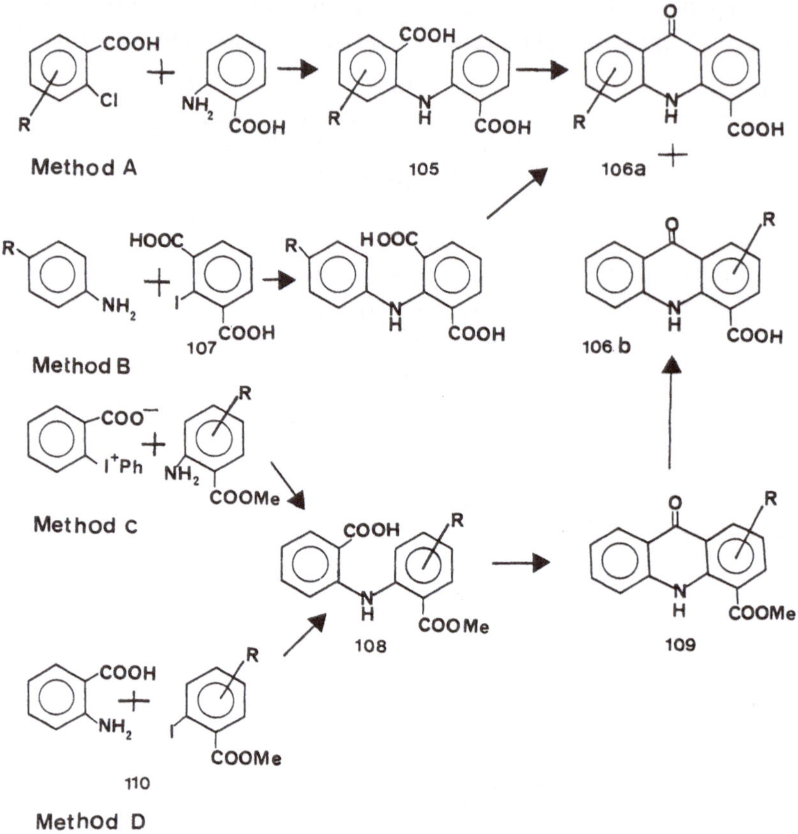

Scheme 1.10

diacid (105) is masked as a carbomethoxy group. Cyclisation with polyphosphate ester then gives the acridone esters (109) unequivocally (Scheme 1.10, Method C). The half esters (108) are prepared by a heavily modified Ullmann reaction, using methyl anthranilates and diphenyliodonium-2-carboxylates[17,127] without an acid acceptor, or methyl iodobenzoates (110) using N-ethylmorpholine as an acid acceptor[129].

1.6.2 Properties

The physiocochemical properties of the parent compound (94) are shown in Table 1.4. Addition of the carboxamide sidechain lowers the chromophore pK_a from the 9.99 recorded[14] for 9-aminoacridine (45) to 8.30, but the compound still exists as the dication under physiological conditions[126]. Crystal structures of the parent compound (94) and the butylcarboxamide (98) as free bases show that the carboxamide groups at the 4-position are approximately planar, and lie coplanar with the acridine ring[130]. This

Table 1.4 Physicochemical properties of acridine-4-carboxamides (**94**) and (**115**).

Name	Aminoacridinecarboxamide (**94**)	DACA (**115**)
Alternative name	NSC 342965	NSC 601316
UV spectrum		
λ_{max}(nm)	407	357
log E	3.97	4.01
pK_a		
Acridine N	8.30	3.54
Dimethylamino N	8.80	8.80
DNA association constant (0.01 M) (M^{-1})	1.58×10^7	2.00×10^6
DNA helix unwinding angle	16°	12°

structure is stabilised by the formation of a hydrogen bond between the acridine ring N10 acceptor and the amide NH donor with the sidechain fully extended (**111**). The length of the C4–C11 bond linking the sidechain to the chromophore is approx. 1.5 Å, suggesting little double bond character[130]. Molecular mechanics calculations confirm that the crystal structure conformation is the preferred one for the free bases, but the barrier to rotation is so low (4.1 kcal/mol) that sidechain rotation about the carboxamide C4–C11 bond is virtually unhindered[130]. The minimum energy conformation computed for the dications is also essentially planar, with a hydrogen bond between the protonated N10 and the amide carbonyl oxygen (**112**).

111   112

Again, the barrier for rotation of the sidechain up to 60° either side of this minimum is low, suggesting that models of DNA binding of these compounds which have the sidechain rotated approx. 30° out of plane with the chromophore (see below) are feasible. The parent compound (**94**) binds to DNA by intercalation, with an unwinding angle (16°) identical to that of 9-aminoacridine, but has a much higher association constant and a preference for binding to GC basepairs[124]. The higher association constant is to be expected for a dicationic compound, but the preferential binding to GC basepairs is structure-dependent, being restricted to compounds bearing a 4-CONH(CH$_2$)$_2$NR^{1}R^2 sidechain, where R^1 and R^2 are groups which permit the nitrogen to be protonated at neutral pH[92].

A detailed study of the dissociation kinetics of (94) and derivatives showed that, while all the compounds dissociate from random sequence DNA by a complex mechanism which involves at least three intermediate forms, derivatives bearing the above sidechains also possess a fourth binding mode of greater kinetic stability. This manifests itself by a fourth, long-lived dissociation transient observed in stopped-flow studies[92], and correlates with *in vivo* antitumour activity within the series. The kinetic data have been interpreted in terms of a molecular model for binding of the 9-aminoacridine-4-carboxamides to DNA. The acridine chromophore is postulated to intercalate from the narrow groove, with the major axis of the acridine lying at an angle to the major basepair axis. With this orientation, the amide NH and protonated $\overset{+}{N}H(Me)_2$ groups of the sidechain are positioned to become hydrogen bond donors to a single DNA acceptor molecule, the O2 of a cytosine base adjacent to the binding site (Figure 1.3).

Although the mechanism of action of the 9-aminoacridinecarboxamides has not been determined, the observation that (94) is an extremely potent inducer of DNA strand breaks [91] suggests that they may act similarly to other DNA intercalating agents on topoisomerase II (see above). Among a small series of compounds studied, DNA breakage ability correlated best with drug/DNA dissociation kinetics, suggesting that long drug residence times at DNA binding sites might be an important determinant of DNA breakage ability[91].

Since the lack of *in vivo* solid tumour activity of (94) and derivatives was considered to be partly due to poor distributive properties of such dicationic species, more weakly basic compounds were evaluated[126]. Bulky electron-withdrawing groups in the 5-position provided much more weakly basic compounds than when the same groups were placed in the electronically equivalent 7-position, presumably due to the additional steric shielding of proton approach to the acridine nitrogen. The acridine pK_a of the 5-SO_2Me

DNA BINDING MODEL

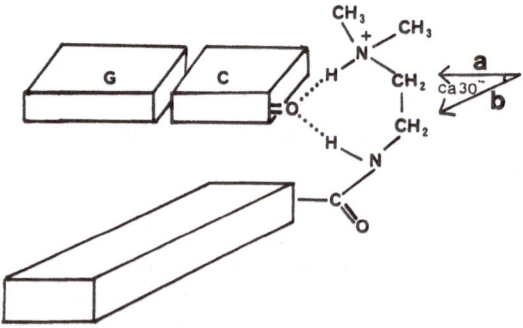

Figure 1.3 Model of aminoacridinecarboxamide (94) binding to DNA. (a) Longitudinal basepair axis; (b) longitudinal drug axis.

derivative (101) was determined to be 5.15, indicating that the acridine would be essentially uncharged at neutral pH. In agreement with the above hypothesis, (101) showed significant *in vivo* solid tumour activity[126]. In further attempts to provide weakly basic derivatives of (94), the resonant 9-amino group was removed to give the acridine-4-carboxamides (114)[131]. These compounds were simply prepared from the corresponding substituted acridone-4-carboxylic acids (106) by reduction with Al/Hg under basic conditions to give the acridine-4-carboxylic acids (113), which were coupled with the appropriate amines to give the carboxamides (114) (Scheme 1.11)

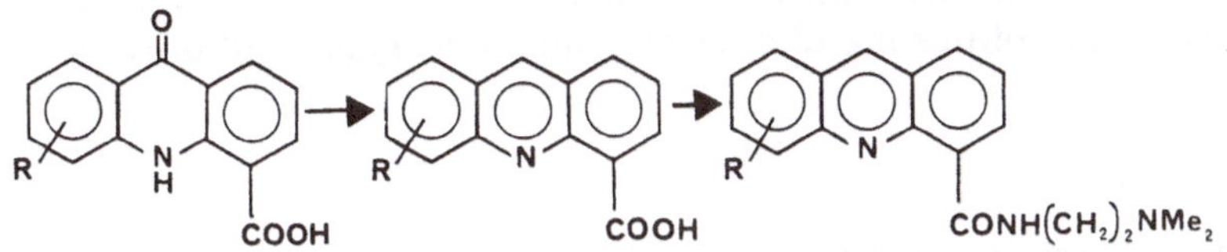

Scheme 1.11

The physicochemical properties of the parent unsubstituted compound (115) (NSC 601316, DACA) are given in Table 1.4. Although the acridine pK_a has been reduced to 3.54 (ensuring an uncharged chromophore), (115) binds to DNA by intercalation only about 10-fold less strongly than (94), and has similar *in vivo* potency and activity against the P388 leukaemia [126]. However, (115) shows remarkable activity against the remotely implanted Lewis lung carcinoma, being repeatedly able to effect virtually 100% cures of the advanced disease[126]. This is substantially better than the activity of (94), amsacrine and adriamycin (which are essentially inactive[126,132]), and even CI-921, which can effect about 50% cures[80]. Many of the substituted acridine-4-carboxamides (114) also show curative activity against the Lewis lung tumour[131].

1.7 Conclusions

Acridine compounds have a long history as therapeutic agents, which continues to the present day. One important reason for this is that the planar, hydrophilic, weakly basic acridine nucleus is a classical DNA intercalating agent, which becomes rapidly but reversibly bound to cellular DNA in a predictable manner. An equally important factor is the stability and accessibility of the acridine nucleus, which makes the synthesis of a wide variety of modified structures possible. The acridine nucleus can therefore be regarded as an ideal 'carrier' for the delivery of various types of functionality to DNA, and all the drugs discussed above can be considered in this light. The special properties of nitracrine (5) and analogues are clearly due to delivery of the redox-active 1-nitro group to the DNA, although in this case the acridine nucleus acts as more than just a carrier molecule. The exceptional topoisomerase inhibitory activity of the 9-anilinoacridines such as amsacrine (35) is almost certainly due to the efficient delivery of the potential quinonediimine system (43) to DNA. Although the diacridines such as (69) and the acridinecarboxamides such as (94) may act solely via their reversible binding to DNA, the geometry and kinetics of this process have been heavily modified by the additional sidechain structures they possess. However it is rationalised, the recent discovery of the structurally simple series of antitumour agents exemplified by (94) and (115) indicate again the remarkable biological activity associated with acridine derivatives, and suggest that further studies of this broad class of compounds might still prove useful.

References

1. W. A. Denny, B. C. Baguley, B. F. Cain and M. J. Waring, 'Anti-tumour Acridines', in *Molecular Aspects of Anticancer Drug Action*, eds. S. Neidle and M. J. Waring, MacMillan, London (1983) pp. 1–34.
2. L. S. Lerman, *J. Mol. Biol.* **3** (1961) 18.
3. L. P. G. Wakelin and M. J. Waring, *Biochem. J.* **173** (1976) 115.
4. S. Neidle and Z. Abraham, *CRC Crit. Rev. Biochem.* **17** (1984) 73.
5. C. Radzikowski, A. Ledochowski, M. Hrabowska, B. Stefanska, B. Horowska, J. Konopa, E. Murowska and M. Urbanski, *Arch. Immunol. Ther. Exp.* **17** (1969) 99.
6. J. Gieldanowski, J. Patkowski, B. Szaga and J. Teodoroczyk, *Arch. Immunol. Ther. Exp.* **20** (1972) 399, 419.
7. M. Warwas, D. Narezewska and W. Dobryszycka, *Arch. Immunol. Ther. Exp.* **25** (1977) 235.
8. S. Schepartz, J. M. Venditti, J. Plowman and M. K. Wolpert, Report on the Programme of the DCT, NCI, National Institute, Bethesda (1981).
9. Z. Mazerska, M. Cholody, J. Lukowicz, B. Wysocka-Skrela and A. Ledochowski, *Arzneimittel-Forschung* **37** (1987) 1276.
10. W. R. Wilson, W. A. Denny, S. J. Twigden, B. C. Baguley and J. C. Probert *Br. J. Cancer* **49**, (1984) 215.
11. W. R. Wilson, W. A. Denny, G. M. Stewart, A. Fenn and J. C. Probert, *Int. J. Radiat. Oncol. Biol. Phys.* **12** (1986) 1235.
12. P. B. Roberts, R. F. Anderson and W. R. Wilson, *Int. J. Radiat. Biol.* **51** (1987) 641.
13. A Ledochowski and B. Stefanska, *Ann. Soc. Chim. Polonorum* **40** (1966) 301.
14. A. Albert, in *The Acridines*, 2nd edn., Edward Arnold, London (1966).
15. A. Albert and E. F. Ritchie, *J. Soc. Chem. Indust.* **60** (1943) 120.
16. A. Kolodziejczyk and A. Arendt, *J. Labelled Compounds* **11** (1975) 385.
17. G. W. Rewcastle and W. A. Denny, *Synthesis* (1985) 220.
18. W. R. Wilson, R. F. Anderson and W. A. Denny *J. Med. Chem.* **32** (1989) 23.
19. D. Chambers and W. A. Denny, *J. Chem. Soc. (Perkin Trans. I)* (1986) 1055.
20. W. A. Denny, D. Chambers, G. M. Stewart and W. R. Wilson, *J. Labelled Compounds Radiopharm.* **22** (1985) 995.
21. W. Cruszecki and E. Borowski, *Ann. Soc. Chim. Polonorum* **42** (1968) 533.
22. S. Skonieczny, *Heterocycles* **14** (1980) 985.
23. A. Ledochowski and J. Zelinski, Polish Pat. 66, 639 (1973) (*Chem. Abstracts* **79** (1973) 31941t).
24. A. Ledochowski and S. Skonieczny, *Roczniki Chemii* **50** (1976) 415.
25. Z. Dauter, M. Bogucka-Ledochowska, A. Hempel, A. Ledochowski and Z. Kosturkiewcz, *Roczniki Chemii* **49** (1975) 859.
26. J. J. Stekowski, P. Kollat, M. Bogucka-Ledochowska and J. P. Glusker, *J. Am. Chem. Soc.* **107** (1985) 2067.
27. P. B. Roberts, W. R. Wilson, L. P. G. Wakelin, R. F. Anderson and W. A. Denny, submitted to *Radiat. Res.*

28. S. Skonieczny and A. Ledochowski, *Heterocycles* **4** (1976) 1693.
29. S. Skonieczny, A. Organiak, A. Snarska, A. Kunikowska, S. Nowak and A. Ledochowski, *Pol. J. Chem.* **52** (1978) 2125.
30. M. Weltrowski and A. Ledochowski, *Pol. J. Chem.* **52** (1978) 215.
31. M. Weltrowski, A. Ledochowski and P. Sowinski, *Pol. J. Chem.* **55** (1981) 2309.
32. W. R. Wilson and R. F. Anderson, *Int. J. Radiat. Oncol. Biol. Phys.* **16** (1989) 1001.
33. J. Chodkowski and W. Kiwak, *Roczniki Chemii* **47** (1973) 157.
34. L. R. Ferguson, P. W. Turner and W. A. Denny, *Mutat. Res.* **187** (1986) 1.
35. W. A. Denny, G. J. Atwell, R. F. Anderson and W. R. Wilson. *J. Med. Chem.* in press.
36. K. Pawlak, J. W. Pawlak and J. Konopa, *Cancer Res.* **44** (1984) 4289.
37. M. Hrabowska, A. Ledochowski, B. Horowska, J. Konopa and K. Onoszaka, *Arch. Immunol. Ther. Exp.* **19** (1971) 879.
38. B. Wysocka-Skrela and A. Ledochowski, *Roczniki Chemii* **55**, (1981) 1735.
39. G. R. Clark, S. Hall, W. A. Denny and G. M. Stewart, *Acta Crystallogr.* **C42** (1986) 628.
40. W. A. Denny, B. F. Cain, G. J. Atwell, C. Hansch, A. Panthananickal and A. Leo, *J. Med. Chem.* **25** (1982) 276.
41. S. S. Legha, J. U. Gutterman, S. W. Hall, R. S. Benjamin, M. A. Burgess, M. Valdivieso and G. P. Bodey, *Cancer Res.* **38** (1978) 3712.
42. F. Cabanillas, S. S. Legha, G. P. Bodey and E. J. Freireich, *Blood* **57** (1981) 614.
43. R. de Jager, P. Seigenthaler, F. Cavallo, O. Klepp, V. Bramwell, R. Joss, P. Alberts, M. von Glabbeke, J. Reynard, M. Rozencweig and H. Hansen, *Eur. J. Cancer Clin. Oncol.* **19** (1985) 289.
44. H. J. Lawrence, C. A. Ries, R. D. Reynolds, J. P. Lewis, M. M. Koretz and F. M. Torti, *Cancer Treatment Rep.* **66** (1982) 1475.
45. R. Zittoun, *Cancer Treatment Rep.* **69** (1985) 1447.
46. G. W. Rewcastle, B. C. Baguley and B. F. Cain, *J. Med. Chem.* **25** (1982) 1231.
47. B. C. Baguley, *Drugs Today* **20** (1984) 237.
48. F. Reverdin and K. Widmer, *Ber.* **46** (1913) 4066.
49. B. F. Cain, G. J. Atwell and W. A. Denny, *J. Med. Chem.* **18** (1975) 1110.
50. W. A. Denny, G. J. Atwell and B. C. Baguley, *J. Med. Chem.* **26** (1983) 1625.
51. D. Hall, D. A. Swann and T. N. M. Waters, *J. Chem. Soc. (Perkin I)* (1974) 1334.
52. J. M. Karle, R. L. Cysyk and I. L. Karle, *Acta Crystallogr.* **B36** (1980) 3012.
53. W. A. Denny, G. J. Atwell and B. F. Cain, *J. Med. Chem.* **21** (1978) 5.
54. M. J. Waring, *Eur. J. Cancer* **12** (1976) 995.
55. W. R. Wilson, B. C. Baguley, L. P. G. Wakelin and M. J. Waring, *Mol. Pharmacol.* **20** (1981) 404.
56. T. D. Sakore, B. S. Reddy and H. M. Sobell, *J. Mol. Biol.* **135** (1979) 763.
57. W. A. Denny and L. P. G. Wakelin, *Cancer Res.* **46** (1986) 1717.
58. R. F. Anderson, J. E. Packer and W. A. Denny, *J. Chem. Soc. (Perkin II)* (1984) 49.
59. J. L. Jurlina, A. Lindsay, J. E. Packer and W. A. Denny, *J. Med. Chem.* **30** (1987) 473.
60. A. Wong, H-Y. Cheng and S. T. Crooke, *Biochem. Pharmacol.* **35** (1986) 1071.
61. D. D. Shoemaker, R L. Cysyk, S. Padmanabhan, H. B. Bhat and L. Malspeis, *Drug. Met. Dispos* **10** (1982) 35.
62. D. D. Shoemaker, R. L. Cysyk, P. E. Gormley, J. J. V. DeSouza and L. Malspeis, *Cancer Res.* **44** (1984) 1939.
63. R. F. Anderson, J. F. Packer and W. A. Denny, *J. Chem. Soc. (Perkin II)* **26** (1988) 191.
64. H. H. Lee, B. D. Palmer and W. A. Denny, *J. Org. Chem.* **53** (1988) 6042.
65. A. Wong, C-H. Huang and S. T. Crooke, *Biochemistry* **23** (1984) 2939, 2946.
66. B. F. Cain, W. R. Wilson and B. C. Baguley, *Mol. Pharmacol.* **12** (1976) 1027.
67. W. R. Wilson, B. F. Cain and B. F. Baguley, *Chem.-Biol. Interact.* **18** (1977) 163.
68. A. Wong, C-H. Haung, S-M. Hwang, A. W. Prestakyo and S. T. Crooke, *Biochem. Pharmacol.* **35** (1986) 1655.
69. M. N. Khan and L. Malspeis, *J. Org. Chem.* **47** (1982) 2731.
70. W. A. Denny, G. J. Atwell and B. F. Cain, *J. Med. Chem.* **23** (1979) 1453.
71. L. F. Liu, *Crit. Rev. Biochem.* **15** (1983) 1.
72. L. A. Zwelling, *Cancer Met. Rev.* **4** (1985) 263.
73. E. M. Nelson, K. M. Tewey and L. F. Liu, *Proc. Natl. Acad. Sci. (USA)* **81** (1984) 1361.
74. T. C. Rowe, G. L. Chen, Y.-H. Hsiang and L. F. Liu, *Cancer Res.* **46** (1986) 2021.

75. L. A. Zwelling, S. Michaels, D. Kerrigan, Y. Pommier and K. W. Kohn, *Biochem. Pharmacol.* **31** (1982) 3261.
76. K. M. Tewey, G. L. Chen, E. M. Nelson and L. F. Liu, *J. Biol. Chem.* **259** (1984) 9182.
77. B. F. Issell, *Cancer Treat. Rev.* **7** (1980) 73.
78. W. A. Denny, G. J. Atwell and B. C. Baguley, *J. Med. Chem.* **26** (1983) 1619.
79. W. A. Denny, G. J. Atwell and B. C. Baguley, *J. Med. Chem.* **27** (1984) 363.
80. B. C. Baguley, W. A. Denny, G. J. Atwell, G. J. Finlay, G. W. Rewcastle, S. J. Twigden and W. R. Wilson, *Cancer Res.* **44** (1984) 3245.
81. K. Soda and K. Nishiide, Jap. Patent 66 278 818 (1969) (*Chem. Abstracts* **72** (1970) 31645e).
82. J. M. S. Gagan, in *Chemistry of Heterocyclic Compounds*, Vol. 9, *Acridines*, ed. R. M. Acheson, Wiley, New York.
83. J. R. Hardy, V. J. Harvey, J. W. Paxton, P. C. Evans, S. Smith, A Grillo-Lopez, W. Grove and B. C. Baguley, *Cancer Res.* (1988) **48** 6593.
84. G. J. Atwell, G. W. Rewcastle, W. A. Denny, B. F. Cain and B. C. Baguley, *J. Med. Chem.* **27** (1984) 367.
85. G. J. Atwell, B. C. Baguley, G. J. Finlay, G. W. Rewcastle and W. A. Denny, *J. Med. Chem.* **29** (1986) 1769.
86. G. J. Atwell, G. W. Rewcastle, B. C. Baguley and W. A. Denny, *J. Med. Chem.* **30** (1987) 652.
87. J.-B. Le Pecq, N.-D. Xuong, G. Gosse and C. Paoletti, *Proc. Natl. Acad. Sci. (USA)* **71** (1974) 5078.
88. B. C. Baguley, W. A. Denny, G. J. Atwell and B. F. Cain *J. Med. Chem.* **24** (1981) 520.
89. S. I. Fink, A. Leo, M. Yamakawa, C. Hansch and F. R. Quinn, *Farmaco. Ed. Sci.* **35** (1980) 965.
90. W. Muller and D. M. Crothers, *J. Mol. Biol.* **25** (1968) 251.
91. W. A. Denny, I. A.G. Roos and L. P. G. Wakelin, *Anticancer Drug Design* **1** (1986) 141.
92. L. P. G. Wakelin, G. J. Atwell, G. W. Rewcastle and W. A. Denny, *J. Med. Chem.* **30** (1987) 855.
93. J. Feigon, W. A. Denny. W. Leupin and D. R. Kearns, *J. Med. Chem.* **27** (1984) 450.
94. P. H. von Hippel and J. D. McGhee, *Annu. Rev. Biochem.* **41** (1972) 231.
95. N. Capelle, J. Barbet, P. Desson, S. Blanquet, B. P. Roques and J.-B. LePecq, *Biochemistry* **18** (1979) 3354.
96. L. P. G. Wakelin *Med. Res. Rev.* **6** (1986) 275.
97. W. A. Denny, G. J. Atwell, B. C. Baguley and L. P. G. Wakelin, *J. Med. Chem.* **28** (1985) 1568.
98. R. K. Johnson and W. S. Howard, *Eur. J. Cancer Clin. Oncol.* **18** (1982) 179.
99. A. Goldin, J. M. Venditti, J. S. McDonald, F. M. Muggia, J. E. Henney and V. T. DeVita *Eur. J. Cancer* **17** (1981) 129.
100. Y. A. Beltagy, W. Waugh and A. J. Repta, *Drug Dev. Ind. Pharmacol.* **6** (1980) 411.
101. T.-K. Chen, R. M. Fico and E. S. Canellakis, *J. Med. Chem.* **21** (1978) 868.
102. D. P. Kelly, P. O.-L. Mack, R. F. Martin and L. P. G. Wakelin, *Int. J. Peptide Protein Res.* **26** (1985) 400.
103. B. Stefanska and A. Ledochowska, *Roczniki Chemii* **46** (1972) 1637.
104. E. S. Canellakis, Y. H. Shaw, W. E. Hanners and R. A. Schwartz, *Biochim. Biophys. Acta* **418** (1976) 277.
105. J. B. Hansen, E. Langvad, F. Frandsen and O. Buchardt, *J. Med. Chem.* **26** (1983) 1510.
106. C. Courseille, F. Leroy, M. Hospital and and J. Barbet, *Acta Crystallogr* **B33** (1977) 1565; C. Courseille, F. Leroy, B. Buseta and J. Barbet, *ibid.* **B33** (1977) 1570.
107. J. Barbet, P. B. Rocques, S. Combrisson and J.-B. Le Pecq, *Biochemistry* **15** (1976) 2642.
108. L. P. G. Wakelin, N. Romanos, T.-K. Chen, D. Glaubiger and E. S. Canellakis, *Biochemistry* **13** (1978) 5057.
109. P. H. von Hippel and J. D. McGhee, *Annu. Rev. Biochem.* **41** (1972) 231.
110. N. Assa-Munt, W. A. Denny, W. Leupin and D. R. Kearns, *Biochemistry* **24** (1985) 1441.
111. R. H. Schafer and M J. Waring, *Biopolymers* **21** (1982) 2279.
112. G. J. Atwell, G. M. Stewart, W. Leupin and W. A. Denny, *J. Am. Chem. Soc.* **107** (1985) 4335.
113. R. G. McR. Wright, L..P. G. Wakelin, A. Fieldes, R. M. Acheson and M. J. Waring, *Biochemistry* **19** (1980) 5825.
114. J-B. LePecq, M. Le Bret, J. Barbet and P. B. Roques, *Proc. Natl. Acad. Sci. (USA)* **72** (1975) 2915.

115. N. Assa-Munt, W. A. Denny, W. Leupin and D. R. Kearns, *Biochemistry* **24** (1985) 1449.
116. T.-K. Chen, R. M. Fico and E. S. Canellakis, *J. Med. Chem.* **21** (1978) 868.
117. A. Adams, B. Jarrot, B. C. Elmes, W. A. Denny and L. P. G. Wakelin, *Mol. Pharmacol* **27** (1985) 480.
118. J. B. Hansen, T. Koch, O. Buchardt, P. E. Neilson, B. Norden and M. Wirth, *J. Chem. Soc. Chem. Commun.* (1984) 509.
119. G. J. Atwell, W. Leupin, S. J. Twigden and W. A. Denny, *J. Am. Chem. Soc.* **105** (1983) 2913.
120. W. A. Denny, G. J. Atwell, G. Willmott and L. P. G. Wakelin *Biophys. Chem.* **22** (1985) 17.
121. B. Gaugain, J. Markovits, J.-B. Le Pecq and P. B. Rocques *FEBS Le tt.* **169** (1984) 123.
122. G. J. Atwell, B. C. Baguley, D. Wilmanska and W. A. Denny, *J. Med. Chem.* **29** (1986) 69.
123. J. B. Hansen and O. Buchardt, *J. Chem. Soc. Chem. Commun.* (1983) 162.
124. G. J. Atwell, B. F. Cain, B. C. Baguley, G. J. Finlay and W. A. Denny, *J. Med. Chem.* **27** (1984) 1481.
125. G. W. Rewcastle, G. J. Atwell, D. Chambers, B. C. Baguley and W. A. Denny, *J. Med. Chem.* **29** (1986) 472.
126. W. A. Denny, G. J. Atwell, G. W. Rewcastle and B. C. Baguley, *J. Med. Chem.* **30** (1987) 658.
127. G. M. Stewart, G. W. Rewcastle and W. A. Denny, *Aust. J. Chem.* **37** (1984) 1939.
128. G. W. Rewcastle and W. A. Denny, *Synthesis* (1985) 217.
129. G. W. Rewcastle and W. A. Denny, *Synthetic Comm.* **17** (1987) 309.
130. B. D. Hudson, R. Kuroda, W. A. Denny and S. Neidle, *J. Biomol. Struct. Dynamics* **5** (1987) 145.
131. G. J. Atwell, G. W. Rewcastle, B. C. Baguley and W. A. Denny, *J. Med. Chem.* **30** (1987) 664.
132. B. C. Baguley, A. R. Kernohan and W. R. Wilson, *Eur. J. Cancer Clin. Oncol.* **19** (1983) 1607.

2 Antitumour anthracyclines

A. SUARATO

2.1 Introduction and historical development

The first reports on structure and chemotherapeutic potential of two 'red pigments' belonging to the anthracycline class of natural products, daunorubicin and doxorubicin, were published 20 years ago. Since then many other anthracyclines have been studied, but it was only the discovery of the potent antineoplastic activity of doxorubicin and its subsequent fast approval in 1974 by the U.S. Federal Drug Administration (FDA) as a prescription drug which triggered off the attention of the scientific community. The search for other anthracycline analogues with better therapeutic indices than doxorubicin has been tenaciously pursued in many laboratories around the world. The therapeutic efficacy of doxorubicin was limited by toxic side effects, mainly cardiotoxicity and myelosuppression. The research efforts have brought the discovery of other clinically useful anthracyclines, such as aclacinomycin A and carminomycin, isolated from natural sources, and the discovery, in synthetic laboratories, that minor structural changes in the lead compound doxorubicin might significantly improve the therapeutic index.

The few anthracyclines, beside doxorubicin, that have passed through the preclinical and clinical investigations are described in this chapter, with the aim of comparing their characteristics in a multidisciplinary fashion, i.e. pharmacological, chemical and metabolic. In particular two drugs, epirubicin and 4-demethoxydaunorubicin, which chemically differ very little from the parent, and aclacinomycin A, which probably represents the most important anthracycline discovered in fermentation broths since daunorubicin and doxorubicin, are considered. Several tables summarising most of the reactions in which the anthracyclines are involved are included, as these may be helpful to the reader for a better understanding of the chemical potential of the anthracyclines.

The study of anthracyclines as a class of naturally occurring compounds began in the 1950s with the work of Brockmann and Brown, who elucidated the structure of the red-orange pigments isolated from cultures of different *Streptomyces* sp[1]. The aglycone moieties of these compounds were identified as anthracyclinones belonging to the types known as rhodomycinone, isorhodomycinone and pyrromycinones, whose colour was due to the presence of a polyhydroxyanthraquinone chromophore (Figure 2.1).

Early in the 1960s, in the laboratories of Farmitalia (Italy)[2] and indepen-

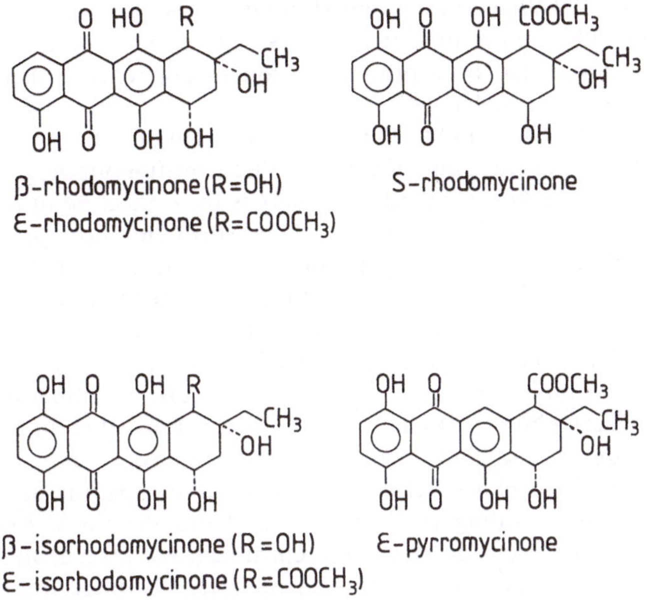

β-rhodomycinone (R=OH)
ε-rhodomycinone (R=COOCH₃)

S-rhodomycinone

β-isorhodomycinone (R = OH)
ε-isorhodomycinone (R = COOCH₃)

ε-pyrromycinone

Figure 2.1 Anthracyclines isolated and studied by Brockman and Bauer.

dently in those of Rhone-Poulenc (France)[3], a new anthracycline, named daunorubicin (1), was isolated. This compound showed the same typical hydroxyanthraquinone chromophore system as the other anthracyclines, and was characterised by the presence of an acetyl group in the alicyclic ring and of a unique amino sugar, daunosamine[4]. Daunorubicin was shown to be endowed with remarkable antitumour activity[5]. Later on, doxorubicin (2) was discovered in the Farmitalia laboratories[6]. The biological activity of these antitumour anthracyclines is related to their ability to complex with DNA with the consequent inhibition of replication and transcription. Note: The rings are indicated by the letters A, B, C and D, starting from the alicyclic ring;

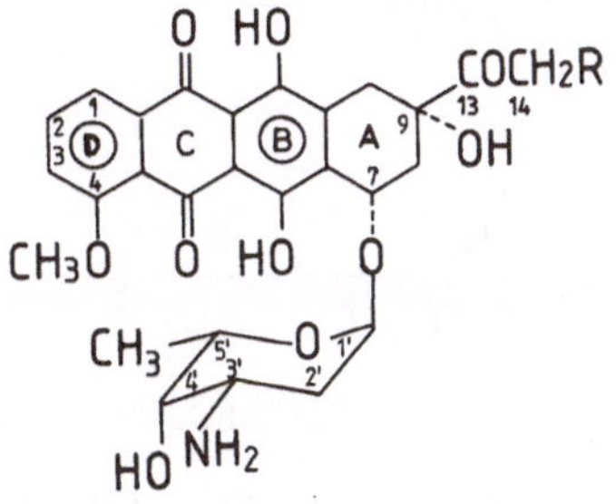

Daunorubicin (1 R = H)
Doxorubicin   (2 R = OH)

the numbering of the carbon atoms starts from the left side and the carbon bearing the sidechain is indicated as C-9. The carbon atoms in the sugar moiety are designated by a prime notation to differentiate them from the carbon atoms of the aglycone[1].

At present, doxorubicin has the widest spectrum of antitumour activity of all antineoplastic agents and is the most utilised antitumour drug worldwide[7]. A cumulative dose-limiting cardiac toxicity is the crucial point in its clinical use[8]. The awareness of the toxic effects of these drugs has prompted the search for new analogues with reduced cardiotoxic potential and a more favourable therapeutic index in comparison with daunorubicin and doxorubicin[9]. Hundreds of anthracyclines have been isolated from culture broths or synthesised by several methods: (a) partial synthesis, (b) total synthesis, (c) biosynthesis, (d) microbial transformation[10]. So far, a number of anthracyclines have been chosen for clinical evaluation. Many of them have already been discontinued since they do not offer any advantage over doxorubicin[7,11]. The main anthracyclines selected for clinical evaluation are shown in Figures 2.2–2.6 according to their structural modifications. Among those mentioned, doxorubicin, daunorubicin, epirubicin and idarubicin represent at the moment the most extensively studied and clinically interesting derivatives. This review focuses on a description of the chemistry, physical aspects, reactivity and metabolism of these drugs.

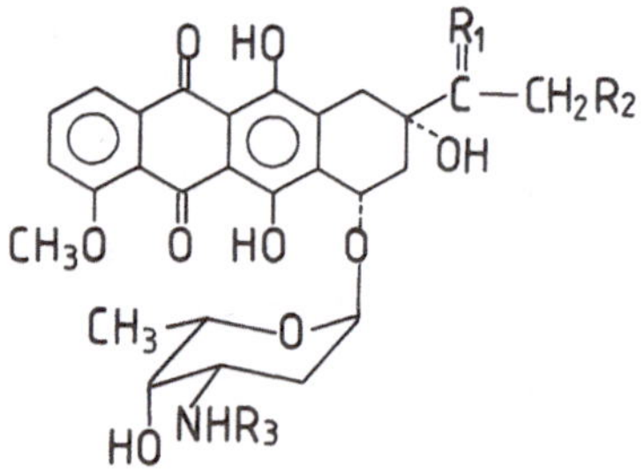

Compound	R_1	R_2	R_3	Ref.
Rubidazone	$=N-NHCOC_6H_5$	H	H	[20]
Detorubicin	O	$OCOCH(OC_2H_5)_2$	H	[21]
AD–32 14-valerate	O	$OCO(CH_2)_3CH_3$	$COCF_3$	[22]

Figure 2.2 Daunorubicin derivatives modified in the chromophore ring system.

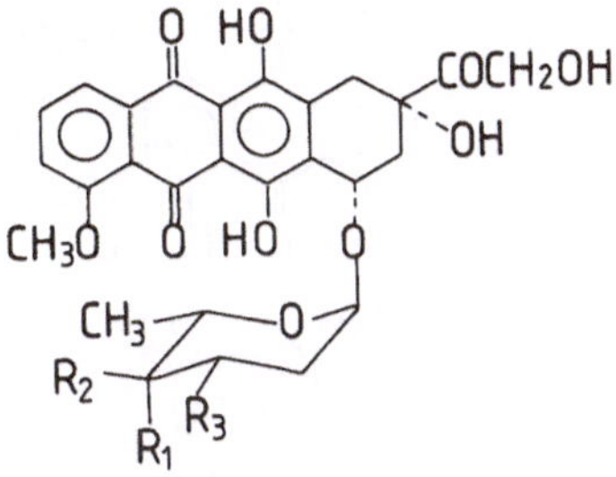

Compound	R_1	R_2	R_3	Ref.
Doxorubicin	OH	H	NH_2	[1]
4'-epidoxorubicin (Epirubicin)	H	OH	NH_2	[15]
4'-deoxydoxorubicin	H	H	NH_2	[16]
4'-jodo-4'-deoxy- doxorubicin	J	H	NH_2	[17]
4'-O-tetrahydro- pyranyladriamycin (β-isomer)	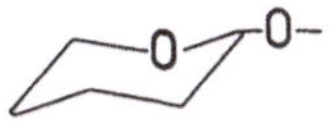 (tetrahydropyranyl-O-)	H	NH_2	[18]
3'-desamino-(3"- cyano-4"-morpholynyl)- doxorubicin	OH	H	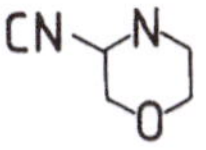 (3-cyano-4-morpholinyl)	[19]

Figure 2.3 Doxorubicin derivatives substituted on the sugar moiety.

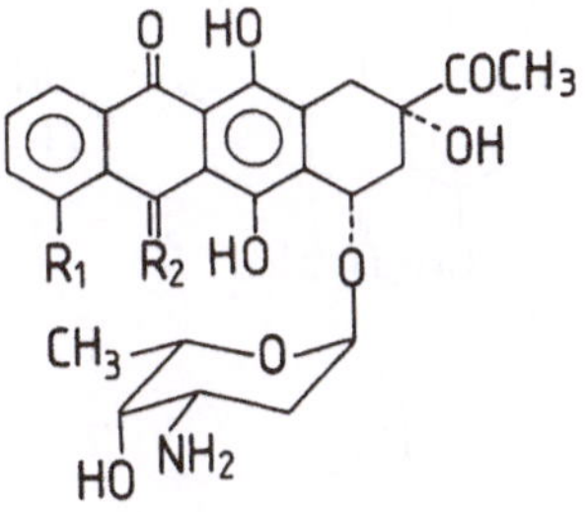

Compound	R_1	R_2	Ref.
Daunorubicin	OCH_3	O	[2]
Carminomycin	OH	O	[12]
4-demethoxydauno- rubicin (Idarubicin)	H	O	[13]
5-iminodaunorubicin	OCH_3	NH	[14]

Figure 2.4 Daunorubicin and doxorubicin derivatives substituted on the sidechain.

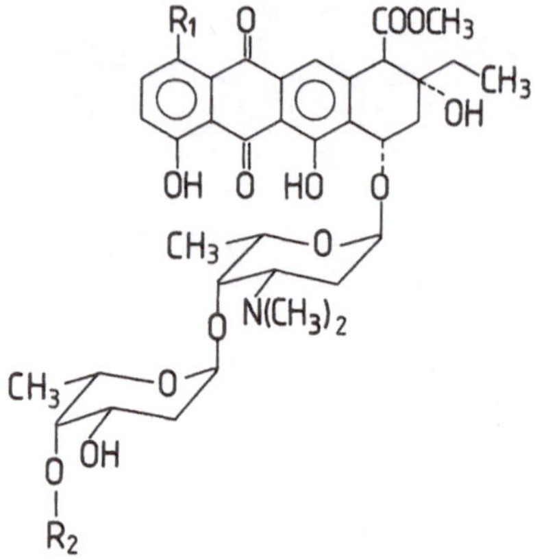

Compound	R₁	R₂	Ref
Aclacynomycin A	H	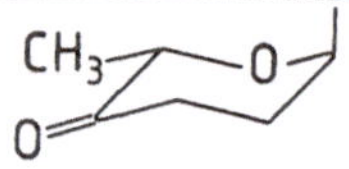	[23]
Marcellomycin	OH	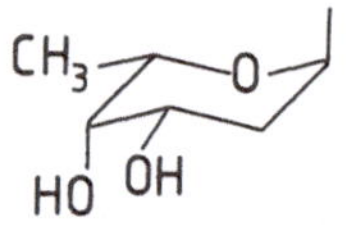	[24]

Figure 2.5 Antitumour anthracyclines having the aglycone pyrromycinone linked at C-7 to a trisaccharide chain.

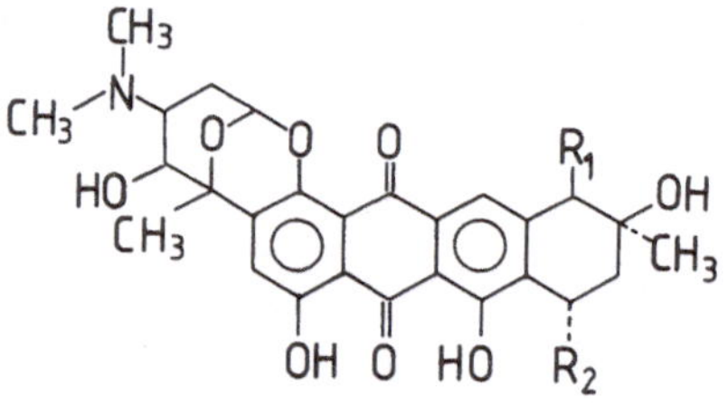

Compound	R₁	R₂	Ref.
Nogalamycin	CO_2CH_3	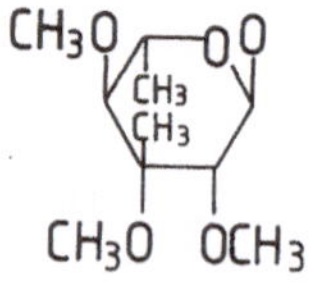	[25]
7-CON-O-Methyl-nogarol (Menogaril)	H	OCH_3	[26]

Figure 2.6 Anthracyclines having an amino sugar residue linked to the D ring.

2.2 Doxorubicin

Doxorubicin was first isolated from cultures of a *Streptomyces* strain (*S. peucetius* var. *caesius*), a variety derived from the original *S. peucetius*[6]. Doxorubicin is the 'drug' against which all other anthracyclines are compared, experimentally and clinically. The difference between the chemical structure of doxorubicin and daunorubicin lies in a single hydroxyl group. Despite this minor chemical difference, there is a marked difference in their antitumour efficacy. While daunorubicin has little activity against solid tumours[27], doxorubicin is active against carcinomas and sarcomas[11]. Acid hydrolysis of doxorubicin affords the aglycon adriamycinone and the amino sugar daunosamine, 3-amino-2,3,6-trideoxy-L-*lyxo*-hexopyranose.

Adriamycinone   Daunosamine

The presence of this particular amino sugar residue is an important requirement for bioactivity, since a biological action for the isolated aglycone moiety has never been reported. The amino group of the glycoside confers good water solubility at neutral pH, as it is in the cationic form. The six chiral centres of doxorubicin are, respectively, 7(*S*) and 9(*S*) for the aglycone and 1′(*R*), 3′(*S*), 4′(*S*) and 5′(*S*) for the sugar corresponding to the L-*lyxo* configuration. The basic C-3′ amino group and the slightly acidic phenolic groups present in doxorubicin, as well as in its congeners, make this molecule soluble in both acidic and basic media.

2.2.1 *Microbiological isolation*

Doxorubicin can be obtained by aerobic fermentation of *Streptomyces peucetius* var. *caesius*. However, the isolation in pure form is difficult, due its low stability in aqueous solution at pH values above neutrality and to the concomitant presence, in the same culture broth, of other anthracyclines having the same chromatographic properties. Doxorubicin is recovered from the mycelium and the fermentation broth by extraction with acidic acetone and purification by partition chromatography on a column of cellulose buffered with pH 5.4 M/15 phosphate. The antibiotic is recovered from the eluate in l-butanol saturated with the pH 5.4 buffer, followed by re-extraction with methylene chloride at pH 8.6. The organic solution is concentrated and doxorubicin crystallised as the hydrochloride by adding an equivalent of hydrogen chloride in anhydrous methanol. Final purification is performed by crystallisation from a methanol/1-propanol mixture[6].

2.2.2   Chemical preparation

Doxorubicin is obtained by reacting daunorubicin hydrochloride (1) in a solvent mixture of methanol and dioxane with a chloroform solution of bromine, forming 14-bromo-daunorubicin (3) which is hydrolysed with an aqueous solution of sodium formate under a nitrogen atmosphere. The mixture is acidified to pH 3 and extracted with methylene chloride; the aqueous solution is brought to pH 8.6 and extracted with methylene chloride. The organic extract is dried over anhydrous sodium sulphate, concentrated, treated with methanolic hydrogen chloride, and then diluted with diethylether. The precipitate formed is doxorubicin hydrochloride (2), which is purified by crystallisation from a mixture of methanol and 1-propanol[28]. The reaction pathway is summarised in Scheme 2.1.

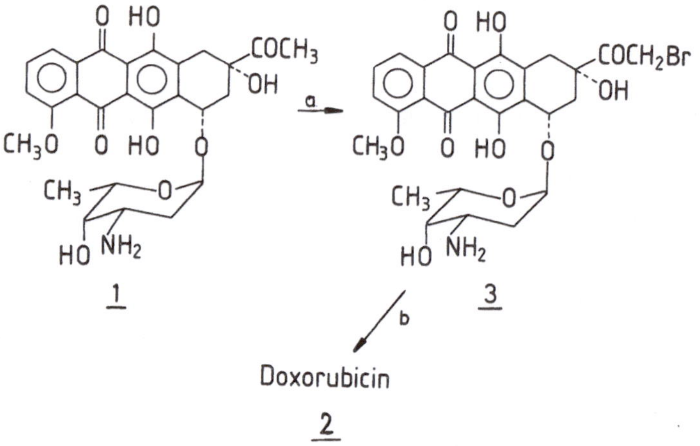

Scheme 2.1 Synthetic pathway for doxorubicin. (a) Br_2, CH_3OH-dioxane; (b) HCOONa.

2.2.3   Analytical profile

According to Chemical Abstracts, doxorubicin is chemically named (8S, 10S)-10-[3-amino-2,3,6-trideoxy-α-L-*lyxo*-hexopyranosyl)oxy]-7,8,9,10-tetrahydroxy-6,8,11-trihydroxy-8-hydroxyacetyl-1-methoxy-5,12-naphtha-cenedione hydrochloride. Other name: (7S, 9S)-9-hydroxyacetyl-4-methoxy-7,8,9,10-tetrahydro-6,7,9,11-tetrahydroxy-7-O-(2,3,6-trideoxy-3-amino-α-*lyxo*-hexopyranosyl)-5,12-naphthacenedione hydrochloride[29].

$$\underline{2}$$

$C_{27}H_{29}NO_{11}$ mol.wt. 543.5
Hydrochloride salt (CAS 25316-40-9)
$C_{27}H_{29}NO_{11} \cdot HCl$ mol.wt. 580.0
Melting point[30]   m.p. 205°C
Optical rotation[31] $[\alpha]_D^{25} + 225°$
Ionization constant[32] pK_a 8.22
Apparent partition coefficient[33] P_{app} 0.52

2.2.3.1 *Appearance.* The hydrochloride salt is a red, free-flowing crystalline powder. The pharmaceutical form, containing lactose, is a lyophilised porous cake.

2.2.3.2 *Solubility.* Doxorubicin hydrochloride is readily soluble in water, methanol, acetonitrile and tetrahydrofuran, but only slightly soluble or insoluble in less polar solvents.

2.3 Epirubicin (4'-epidoxorubicin)

Epirubicin 7 is a new anthracycline antibiotic in which the stereochemistry at C-4' of daunosamine has been inverted. Thus the amino sugar is in the L-*arabino* configuration instead of the L-*lyxo* configuration present in doxorubicin[15,34]. The hydroxyl group on the sugar moiety, possessing the stable $^1C_4$ conformation, has an equatorial orientation. This molecular modification does not substantially change the antitumour efficacy, as compared with that of doxorubicin[35]; however, the analogue appeared to be significantly better tolerated and less cardiotoxic than the parent drug[36]. The lower incidence of toxic side effects of epirubicin in humans may be related to the ability of the drug to be a substrate for human β-glucuronidase; in fact the main urinary metabolite of the drug in humans is the corresponding 4'-O-β-glucuronide[37].

2.3.1 *Preparation*

The process for manufacturing epirubicin, illustrated in Scheme 2.2, is based on two key reactions; an oxidation followed by a regio- and stereoselective reduction, which permits the inversion of configuration at C-4' of N-trifluoroacetyldaunorubicin. The starting material (4), obtained from the natural antibiotic daunomycin by a common N-trifluoroacetylation, is subjected at very low temperature ($-70°C$) to an oxidative reaction with dimethylsulfoxide activated by trifluoroacetic anhydride and followed by addition of a tertiary amine base affording the ketone (5). Treatment of (5) with an equivalent of sodium borohydride at $-30°C$ affords 4'-epi-N-trifluoro-acetyldaunorubicin (6)[15b] from which epirubicin (7) is obtained by the method used for the chemical transformation of daunorubicin to doxorubicin[28].

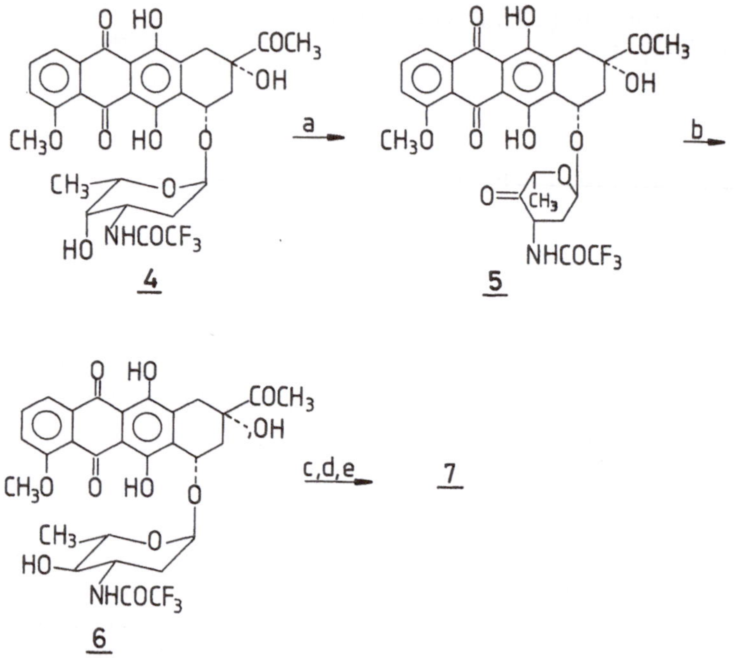

Scheme 2.2 Preparation of epirubicin, reagents and conditions. (a) $(CH_3)_2SO$, $(CF_3CO)_2O$, Et_3N, CH_2Cl_2, $-70°C$; (b) $NaBH_4$, CH_3OH, $-70°C$; (c) 0.1 N NaOH, 0°C; (d) Br_2, CH_3OH-dioxane, HBr; (e) HCOONa.

2.3.2 Analytical profile

According to Chemical Abstracts, epirubicin is named (8S-cis)-10-[(3-amino-2,3,6-trideoxy-α-arabino-hexopyranosyl)oxy]-7,8,9,19-tetrahydro-6,8,11-trideoxy-8-(hydroxyacetyl)-1-methoxy-5,12-naphthacenedione hydrochloride. It is also named: (7S, 9S)-9-hydroxyacetyl-4-methoxy-7,8,9,10-tetrahydro-6,7,9,11-tetrahydroxy-7-O-(2,3,6-trideoxy-3-amino-α-arabino-hexopyranosyl)-5,12-naphthacenedione hydrochloride[34].

$$\underline{7}$$

$C_{27}H_{29}NO_{11}$ mol.wt. 543.5
Hydrochloride salt (CAS 56390-09-1)
$C_{27}H_{29}NO_{11} \cdot HCl$ mol.wt. 580.0
Melting point[30]   m.p. 173°C
Optical rotation[31] $[\alpha]_D^{25} + 314°$
Ionization constant[32] pK_a 7.7
Apparent partition coefficient[33] P_{app} 1.14

2.3.2.1 *Appearance.* The hydrochloride salt is a red, free-flowing crystalline powder.

2.3.2.2 *Solubility.* Epirubicin hydrochloride is soluble in water and in methanol, slightly soluble in ethanol, and practically insoluble in acetone or methylene chloride.

2.4 Idarubicin (4-demethoxydaunorubicin)

The anthraquinone chromophore system is an important structural feature for all the anthracyclines. A wide range of oxygenation patterns appears to be compatible with biological activity[1]. Idarubicin (4-demethoxydaunorubicin) (13) is a totally synthetic anthracycline related to daunorubicin (Figure 2.2)[13]; the only change involves the lack of the methoxy group in position C-4 of the aglycone system. Pharmacologically the compound is characterised by a higher potency than that of the parent compound daunorubicin[38] and by a lower cardiotoxic index with respect to doxorubicin[38]. Idarubicin shows important antileukaemic activity in adult acute non-lymphocytic leukaemia, acute lymphocytic leukaemia, chronic myelogenous leukaemia and in paediatric leukaemias. There is also evidence of antileukaemic activity for idarubicin administered by the oral route[39]. An important aspect of the metabolic fate of idarubicin is the extensive conversion in humans to the corresponding 13-dihydro derivative (idarubicinol)[40] which exhibits remarkable antitumour activity in experimental murine systems.

2.4.1 *Preparation*

The process for manufacturing idarubicin is shown in Scheme 2.3 and is carried out by performing the total synthesis of the corresponding 4-demethoxydaunomycinone (**10**) followed by coupling with the natural amino sugar, daunosamine. Among several synthetic methods described for the preparation of (**10**)[41], the practical one, used for large-scale preparation, is largely based on the work of Wong[42]. The optically resolved $(-)$-1,4-dimethoxy-6-hydroxy-6-acetyltetralin (**8**)[13,43] is heated to melting with phthalic anhydride in the presence of aluminium chloride and sodium chloride to give 4-demethoxy-7-deoxydaunomycinone (**9**) in good yield without apparent racemisation[13]. Conversion to the corresponding C-13 cyclic acetal, followed by treatment with bromine and 2,2'-azo-bisisobutyronitrile, affords 4-demethoxydaunomycinone (**10**) after hydrolytic work up. Glycosidation of the latter with N,O-ditrifluoroacetyldaunosaminyl chloride (**11**) in the presence of silver triflate, followed by methanolysis, affords 4-demethoxy-N-trifluoroacetyldaunorubicin (**12**) from which idarubicin is obtained by mild alkaline hydrolysis. Treatment of the free base with methanolic hydrogen chloride gives idarubicin hydrochloride in acceptable pharmaceutical form.

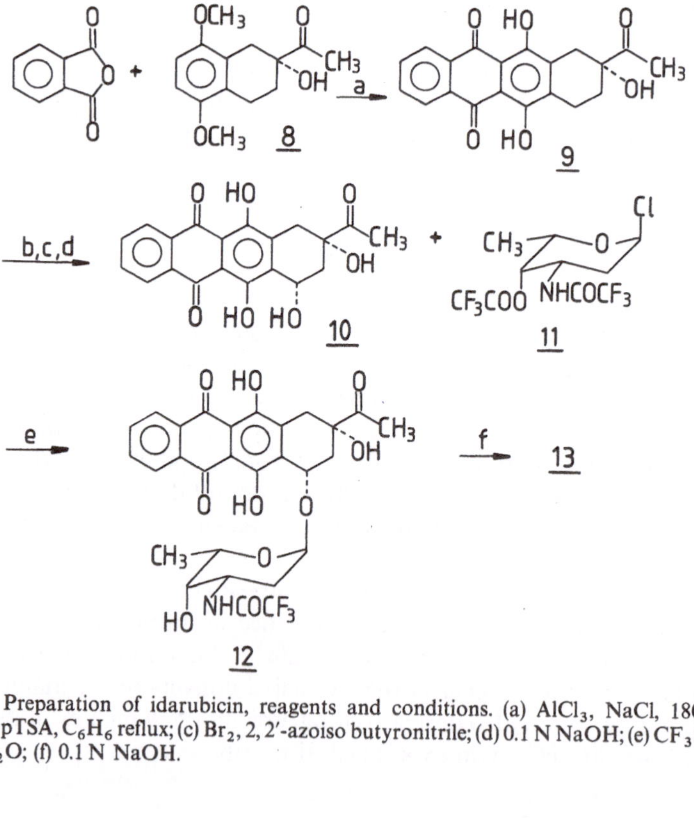

Scheme 2.3 Preparation of idarubicin, reagents and conditions. (a) AlCl$_3$, NaCl, 180°C; (b) (CH$_2$OH)$_2$, pTSA, C$_6$H$_6$ reflux; (c) Br$_2$, 2, 2'-azoiso butyronitrile; (d) 0.1 N NaOH; (e) CF$_3$SO$_2$Ag, CH$_2$Cl$_2$/Et$_2$O; (f) 0.1 N NaOH.

2.4.2 Analytical profile

According to Chemical Abstracts, idarubicin hydrochloride is best chemically named: (7-cis)-9-acetyl-7-[(3-amino-2,3,6-trideoxy-α-lyxo-hexopyranosyl)-oxy]-7,8,9,10-tetrahydro-8,9,11-trideoxy-5,12-naphthacenedione hydrochloride. Other name: (7S,9S)-9-acetyl-7,8,9,11-tetrahydro-6,7,9,11-tetrahydro-xy-7-O-(2,3,6-trideoxy-3-amino-α-lyxo-hexopyranosyl)-5,12-naphthacenedione hydrochloride.

13

$C_{26}H_{27}NO_9$	mol.wt. 497
Hydrochloride salt (CAS 57852-57-0)	
$C_{26}H_{27}NO_9 \cdot HCl$	mol.wt. 533.5
Melting point[30]	m.p. 173°C
Optical rotation[31]	$[\alpha]_D^{25} + 178°$
Ionization constant[32]	pK_a 8.5 (water)
Apparent partition coefficient[33]	P_{app} 7.6

2.4.2.1 *Appearance.* Orange-red powder.

2.4.2.2 *Solubility.* Slightly soluble in water, sparingly soluble in non-polar organic solvents.

2.5 Spectroscopic properties

2.5.1 Infrared spectra

In Table 2.1 the main absorption bands[44] of the infrared spectra of doxorubicin[29], epirubicin[34] and idarubicin as their hydrochloric salts, recorded from a KBr pellet (0.4%) on a Perkin-Elmer 457 grating spectrophotometer, are reported.

2.5.2 Mass spectra

The mass spectrum of the anthracyclinone can be recorded by electron-impact ionisation[45], but this technique cannot be used to obtain the spectra of the

Table 2.1 Infrared spectra of doxorubicin (D), epirubicin (E), idarubicin (I).

IR absorption band (cm^{-1})	Compounds	Assignments
3560–3160	D, E, I	O–H streching (hydrogen bonded)
3160–2300	D, E, I	NH$_3$ and O–H streching (hydrogen bonded)
1724	D, E	C=O (ketone)
1710	I	C=O (ketone)
1613 and 1580	D, E	C=O stretching (intra-hydrogen bonded quinone)
1620 and 1585	I	C=O streching (intra-hydrogen bonded quinone)
1420–1340	D, E, I	C–O stretching and O–H deformation
1282	D, E	C–O–C stretching (ether)
1270–1230	D, E, I	C–O stretching and O–H deformation
1115	D, E, I	C–O (tertiary alcohol)
1071	D, E, I	C–O (secondary alcohol)
1008	D, E	C–O (primary alcohol)
730	D, E, I	Out-of-plane deformation of the aromatic C–H

Table 2.2 ^{1}H-NMR data of doxorubicin[29] and epirubicin[34] hydrochloride in DMSO-d_6 solution, trimethylsilane (TMS) is the internal reference.

Doxorubicin				Epirubicin		
J^a	Multiplicity^b	Position (ppm)	Proton	Position (ppm)	Multiplicity	J^a
6.5	d	1.15	CH$_3$-5′	1.20	d	6.2
			H-2′ax	1.75	bt	12
	m	1.77	H-2′eq	2.05	bd	12
	m	2.15	H$_2$-8	2.14	bs	
10.0	bs	4.90	H$_2$-10	2.95	s	
	m	3.31	H-3′ax			
			H-4′ax	3.09	bs	
6.0	bs	3.62	H-4′eq			
	s	3.94	CH$_3$O	3.97	s	
6.5	dq	4.14	H-5′	3.97	m	
	s	4.57	H$_2$-14	4.55	d	6
	s	2.92	H-7	4.94	t	
7.0	bs	5.25	H-1′	5.26	bt	
	bs	4.46	OH-9	5.48	bd	
			OH-4′	5.77	s	6
			H-2	7.68	d	
7.0	t	7.53	H-3			
			H-4	7.91	m	
	s	13.85	OH-11	13.24	s	
	s	16.08	OH-6	14.03	s	
			H-4			
7.0	d	7.79	H-2			

^{a} Or W_H (Hz).

^{b} s, singlet; d, doublet; t, triplet; m, multiplet; b, broad; bs, broad signals; dq, double quartet; dd, double doublet.

glycosides. The intact molecule can be examined by field desorption ionisation mass spectrometry[29]. The major fragmentation peaks have m/e corresponding to M + 1, molecular ion, aglycone and aglycone aromatised on the A ring (7,9-bisanhydroanthracyclinone).

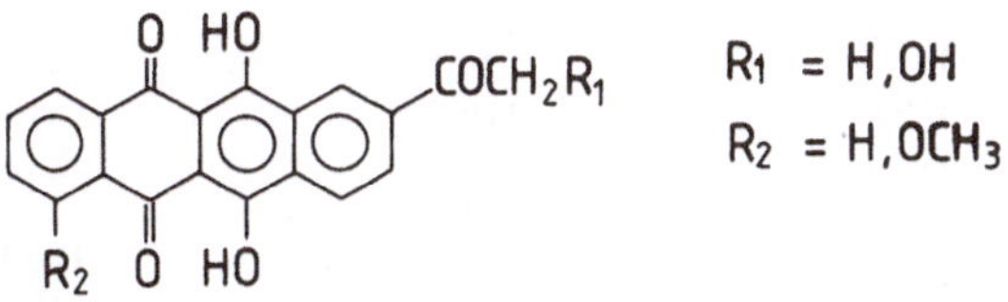

R₁ = H,OH
R₂ = H,OCH₃

7,9-bisanhydroanthracyclinone

2.5.3 Nuclear magnetic resonance spectra

The attribution of the chemical shift values for the ^{1}H-NMR spectrum of doxorubicin, epirubicin and idarubicin was made on the basis of the data for daunorubicin[4]. The assignments of the chemical shift values are reported in Tables 2.2 and 2.3. The assignments of the chemical shift values of the ^{13}C-NMR are given in Table 2.4. Proton and carbon NMR were recorded at 200 MHz on a Varian XL-200 spectrometer.

Table 2.3 ^{1}H-NMR data of idarubicin hydrochloride in DMSO-d_6 solution, TMS is the internal references.

Proton	ppm	Multiplicity	J or W_H (Hz)
CH$_3$-5′	1.13	d	6.6
H$_3$-14	2.25	s	
H-2′ax	1.88	ddb	12.0
H-2′eq	1.69	td	12.0
H-8ax	2.09	dd	14.5
H-8eq	2.15	dd	14.5
H-10ax	2.81	d	18.5
H-10eq	2.92	d	18.5
H-4′	3.55	m	6
H-3′	3.39	m	12
H-5′	4.19	qb	6.6
H-7	4.93	dd	5.5
H-1′	5.28	d	3
OH-4′	5.46	d	6
OH-9	5.54	bs	
NH$_3$	7.84	bs	
H-2			
H-3	8.02–7.95	m	
H-1			
H-4	8.03–8.25	m	
OH-11	13.31	s	
OH-6	13.52	s	

[a]Abbreviations are as indicated for Table 2.2

Table 2.4 ^{13}C-NMR data of doxorubicin[29], epirubicin[34] and idarubicin hydrochloride in D_2O solution. Values (ppm) are referred to TMS.

Carbon atom	Doxorubicin hydrochloride	Epirubicin hydrochloride	Idarubicin hydrochloride
1	119.4	119.5	126.6
2	136.6	136.7	135.5
3	119.4	119.5	135.6
4	160.2	160.2	126.6
OCH$_3$	56.4	56.4	
5	185.2	(185.4)	(186.3)
6	155.4	155.4	156.6
7	68.4	68.4	69.9
8	35.3	35.3	35.9
9	75.7	75.9	75.0
10	32.2	32.3	31.8
11	153.8	153.9	155.2
12	185.2	(185.2)	(186.1)
4a	119.4	118.4	132.7
5a	110.3	110.3	110.0
6a	(133.7)[a]	(133.9)	136.4
10a	(133.7)	(133.3)	135.0
11a	110.3	110.3	110.0
12a	(133.4)	(133.6)	132.7
1′	98.8	98.2	99.3
2′	27.6	38.3	28.3
3′	46.9	49.7	49.7
4′	67.1	72.3	66.2
5′	66.2	69.4	66.2
6′	15.8	16.5	16.8
13	214.5	214.5	212.0
14	64.4	64.5	24.3

[a] Similar values in ( ) may be interchanged.

2.5.4 Ultraviolet and visible spectra

The ultraviolet and visible spectra of doxorubicin, epirubicin and idarubicin as their hydrochloride salts in methanol containing 0.01 N HCl are shown in Figure 2.7. Those of doxorubicin and epirubicin are comparable, due to the presence in the molecules of the same chromophore; that of idarubicin is

Table 2.5 Ultraviolet and visible molecular absorptivities of doxorubicin and epirubicin hydrochloride.

Wavelength (nm)	e	$E^{1\%}_{1cm}$
233	38150	658
253	23500	440
290	8400	145
477	13050	225
495	13000	224
530	7200	124

Table 2.6 Ultraviolet and visible molecular absorptivities of idarubicin hydrochloride.

Wavelength (nm)	e	$E_{1cm}^{1\%}$
253	37840	709
290	9670	181
460	9276	174
481	10410	195
519	6630	124

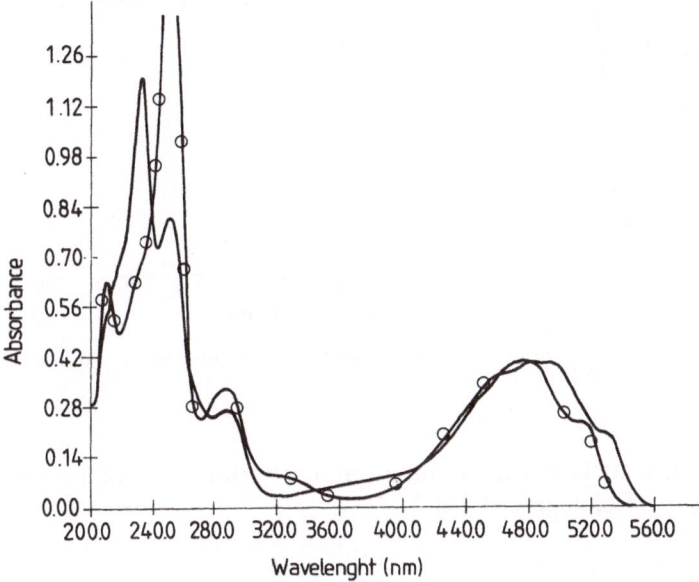

Figure 2.7 Ultraviolet and visible spectra of doxorubicin and/or epirubicin (——) and idarubicin hydrochloride (o——o) in methanol.

slightly different. The molecular absorptivities and $E_{1cm}^{1\%}$ values are reported in Tables 2.5 and 2.6.

2.5.5 *Fluorescence spectra*

In Table 2.7 excitation and emission wavelength of daunorubicin, epirubicin and idarubicin determined using a Perkin-Elmer MPF-44 spectrofluorometer are reported.

Table 2.7

Compound	Excitation (nm)	Emission (nm)
Doxorubicin	495	595
Epirubicin	495	595
Idarubicin	468	572

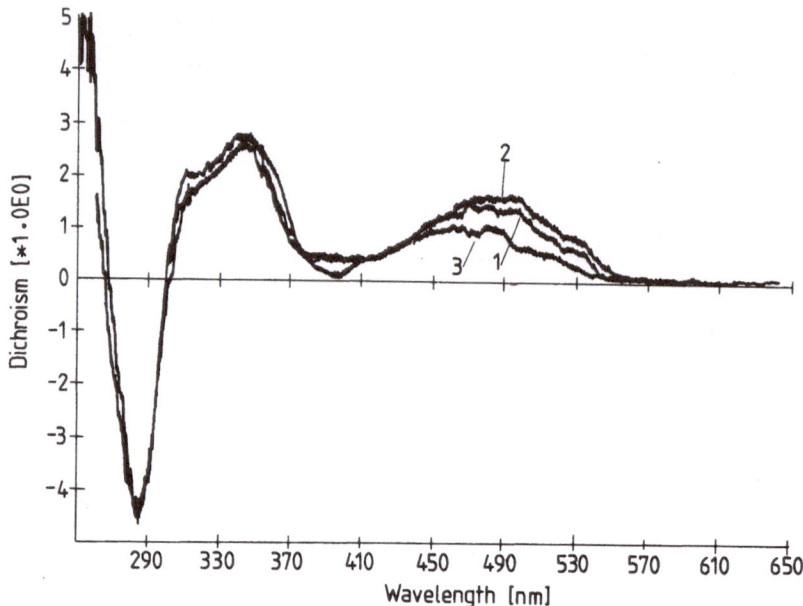

Figure 2.8 Circular dichroism curves of doxorubicin (1), epirubicin (2) and idarubicin (3) in methanolic solution, 0.01 M HCl.

Fluorometric methods can be employed to determine the total anthracycline components in biological fluids[46].

2.5.6   *Circular dichroism*

The chiral centres at C-7 and C-9 in the anthracyclines described are responsible for the Cotton effects seen at 345 and 285 nm[47]. The circular dichroism curves of methanolic solutions of doxorubicin, epirubicin and idarubicin, determined using a Jobin-Yvon Mark V dichrograph, are shown in Figure 2.8. The curves are similar, indicating identical configuration at C-7 and C-9.

2.6   Stability and reactivity

Doxorubicin, epirubicin and idarubicin are stable in acidic aqueous solution in the pH range 3.0–6.3, for a few days at 4°C in the dark. At room temperature and upon exposure to light, hydrolysis of the glycosidic linkage may occur and the aglycone may be released. Also, aqueous solutions of the drugs decompose if they are exposed to daylight. The presence of the hydroxyacetyl chain in doxorubicin and epirubicin make these compounds very unstable in aqueous alkaline solutions. As the pH is increased from 6.3 to 12, there is formation of a

complex mixture of pigmented compounds with a wide range of chromatographic polarities.

The identification of these mixtures has not been accomplished. Also
idarubicin decomposes in alkaline aqueous solutions but, after the first 3 h at
room temperature, a product of the decomposition has been identified as 4-
demethoxy-7-epi-daunomycinone[48]. Doxorubicin, epirubicin and idarubicin

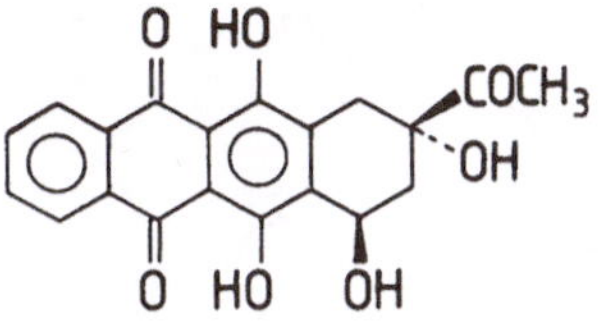

4-demethoxy-7-epi-daunomycinone

hydrochloride, as well as the lyophilised compositions containing lactose, are
very stable for a long period of time in the solid state at room temperature if
dry and if stored in well-closed containers.

Very few reactions may be performed on doxorubicin and epirubicin
without furnishing degradation products. Among these reactions are: preparation of the C-14 ^{14}C-labelled derivatives; hydrazone derivatives at the 13-
carbonyl group[20]; 5-imino-doxorubicin[63] and acyl derivatives at the 3'-
amino group of the sugar moiety[64]; 3'-deamino-3'-morpholinyl derivatives[19];
4'-dihydropyranyl derivatives[18]. These reactions are summarised in
Scheme 2.4 (the numbers in square brackets apply to references).

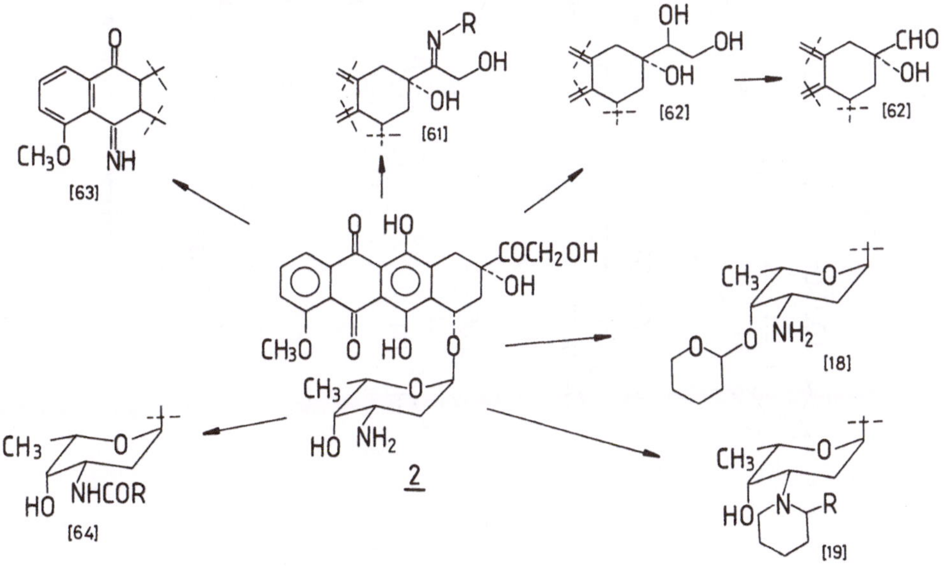

Scheme 2.4 Chemical modifications of doxorubicin.

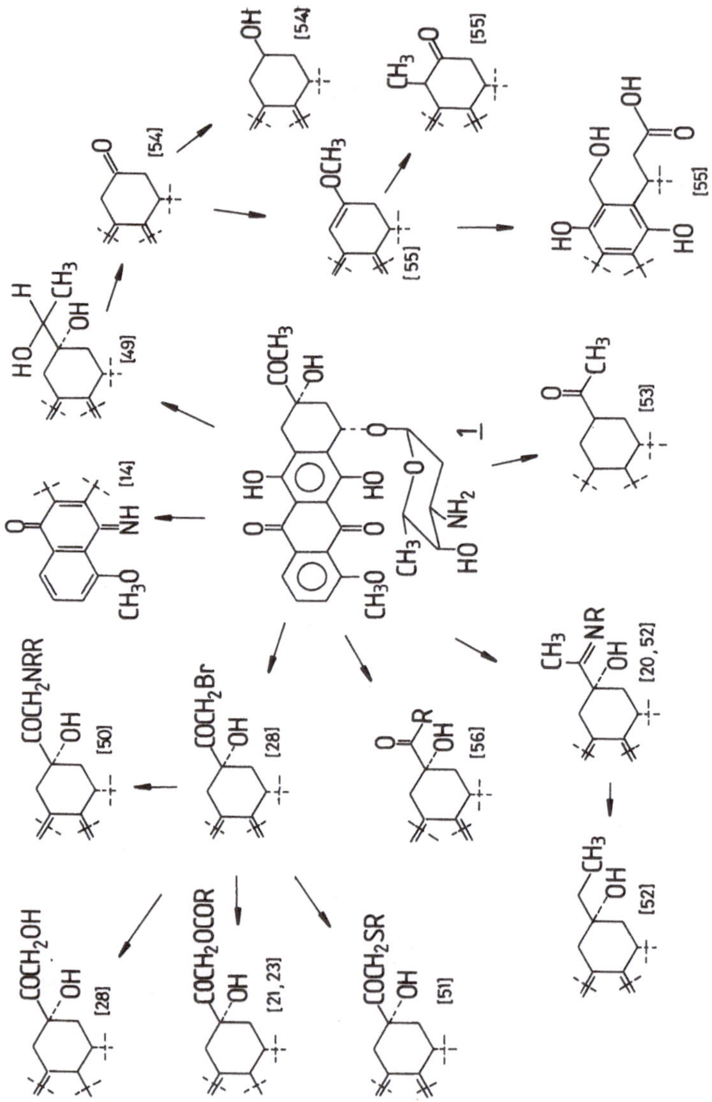

Scheme 2.5 Chemical modification to the aglycon moiety of daunorubicin.

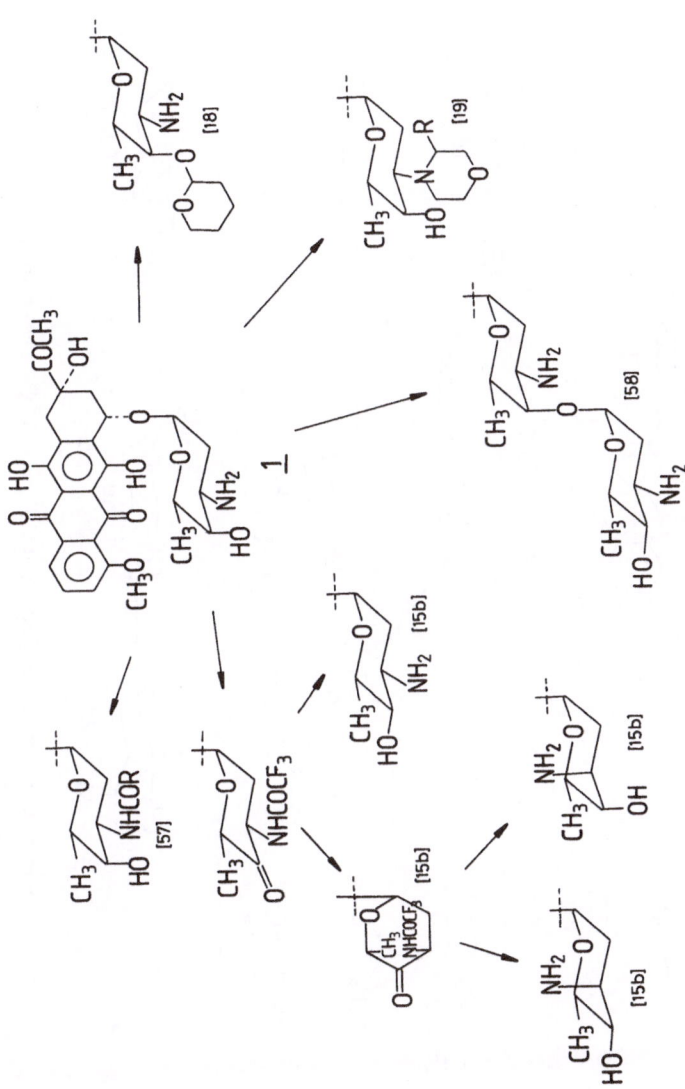

Scheme 2.6 Chemical modification on the sugar moiety of daunorubicin.

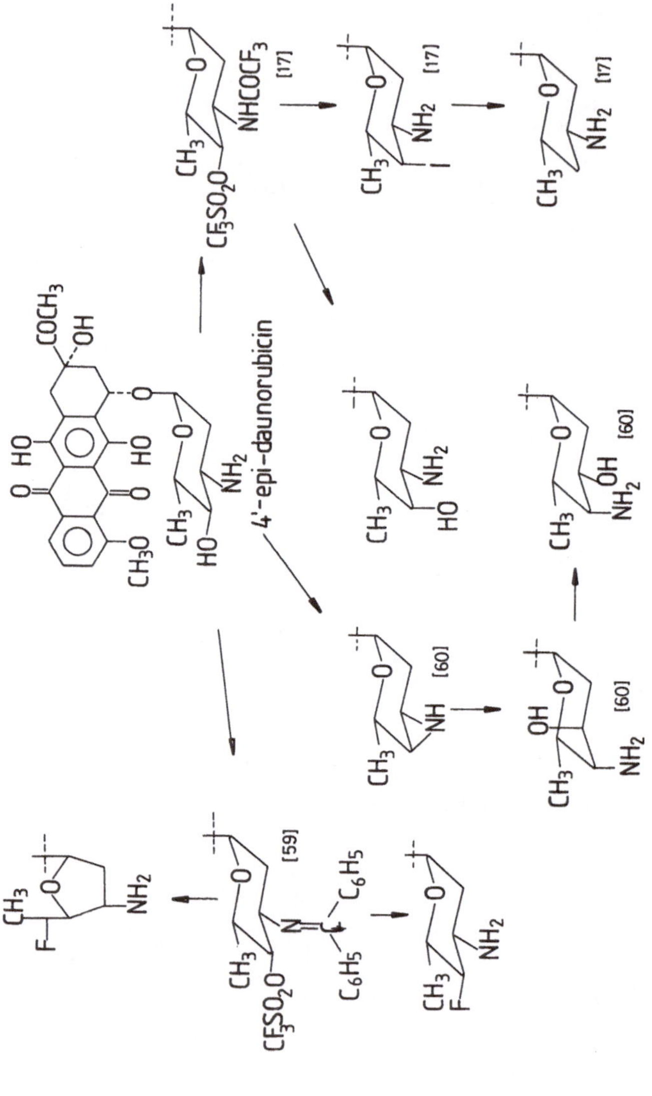

Scheme 2.7 Chemical modifications at C-4′ and C-3′ of 4′-epidaunorubicin.

On the other hand, many transformations have been carried out on the structure of daunorubicin without loss of the sugar residue. The lack of the hydroxyl function in the sidechain makes this compound and its cogeners chemically more stable than doxorubicin. In Scheme 2.5 some of the modifications carried out on the aglycon portion of daunorubicin are shown. In Scheme 2.6 the modifications on the sugar moiety are presented.

In Scheme 2.7 reactions involving the C-4′ position of 4′-epidaunorubicin are shown. The equatorial hydroxyl group is an important feature in this structure. Its activation as an O-trifluoromethanesulfonate leads to several products depending on the protective group of the 3′-amino function. Idarubicin is a potential substrate for the chemical transformations shown in Schemes 2.6 and 2.7.

2.7 Metabolism

2.7.1 Synthesis of C-14 ^{14}C-labelled derivatives

The synthesis of radiolabelled anthracyclines is an important prerequisite for the study of the metabolism of anthracyclines in laboratory animals. The synthesis of [^{14}C]doxorubicin and epirubicin has been described [29,34]. The reaction pathway for the preparation of [14-^{14}C]doxorubicin is illustrated in Scheme 2.8. The hydroxyacetyl sidechain is reduced to the dihydroxy derivative (14), which is oxidatively cleaved to give the aldehyde (16) protected on the amino group. The [14-^{14}C]acetyl group is restored by using [^{14}C]diazomethane. [14-^{14}C]Daunorubicin (17) is converted to [14-^{14}C]doxorubicin by standard methods. Other syntheses of [14-^{14}C]doxorubicin have been reported by Malspeis[65] and Chen[66].

2.7.2 Metabolism in patients

Although little difference appears in the antitumour efficacy in animals between epirubicin and doxorubicin, the differences in molecular structure of these drugs seem to cause differences in their pharmacokinetics. Studies in both animals and patients have demonstrated that the metabolism and excretion of epirubicin are different from those of doxorubicin. The use of the high-pressure liquid chromatography technique has proved to be the method of choice for quantitative separation of the drug and its metabolites in extracts of urine and plasma[67]. According to this methodology, doxorubicin and 13-dihydrodoxorubicin (18) (doxorubicinol) were found to be the major components of urine of patients[69]. The reduction of the 13-carbonyl group is catalysed by a very ubiquitous enzyme named 'daunorubicin reductase'. Another metabolite is 7-deoxy-adriamycinone (20) which represents the product of the reductive cleavage of the daunosamine moiety. These products are further metabolised as shown in Scheme 2.9. Epirubicin is more extensively

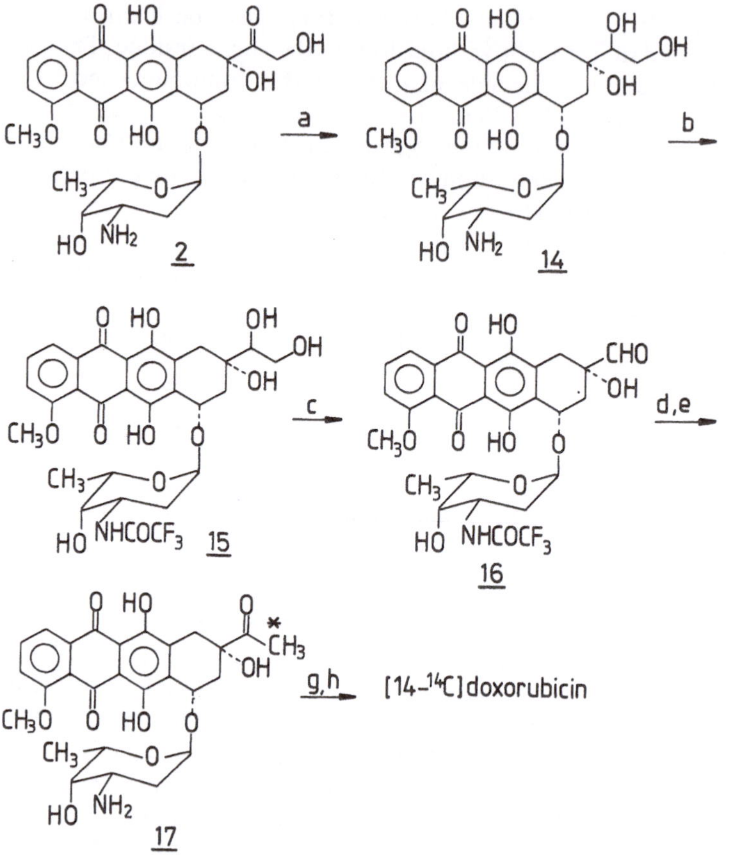

Scheme 2.8 Preparation of [14-^{14}C]doxorubicin, reagents and conditions. (a) NaBH$_4$, CH$_3$OH-dioxane, 0°C; (b) (CF$_3$CO)$_2$O, CH$_2$Cl$_2$, 0°C; (c) HIO$_4$, acetone/water; (d) [^{14}C]CH$_2$N$_2$, CH$_2$Cl$_2$, 0°C; (e) 0.1 N NaOH, 0°C; (f) HCl/CH$_3$OH; (g) Br$_2$, CH$_3$OH-dioxane; (h) HCOONa.

metabolised then doxorubicin. The formation of large quantities of metabolites resulting from glucuronidation of the 4′-hydroxyl group in the sugar moiety may account for its lower toxicity in comparison with doxorubicin[35]. The glucuronides formed, 4′-O-(β-glucuronyl) epirubicin (**25**) and its reduced form derived from 13-dihydroepirubicin (**26**) (epirubicinol), are extracted from urine (Scheme 2.10). Formation of glucuronides facilitates the excretion process of many substances[70]. In contrast, doxorubicin is not metabolised into such glucuronides. This different pattern in terms of glucuronide conjugation can be correlated with the different orientation of the C-4′ hydroxyl group, which is equatorial in epirubicin and axial in doxorubicin.

The presence of the *cis* equatorial 3′-amino group in the structure of doxorubicin can hinder the glucuronyl transferase. The most important metabolic species of idarubicin is the reduced carbonyl form, 13-dihydro-

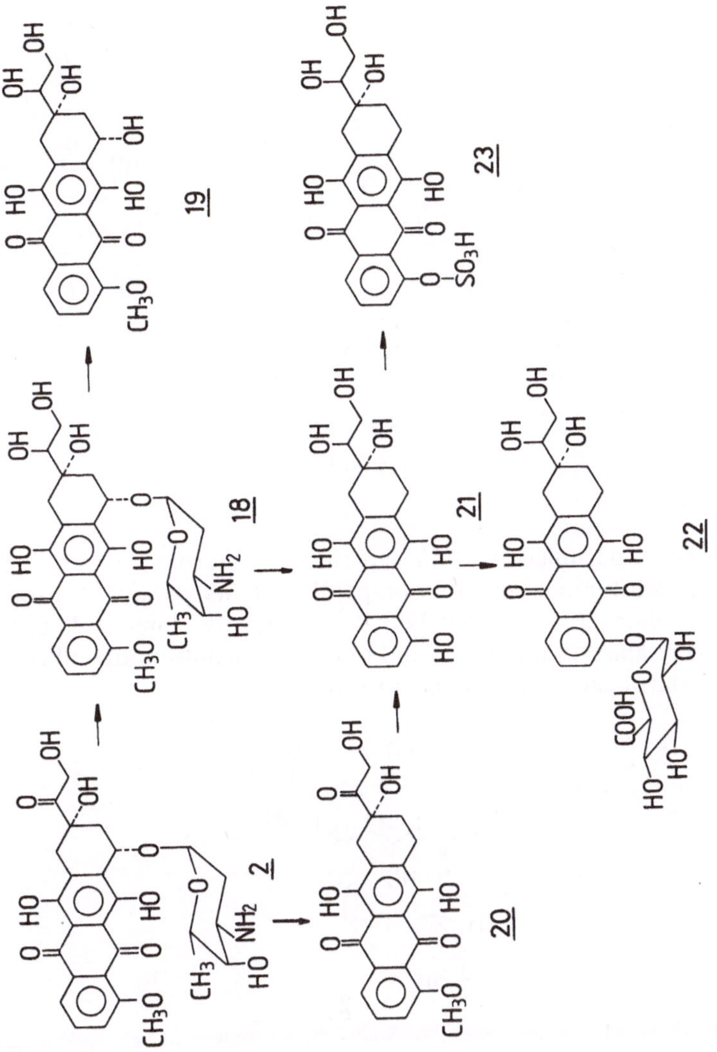

Scheme 2.9

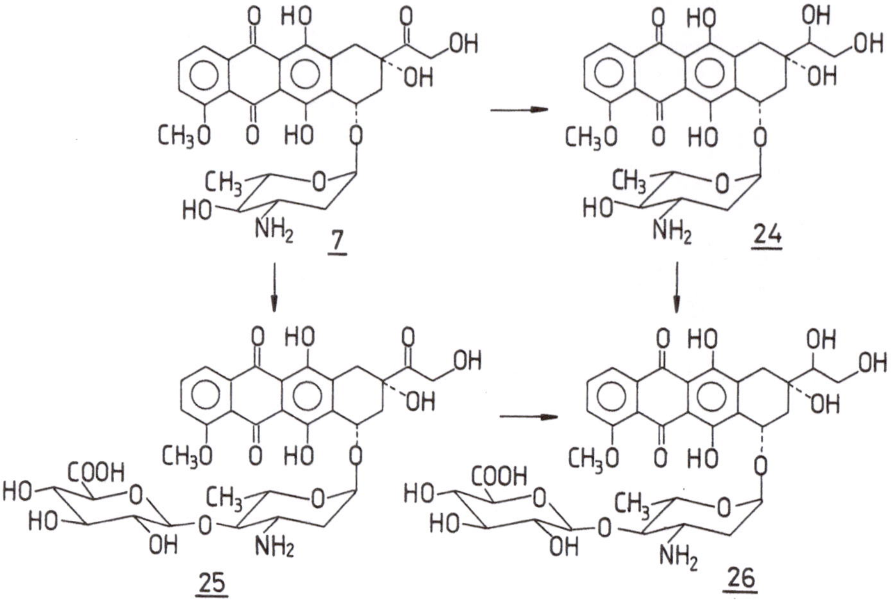

Scheme 2.10

idarubicin (**27**) (idarubicinol)[71]. In experimental murine systems, idarubicinol exibits antitumour activity comparable to its parent[40b]. From the experimental data, it seems that (**27**) is only slowly eliminated, probably because it is retained in the tissues and tumour. The antitumour efficacy and potency of idarubicin may be related to this property.

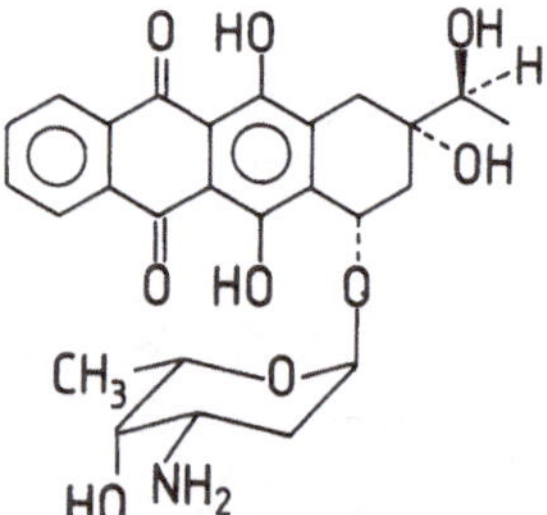

2.8 Method of analysis

2.8.1 *Spectroscopic analysis*

The ultraviolet and visible absorption maxima (Tables 2.5 and 2.6) can be used for the quantification of anthracyclines in dosage forms. The fluorescence

Table 2.8 R_f values on TLC[a].

Compound	A[b]	B[b]
Doxorubicin	0.30	
Epirubicin	0.34	
Idarubicin	0.42	
13-Dihydrodoxorubicin	0.18	
13-Dihydroepirubicin	0.23	
13-Dihydroidarubicin	0.32	
Adriamycinone		0.73
4-Demethoxydaunomycinone		0.83
13-Dihydroadriamycinone	0.65	0.50
7-Deoxyadriamycinone		0.72
7-Deoxy-4-demethoxydaunomycinone		0.89
7-Deoxy-13-dihydroadriamycinone	0.78	0.63

[a] Silicagel F 254 Merck Plates, thickness 0.25 nm.
[b] The following solvent systems were used:
(A) methylene chloride/methanol/acetic acid/water (80 : 20 : 7 : 3, by vol.) (B) methylene chloride/methanol/water (100 : 20 : 10 by vol.).

properties of those drugs (Table 2.7) can be used for their determination at low concentration[46].

2.8.2 Thin-layer chromatography

In Table 2.8 are listed the solvent systems and the R_f values of doxorubicin, epirubicin, idarubicin and their related compounds.

2.8.3 Liquid chromatography

A number of liquid chromatographic methods for the determination of the anthracyclines have been described[34,40]. An interesting procedure has been reported for the separation of the diastereoisomeric mixture of the chemically reduced 13-carbonyl group of anthracyclines and their comparison with the product of the enzymatic reduction (18), (24) and (27)[72].

2.8.4 Determination of anthracyclines in biological fluids

Several methods have been described for recovering the intact drugs[67] and their metabolites[29,34,40b] in biological fluids and for their quantification[46], as well as for determining their tissue distribution[68].

2.9 Aclacinomycin A

Aclacinomycin A (28) is a clinically important antitumour agent belonging to the anthracycline group isolated by Oki et al. from a culture of Streptomyces galilaeus MA144-M1 in 1975[73].

The compound bears several structural modifications, in comparison to the anthracyclines previously described[74]. It consists of the tetracyclic quinoid aglycone aklavinone linked at C-7 to a trisaccharide composed of the amino sugar L-rhodosamine[75] and two other deoxy-sugars, 2-deoxy-L-fucose[76] and L-cinerulose A[77]. Aklavinone differs from daunomycinone by the lack of the C-11 phenolic group, the presence of a C-10 carboxymethyl group and a hydroxyethyl group instead of the hydroxyacetyl sidechain. The absolute configuration of the aglycone is $7S,9R$ and $10R$[78].

The ability of aclacinomycin A to bind to DNA is similar to that of doxorubicin, however, aclacinomycin A is classified as one of the class II anthracyclines which mainly inhibit RNA synthesis, in contrast to class I anthracyclines which show similar inhibition of DNA and RNA synthesis and which include doxorubicin[79].

The drug shows antitumour activity superior to that of doxorubicin in some experimental models such as human xenografts of gastric cancer, CD mouse mammary carcinoma and colon 38[80]. It is also active against Ehrlich tumours, Lewis lung carcinoma and B16 melanoma, while it is less active against L1210 and P388 leukaemias[81].

Toxicological studies performed in hamsters indicate that aclacinomycin A produces a milder cardiac toxicity than does doxorubicin. In addition, there are indications that myocardial changes are generally reversible[82]. The initial clinical study was conducted by Furue et al. in 1977–1978[83] and large-scale phase I–II studies have been performed subsequently[84]. In patients, aclacino-mycin A is active against leukaemias[85], but its response rate in solid tumours is much lower than that of doxorubicin. Haematological toxicity represents the dose-limiting factor for aclacinomycin A. Gastrointestinal symptoms such as nausea and vomiting occur in the majority of the patients but they are generally milder than with doxorubicin.

2.9.1   Microbiological isolation

Besides aclacinomycin A, many other anthracycline derivatives belonging to the aklavinone and ε-pyrromycinone were isolated from the fermentation broth of S. galilaeus MA144-M1. The compounds were identified by Oki et al. and divided into two groups by colour[86]. The first group consisted of yellow-coloured pigments, named aclacinomycins, having aklavinone as the aglycone and the second group consisted of red-coloured pigments having ε-pyrromycinone as the aglycone. The aglycones were glycosidically linked with some of the following sugars: L-rhodosamine, L-cinerulose A, L-cinerulose B, 2-deoxy-L-fucose, L-amicetose, L-rhodinone, D-cinerulose A, L-aculose, L-daunosamine, N-monomethyl-L-daunosamine. To perform the isolation of the compounds, the harvested broth of S. galilaeus MA144-M1 is filtered and the pigments are extracted separately from the mycelium and from the filtrate. After evaporating the solvents, the extracts are combined and the pigments re-extracted with toluene, then the glycosides are extracted into acetate buffered water at pH 3.5 and removed from the aglycones.

To separate the yellow from the red pigments a cupric sulphate solution is added to the acetate buffer layer at pH 5.0. The resulting precipitate of the red anthracycline-copper complex is separated off and the yellow pigments in the solution are extracted into toluene, washed with aqueous EDTA solution and fractionated by silicic acid column chromatography. Aclacinomycin A is obtained after further chromatographic separation of the yellow pigments.

2.9.2 *Analytical profile*

According to Chemical Abstracts, aclacinomycin A is named: 2-ethyl-1,2,3,4,6,11-hexahydro-2,5,7-trihydroxy-6,11-dioxo-4-[[2,3,6-trideoxy-4-*O*-[2,6-dideoxy-4-*O*[(2*R-trans*)-tetrahydro-6-methyl-5-oxo-2*H*-pyran-2-yl]-α-L-*lyxo*-hexopyranosyl]-3-(dimethylamino)-α-L-*lyxo*-hexopyranosyl]oxy-1-naphthacenecarboxylic acid methyl ester.

$C_{42}H_{53}NO_{15}$ mol.wt. 811.9
Registry number   57576-44-0
Melting point[86]   151–153°C
Optical rotation[86] $[\alpha]_D^{25} - 11.5°$ (c 1.0, $CHCl_3$)

2.9.2.1 *Appearance.* Aclacinomycin A is a yellow crystalline powder.

2.9.2.2 *Solubility.* Aclacinomycin A is soluble in methanol, ethanol and sparingly soluble or insoluble in water. The drug is stable in aqueous acidic solution, but unstable at basic pH.

Table 2.9 Ultraviolet and visible absorptions of aclacinomycin A.

Methanol		0.1 N HCl		0.1 N NaOH	
λ_{max}(nm)	$E^{1\%}$	λ_{max}(nm)	$E^{1\%}$	λ_{max}(nm)	$E^{1\%}$
229.5	550	229.5	571	239	450
259	326	258.5	338	287	113
289.5	135	290	130	523	127
431	161	431	161		

2.9.3 *Spectroscopic properties*

2.9.3.1 *Infrared spectra.* The main signals recorded in KBr are: hydroxy groups (3400–3300 cm^{-1}); ester carbonyl (1740 cm^{-1}); non-chelated carbonyl (1670 cm^{-1}); chelated carbonyl (1620 cm^{-1})[86].

2.9.3.2 *Ultraviolet and visible spectrum.* The absorption spectra of aclacinomycin A recorded in methanol, aqueous 0.1 N HCl and 0.1 N NaOH show the maximum values reported in Table 1.9.

2.9.3.3 *Mass spectra.* Field desorption mass spectroscopy (FDMS) of the triglycoside antibiotic aclacinomycin A shows a clearly detectable protonated molecular ion of m/z 812. The fragmentation pattern is similar to that of other anthracyclines. Additional information can be gained from peaks corresponding to the cleavage of the glycosidic bonds; these occur at m/z 700, 569 and 412 and are useful in confirming the carbohydrate sequence. Furthermore it is noteworthy that in FDMS, an ion at m/z 418 is present, which corresponds to the glycosidic part of the molecule[87].

2.9.3.4 *Nuclear magnetic resonance spectra.* Proton and carbon magnetic resonance studies of aclacinomycin A indicate that the configuration at the glycosidic linkages at C-1′, C-1″ and C-1‴ are α. Conformational analysis of the methyl glucoside of the disaccharide 2-deoxyfucose and cinerulose A suggests that L-cinerulose A in aclacinomycin A takes a flatter shape[86].

2.9.4 *Metabolism*

Metabolic pathways of aclacinomycin A, examined with rat liver homogenates in the presence of NADPH, proceed in two main directions[88]. Under aerobic conditions aclacinomycin A is converted to the glycosidic metabolites M_1 and N_1 in which the terminal keto-sugar, cinerulose A, is reduced. Subsequent cleavage of aclacinomycin's sugars produces first the disaccharide metabolite S_1, and then the monosaccharide aklavin. Under anaerobic conditions, aclacinomycin A is degraded to the aglycone-type metabolites C_1

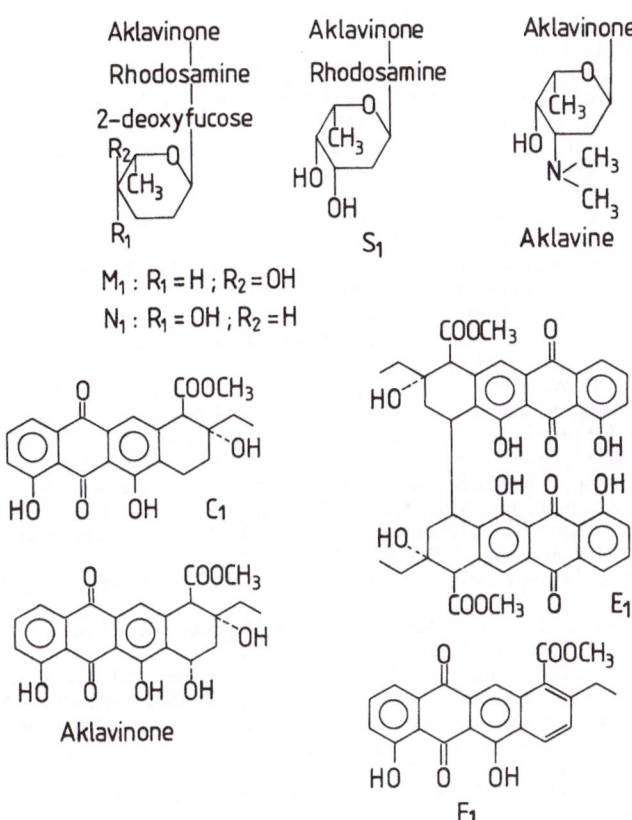

Figure 2.9 Products of aclacinomycin A hydrolysis or metabolism under aerobic and anaerobic conditions.

and E_1 through reductive glycosidic cleavage at the C-7 position. Finally, hydrolysis of the C-7 glycosidic linkage can afford the aglycones, aklavinone and bisanhydroaklavinone F_1[89] (Figure 2.9).

References and notes

1. F. Arcamone, 'Doxorubicin', in *Medicinal Chemistry*, Vol. 17, Academic Press, New York (1981).
2. A. Grein, C. Spalla, A. Di Marco and G. Canevazzi, *G. Microbiol.* **11** (1983) 109.
3. M. Dubost, P. Ganter, R. Maral, L. Ninet, S. Pinneret, J. Preud'Homme and G.H. Werner, *C.R. Acad. Sci.* **257** (1963) 1813.
4. F. Arcamone, G. Cassinelli, G. Franceschi, P. Orezzi and S. Penco, *Gazz. Chim. Ital.* **100** (1970) 949.
5. A. Di Marco, M. Gaetani, L. Dorigotti, M. Soldati and O. Bellini, *Tumori* **49** (1963) 203.
6. F. Arcamone, G. Cassinelli, G. Fantini, A. Grein, P. Orezzi, C. Pol and C. Spalla, *Biotechnol. Bioeng.* **11** (1969) 1101.

7. R. B. Weiss, G. Sarosy, K. Clagett-Carr, M. Russo and B. Leyland-Jones, *Cancer Chemother. Pharmacol.* **18** (1986) 185.
8. (a) M. D. Green, J. L. Speyer and F. M. Muggia, *Eur. J. Cancer Clin. Oncol.* **20** (1984) 293; (b) D. H. Von Hoff, M. W. Layard, P. Basa, H. L. Davis Jr., A. L. Von Hoff, N. Rozencweig and F. M. Muggia, *Ann. Intern. Med.* **91** (1979) 710.
9. M. B. Naff, J. Plowman and V. L. Narayanan, in *Anthracycline Antibiotics*, ed. H. S. El Kadem, Academic Press, New York (1982).
10. F. Arcamone, *Cancer Res.* **45** (1985) 5995.
11. F. M. Muggia, C. W. Young and S. K. Carter (eds.), *Anthracycline Antibiotics in Cancer Therapy*, Martinus Nijhoff, The Hague (1982).
12. (a) M. G. Brazknikova, V. B. Zbarskiy, M. K. Kudinova, L. I. Murav'yeva, V. I. Potomar-enko and N. P. Potopova, *Antibiotiki* (Moskow) **18** (1973) 678; (b) L. H. Baker, D. H. Kessel, R. L. Lamis, S. D. Reich, M. D. De Furia and S. T. Crooke, *Cancer Treat. Rep.* **63** (1979) 899.
13. F. Arcamone, L. Bernardi, B. Patelli, A. Di Marco, A. M. Casazza, G. Pratesi and P. Reggiani, *Cancer Treat. Rep.* **60** (1976) 829.
14. G. L. Tong, D. W. Henry and E. M. Acton, *J. Med. Chem.* **22** (1979) 36.
15. (a) F. Arcamone, S. Penco, A. Vigevani, S. Redaelli, G. Franchi, A. Di Marco, A. M. Casazza, T. Dasdia, F. Formelli, A. Necco and C. Soranzo, *J. Med. Chem.* **18** (1975) 703; (b) A. Suarato, S. Penco, A. Vigevani and F. Arcamone, *Carbohydr. Res.* **98** (1981) C1.
16. (a) F. Arcamone, S. Penco, S. Redaelli and S. Hanessian, *J. Med. Chem.* **19** (1976) 1424; (b) S. Penco, F. Arcamone and A. Di Marco, U. S. Patent 4, 067, 969 (January 10, 1978); (c) A. Suarato, S. Penco and F. Arcamone, British Patent Appl, 8031382 (September 29, 1980); (d) F. Arcamone, *Adv. Med. Oncol. Res. Educ.* **5** (1979) 21.
17. A. Suarato, S. Penco, F. Arcamone and A. M. Casazza, U. S. Patent 4, 438, 105 (March 20, 1984).
18. H. Umezawa, Y. Takahashi, M. Kinoshita, H. Naganawa, T. Masuda, M. Ishizuka, K. Tatsuta and T. Takeuchi, *J. Antibiot.* **32** (1979) 1082.
19. (a) E. M. Acton, G. L. Tong and C. W. Mosher, *J. Med. Chem.* **27** (1984) 638; (b) E. M. Acton, G. L. Tong, D. L. Taylor, J. A. Filppi and R. L. Walgemuth, *J. Med. Chem.* **29** (1986) 1225.
20. (a) G. Jolles, Ger. Patent 1, 803, 892 (May 29, 1969) (*Chem. Abstracts* **71** (1969) 70907; (b) R. K. Johnson, A. A. Ovejera and A. Goldin, *Cancer Treat. Rep.* **60** (1976) 99; (c) R. Maral, *Cancer Chemother. Pharmacol.* **2** (1979) 31.
21. R. Maral, J. B. Ducer, D. Farge, G. Rousinet and D. Reisdorf, *C.R. Acad. Sci. Paris* (D) **286** (1978) 443.
22. (a) M. Israel and E. J. Modest, U. S. Patent 4, 035, 566 (1977); (b) L. M. Parker, M. Hirst and M. Israel, *Cancer Treat. Rep.* **62** (1978) 119; (c) R. H. Blum, M. B. Garmick, M. Israel, G. P. Canellas, I. C. Henderson and E. Frei III, *Cancer Treat. Rep.* **63** (1979) 919; (d) M. Israel and G. Patti, *J. Med. Chem.* **25** (1982) 187.
23. T. Oki, Y. Matsuzawa, A. Yashimoto, K. Numata, I. Kitamura, S. Hori, A. Takamatsu, H. Umezawa, M. Ishizuka, H. Naganawa, H. Suda, M. Hamada and T. Takeuchi, *J. Antibiot.* **28** (1975) 830.
24. (a) D. E. Nettleton Jr., T. W. Doyle and W. T. Bradner, U. S. Patent 4, 123, 608 (Oct. 31, 1978); (b) W. T. Bradner and M. Misiek, *J. Antibiot.* **30** (1977) 519.
25. (a) L. D. Zelezmick and C.M. Sweeney, *Arch. Biochem. Biophys.* **120** (1967) 292; (b) P. F. Wiley, F. A. Mac Keller, E. L. Caron and R. B. Kelly, *Tetrahedron Lett.* (1968) 663; (c) P. F. Wiley, R. B. Kelly, E. L. Caron, V. H. Wiley, J. H. Johnson, F. A. Mac Keller and S. A. Mizsak, *J. Am. Chem. Soc.* **99** (1972) 542.
26. (a) P. F. Wiley, J. H. Johnson and D. J. Hauser, *J. Antibiot.* **30** (1977) 628; (b) G. L. Neil, C. L. Blowers, D. J. Houser, J. H. Johnson and P. F. Wiley, *Proc. Am. Assoc. Cancer Res.* **18** (1977) 2.
27. R.B. Weiss and S. Bruno, *Cancer Treat. Rep.* **65** (Suppl. 4) (1981) 25.
28. F. Arcamone, G. Franceschi and S. Penco, U. S. Patent 3, 803, 124 (April 9, 1974).
29. A. Vigevani, M. J. Williamson, *Doxorubicin, Analytical Profiles of Drug Substances*, Vol. 9, ed. K. Florey, Academic Press, New York (1980).
30. Due to the decomposition, the melting temperature is affected by heating rate.
31. Determined in methanol (0.1%) at 589 nm using a Perkin-Elmer Model 241 MC polarimeter.
32. Titration of an aqueous solution of the compound with N/20 sodium hydroxide. Solutions of anthracycline show indicator-like properties, turning from orange-red to blue-violet about pH 9.

33. Partition coefficient between 1-octanol and Tris buffer (pH 7) with constant ionic strength ($I = 0.1$) at room temperature after shaking for 15 h.
34. F. Arcamone, G. Cassinelli, S. Penco, G. P. Vicario and A. Vigevani, in *Epirubicin*, ed. G. Bonadonna, Masson (Milan) (1984) pp. 3–28.
35. F. Ganzina, *Cancer Treat. Rev.* 10 (1983) 1.
36. F. Ganzina, N. Di Pietro and O. Magni, *Tumori* 71 (1985) 233.
37. (a) P. Deesen and B. Leyland-Jones, *Drug Metab. Dispos.* 12 (1984) 9; (b) G. Cassinelli, E. Configliacchi, S. Penco, G. Rivola, F. Arcamone, A. Pacciarini and L. Ferrari, *Drug Metab. Dispos.* 12 (1984) 506.
38. F. Ganzina, M. A. Pacciarini and N. Di Pietro, *Invest. New Drugs* 4 (1986) 85.
39. (a) A. DiMarco, A. M. Casazza and G. Pratesi, *Cancer Treat Rep.* 61 (1977) 893; (b) F. Pannuti, C. M. Camaggi, E. Stropcchi, R. Comparsi, B. Angelelli and M. A. Pacciarini, *Cancer Chemother. Pharmacol.* 16 (1986) 295.
40. (a) E. Moro, V. Bellotti, M. G. Jannuzzo, S. Stegnjich and G. Valzelli, *J. Chromatogr. Biomed Appl.* 274 (1983) 281; (b) T. Bordoni, B. Barbieri, C. Geroni, G. Cassinelli, M. Grandi and F. C. Giuliani, 7th Int. Symp. on Future Trends in Chemotherapy, Tirrenia, Italy, Abst. 107 (1986).
41. (a) R. J. Stoodley, The anthracycline antibiotics: some synthetic endeavours, Second SCI-RSC Medicinal Chemistry Symposium (1984) pp. 134–147; (b) M. Suzuki, T. Matsumoto, R. Abe, Y. Kimura and S. Terashima, *Chem. Lett.* (1985)57; (c) M. J. Broadhurst, C. H. Hassal and G. J. Thomas, *Chem. Ind. (London)* (1985) 106.
42. C. M. Wong, D. Popien, R. Schwenk and T. Raa, *Can. J. Chem.* 49 (1971) 2712.
43. (a) M. Suzuki, Y. Kimura and S. Terashima, *Chem. Lett.* (1985) 367; (b) M. Sodeoka, T. Iimori and M. Shibosaki, *Tetrahedron Lett.* (1985) 6497.
44. F. Arcamone, *Topics in Antibiotics Chemistry*, Vol. 2, ed. P. Sammes, Ellis Harwood, Chichester (1978) pp. 102–239.
45. F. Arcamone, G. Franceschi, S. Penco and A. Selva, *Tetrahedron Lett.* (1969) 1007.
46. L. Dusonchet, N. Gebbia and F. Gerbassi, *Pharmacol. Res. Commun.* 3 (1971) 55.
47. H. Brockman, H. Brockman Jr. and J. Niemeyer, *Tetrahedron Lett.* (1968) 4719.
48. A. Suarato, unpublished results.
49. J. Jolles and G. Ponsinet, Ger. Patent 2, 206, 690 (July 27, 1972).
50. F. Arcamone, L. Bernardi, B. Patelli and A. Di Marco, Ger. Patent 2, 557, 537 (July 8, 1976).
51. F. Arcamone, L. Bernardi and B. Patelli, Ger. Patent 2, 713, 745 (Oct. 13, 1977).
52. (a) T. H. Smith, A. N. Fujiwara and D. W. Henry, *Am. Chem. Soc. 172nd Meet. Abstr. Medi* (1976) 88; (b) T. H. Smith, A. N. Fujiwara and D. H. Henry *J. Med. Chem.* 21 (1978) 280.
53. S. Penco, G. Franchi and F. Arcamone, Belg. Patent, 876, 100 (Nov. 8, 1979).
54. S. Penco, F. Angelucci, A. Vigevani, E. Arlandini and F. Arcamone, *J. Antibiot.* 30 (1977) 764.
55. S. Penco, F. Angelucci, F. Gozzi, G. Franchi, B. Gioia, A. Vigevani and F. Arcamone, *Int. Symp. Chem. Nat. Prod*, 11th Golden Sands Bulg, 4, Part 1, 448 (1978).
56. T. H. Smith, A. N. Fujiwara and D. W. Henry, *J. Med. Chem.* 22 (1979) 40.
57. D. W. Wilson, D. Grier, R. Reimer, J. D. Bauman, J. F. Preston and E. J. Grabbay, *J. Med. Chem.* 19 (1976) 381.
58. S. Penco, G. Franchi and F. Arcamone, Ger. Patent 2, 751, 395 (May 18, 1978).
59. A. Surato, S. Penco and M. Caruso, G. B. Pat. 2, 159, 518A (December 4, 1985).
60. A. Bargiotti, M. Caruso, A. Suarato, S. Penco and F. Giuliani, E. Patent 199920 (December 10, 1986).
61. D. W. Henry and G. L. Tong, U. S. Patent 4, 112, 217 (September 5, 1978).
62. (a) G. P. Vicario, S. Penco and F. Arcamone, U. S. Patent 4, 211, 864 (July 8, 1980); (b) S. Penco, G. P. Vicario, F. Angelucci and F. Arcamone, *J. Antibiot.* 30 (1977) 773.
63. E. M. Acton and G. L. Tong, *J. Med. Chem.* 24 (1981) 669.
64. P. K. Chakravarty, C. L. Philip, W. J. Michael and J. A. Katzenellenbogen, *J. Med. Chem.* 26 (1983) 638.
65. B. R. Vishnuvajjala, T. Kataoka, F. D. Cazer, D. T. Witiak, and L. Malspeis, *J. Labelled Compd. Radiopharm.* 14 (1978) 111.
66. C. R. Chen, M. Tan Fong, A. N. Fujiwara, D. W. Henry, M. A. Leafer, W. W. Lee and T. H. Smith, *Cancer Res.* 35 (1975) 1175.
67. J. J. Langone, H. van Vunakis, N. Bachur, *Biochem. Med.* 12 (1975) 283.
68. M. Israel, W. J. Pegg, P. M. Wilkinson and M. B. Garnick, *J. Liquid Chromatogr.* 1 (1978) 795.
69. R. Hulhoven and J. O. Desager, *J. Chromatogr.* 125 (1976) 369.

70. R. B. Weenen, J. M. S. Van Maanen and M. M. Planque, *Eur. J. Cancer Clin. Oncol.* **29** (1984) 919.
71. M. Broggini and T. Colombo, *Cancer Treat. Rep.* **68** (1984) 739.
72. S. Penco, G. Cassinelli, A. Vigevani, P. Zini, G. Rivola and F. Arcamone, *Gazz. Chim. Ital.* **115** (1985) 195.
73. (a) T. Oki, Y. Matsuzawa, A. Yoshimoto, K. Numata, I. Kitamura, S. Hori, A. Takamatsu, H. Umezawa, M. Ishizuka, H. Naganawa, H. Suda, M. Hamada and T. Takeuchi, *J. Antibiot.* **28** (1975) 830; (b) T. Oki, N. Shibamoto, Y. Matsuzawa, T. Ogasawara, A. Yashimoto, I. Kitamura, T. Inui, H. Naganawa, T. Takeuchi and H. Umezawa, *J. Antibiot.* **30** (1987) 683.
74. T. Oki, I. Kitamura, Y. Matsuzawa, N. Shibamoto, T. Ogasawara, A. Yoshimoto, T. Inui, H. Naganawa, T. Takeuchi and H. Umezawa, *J. Antibiot.* **32** (1979) 801.
75. H. Brockman, E. Spohler and T. Waehnelt, *Chem. Ber.* **96** (1963) 2925.
76. B. Iselin and T. Reichstein *Helv. Chim. Acta* **27** (1944) 1200.
77. Keller-Schielein and W. W. Richle *Chimia* **24** (1970) 35.
78. (a) J. J. Gordon, *Tetrahedron Lett.* **8** (1960) 28; (b) H. Brockmann, Jr., H. H. Budzikewicz, C. Djerssi, H. Brockmann and J. Niemeyer, *Chem. Ber.* **98** (1965) 1260.
79. S. T. Crooke, V. H. Duvernay and L. Galvan, *Mol. Pharmacol.* **14** (1978) 290.
80. (a) T. Oki, T. Takeuchi, S. Oka and H. Umezawa, in *Recent Results in Cancer Research*, eds. G. Mathe and F. M. Muggia, Springer-Verlag, Berlin, **74** (1980) 207; (b) T. Oki, T. Takeuchi, S. Oka and H. Umezawa, in *Recent Results in Cancer Research*, eds. S. K. Carter, Y. Sakurai and H. Umezawa, Springer-Verlag, Berlin **76** (1981) 21; (c) T. Kubota, Y. Shimosato and Y. Hwa, *Jpn. J. Cancer Chemother.* **5** (1978) 55.
81. S. Fujimoto, J. Inagati, N. Horikoshi and M. Ogawa, *Gann* **70** (1979) 411.
82. D. Dantchev, V. Slioussartchouk and M. Paintrad, *Cancer Treat. Rep.* **63** (1979) 875.
83. H. Furue, T. Komita and I. Nakao, in *Recent Results in Cancer Research*, eds. S. K. Carter, H. Umezawa, J. Douras and Y. Sakurai, Springer-Verlag, Berlin, **63** (1978).
84. (a) K. Yamada, *Jpn. J. Cancer Chemother.* **10** (Part II) (1983) 1030; (b) K. Yamada, T. Nakamura, T. Tsuruo *et al.*, *Cancer Treat. Rev.* **7** (1980) 177.
85. H. Suzuki, K. Kawashima and K. Yamada, *Lancet* (1979) 870.
86. T. Oki, I. Kitamura, Y. Matsuzawa, N. Shibamoto, T. Ogasawara, A. Yoshimoto, T. Inui, H. Naganawa, T. Takeuchi and H. Umezawa, *J. Antibiot.* **32** (1979) 791.
87. B. Gioia, E. Arlandini and A. Vigevani, *Biomed. Mass Spectrom.* **11** (1984) 35.
88. T. Kamijana, T. Oki, T. Inui, T. Takeuchi and H. Umezawa, *Biochem. Biophys. Res. Commun.* **82** (1978) 188.
89. (a) T. Komiyane, T. Oki and T. Inui, *J. Antibiot.* **32** (1979) 1219; (b) T. Ogasawara, Y. Masuda, S. Goto, S. Mari and T. Oki *J. Antibiot.* **34** (1981) 52.

3   Alkylating agents

I. NICULESCU-DUVĂZ, I. BARACU AND A. T. BALABAN

3.1   Introduction

In 1962, Ross[1] defined a biological alkylating agent as a compound which can replace a hydrogen atom by an alkyl group under physiological conditions (pH 7.0–7.4, 37°C, aqueous solution).

Alkylating agents were the first group of chemicals systematically investigated in order to find effective inhibitors against neoplastic cells. They provide almost half of today's clinically useful products (see Table 3.1). Abbreviations used in this chapter are indicated in Table 3.1.

There have been at least two main trends in the early development of alkylating agents: (a) the search for new cytotoxic alkylating functions, which led to the first generation of clinically useful agents; (b) the search for structures with enhanced selectivity towards malignant cells, by exploiting, in the design of the new alkylating agents, some theoretical principles (such as the latent activity concept, carrier principle). This approach led to the second generation of clinically useful drugs: CP, IF, Estracyt, L-PAM, etc. However, it remains questionable whether real cytotoxic selectivity was achieved until now.

At the present time, new approaches such as the use of monoclonal antibodies as selective carriers of alkylating moieties, the development of novel DNA-groove binding alkylators, capable of recognising and binding to specific sequences of this macromolecule, will probably afford a third generation of alkylating agents.

Alkylating agents exhibit a quite paradoxical biological activity, being able to both inhibit tumour cell growth and often to act as potent carcinogens[18, 19]. Chlorambucil, mustard gas, L-PAM, My leran and N, N-bis(2-chloroethyl)-2-naphthylamine were recognised as human carcinogens[19] and there is also enough evidence of the carcinogenic activity of some N-nitrosoureas, cyclic lactones, epoxides, etc. on animals[19].

The antitumour efficacy of alkylating agents can be roughly correlated with their ability to induce lethal mutations and to inhibit DNA synthesis. In this respect the compounds able to cross-link DNA are much more effective than the methylating, ethylating or other singly alkylating agents. In contrast, the latter are more adapted to generate mutations associated with miscoding, which further lead to transformed cells. These data subtly support the well-known bifunctionality concept of Ross[1]. Accordingly the true monofunctional alkylating agents (such as methyl-methane sulphonate, N-methyl-N-

Table 3.1   Alkylating agents of clinical use[a].

Drugs	Abbreviations[b]	Types of responsive malignancies	Ref.[c]
Carmustine	BCNU	Gastric carcinoma, malignant glioma, lymphocytic lymphoma	2
Cyclophosphamide	CP	Head and neck cancer, small and non-small cell lung carcinoma, breast carcinoma, gastric carcinoma, bladder carcinoma, soft-tissue sarcoma, non-Hodgkin's lymphoma	3
Dacarbazine	DTIC	Soft-tissue sarcoma, osteosarcoma, Hodgkin's disease, melanoma	4, 5
Dibromdulcitol	DBD	Squamous cell carcinoma	6
Estracyt		Prostatic carcinoma	7
Iphosphamide	IF	Soft-tissue sarcoma, testicular carcinoma, non-Hodgkin's lymphoma	8, 9
Chlorambucil		Haemoblastoses	10
Lomustine	CCNU	Small cell and non-small cell lung carcinoma, kidney carcinoma, primary and metastatic brain tumours, Hodgkins's disease	11
Mechlorethamine	HN_2	Hodgkin's disease, lymphoid leukaemia	12
Melphalan	L-PAM	Ovarian carcinoma, osteosarcoma, testicular cancer, multiple myeloma	13
Myleran		Chronic myeloid leukaemia	14
Mytomicin C	MMC	Non-small cell lung carcinoma, breast carcinoma, gastric carcinoma	15
Semustine	MeCCNU	Gastric carcinoma, soft-tissue sarcoma	16
Thio-TEPA	TSPA	Breast carcinoma, Hodgkin's disease	17

[a] Only the most clinically widespread alkylating agents are listed.
[b] These abbreviations, commonly used in cancer chemotherapy regimens, are also adopted in this chapter.
[c] Important reviews or clinical trials.

nitrosourea) are cancerogenic. An excellent review on this topic was recently written by Lawley[20].

In this chapter we generally avoid review of the mechanisms of action of alkylating agents. However, selected mention of this topic is sometimes necessary. On such occasions we always implicitly assume that the biochemical interaction which triggers the antitumour effect is DNA alkylation.

It is very difficult to rationalise the chemistry of this class of compounds because of: (i) the large variety of alkylating structures synthesised and tested for antitumour activity; (ii) the extreme diversity of the chemistry involved in such an approach (i.e. preparative chemistry, stereochemistry of intermediates or ultimate alkylators, reactivity, etc.); (iii) in many cases their metabolism and mechanisms of action are unknown.

In this chapter, the chemical features (structure, reactivity, physico-chemical properties, etc.) which are responsible for both the cytotoxicity and the selectivity against cancer cells of the alkylating agents are reviewed.

This chapter is by no means an exhaustive approach; we have tried to present comprehensively the essential chemical features needed for the design of alkylating agents, taking into account the progress made in this area during the last two decades. The preparative chemistry of some clinically used alkylating agents is described.

3.2   Classification of biological alkylating agents

The reaction between an alkylating agent and a substrate takes place according to equation (3.1)

$$R-H + alkyl-X \rightarrow R-alkyl + H^+ + X^- \qquad (3.1)$$

where alkylation could occur at a C, N, O, S, P, etc. Biological alkylating agents react following equation (3.1) at pH 7.0–7.4, 37°C in aqueous media, specifications which drastically restrict the area of biological alkylators, which remains, nevertheless, very large. The main alkylating moieties, as well as the corresponding clinically most useful alkylating agents, are given in Table 3.2.

Generally only the bifunctional alkylating agents are effective inhibitors of malignancies. A larger number of alkylating moieties accumulated on the same molecule does not lead to a proportional increase in the activity[21]. Obviously, the alkylating moieties grafted on a molecule could be identical or different, such as in IF, 3-(2-chloroethyl)-2-(2-methylsulphonyl)-ethylamino-tetrahydro-1,3,2-oxazaphosphorine-2-oxide (**1**), 2-chloroethyl-(methylsulphonyl)methanesulphonate (**2**) (clomesone) etc.

The chemical reactivity of some alkylating moieties can be modulated by their subsequent transformation. A series of such derivatised alkylating groups in summarised in Table 3.3.

A different classification of biological alkylating agents according to their mechanism of action is given in section 3.3.

3.3   Reactivity of alkylating species

Most of the biological alkylating agents react directly with target macro-molecules (DNA, RNA, proteins), both *in vitro* and *in vivo*. According to the nomenclature used for carcinogens they might be termed 'direct alkylators'. During solvolysis (in aqueous media), the alkylating agents are transformed into electrophilic species which interact directly, through a bimolecular process, with target macromolecules.

Both the extent of alkylation of the biomolecules responsible for (or involved in) cell proliferation and the specificity of this reaction depend on the

Table 3.2 Type of biological alkylating agents.

No.	Class of compounds[a]	Chemical structure of typical alkylating compounds	Examples of important clinically useful compounds
1	Alkyl halides	R–X; X = Cl, Br, I; R = alkyl, alyl etc.	DBD[b]; Mielobromol (1,6-dibromo-1,6-dideoxy-D-mannitol)
2	Alkyl phosphates	$(RO)_3PO$, R = alkyl, etc.	
3	Aziridines	R–N⟨ ; R–N⟨R_2/R_1 ; HN⟨R_3; R = alkyl, aryl, quinones, etc.; R_1, R_2 = CH₃, C₂H₅; R_3 = fused heterocycles	TEM (N,N',N''-triaziridinylmelamine); MMC[b]; Diaziquinone (AZQ, 3,6-diaziridinyl-2,5-bis-(carbethoxyamino)-1,4-benzoquinone); Trenimon (2,3,5-tris-aziridinyl-1,4-benzoquinone)
4	Benzo- and naphthoquinones	[naphthoquinone with R_1, R_2]; [benzoquinone with R_1, R_2]; R_1, R_2 = CH₂OH, CH₂X (X = leaving group)	
5	Diazoalkanes	N₂CHR; R = alkyl, amino acid residues, etc.	Azaserine; DON (6-diazo-5-oxo-norleucine); Azotomycin
6	2-Haloethyl sulphides	R–SCH₂CH₂X	
7	2-Haloethylamines[c], 2(3)-halo-propyl-amines (nitrogen mustards)	R = alkyl, CH₂CH₂X; X = Cl, Br, I; R–N(CH₂CH₂X)₂; R–N(CH₂CHX)₂ — CH₃; R–N(CH₂CH₂CH₂X)₂; R = alkyl, aryl, hetanyl, etc.; X = Cl, Br, I	HN₂[b] (N-methyl-N,N-bis-(2-chloroethyl)amine; nitrogen mustard; nor-HN₂(N,N-bis-(2-chloroethyl)amine); L-PAM(4-[N,N-bis-(2-chloroethyl)amino]phenyl-L-alanine); Chloroambucil(4-[N,N-bis(2-chloroethyl)—aminophenyl]butyric acid); Dopan (uracil-mustard; 5-bis-(2-chloroethyl)aminouracil); Alkyrom (IOB 82; 3-[bis-(2-chloroethyl)—amino]-4-methylbenzoic acid)

No.	Class	Structure / Substituents	Examples
8	δ-Lactones; α-methylene-γ-lactones	[γ-lactone ring bearing =CHR₁] $R\!-\!CH\!-\!CH_2$ / $O\!-\!C\!=\!O$ R = alkyl, aralkyl etc. R_1 = H, Cl, Br, CH_3SO_2O, CF_3SO_2O, etc.	
9	Methane sulphonates	CH_3SO_2OR R = alkyl, 1,ω-alkylene, carbohydrate group, etc.	Myleran[b], (1,4-dimethylsulphonyloxybutane); Clomesone (2-chloroethyl-(methylsulphonyl)methanesulphonate); 1,6-dimethylsulpho-nyloxy-1,6-dideoxy-D-mannitol.
10	N-Nitrosoureas	$RNHC(O)N(NO)R_1$ R = alkyl, 2-haloethyl, cycloalkyl, heterocycle, etc. $R_1 = CH_3, CH_2CH_2X$, etc.; X = F, Cl	BCNU[b] (1,3-bis(2-chloroethyl)-1-nitrosourea); CCNU[b] (1-(2'-chloroethyl)-3-cyclohexyl-1-nitrosourea); MeCCNU[b] (1-(2-chloroethyl)-3-(4-methylcyclohexyl)-1-nitrosourea); Chlorozotocin (2-[3-(2-chloroethyl)-3-nitro-soureido]-2-deoxy-D-glucopyranose
11	Oxiranes	$R\!-\!CH\!-\!CH_2$ (epoxide, O) R = alkyl	Teroxirone (α-TGT; 1,3,5-tri-glicidyl-s-triazinetrione); 1,2,5,6-dianhydrogalactitol
12	Piperazinediones	[bis-piperazinedione structure] R = alkyl, 1-ω-alkylene, etc.	Razoxane (ICRF-159; (±)-1,2-bis-(3,5-dioxo-piperazin-1-yl) propane)
13	s-Triazines	[triazine structure] R = N(CH₃)₂, N(CH₂OH)₂, etc. R-N = N-NR₁R₂ R = aromatic or heteroaromatic systems $R_1, R_2 = CH_3, C_2H_5$, etc.	HMM (Hexamethylmelamine) (see Chapter 11)
14	Triazene derivatives		DTIC[b] (3,3-bis(2-chloroethyl)-1-triazeno-4(5)-imidazolyl-5(4)-carboxamide) (see Chapter 6)

[a] In alphabetical order.
[b] For abbreviations see Table 3.1.
[c] For the derivatised alkylating moieties see Table 3.3.

Table 3.3 Derivatised alkylating functions.

No.	Starting alkylating moiety[a]	Derivatised alkylating moiety		
		Class of compound	Structure	Important compounds
1	Aziridines	Aziridinylamides	$R-\overset{\text{O}}{\underset{\parallel}{C}}-N{<}$ (aziridine)	
		(Thio)phosphoramide aziridines	$RR_1-\overset{Z}{\underset{\parallel}{P}}-N\left[\triangle\right]_2$ z=O,S	TSPA[b] (triaziridinyl) thiophosphoramide); TEPA (triaziridinyl-phosphoramide); etc.
		2-Haloethylamides	$RC(O)N(CH_2CH_2X)_2$ R = alkyl, aryl, etc. X = Cl, Br, etc.	
2	2-Haloethylamines (nitrogen mustards)	Phosphoramidic 2-halo-ethylamines	$RR_1N\!\!\diagdown_{\overset{\parallel}{O}}\!\!P\!\!-\!\!N(CH_2CH_2Cl)_2$ $RO\!\!\diagdown_{\overset{\parallel}{O}}\!\!P\!\!-\!\!N(CH_2CH_2Cl)_2$ $(RR_1N)_2\!\overset{}{\underset{\overset{\parallel}{O}}{P}}\!\!-\!\!N(CH_2CH_2Cl)_2$ $\overset{NR_3}{R_2'}\!\!\diagdown_{\overset{\parallel}{O}}\!\!P\!\!-\!\!N(CH_2CH_2Cl)_2$ R, R₁ = alkyl, aryl, etc. R₂ = (CH₂)ₘ, n = 2, 3 or substituted alkylene R₃ = 2-haloethyl, ethyl-aziridinyl, etc.	CP[b]: 2-Bis-(2-chloroethylamino) perhydro-1,3,2-oxazaphosphorin-2-oxide; IF[b]: 3-(2-chloroethyl)-2-(2-chloroethylamino) perhydro-1,3,2-oxazaphosphorin-2-oxide; Thiophosphamide: 2-bis(2-chloroethyl)amino-3-(2-chloroethyl) perhydro-1,3,2-oxazaphorin-2-oxide; CP mustard: bis-(2-chloroethyl) phosphoramide acid)
		Urethane type nitrogen mustards	$ROCON(CH_2CH_2Cl)_2$ R = steroid hormones, carbohydrates, etc.	Estracyt[b] (3-[bis(2-chloroethyl) carbamoyl]oestradiol, 17β-phosphate);
		Triazene nitrogen mustards	$R-N{=}N-N(CH_2CH_2Cl)_2$ R = aryl, substituted aryl, aromatic heterocycles, etc.	4-[1,1-bis-(2-chloroethyl)-triazeno-3-imidazole]-5-carboxamide; Mitozolamide (8-carbamoyl-3-(2-chloroethyl-imidazo 5,1-d-1,2,3,5-tetrazin-4(3H) one) etc.

[a] For the starting alkylating moieties see Table 3.2.
[b] For abbreviations see Table 3.1.

reactivity of the alkylating agents, a parameter thus controlling both the antitumour (or carcinogenic) activity and the toxicity of these compounds.

Ross[22] assumed a proportional relationship between the alkylating activity and the cytostatic action. Bardos et al.[23] could not confirm such a relationship, although in a series of aromatic nitrogen mustards they demonstrated that such a correlation exists between the solvolysis rate and the cytotoxic effect on tumour cells in vitro. Kohn et al.[24] found a positive correlation between the extent of cross-linking and cytotoxicity. A correlation between chemical reactivity and carcinogenic potency was also demonstrated for various carcinogenic alkylating agents[20]. It follows that the chemical reactivity of alkylating agents is of high practical value since this provides a quantitative basis for the design of such compounds. However, the relationship between reactivity and cytotoxicity is not linear. Optimal reactivities, corresponding to a maximal cytotoxic effect, can be determined for various types of alkylating derivatives.

The reactivity of alkylating agents has normally been expressed in terms of: (a) the rate of the hydrolysis reaction (aqueous medium, pH 7.4, 37°C); (b) the half-life ($t_{\frac{1}{2}}$) under the same conditions, or (c) the alkylating activity as determined in standard tests (see section 3.3.1.1.)

Most of the clinically useful alkylating derivatives exhibit, under such conditions, a $t_{\frac{1}{2}}$ in the range of tens of minutes to several hours. However, drugs which react significantly faster ($t_{\frac{1}{2}}$ in the range of seconds) were synthesised for use in some special procedures (intra-arterial or intratumour perfusion, etc.) so that the compound which escapes into the general circulation is rapidly inactivated[25].

Finally, alkylating agents were designed that require metabolic (enzymatic) activation in order to release the 'ultimate alkylator', which is responsible for DNA cross-linking. In the transport form (prodrug) such compounds are chemically inactive, showing very long $t_{\frac{1}{2}}$ (i.e. for CP or IF $t_{\frac{1}{2}} > 7$ days). According to these criteria, the biological alkylating agents may be classified as in Scheme 3.1.

Alkylating agents

directly reacting with target macromolecules
- with high chemical reactivity (i.e compounds especially designed for intra-arterial or intratumoral perfusion)
- with medium chemical reactivity (i.e. Chlorambucil, L-PAM, Lomustine, Carmustine, Thio-TEPA, Myleran, Trenimon)

which need metabolic activation for the reaction with target macromolecules
- with low chemical reactivity (i.e. CP, IF, Estracyt, Mitozolomide, benzo- (and naphtho)quinone derivatives)

Scheme 3.1 Classification of alkylating agents according to their mechanism of action.

The following factors are important in defining and evaluating the reactivity of the direct or the 'ultimate' alkylators:

1. the mechanism of the reaction with the nucleophiles or DNA nucleophilic sites;
2. the electrophilicity of the alkylating species;
3. the softness of the alkylating moieties;
4. the stereochemistry of the electrophilic species;
5. the nature, reactivity and stereochemistry of the DNA nucleophilic site.

All these parameters (and other, less important ones) determine the extent of alkylation of the target molecules, as well as the nature of the alkylated sites.

3.3.1   The reaction mechanisms of the alkylating agents with water and other nucleophiles

The nucleophilic substitution involves the breaking of an old bond and the formation of a new one. The principal mechanistic variations are associated with the timing of these two events.

According to classical concepts, the nucleophilic substitutions can follow a unimolecular (S_N1) (equations (3.2), (3.3)) or a bimolecular (S_N2) (equations (3.4), (3.5)) pathway.

$$\text{Alkyl–X} \xrightarrow{k_1} \text{Alkyl}^+ + \text{X:}^- \text{ (prior ionisation)} \tag{3.2}$$

$$\text{Alkyl}^+ + \text{Nu:}^- \xrightarrow{k_2} \text{Alkyl–Nu for } k_1 \ll k_2 \tag{3.3}$$

$$\text{Alkyl–X} + \text{Nu:}^- \xrightarrow{k_1} [\text{X} \cdots \text{Alkyl} \cdots \text{Nu}] \xrightarrow{k_2} \text{Alkyl–Nu} + \text{X:}^- \tag{3.4}$$

$$d\,[\text{Alkyl–X}]/dt = -k_1\,[\text{Alkyl–X}]\,[\text{Nu:}^-] \tag{3.5}$$

The S_N1 mechanism proceeds through carbocations (3.2) with X the leaving group. For many alkylating agents the factors favouring one or the other of the two mechanisms have been determined[1,22,23,26,27]. In a different approach these mechanisms could be regarded as limiting cases of a common mechanism proceeding through ion pairs[28].

$$\text{Alkyl–X} \underset{k_{-1}}{\overset{k_1}{\rightleftharpoons}} \text{Alkyl}^+\text{X:}^- \quad
\begin{array}{l} \xrightarrow{k_S,\ H_2O} \text{Alkyl–OH} \\[2mm] \xrightarrow[K_2,\ Y:^-]{} \text{Alkyl–Y} \end{array} \tag{3.6}$$

$k_{-1}/k_s \to 0 \to$ resembles S_N1

$k_{-1}/k_s \to \infty \to$ resembles S_N2

In the case of sulphur and nitrogen mustards some mechanistic differences appear because of 'the neighbouring' group effect exerted by the sulphur and nitrogen atoms, respectively, on the reaction centre (see section 3.3.1.1).

The reaction between α-methylene-γ-lactones and nucleophiles occurs by a Michael type addition[29]. The quinone-methide intermediates which probably result from metabolic activation of benzo- and naphthoquinones react with nucleophiles by a similar mechanism. The α-methylene-γ-lactones, substituted with leaving groups, react also by nucleophilic addition-elimination reaction of Michael type[30].

Finally, the diazoalkanes alkylate nucleophiles through a completely different mechanism (namely a carbenoid or radical mechanism), with significant difference in selectivity[27, 31].

The electrophilic species generated during alkylation reactions are usually classical alkyl carbocations (carbenium ions), most often RCH_2^+, CH_3^+, $C_2H_5^+$, $C_6H_5CH_2^+$, etc., but also cyclic ions resulting from the solvolysis of the nitrogen mustards or aziridines (7), sulphur mustards (8), oxiranes (9) etc. The formation of chloronium ions (10) has also been reported, for instance, during the non-enzymatic hydrolysis of the N-nitrosourea derivatives[32, 33].

The reactivity of the aliphatic carbocations, calculated from the dissociation energy of compounds RX to form the corresponding carbenium ions, R^+, varies in the following order:

$$CH_3^+ > C_2H_5^+ > n\text{--}C_3H_7^+ > i\text{--}C_3H_7^+ > (CH_3)_3C^+ > CH_2{=}CH\text{--}CH_2^+$$
$$> C_6H_5CH_2^+ > (C_6H_5)_3C^+$$

The difference between the formation energies of the first and last carbocation in this series is approximately 100 kcal/mol. However, this order of reactivity is valid for an S_N1 reaction mechanism, where the formation of the carbocation precedes its reaction with the nucleophile. In contrast, for the

Table 3.4 Kinetic data, alkylating activities, Swain–Scott constants (s), reaction mechanisms and lipophilicities for some alkylating agents.

No.	Class of compounds	Compound	Hydrolysis $k_1(h^{-1})$	Hydrolysis $t_{\frac{1}{2}}(h)$	Alkylating activity	s^a	Mechanism	log P	Ref.
1	Alkyl halides	Methyl iodide	—	—	62	1.20	S_N2	1.69,1.80	1,49
		Methyl bromide	5.9×10^{-3}	1.16×10^2	—	1.00	S_N2	1.38[i]	1
		tert-Butyl bromide	2.8×10^{-1c}	2.47	—	—[d]	S_N1	2.69[i]	37
		Benzyl chloride	—	—	40	0.87	S_N2	2.30	1,49
		DBD[b]	—	—	—	—	S_N2	-0.29	39
2	Alkyl phosphates	Trimethyl phosphate	—	—	—	0.80	S_N2	-0.52	38
3	Aziridines[e]	TSPA[b]	—	—	80	1.10	S_N2	0.92	49
		TEM[b]	—	—	6.2[f]	—	S_N2	—	
4	2-Haloethyl amines, derivatised 2-haloethyl amines	nor-HN$_2$[b]	0.11	6.33	100[h]	1.26	S_N2	1.67	1,22,45
		HN$_2$[b]	2.77	0.25	100[h]	1.18	S_N2	1.97	1,22,45
		L-PAM[b]	0.52	1.33	110[c]	1.39	S_N1	-1.71[i]	1,44
		Chlorambucil	1.39	0.50	100[h]	1.26	S_N1	-0.60[i]	1,44
		CP	0.00	0.00	1.3[h]	—	e.a.[n]	-1.15	43,45
		CP mustard[b]	—	—	—	—	S_N2	—	
		Spirohydantoin mustard	—	—	—	—	S_N2	2.47	38,46
5	2-Haloethyl sulphides	2-Chloroethyl sulphide	9.00	7.7×10^{-2}	—	—	S_N1	2.87[i]	22
		2-Chloroethyl sulphide cation	—	—	—	0.95	S_N2	—	22
		4 2,3-Bis(2-bromo-ethylthio)-n-pro-pyloxy benzoic acid	5.54×10^2	1.25×10^{-3}	—	—[g]	S_N2	—	25
6	Lactones	Propiolactone	0.68	1.00	—	0.77	S_N2	-0.31[l]	47
7	Methane-sulphonates[m]	Methylmethane-sulphonate	0.72×10^{-1}	9.5	60	0.83,0.87	S_N2	-0.07	20,49
		Ethylmethane-sulphonate	0.6×10^{-1}	1.15	15	0.67	S_N2^p	0.43	20,49

#		Compound							
		Isopropylmethane-sulphonate	3.1	0.22	—	0.29	S_N1[p]	0.73	20
		Myleran	0.36	1.95	20	0.90	S_N2	0.68	1
		Erythritol-1,4-dimethanesulphonate	0.94	0.73	—	—	S_N2	—	1
		2-Chloroethyl-(methylsulphonyl)-methanesulphonate	—	—	—	—	S_N2	-0.05	
8	N-Nitroso-ureas	BCNU[b]	0.96	0.72	1.4[k]	0.94	S_N1	1.53	50
		CCNU[b]	0.79	0.88	0.52[k]	—	S_N1	2.83	50
		N-Methyl-N-nitroso-urea	2.74	0.25	0.36[k]	0.42	S_N1	-0.59[l]	20,50
		N-(2-Chloroethyl)-N-nitrosourea	5.78	0.12	2.00[k]	0.71	S_N1	-0.57	50
			3.15×10	2.20×10^{-2}	3.00[k]				
9	Oxiranes	Styrene oxide	—	—	—	—[g]	S_N1/S_N2	1.28[l]	41
		Glycidol	6.86×10^{-3}	1.01×10^2	—	0.96	S_N2	-2.09[l]	39,42
		1,2;5,6-Diepoxy-hexane	2.30×10^{-3}	3.01×10^2	—	—	S_N2	-1.67[l]	1

[a] Substrate (Swain–Scott) constant.
[b] For abbreviations see Tables 3.1–3.3.
[c] Calculated using data from Ref. 37.
[d] $s = 0.40$ for tert-butylchloride[34]
[e] log P of aziridine is 0.18[40]
[f] Ref. 26 determined at 80°C.
[g] $s = 0.23$ for epichlorhydrin.
[h] According to Ref. 43.
[i] Calculated using equation $\log P_{benzene} = 1.015 \log P_{octanol} - 1.042$.
[j] $\log P = -0.57$[48]
[k] According to Ref. 50.
[l] Calculated values using Ref. 121.
[m] Of methanesulphonyloxy group is -1.77.
[n] Enzymatic activation.
[p] Mainly.

one-step, S_N2 alkylation reactions, an opposite order of reactivity was found, benzylating derivatives reacting faster than methylating ones.

Since primary carbenium ions are too energy rich, they are not formed and the mechanism which would afford such ions changes to S_N2 which involves a quasi-penta-coordinated carbon atom. Only tertiary, benzylic or allylic carbenium ions are stable enough to be formed in S_N1 reactions.

A knowledge of the reaction mechanism of the alkylating agents is of practical value for their design. Thus, the alkylating efficiency of compounds reacting by an S_N1 mechanism (in contrast with those alkylating via S_N2) is independent of the concentration of nucleophilic centres in the medium or in the target. In contrast, their rate of ionisation is largely dependent on the dielectric constant of the medium, occurring faster in aqueous solutions. The rate-reducing effect of the common ion could be noted for this mechanism. On the other hand, agents reacting by the S_N1 mechanism generally exhibit lower sensitivity to the nucleophilicity of the substrate, because of the high reactivity of the carbenium ion intermediates. Obviously, these comments do not apply to the alkylating agents which undergo enzymatic activation.

The reaction mechanism also determines, to a certain extent, the selectivity for the DNA nucleophilic sites. Thus, most of the agents alkylating by S_N1 or S_N2 mechanisms attack the nucleotides in order of their increasing basicity: $G > A > C \gg T(U)$, whereas those alkylating by a carbenoid or radical mechanism exhibit a reverse order of attack for guanosine[27,31]. The α-methylene-γ-lactones prefer to react with thiol groups of proteins (see section 3.5.9).

The data concerning the reaction mechanisms of some commonly used alkylating agents (excepting those which require enzymatic activation) are given in Table 3.4.

3.3.1.1 *Kinetic criteria for the recognition of the reaction mechanisms.* Kinetic measurements are needed to establish reaction order. However, this methodology is often unsatisfactory in delineating the reaction mechanism.

Early concepts about the biological alkylating agents assumed that the alkylation of water molecules parallels (mechanistically and kinetically) that of nucleic acids, hence the hydrolysis (solvolysis) rate would be a suitable parameter for a quantitative correlation of the effectiveness of these alkylators. Thus, the hydrolysis rates for hundreds of alkylating agents were measured using standard conditions of hydrolysis (Me_2CO/H_2O, 1:1, 66°C, 30 min)[1].

The kinetics obtained for the hydrolysis of sulphur and nitrogen mustards under these conditions suggested that the mechanisms depend on the basicity of the hetero atom and that, from this point of view, aliphatic nitrogen mustards behave differently from aromatic ones[1,22,27]. Aliphatic nitrogen mustards cyclise rapidly to an aziridinium ion, which can be titrated (with thiosulphate) or detected by ^{1}H-NMR[22,27]. This ion further reacts with nucleophiles according to an S_N2 mechanism.

For instance, one arm of HN_2 cyclises in 1–2 min at 37°C, while the half-life of this ion is around 1 week under the same conditions.

$$RR_1NCH_2CH_2Cl \rightarrow RR_1N\!\!\triangleleft\, Cl^- \xrightarrow{\ Nu:^-\ } RR_1NCH_2CH_2Nu + Cl^-$$

$$\quad\quad\mathbf{11}\quad\quad\quad\quad\quad\mathbf{12}\quad\quad\quad\quad\quad\quad\mathbf{13}$$

Nor-HN_2, having nitrogen basicity between that of aliphatic and aromatic nitrogen mustards, yields an unionised N-substituted aziridine[22] under these conditions.

$$HN(CH_2CH_2Cl)_2 \rightarrow ClCH_2CH_2N\!\!\triangleleft + HCl$$

$$\quad\quad\mathbf{14}\quad\quad\quad\quad\quad\quad\mathbf{15}$$

Finally, if the reaction affording the cyclic intermediate is very slow compared to that in which the aziridine ring is opened, then this intermediate does not accumulate and the kinetics become first-order, consistent with an S_N1 mechanism[27]. Accordingly, sulphur mustard (less basic than HN_2 or nor-HN_2). shows hydrolysis kinetics supporting ionisation as the rate-controlling step, as required by the S_N1 mechanism[1,22,34]. There are no data to support the accumulation of the cyclic sulphonium ion intermediate (**16a**) which has a real carbonium character[35], and reacts further with nucleophiles according to a bimolecular mechanism.

$$RSCH_2CH_2Cl + H_2O \longrightarrow R\overset{\oplus}{S}\!\!\triangleleft\, Cl^- \xrightarrow{\ H_2O\ } RSCH_2CH_2OH$$

$$\quad\quad\mathbf{16}\quad\quad\quad\quad\quad\quad\mathbf{16a}\quad\quad\quad\quad\quad\mathbf{17}$$

A similar behaviour could be assigned to the aromatic nitrogen mustards which are even less basic. The rate-determining step is the formation of the solvated carbocation (**18a**) in equilibrium with the aziridinium ion (**18b**)[36].

$$Aryl-N(CH_2CH_2Cl)_2 \longrightarrow \left[Aryl-N\overset{\oplus}{\underset{CH_2CH_2Cl}{\overset{CH_2}{\underset{CH_2}{\diagup}}}}\right]Cl^{\ominus}\!.H_2O \rightleftharpoons \left[Aryl-\overset{\oplus}{N}\overset{\diagdown}{\underset{CH_2CH_2Cl}{}}\right]Cl^{\ominus}$$

$$\quad\quad\mathbf{18}\quad\quad\quad\quad\quad\quad\mathbf{18a}\quad\quad\quad\quad\quad\quad\mathbf{18b}$$

Therefore, the passage from strongly basic nitrogen mustards to weakly basic (aromatic) ones or to sulphur mustard gas does not alter the alkylation mechanism, which remains the nucleophilic attack of the cyclic intermediate. It also indicates that the basicities of the precursor amines are useful parameters in estimating the reactivity of the corresponding nitrogen mustards, because they may be satisfactorily correlated with hydrolysis rates (determined as $\log k_{66}$ in Ross's standard test). Thus, for a series of eleven *m*-

substituted nitrogen mustards equation (3.7) is obtained[51]

$$\log k_{66} = -6.62 + 0.58\,pK_a \qquad (3.7)$$
$$n = 10, \qquad r = 0.949, \qquad s = 0.18$$

For aromatic nitrogen mustards with several substituents on the aromatic nucleus, or when *ortho* substituents (with respect to the nitrogen mustard moiety as in compounds (**19**)) are present, such correlations failed. This failure could be explained by the different extent to which these various *ortho*-substituents affect the reaction centre by producing a hindrance to planarity which increases the basicity of the nitrogen mustard moiety. However, for homologous series in which the *ortho* effect remains constant (as in compounds (**19**) with $R = CH_3$ and various R_1) linear hydrolysis rate–basicity relationships could be obtained (3.8)[52]

$$\log k_{66} = -5.58 + 0.678\,pKa \qquad (3.8)$$
$$n = 9, \qquad r = 0.968, \qquad s = 0.288$$

$R = CH_3, C_2H_5, OCH_3$, etc; $R_1 = H, CO_2H$

Similar equations can also be derived for the nitrogen mustard group attached to other aromatic systems[53].

Table 3.4 summarises the hydrolysis rate constants (determined under physiological conditions (pH 7.0–7.4, 37°C, various buffers), as well as the half-lives of some important alkylating agents.

Finally, we want to emphasise that the hydrolysis rate constants are of poor predictive value for the ability of the compounds to alkylate DNA, because:

(a) solvolytic reactions are unsuitable for determining the reaction mechanisms, as the solvent effects often produce misleading kinetics; the neighbouring group effect also contributes to this situation;

(b) the data thus obtained are useful only to compare the reactivities of the derivatives belonging to the same homologous series;

(c) water, being a good solvent and also a hard Lewis base, does not represent a satisfactory model for the interaction of DNA with alkylating agents.

The structure and the concentration of the final hydrolysis products also afford some clues to the reaction mechanism. Thus, the alkaline hydrolysis of 1-diethylamino-2-chloropropane (**20**) which occurs with the rearrangement of the methyl group from the halogenated sidechain, is consistent with an S_N2 mechanism involving a cyclic intermediate.

$$(C_2H_5)_2NCH_2CHCl \longrightarrow (C_2H_5)_2\overset{\oplus}{N}\!\! \diagdown \quad Cl^- \xrightarrow{OH^-} (C_2H_5)_2NCHCH_2OH$$
$$\underset{CH_3}{|} \qquad\qquad\qquad \underset{CH_3}{|} \qquad\qquad\qquad\qquad\qquad \underset{CH_3}{|}$$

20   **21**   **22**

Similar effects are observed during the hydrolysis of substituted aziridines or oxiranes. In contrast, such transpositions have never been observed during the solvolysis of aromatic nitrogen mustards[22].

In order to avoid the previously mentioned disadvantages of the solvolytic reactions, and also to find a parameter which could better define the reactivity of alkylating agents, the determination of the rate constant for the reaction with nucleophiles other than water was proposed (i.e. p-nitrobenzylpyridine (NBP)[54], 4-pyridine-aldehyde-2-benzothiazolylhydrazone (PBH)[55,56], 2,4-dichlorothiophenol, etc.). Among these reagents, the most widely used (also for analytical purposes) was NBP. The reactivity of NBP toward alkylating species results from the nucleophilicity of the unshared electron pair of the pyridine nitrogen. Using this reagent the reactivity of an alkylating agent is defined as the time-dependent alkylation of NBP in the presence of competing nucleophiles such as water or other solvent molecules[26,43,57].

The concentration of alkylated NBP, determined under standard conditions, allows the calculation of the 'alkylating activity' of the investigated derivative. This procedure can be applied to any chemical class of biological alkylating agents, excepting those which require enzymatic activation. For the latter, the alkylating activity can be determined following their activation.

Usually the reaction mechanism of NBP alkylation is of S_N2 type[23,26]. However, first-order kinetics have also been observed, but in such cases the observed NBP alkylation rate only reflects the rate of the initial ionization step in forming the aziridinium intermediate.

As expected, there is not always a parallel between the hydrolysis rate constants and the alkylating activity. For instance, there are alkylating agents with relatively large hydrolytic half-lives which exhibit a significant alkylating activity (i.e. NH_2, or nitrogen mustard-N-oxide, see Table 3.4).

The alkylating activities of some clinically useful alkylating agents are summarised in Table 3.4. Unfortunately, since various authors have used different experimental conditions for determining this parameter, not all the figures reported in Table 3.4 are comparable.

NBP has also been used in determining the hydrolysis rates of alkylating agents, by measuring the decrease in alkylating activity in aqueous solutions in the absence of nucleophilic partners other than water[58]. More recently, nucleophilic selectivity ratios of clinical alkylating agents have also been measured by NBP competition[49]. This last procedure is very useful for the determination of the Swain–Scott, s constant (see section 3.3.2.1) and also for the rapid identification of protective nucleophile-clinical alkylating agent combinations[49].

The alkylating activity offers a more accurate description for the reactivity of the alkylating agents. Nevertheless, this parameter does not entirely reflect the behaviour of alkylating agents against target macromolecules.

3.3.1.2 *Correlation of the reaction rate constants with electronic and structural parameters.* The correlation of reaction rate constants with electronic parameters could provide indications about the alkylation reaction mechanism for a homologous series. This mechanism could be inferred from the influence exerted by electron-releasing or electron-withdrawing substituents on the reaction centre. Generally, Hammett constants (σ, σ^+, or σ^-) were used to describe this influence. The Taft steric parameters have also been used. Two examples are given.

(a) In a series of monosubstituted aromatic nitrogen mustards, a linear relationship can be shown between the hydrolysis rates determined in Me_2CO/H_2O (1:1, 66°C, 30 min and expressed as $\log k_{66}$) and σ constants of the substituents[59, 60]. The formation of an aziridinium intermediate would be highly dependent on the electron density on the nitrogen, and hence, associated with a large ρ for the σ Hammett parameter. The following equations were obtained ((3.9), (3.10)) which are consistent with this assumption

$$\log k_{66} = -1.84(\pm 0.40)\sigma_p - 4.02(\pm 0.08) \tag{3.9}$$
$$n = 11, \quad r = 0.961, \quad s = 0.116$$
$$\log k_{66} = -2.07(\pm 0.27)\sigma_m - 3.80 \tag{3.10}$$
$$n = 9, \quad r = 0.947, \quad s = 0.150, \quad (F = 61.68)$$

The correlation remains satisfactory even for the unified sets of *meta-* and *para*-congeners

$$\log k_{66} = -1.79(\pm 0.13)\sigma_i - 3.94 \tag{3.11}$$
$$n = 20, \quad r = 0.959, \quad s = 0.173, \quad (F = 204.76)$$

These equations suggest an S_N1 mechanism, since the electron-releasing substituents enhance the hydrolysis rate, while electron-withdrawing ones have an opposite effect.

(b) A second example is provided by a study performed by Sugiura et al.[41]

on the reaction between styrene oxides and nucleophiles. They showed that ρ (the slope of Hammett's equation) is 0.87 for the 'normal' reaction of styrene oxides with benzylamine, while ρ for the 'abnormal' reaction, with the same amine, is -1.15. Thus, the negative slope suggests an S_N1 mechanism for the formation of the abnormal product, and the positive slope an S_N2 pathway for the formation of the normal one.

26   **27, normal**   **28, abnormal**

3.3.1.3 *The effect of the reaction medium.* Suggestions regarding the alkylation reaction mechanism could also be obtained by investigating the behaviour of alkylating agents in various solvolytic media.

Thus, for halogenated alkylating agents, i.e. for nitrogen and sulphur mustards, the kinetic data obtained for Cl^- elimination in aqueous solutions could be well correlated with the hydrolysis mechanism. Thus, for an S_N2 reaction there is no parallelism between the concentration of Cl^- and that of H^+, because the rate-determining step is the reaction between the ionised aziridinium intermediate and the nucleophile. This results in the H^+ concentration (acidity) being significantly lower than that of Cl^-, because Cl^- ions stabilise the cyclic intermediate. The same is true for mustard gas. Therefore, these reactions are relatively insensitive to pH variations in the range of 5–10. If the hydrolysis of aliphatic nitrogen mustard occurs in an unbuffered medium it usually stops after the elimination of one Cl^- equivalent/mol. In buffered solutions the hydrolysis proceeds quantitatively[22].

For aromatic nitrogen mustards, where the formation of the ionic intermediates is the rate-determining step and the low basicity of the amines precludes their conversion to salts, a parallelism between $[Cl^-]$ and $[H^+]$ is observed. The hydrolysis takes place quantitatively in both buffered and unbuffered media and is pH sensitive[22,45].

From a practical point of view the Cl^- elimination is not particularly suited to the determination of the reactivity of the halogenated alkylating agents, because for nitrogen mustards reacting through an S_N2 pathway, it might simulate too high reactivity rates[45].

3.3.2 *The electrophilicity of alkylating agents*

Electrophilic species do not usually pre-exist in the reaction media, but are formed *in situ* by reversible or irreversible reactions. A quantitative estimation of reagent electrophilicity is a difficult task, because it depends on several factors, such as the magnitude of its electron deficiency, the ease of its formation, its stability in solution, its stereochemistry and finally the nature of the reacting nucleophile. We discussed previously how the alkylating activity (determined by NBP procedure) satisfactorily parallels the reactivity of alkylating agents and hence their electrophilicity. However, two more aspects must be taken into account in order to define this concept accurately. These are:

(a) the effect of the nucleophiles on the reaction rate:
(b) the stereochemistry of the electrophilic species.

3.3.2.1 *The effect of nucleophiles on the rate of alkylation reactions.* The rate of the alkylation reaction depends on the nucleophilicity of the substrate. Swain *et al.*[42, 61] expressed the nucleophilicity, n, of a nucleophilic reagent, by a linear free energy relationship

$$\log k/k_0 = ns \qquad (3.12)$$

where k is the second-order rate constant for the reaction between the considered nucleophile and the alkylating agent, k_0 is the rate constant for the corresponding reaction with a standard nucleophile at 25°C (water), s is the Swain–Scott substrate constant and n is the nucleophilic constant. The value of s was taken as 1.00 for methyl bromide (the standard reference). Taking into account this standard, Swain and Scott[42] computed a scale of nucleophilicities, assuming for water $n_{H_2O} = 0.00$. Using this scale, they further calculated a series of substrate constants, s. The preferential reaction of an alkylating agent with the stronger nucleophile relative to a weaker one, when both are present in excess, will rise steeply with increasing s values. In other words, the nucleophilic selectivity increases in the same direction as the s parameter[49].

These parameters are utilised in estimating the extent of alkylation at various nucleophilic centres in DNA, by means of the competition factor F_y [27,34]. This factor is related to s and n by

$$\log F_y = \log\left(\frac{k}{k_o}\right) - \log H_2O \qquad (3.13)$$

where k and k_o have the same significance as for (3.12). Replacing these values from (3.12) we obtain

$$\log F_y = sn - 1.74 \qquad (3.13a)$$

A series of values of n for nucleophiles of biological interest are given by Ross[1] and Lawley[20].

In fact the Swain–Scott substrate constant affords a quantitative basis for the nucleophilic selectivity of alkylating agents. The importance of this concept for the design of effective alkylating agents was recognised early by Ross[22]. In particular, it has been stressed that low nucleophilic selectivity may result in greater alkylation of weaker nucleophilic sites[49].

An alternative way to express the nucleophilic selectivity of alkylating agents is the softness concept in Pearson's hard-and-soft-acids-and-bases (HSAB) theory. A reaction between an electrophile (electron acceptor) and a nucleophile (electron donor) could also be regarded as a reaction between a Lewis acid and a Lewis base. The equilibrium constant, K, of such a reaction is determined by the strength factor, S. However, S does not always satisfactorily express K, so that, according to Pearson's equation, the introduction of a supplementary parameter is required in order to obtain a good correlation[62,63,64].

$$\log K = S_A S_B + \sigma_A \sigma_B \qquad (3.14)$$

where σ_A, σ_B are called softness.

Some attempts to define softness were made by Pearson[63], Klopman[65], etc., but a general consensus was not achieved. The softness of an acid or a base is characterised by a large atomic radius, a small effective nuclear charge and a high polarisability, whereas hardness implies all the opposite properties. According to these properties nucleophiles and electrophiles could be classified as hard or soft[62-66].

In 1963, Pearson[62] expressed the HSAB principle, which stated that 'hard acids (electrophiles) prefer to bind to hard bases, while soft acids prefer to bind to soft bases'. This statement is an extremely powerful qualitative rule, correlating the chemical behaviour of alkylating agents with cellular nucleophilic sites.

Although an exact softness scale is not available, empirical systematisation of various acids and bases was undertaken. The alkylating agents (carbon Lewis acids) are relatively soft. The hardness sequence of several carbenium ions decreases in the following order

$$RCO^+ > CN^+ > C_6H_5^+ > t\text{-}Bu^+ > i\text{-}Pr^+ > (Et^+ > Me^+)$$

Replacement of the hydrogen atoms of CH_3^+ by electronegative groups would certainly harden the cation (i.e., $RCO^+ > RCOCH_2^+ > CH_3^+$; RCO^+ being a borderline acid). Since H^+ is one of the softer acids, CH_3^+ represents the extreme limit of the softness class of the carbon Lewis acids bearing a positive charge. The corresponding carbene is even softer[66].

A correlation between softness and Swain's s constant may be made. Generally the soft alkylating agents are those with a higher s factor. An approximative scale of softness for Lewis bases (or nucleophiles) is given in Table 3.5.

3.3.2.2 *The stereochemistry of the electrophilic reagents.* The stereochemistry, the van der Waals volume, and the degree of hydration of the electrophilic species of alkylating agents represent important factors in determining the alkylation extent and specificity of the macromolecules and hence their biological properties.

Table 3.5 Classification of bases according to softness.

Softness	Bases
Hard	NH_3, Alk-NH_2, H_2O, HO^-, ROH, RO^-, $CH_3CO_2^-$, CO_3^{2-}, F^-
Borderline	Aryl-NH_2, N-piperidines, N-purines, Br^-, etc.
Soft	$H:^-$, organic anions, olefines, R_2S, RSH, RS^-, $S_2O_3^{2-}$, I^-

As they are usually carbocations of the R_3C^+ type (carbenium ions) with sp^2 hybridisation, their structure is mostly planar[67]. Hence, the alkylation reactions of the classical carbenium ions (as such or symmetrically solvated) are not stereospecific. However, the reactions of ion pairs or other ionic aggregates could occur stereospecifically, due to the chirality of these species. The chloronium ion (**10**) possessing a cyclopropene-like structure, allows geometric isomerism. The ring opening theoretically leads to diastereoisomers if the carbons are suitably substituted, such reactions exhibiting high stereospecificities. Such ions could be generated during the non-enzymatic decomposition of N-(2-chloroethyl)-N-nitrosourea derivatives. Their formation is consistent with some abnormal reaction products found during their hydrolysis[32].

3.3.3 *Formation of DNA adducts*

An important and often neglected aspect in the design of alkylating agents is that of the accessibility and reactivity of the nucleophilic sites in the DNA molecule.

3.3.3.1 *The reactivity of the DNA macromolecule.* Unusual electronic interactions in a DNA macromolecule lead to new qualitative effects which also affect the reactivity of the nucleophilic sites of the purine and pyrimidine bases.

Steric hindrance is one of the factors which drastically limits the access of exogenous electrophiles to some nucleophilic sites. Thus, whereas the phosphodiester groups and the sugar hydroxyls remain largely exposed to electrophilic attack, some nucleophilic centres of the purine and pyrimidine bases are partially or totally hindered, either by (a) their involvement in Watson–Crick hydrogen bonding, or (b) their position in the interior of the DNA double helix and the superposition of the electrostatic potential* associated with the phosphate groups and bases themselves. This fact greatly

---

*The electrostatic potential of a macromolecule is defined as an expression of "the molecular reality, closely related to what a reactant feels upon approaching the substrate at not too close a distance (> 5 Å)"; this concept developed by Pullman[68] is particularly well suited for the study of the interaction of various electrophiles with DNA. The quantum mechanical approach of this concept affords a unitary criterion for estimating the reactivity of the various nucleophilic centres in DNA.

reduces their nucleophilic reactivity. Theoretically, only nucleophilic centres situated in the major or minor grooves or in the walls of the double helix remain accessible to electrophilic attack.

The following nucleophilic centres contribute to hydrogen bonding: adenine, N1 and 6-NH_2; guanine, N1, 2-NH_2 and O6; thymine, N3 and O4; cytosine, O2, N3, 4-NH_2. From these, only the N3 and O6 groups in guanine remain more or less readily accessible. However, in some cases (i.e. DNA alkylation with BCNU) the alkylation of cytosine appears to be an important reaction[20].

The role of the overall structure of the DNA macromolecule upon the reactivity of its nucleophilic sites has recently become accessible to quantum mechanical exploration through the use of the concept of the electrostatic molecular potential[68].

In this concept the most prominent potential minima (calculated for all non-hydrogen atoms) are associated with high nucleophilic reactivity. The deepest minimum is generally associated with the N7 of guanine. One of the most striking results of these computations is the progressive increase of the absolute value of the potential minima in the series: single bases < nucleosides < nucleotides ≪ single helix < double helix. This phenomenon may be explained by the penetration and overlapping of the strong electrostatic potential of the phosphate groups with that of the purine and pyrimidine bases when these are arranged in a regular structure.

The involvement of ring nitrogen atoms (i.e. N1(A), N3(C), etc.) in hydrogen bonding in the double helix produces a very strong depletion of the molecular potential minimum[68]. These facts have the following major practical consequences regarding the nucleophilic reactivity of DNA sites.

(a) The increase in the net nucleophilic reactivity of some centres (with respect to the reactivity of the same centres in bases, nucleosides, nucleotides). Thus:

—the alkylation by ethylaziridinium ions of the N7 of guanine proceeds at a 50-fold higher rate in native DNA and a 7-fold higher rate in denatured DNA than in monomeric guanosine[69]. A similar enhancement of the reactivity was shown during adenine alkylation by the same reagent[70];

—reagents such as N'-methyl-N'-nitro-N-nitrosoguanidine, N-methyl-N-nitrosourea, dimethylsulphate, etc., cannot methylate guanosine. However, the same reagents do alkylate the guanine residues (at N3 and N7) in synthetic polynucleotides (i.e. polyG, polyGC, etc.) and DNA[71].

(b) Another significant result of this approach is the possible reversal of the relative ordering of related nucleophilic centres in going from single bases to the double helix. Thus, the order of nucleophilicity of various centres in double helical DNA is as follows: N7(G) > N3(G) > N3(A) > O2(T) >

O6(G), etc.; different from that of the monomeric bases: N7(G) > N3(C) > O2(C) > O6(G) > N1(A) > N7(A), etc.[68].

These theoretical predictions are in surprisingly good agreement with experimental data.

Other potential alkylation sites in DNA macromolecules that are often neglected are the phosphate groups. Because of the overlapping of the electrostatic effects, in the DNA double helix, the minima associated with phosphate groups increase considerably[72]. Therefore, they become more nucleophilic, as was shown experimentally. Thus the DNA reaction with ethyl sulphate or ethyl-N-nitrosoamine affords a significant percentage (16% and 70%, respectively) of phosphoric triesters. A similar amount of phosphate groups are alkylated by N-(2-chloroethyl)-N-nitrosourea[73, 74].

3.3.3.2 Reaction products of alkylating agents with DNA: structure and biological significance.

The reaction products of alkylating agents with DNA have been characterised in many cases and their properties extensively studied. Some excellent reviews of this topic are available[71,75,76].

However, there are alkylating agents for which the structure of the adducts* thus formed is largely unknown. As a general orientation it may be considered that when the electrophilic reactivity of alkylating agents increases (or their corresponding substrate constant, s, decreases), reactions with the less nucleophilic sites become more prominent. The HSAB principle affords the same conclusion (see section 3.3.2.1).

The reaction between alkylating agents and DNA yields addition products, which either involve a single purine or pyrimidine nucleus (i.e. for methylating, ethylating agents) or link together two such moieties, especially in the case of bifunctional alkylating agents (cross-linking). These data, as well as the possible biological significance of the adducts, are given in Table 3.6.

The first situation is typical for monofunctional agents, but bifunctional agents can also lead to monoalkylated bases (for instance mustard gas, HN_2, N-nitrosoureas, etc.)[74,91,94]. In certain cases, cyclic adducts such as (29) and (30) also appear, as for instance, during DNA reaction with N-(2-chloroethyl)-N-nitrosourea, δ-propiolactone, etc.[92, 99]. Their biological significance has not yet been elucidated.

29, 1N6A   30, 3N4C

---

*The term 'adduct' is used to define any covalent reaction product formed during interactions between DNA and alkylating agents.

Table 3.6 Types of adducts induced by alkylating agents and their biological significance.

No.	Alkylating agent	Reaction product of the DNA alkylation	Biological significance of the induced lesion
1	Monofunctional alkylating agents: alkyl halides, 2-haloalkylamines or sulphides, methanesulphonates, aziridines, oxiranes, triazenes, etc.	a. Alkylated nucleotides	Possible miscoding mutagens, depurination (error prone repair) (e.g. O6(G), O4(T)); single strand breaks; could trigger cellular transformation and carcinogenic effects; low cytotoxicity; multiple repair possibilities
2	Bifunctional alkylating agents: nitrogen and sulphur mustards, aziridines, oxiranes, methanesulphonates, nitrosoureas, etc.	a. Cyclic adducts (e.g. nitrosoureas)	Unknown, could represent intermediates in the formation of intra- or interstrand cross-links
		b. Adducts formed by the cross-linking (intra- or interstrand) of two nucleotides	Strong cytotoxic character by blocking DNA synthesis; repair occurs with low efficiency
		c. Adducts formed by the cross-linking between a nucleotide and a protein	Not yet elucidated

Bifunctional alkylating agents are able to induce several types of cross-linkings, namely: (a) intrastrand (both alkylated sites belonging to the same strand); (b) interstrand; (c) DNA-protein cross-links[75,76,100]. The structure of some of the DNA-alkylating agents' reactions products are given in Table 3.7.

Interstrand cross-links are especially related to the cytotoxicity of alkylating agents[76,79,89]. It has been suggested that the lethal effect of bifunctional alkylating agents is due to the blocking of DNA replication by these cross-links[101] and also by a significantly lower efficiency of the removal of these adducts by the repair systems.

There are some apparent exceptions to these general principles. Interesting cases are those of DTIC or CB 1954 (5-(aziridin-l-yl)-2, 4-dinitrobenzamide), which do not seem to fulfill the bifunctionality requirement. In the first case, DNA cross-links have been detected: they could be assigned to the CH_2O released during DTIC metabolism[102]. In the second case, the product undergoes metabolic activation leading to a bifunctional alkylating compound, which is able to alkylate DNA.

Many other effects of alkylation (or cross-linking) can be noted at DNA level, such as depurination, depyrimidation, alkali-labile DNA breaks, which were mentioned for N-nitrosoureas[76], nitrogen mustards[89]. Their contri-

Table 3.7 Interactions of alkylating agents with DNA: reaction sites.

No.	Class of compounds	Compounds	Reaction sites	Ref.
1	Alkyl phosphates	Trimethylphosphate	O6G	82
2	Aziridines	TSPA	7G	78
		TEM	7G,7G × 7G[a]	
3	2-Haloethyl chlorides	nor-HN$_2$	7G,7G × 7G[a]	84,85
		HN$_2$	7G,7G × 7G[a]	84,85
			7G × 3C[a]	89,91
		CP	?	86,87
		L-PAM	?	89,90
		Chlorambucil	?	88
4	2-Haloethyl sulphides	Bis-(2-chloroethyl)-sulphide	7G,7G × 7G[a]	20
		(2-Chloroethyl)-ethylsulphide	7G,3A	20
5	Lactones	Propiolactone	7G,1A,3C,1N6A[b]	92,93
6	Methanesulphonates	Methylmethane-sulphonate	7G,7G × 7G[a]	73,77
		Ethylmethanesul-phonate	7G,3A,1A	73
		2-Chloroethyl-(methyl-sulphonyl)-methane-sulphonate	7G,3,7G × 7G[a] 7G × 3C[a],P	94,95
7	N-Nitrosoureas	BCNU	7G × 7G[a],7G,N6A, N4C,O6G,7G,3C, 1N6A[b],3N4C[b],O6G	81,96, 97,99
		CCNU	7G,3C,3N4C[b]	97,99
		N-Me-N-nitrosourea	7G,3A,1A,3C,7A,P	71,73
		N-Et-N-nitrosourea	O6G,7G,P	71
		N-(2-chloroethyl)-N-nitrosourea	7G[c],7G × 7G[a], 7G × 3C,P	74
8	Oxiranes	Dianhydrogalactitol	7G	83
9	Triazenes	DTIC	7G	98

[a]Cross-links.
[b]Cyclic adducts.
[c]7-(2-Hydroxyethyl)-deoxy-G, 7-(2-chloroethyl)-deoxy-G and ring opened 7-(2-hydroxyethyl)-G are formed.

butions to the overall cytotoxic effects may not be accurately estimated. Regional specificity in DNA alkylation has been reported[75].

3.3.4 *Structural modulation of alkylating agent reactivity*

The antitumour effectiveness of alkylating agents can be improved by increasing either their cytotoxicity, or their selectivity towards malignant cells.

A quantitative expression of these two parameters (according to Brock) for several alkylating agents is given in Table 3.8[103].

Cytotoxicity seems to be easier to handle (see also Table 3.9), because it is closely related to: (a) the chemical reactivity of the alkylating moiety; (b) the bifunctional character of the alkylating agent. The monofunctional agents are

Table 3.8 Cytotoxic activity and specificity of some nitrogen mustards[103].

No.	Compound	Cytotoxic[a] activity (ED_{50}/mol)	Cytotoxic specificity[b] $(ED_{50}/mol\,min^{-1}\,10^{-2})$	$\dfrac{LD_{50}{}^{c}}{CD_{50}}$
1	$HN_2{}^e$	138	40	4.4
2	Chlorambucil	14	10	7.3[d]
3	nor-$HN_2{}^e$	1.4	10	2.5
4	HN_2-N-oxide	1.5	2	12
5	CP-mustard	7.5	8	3.5
6	4-HO-CP	63	225	120.0
7	4-HO-IF	63	700	80.0[d]

[a]Determined according to Schmähl and Druckrey[103b].
[b]Cytotoxic specificity = cytotoxic activity/alkylating activity, determined by NBP test.
[c]Therapeutic index, $LD_{50} = 50\%$ lethal dose determined on Yoshida ascites sarcoma cells in rats, $CD_{50} = 50\%$ curative dose determined in the same system[43].
[d]Calculated from Ref. 103a.

generally significantly less toxic. The increase in the number of alkylating units does not increase the cytotoxicity proportionally[21].

Very reactive and hence cytotoxic alkylating agents may be designed. Such compounds could be used for intratumoral or intra-arterial perfusions, because of their very short-lives. Compounds (**31**) and (**32**) fulfill these requirements[24], having extremely short $t_{1/2}$ (in physiological medium, pH 7.4, 37°C).

$$HOOC\!-\!\bigcirc\!-\!OCH_2\,CHSCH_2CH_2\,Br \qquad \left[Cl\,CH_2CH_2S(CH_2)_3\,CONHCH_2 \right]_2$$
$$\qquad\qquad\quad |\;CH_2SCH_2CH_2Br$$

$$31,\; t_{0.5} = 4\text{-}5\;sec \qquad\qquad 32,\; t_{0.5} = 1.3\;sec.$$

Based on Table 3.8, the cytotoxic specificity is defined as the biological efficacy of the alkylating agents responsible for the cytotoxic effects. However, the design of highly specific compounds against malignant cells is much more difficult, because it requires, for instance, some reproducible biochemical difference between the normal and tumour cells to be exploited. These differences could be related to enzyme systems, metabolic behaviour, genetic material, etc. Unfortunately, although several such features have been demonstrated[104], they present only a qualitative character. However, some of these peculiarities have been exploited and will be further discussed. For instance, many antitumour agents have been designed based on the assumption that the proliferative kinetics of the cancer cells is faster than that of normal cells. This approach was fairly successful and many substances were obtained which are highly cytotoxic against rapidly proliferating cells. Unfortunately, there are also normal cells with rapid proliferation (bone marrow, gastric epithelium, etc.), which limits their use.

Another problem is to design alkylating agents possessing appropriate properties in order to achieve the desired selectivity. Some of these requirements are:

—a suitable reactivity (usually a low one), so that the drug would be able to reach the target in a sufficient concentration;

—a structure able to achieve a selective transport of the alkylating moieties into malignant cells.

—a structure which could be activated specifically inside the malignant cells in order to release an effective alkylator there.

In order to obtain molecules which fulfill, at least partially, these requirements, several theoretical principles have been stated. They have led, in many cases, to clinically useful compounds, despite the fact that often their mechanism of action does not confirm the working hypothesis. Some of these trends are briefly outlined:

(a) The search for new alkylating moieties (quinone-methides, α-methylene-γ-lactones, etc.) or for unusual combinations of alkylating moieties grafted on the same molecule: 2-chloroethyl(methylsulphonyl)-methanesulphonate[105], 5-(aziridin-l-yl)-2,4-dinitrobenzamide[106], amino acids carrying both a nitrogen mustard and an N-(2-chloroethyl)-N-nitrosourea[107], CP or IF derivatives with aziridinyl or methane-sulphonyl groups[108] could be cited.

(b) The latent activity concept: the idea of using a highly reactive drug in a chemically inactive form (prodrug or transport form) able to undergo metabolic activation was put forward by Druckrey [103b, 109] and less explicitly by Seligman[110]. This concept contributed to the design of latent phosphoramide nitrogen mustards, which finally led to a series of cyclic congeners with outstanding antitumour properties. CP is the prototype of these compounds, although its mechanism of action does not entirely correspond to the initial hypothesis. CP, which is chemically inactive ($t_{1/2} > 7$ days), is activated by microsomal enzymes (see section 3.6.1). A similar behaviour is exhibited by IF, Trophosphamide, CB 1954, triazene nitrogen mustards (BTIC) and Estracyt.

According to their mechanism of activation, latent alkylating agents could be classified as:

—compounds activated by oxidation: enzymatic systems such as cytochrome P-450 dependent monooxygenases (usually involved in the xenobiotics metabolism) cause activation of CP, IF and triazenes;

—compounds activated by reduction: several enzymes for instance nitrore-ductases (i.e. CB 1954, see section 3.6.2)[111] or other NADPH dependent reductases (Mitomycine C, benzo- and naphthoquinones) are involved in this type of activation. In the case of naphtho- and benzoquinone, a selectivity against hypoxic cells is to be expected, because of their higher reducing potential as compared to normal cells[112];

—compounds activated by enzymatic hydrolysis: urethane nitrogen mustards (i.e. Estracyt) belong to this class.

(c) The carrier concept: this concept is based on the properties of certain molecules to accumulate in cancer cells. This idea is not new and many attempts were made to use suitable structures (based on active transport potentialities or on preferential uptake of chemicals by a particular tissue), in order to improve the specificity of the alkylating agents. This was a relatively fruitful approach, which led to very active compounds such as L-PAM, DBD, Estracyt, although the improvement in selectivity was rather disappointing.

Attempts have been made to use amino acids, steroids, carbohydrates, dyes and macromolecules, as carriers for alkylating moieties. Two new interesting developments of this concept have recently emerged, namely:

—the chemoimmunotherapy approach (section 3.4.2) which uses mono- or polyclonal antibodies as specific carriers of alkylating moieties;
—the use of specific groove binding DNA compounds[113,114]. This could be a very interesting way of progress, taking into account the oncogene theory.

The combination of both latent activity and carrier concepts led to a very interesting compound, namely 3-N,N-bis-(2-chloroethyl)-carbamoyl oestradiol (Estracyt, see section 3.4.2 (**39**)), with fairly good clinical activity against prostate carcinoma.

Several other concepts have been claimed for the design of alkylating agents (e.g. 'soft' and 'hard' alkylating agents, the principle of latent inactivity[1,118]), but until now, they have not led to useful results.

Other approaches can also be used in order to increase the effectiveness of alkylating agents, namely:

—tumour cell sensitisation: for instance, the prior administration of a methylating agent with low toxicity is able to decrease the cellular levels of alkylguanine transferase (AGT) (which is responsible for some types of repair processes) in tumour cells, thus significantly increasing the sensitivity of these cells to 2-haloethyl-N-nitrosoureas[115];
—normal cell protection: a good example is the development of sodium 2-mercaptoethanesulphonate (Mesna) in order to reduce CP toxicity to bladder[116]. Another example is the use of 5,5-N-dimethylamidine as protection against HN_2 toxicity, by blocking its cellular uptake[117].

Both latent activity and carrier concepts are based on the possibility of modulating the chemical reactivity of the alkylating moieties (see Table 3.9).

3.4 Chemical features involved in the transport and uptake of alkylating agents

On their long way from the site of administration to the target, alkylating agents have to cross several barriers. The cellular uptake of alkylating agents,

Table 3.9 Modulation of chemical reactivity in the alkylating agents.

No.	Trend in reactivity modification	Chemical features involved in reactivity modification	Alkylating moieties submitted to modification	Resulting compounds
1.	Enhancement of reactivity	1. Increase of the basicity of the reaction centre by:		
		— sidechain substitution with electron releasing (+I) substituents	Sulphur mustards, nitrogen mustards, aziridines, methanesulphonates, etc.	HN_2 from nor-HN_2
		— convenient substitution of aromatic nucleus with electron releasing substituents	Aromatic nitrogen mustards	Aminophenyl mustard from phenyl mustard
		— ortho effect	Aromatic nitrogen mustards	IOB 82 (75) from m-(2-chloroethyl) aminobenzoic acid
		— substitution of halogens Cl, I, Br	Nitrogen mustards	Alkylating agents used for perfusions
		2. Suitable substitution of reaction centre in order to change the reaction mechanism from S_N2 to S_N1	Aziridines, oxiranes methanesulphonates	Compound (116) 2,5-dimethylsulphonyl-oxy-n-hexane
2.	Decrease of reactivity	1. Decrease of basicity of the reaction centre:		
		— substitution of the reaction centre with electron withdrawing substituents (derivatisation of the alkylating moiety)	Aziridines, nitrogen mustards	CP, IF, Estracyt, DTIC, Mitozolomide
		— aromatic nucleus substituted with electron withdrawing substituents	Aromatic nitrogen mustards	Nitrogen mustards derived from azobenzene

as well as their translocation to the nucleus, are complex processes which are favoured by various chemical features of the molecule.

The *in vivo* distribution of alkylating agents is comprehensively discussed in a number of pharmacokinetic studies. We will mention only that, in the circulation, alkylating agents and/or their metabolites are transported both free and absorbed to serum proteins (e.g. albumin). Conflicting data are available about the effect of this absorption upon their activity. Increases (e.g. for aromatic nitrogen mustards)[119], as well as decreases (e.g. for *N*-nitrosoureas)[120], in the half-lives of alkylating agents have been reported.

Literature data regarding the uptake of alkylating agents and the mechanism related to this phenomenon are rather poor (Table 3.10).

3.4.1 *Lipophilicity*

Lipophilicity, expressed as $\log P$ (partitioning coefficient of a compound between *n*-octanol and water)[121] or in terms of the substituent constant, π, is a parameter which expresses the ability of a substance to accumulate either in lipid or aqueous medium. This is a very important parameter for the *in vivo* pharmacokinetic behaviour of a compound. Usually, diffusion across biolo-

Table 3.10 Cellular uptake of alkylating agents.

No.	Uptake mechanism	Alkylating agents	Required chemical features
1.	Passive diffusion	*N*-Nitrosoureas, latentiated nitrogen mustards, aziridines, methane sulphonates, oxiranes, triazenes (?), etc.	Optimum lipophilicity, higher for solid tumours ($\log P_0 < 3$) as for leukaemia ($\log P_0 < 0$)[60,123,124]; favoured by low molecular weights and unionised forms[125].
2.	Facilitated diffusion	L-PAM (?), Chlorambucil, carboxy CP	Chemical structure and stereochemistry of the molecule in order to fit a membrane receptor; e.g. choline-like structures for aliphatic nitrogen mustards[125,126], amino acid-like structure for aromatic ones[127,128]
3.	Activated transport	HN_2[a], L-PAM[b]	
4.	Pinocytosis	Alkylating moieties linked to macromolecular carriers[c]	Large molecular weights[119,129]

[a]Transported by choline carriers.
[b]Transported by two amino acid transport systems.
[c]The combination of macromolecules with alkylating agents could be divided into two groups: alkylating agents absorbed on proteins and covalently linked (directly or by a carbodiimide procedure) to proteins[129].

gical membranes requires an optimum lipophilicity, $\log P_0$. A lipophilicity which is either too low or too high hinders a substance entering or leaving the lipid barrier, respectively.

Because the lipophilicity is an additive property, it can be modulated. It is relatively easy to design an alkylating agent with a given lipophilicity both by choosing the appropriate alkylating moiety and a suitable carrier[122].

The most striking example of the importance of this parameter for biological activity is offered by N-nitrosoureas. Thus, it was demonstrated that their activity against L 1210 leukaemia depends on the lipophilicity according to parabolic function[123]

$$\log \frac{1}{C} = 4.53 - 0.069 \log P - 0.057 \, (\log P)^2 \qquad (3.15)$$

$$n = 22, \qquad r = 0.922, \qquad s = 0.163$$

By derivation of this function, an optimum lipophilicity, $\log P_0 = -0.60$, is obtained. Similar approaches indicate that the optimum lipophilicity is higher for solid tumours (Lewis lung carcinoma)[124].

The very important property of some anticancer compounds to cross the blood-brain barrier is also dependent on the lipophilicity, the optimum $\log P$ being in the range 0–3. On this basis, N-(2-chloroethyl)-N-nitrosoureas able to kill leukaemic cells in the brain were obtained (see also section 3.7.1 and Table 3.10).

The effect of this parameter on active transport is much less important because, in this case, the determining parameters are the structural features involved in the interaction of the drug with the membrane receptors.

The $\log P$ values of some important alkylating agents are given in Table 3.4.

3.4.2   Molecular structure

Molecular structure and often stereochemistry are very important elements both for the cellular permeation of the alkylating agent and for its transport across the cytoplasm. In this respect, molecules to which the alkylating moieties are grafted could be regarded as carriers.

Two main types of carriers are as follows:

(a) Xenobiotic structures: such carriers are important when we try to design compounds exhibiting some preselected physico-chemical properties, for instance, to design latent alkylating agents able to cross the blood-brain barrier. Thousands of alkylating agents, some of them clinically useful, were synthesised on this basis (e.g. HN_2, Chlorambucil, CP, IF, BCNU, CCNU, Me-CCNU, 2-chloroethyl-(methanesulphonyl)-methanesulphonate, Myleran, TSPA);

(b) Physiological structures: such carriers are believed to permeate cell membranes by specific transport processes, thus accumulating in tumour cells.

In order to avoid the alteration of the carrier properties of such molecules, the design of the corresponding alkylating agents has to fulfill at least three requirements, namely:

— $\log P$ of the alkylating moieties must be as close as possible to 0.00;

— low reactivity of the alkylating moiety is required;

— the linking of the alkylating moiety to the carrier must not affect or hinder those structural features (of the carrier) which are involved in active transport.

Almost all types of physiological molecules (carbohydrates, lipids, amino acids, peptides and proteins, steroids, purine and pyrimidine bases, mono- and polyclonal antibodies) have been used for this purpose. Because of the numerous data, it is impossible to review exhaustively all the structures used as carriers. We shall present only some significant examples which led to the synthesis of clinical products.

Carbohydrates were used in combination with almost every kind of alkylating moiety, in the hope that they would be used as effective substrates by the malignant cells having a different energetic metabolism. Among the clinically useful products we mention: DBD, 1,6-bis-(2-chloroethylamino)-1, 6-dideoxy-D-mannitol (Degranol); 1,6-dianhydrogalactitol. No transport involvement in the antitumour activity of these derivatives could be demonstrated.

Amino acids and peptides: some amino acids are specifically accumulated in cells, leading to an intracellular concentration which may be five-fold higher than the extracellular concentration. This fact combined with the faster proliferation of malignant cells led to the proposal of these structures as suitable carriers for alkylating moieties.

Accordingly, hundreds of alkylating agents derived from amino acids were synthesised, among which, L-PAM, Sarcolysin (D, L-PAM), *meta*-L-PAM (as in Peptichemio), proved to be of clinical use, although their cytotoxic selectivity remained disappointing.

In an attempt to decrease the toxicity of these derivatives, a series of peptides carrying alkylating moieties were also synthesised and tested[129,130,131].

The most extensively studied of these compounds was L-PAM. Even in this case there is no proof that the carrier hypothesis works. However, the difference of activity noted between the PAM enantiomers (L-PAM being twice as active as D-PAM)[132] could be ascribed to a different uptake by cancer cells. Such activity differences between diastereoisomers were found also with dipeptides. However, unexpectedly, sometimes DD-diastereoisomers are more effective then LL ones[130].

Esterification of the carboxylic group of amino acids or peptides does not significantly affect their activity. In contrast, alteration of the amine group drastically influences the biological properties of these compounds[130,131].

Purine and pyrimidine bases, nucleosides and nucleotides: the use of this class of compounds as carriers was suggested following the idea that the proliferative kinetics of malignant cells is faster than that of normal cells so that they would require more purines and pyrimidines for DNA synthesis.

Among the alkylating agents derived from purines or pyrimidines the best known is uracil mustard (33), which, however, did not gain wide clinical use because of its side effects (toxicity and carcinogenicity). Recently, the synthesis of new types of phosphoramide nitrogen mustards (34, 35) derived from nucleotides has been reported[133,134].

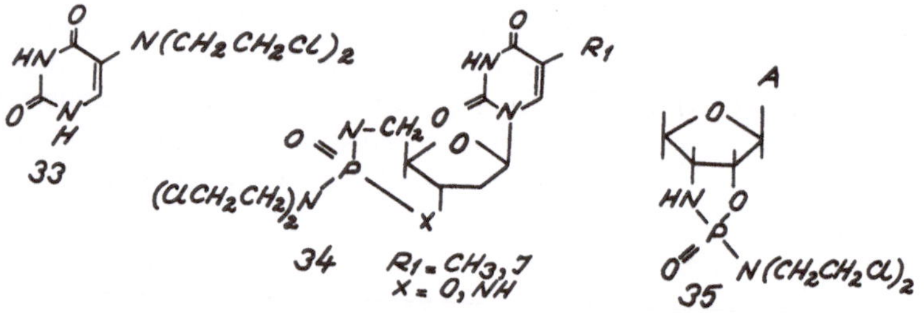

Steroids: the steroid hormones are of particular interest as carriers, because:

—there are several types of cancers (approx. 30–50% of all cancer patients) whose growth is very dependent on the extra- and intracellular concentration of certain types of steroids or other hormones (e.g. prostate and breast carcinomas);

—although the mechanism of steroid hormone uptake is poorly understood[125], a positive correlation was demonstrated between the concentrations of oestrogen (ER) and progesterone receptors (PR) in cancer cells and their sensitivity to hormonal treatment.

These special features were used (but not fully exploited) in the design of alkylating agents which are effective in hormone-dependent cancers. Many attempts to attach alkylating moieties (latent or not latent) to steroid hormones have been reported, several of them before the discovery of hormone protein receptors[135,136].

The fact that natural and synthetic oestrogens can be used by themselves to control some types of malignancies stimulated interest in this area. Thus, early attempts[137] led to compounds (36, 37) which, in spite of their cytotoxicity, demonstrated no selectivity against hormone-dependent cancers. The next step was the synthesis of oestradiol mustard (38) which, despite some affinity for ER, also proved to be rather unspecific[138].

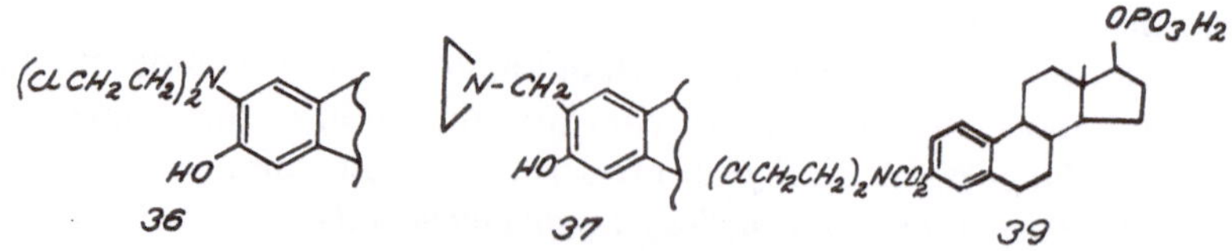

More recently, the synthesis of latent nitrogen mustards (i.e. urethane type nitrogen mustards) derived from natural and synthetic oestrogens was proposed[139,140]. In our laboratory 3-N,N-bis-(2-chloroethyl)-carbamoyl-17β-oestradiol, Estramustine[141], was synthesised, which was independently and simultaneously obtained by Leo[142] in a soluble form (39) (Estracyt). This approach was more fruitful; Estracyt is highly effective in the treatment of prostate carcinoma. The mechanism of action of this compound is not yet elucidated; it probably acts by releasing the alkylating moiety in the target cells. It possesses a certain affinity for ER (especially in dephosphorylated

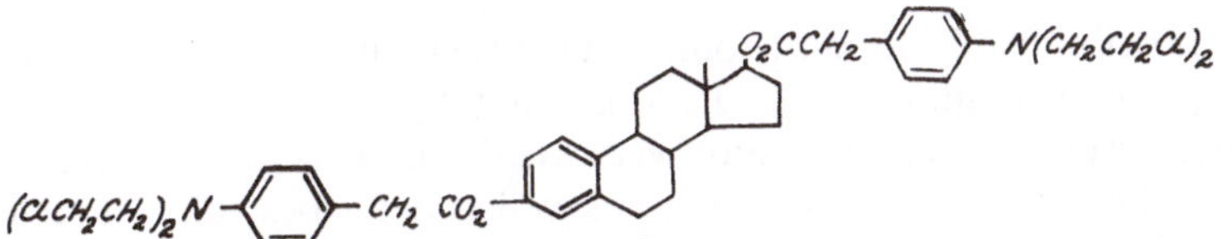

38

form)[135] and a high affinity for a prostatic protein, which probably acts as a carrier, explaining the cytotoxic selectivity[144]. Substitution of the 17β position of oestradiol with urethane nitrogen mustard (40)[141] or with N-2-chloroethyl-N-nitrosoureyl group (41)[143] did not afford clinically interesting products.

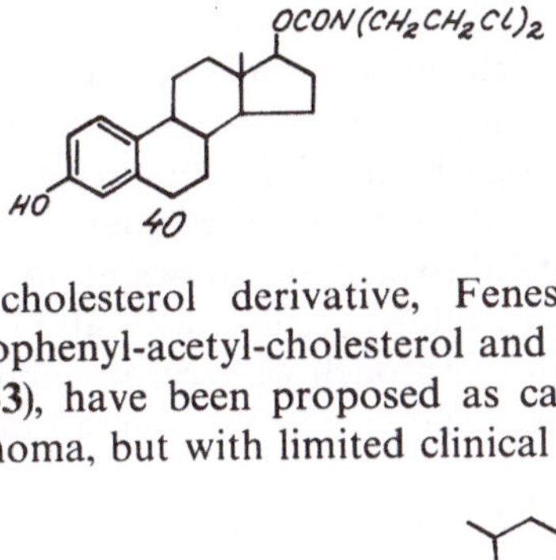

40

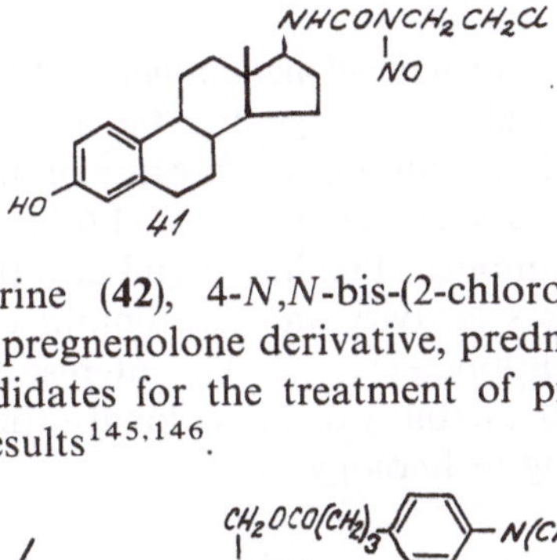

41

A cholesterol derivative, Fenesterine (42), 4-N,N-bis-(2-chloroethyl)-aminophenyl-acetyl-cholesterol and a pregnenolone derivative, prednimustine (43), have been proposed as candidates for the treatment of prostate carcinoma, but with limited clinical results[145,146].

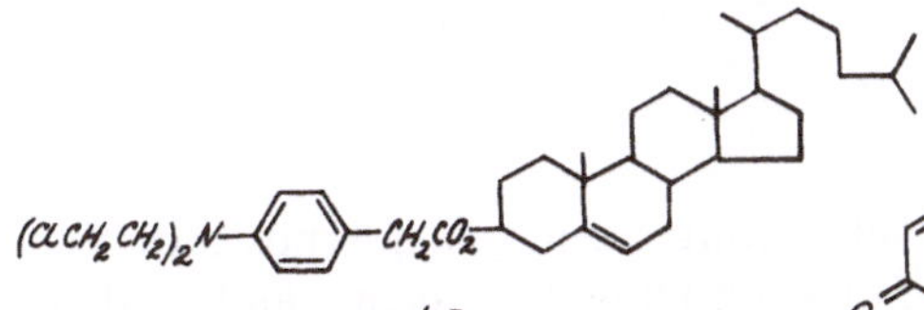

42

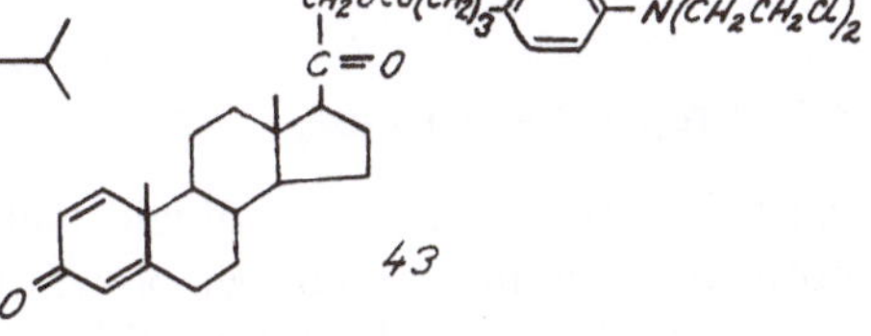

43

When the design of alkylating agents is undertaken on the basis of the interaction between receptors and steroid hormones, the main problem is to choose the most suitable carrier for a selected cytotoxic moiety and also to determine to which part of its molecule these groups could be linked, in order to preserve the alkylating steroid's affinity for the receptor. For instance, the structural requirements for oestrogen-ER interactions were established using a quantitative structure-activity relationship (QSAR) analysis on two series of synthetic and natural oestrogens[147], leading to the following conclusions:

—the 3 and 17β positions (or the 4,4' positions, respectively) are deeply involved in oestrogen-ER interactions, therefore they have to be free in alkylating oestrogens;

—the attachment of the alkylating moieties could be done in the 7, 12 and 17α positions. Substitution of the aromatic ring would lead to ineffective derivatives;

—for hexoestrol and stilboestrol the favourable positions for linking alkylating moieties are located in the sidechains[147].

The possibility of linking a bulky substituent in the 7 position without affecting the ER affinity was demonstrated by Bucourt et al.[147a]. More recently, derivatives of oestradiol were synthesised, with alkylating groups linked in the 6[148] and 7[149a] positions, as well as of hexoestrol, with alkylating moieties located on the sidechains (**44**)[149b].

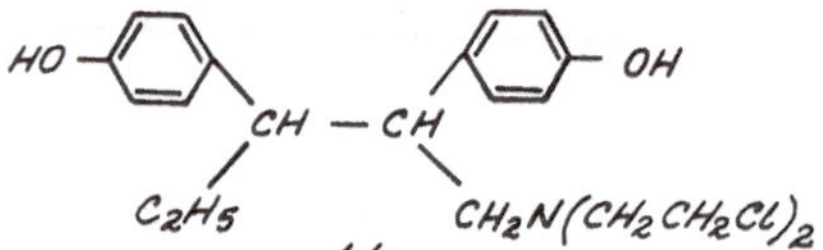

Poly- and monoclonal antibodies. One of the most promising approaches to selective alkylating agents is the use of the targeting properties of the (poly- or monoclonal) antibodies, because of their very high affinity for the cell surface tumour-associated antigens (TAA)[150a,b]. A number of alkylating agents (Chlorambucil, Trenimon) and also protein toxins have shown an increased tumour selectivity and antitumour effects, after coupling with appropriate TAA-antibodies[151,152]. The interest in this approach was recently enhanced by the availability of monoclonal antibodies and also by the progress made in coupling technology[153].

3.5 Preparative chemistry

An exhausive review of biological alkylating agents is not possible, due both to the immense number of synthesised alkylating agents and to their structural diversity. Therefore, this section is mostly concerned with three aspects:

—the preparative chemistry of the main alkylating functions;

—the synthetic pathways employed to obtain some important clinically useful compounds and some of their metabolites;

—particular chemical and stereochemical features, playing a special role in their mechanism of action.

3.5.1 Sulphur mustards

Mustard gas was the first compound which showed inhibitory activity in rat skin cancerogenesis tests[154], due to its strong cytotoxic properties. Its cytotoxicity has been further correlated with its ability to alkylate nucleic acids[20]. Sulphur mustard is prepared by conventional procedures involving either halogenation of the bis-(2-hydroxyethyl)-sulphide (45) with HX or POX$_3$ (X = Cl, Br), SOCl$_2$ or by treating ethene with sulphur dichloride

$$S(CH_2CH_2OH)_2 + SOCl_2 \longrightarrow S(CH_2CH_2Cl)_2 \longleftarrow S_2Cl_2 + CH_2{=}CH_2$$
$$\textbf{45} \qquad\qquad\qquad \textbf{46}$$

Thiodiglycol (45) may be prepared in large amounts by the addition of 2-chloroethanol to sodium sulphide or, even better, by the reaction of ethylene oxide with hydrogen sulphide.

Highly reactive bifunctional alkylators belonging to this class have been synthesised by introducing a spacer between two 2-chloroethyl-thio groups,

$$ClCH_2CH_2S(CH_2)_nSCH_2CH_2Cl$$
$$\textbf{47}$$

such as in compounds (47) ($n = 1$–10). Their vesicant power increases with n until $n = 3$, then decreases as n increases from 4 to 10[155]. The synthetic path which led to the very reactive sulphur mustard (31) ($t_{1/2} = 4.8$ s, 37°C, pH 7.4)[156] used for intra-arterial or intratumoral infusions is given in Scheme 3.2.

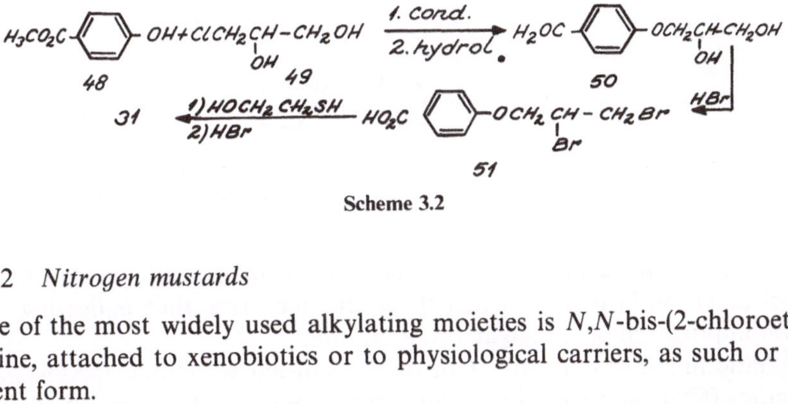

Scheme 3.2

3.5.2 Nitrogen mustards

One of the most widely used alkylating moieties is N,N-bis-(2-chloroethyl) amine, attached to xenobiotics or to physiological carriers, as such or in a latent form.

The general way to prepare aliphatic or aromatic nitrogen mustards is as follows:

$$R-NH_2 \longrightarrow RN(CH_2CH_2OH)_2 \longrightarrow RN(CH_2CH_2X)_2$$
$$\textbf{52} \qquad\qquad \textbf{53} \qquad\qquad\qquad \textbf{54}$$

The N,N-bis-(2-hydroxyethyl)precursors (53) are usually obtained by treating the corresponding primary amines with 2-chloroethanol, or better with ethylene oxide in acetic acid medium. Their halogenation has been carried out with a number of reagents such as phosphorus trichloride or tribromide (PCl_3, PBr_3), phosphoryl chloride or bromide ($POCl_3$, $POBr_3$) or thionyl chloride ($SOCl_2$). The introduction of iodine is achieved by boiling the corresponding alkyl halides (**54**) (X = Cl, Br) with sodium iodide in dry acetone[157].

$$54 + NaI \longrightarrow RN(CH_2CH_2I)_2$$
$$(X = Br) \qquad\qquad 55$$

3.5.2.1 *Aliphatic nitrogen mustards.*   Nitrogen mustard was probably one of the first synthetic compounds tried in cancer chemotherapy[158]. Nor-nitrogen mustard (nor-HN_2), HN_2 and *N,N,N*-tri-(2-chloroethyl) amine (Sinalost) proved to be effective inhibitors of animal and human tumour growth. Among them, HN_2 is still of clinical use (see Table 3.1).

Halogenation of mono-, di- or triethanolamine affords the desired compounds. A large number of nitrogen mustards with a wide range of chemical reactivities were synthesised by varying the halogens or their reactive sidechains (e.g. *n*-propyl, isopropyl)[1]. Among the large number of aliphatic nitrogen mustards investigated, some 2-chloroethyl-amino derivatives of sugars proved to have clinical activity. The best known is 1,6-bis-(2-chloroethylamino)-1,6-dideoxy-D-mannitol (**56**) (Degranol)[159].

Animal experiments performed with diastereoisomeric analogues (sorbitol, dulcitol or iditol) have shown that these are ineffective, thus indicating the decisive importance of the sugar configuration.

Aliphatic nitrogen mustards with rather unusual structures, for instance spirazidine (**57a**) and prospidine (**57b**), were prepared and underwent clinical trials in the Soviet Union[160].

The Mannich reaction proved to be extremely useful in the synthesis of nitrogen mustards linked by methylene bridges to the aromatic nucleus. Nitrogen mustards derived from oestradiol[161], uracil (**59**) (Z = O) and thiouracil (**59**) (Z = S) were thus prepared[162]. A structurally related derivative, 2,4-diamino-6-bis-(2-chloroethylamino)-methylpteridine, highly effective against L 1210 leukaemia was obtained via the corresponding 2,4-diamino-6-halomethylpteridine[163].

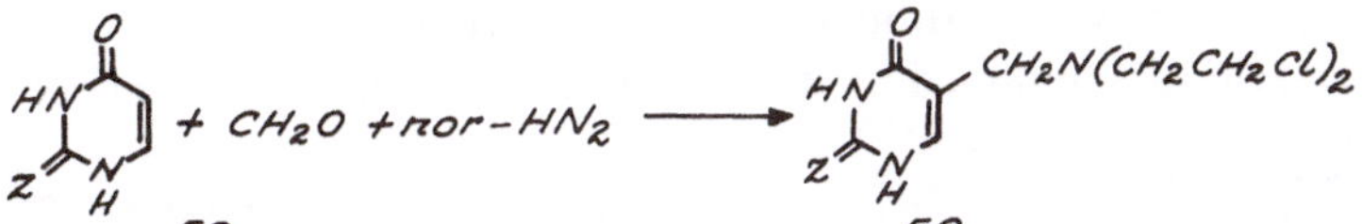

$$58 \quad + CH_2O + nor-HN_2 \longrightarrow 59$$

The synthesis of a novel nitrogen mustard with a spiro structure, designed to cross the blood-brain barrier ($\log P = 2.47$), was also reported. The synthetic path which affords this spirohydantoin mustard (62) is as follows[46].

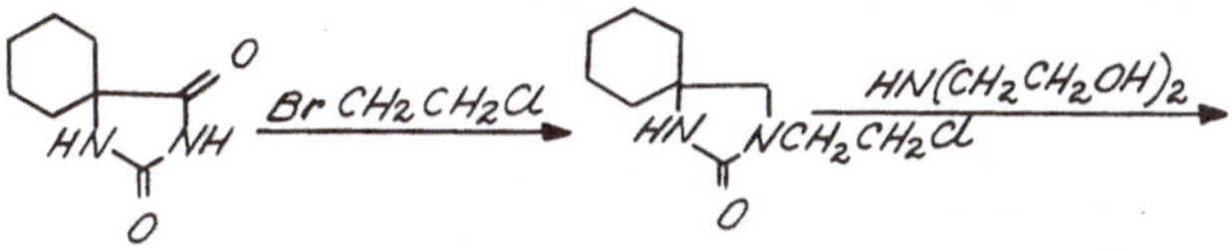

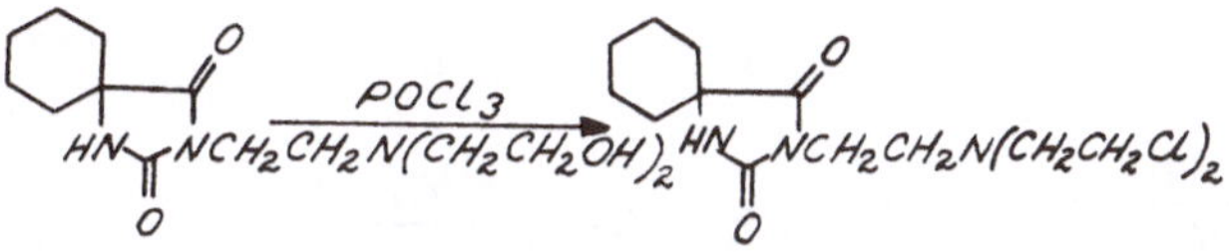

3.5.2.2 *Aromatic nitrogen mustards.* These are the most largely represented category of nitrogen mustards, due to the high number of aromatic or heteroaromatic systems used for this purpose.

Chlorambucil was synthesised by Everett *et al.*[164] according to the following scheme.

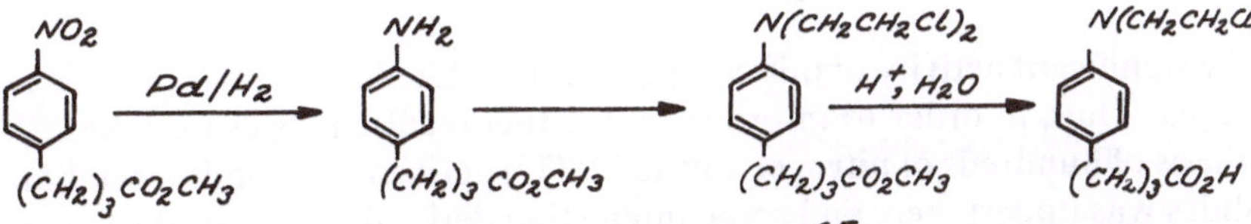

Direct nitration of 4-phenylbutyric acid affords good yields of the intermediate (63). Synthesis of the nitrogen mustard moiety is carried out by the procedures discussed previously. The final hydrolysis performed in concentrated hydrochloric acid led to the free acid (66). The synthesis of highly radioactive tritiated chlorambucil was done by catalytic (PtO_2/Adams)

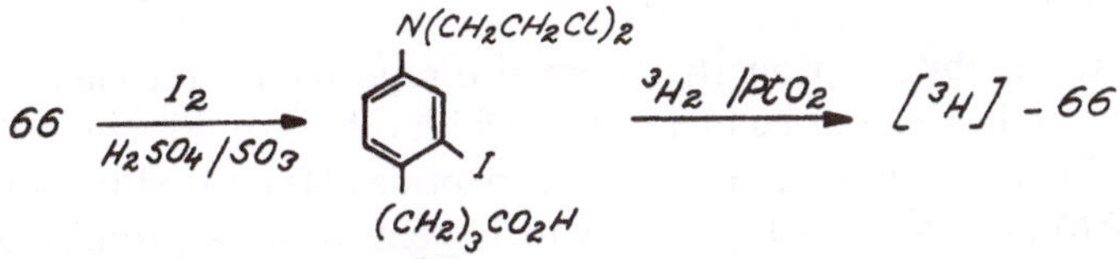

tritiodeiodination with tritium of 4-[bis-(2-chloroethyl)amino]-2-iodophenyl-
n-butyric acid (67) obtained by direct iodination of chlorambucil with iodine
in $H_2SO_4 + SO_3$ or oleum[165].

L-PAM, 4-[N,N-bis-(2-chloroethyl)amino]phenyl-L-alanine (72) was
reported in 1953 by Bergel and Stock[166] simultaneously with its corresponding racemic isomer obtained by Larionov et al.[167]. The L-enantiomer proved
to be twice as effective as the D. The synthesis of L-PAM is described. A suitable
synthetic approach was devised using L-phenylalanine as starting material
which avoided racemisation. The synthesis of the tritiated L-PAM has also
been described[169].

The significant activity of this compound stimulated other investigations in
this area. Thus, in order to improve on the therapeutic index of L-PAM, the
synthesis of hundreds of nitrogen mustards (73) derived from amino acids and
peptides was undertaken, and some interesting derivatives emerged from this
approach[168-171]. Significant differences in the activity of the various diastereoisomers were found in these series also[130].

Significant inhibitory properties were also reported for the *meta* isomer of
L-PAM, which has been used in a mixture with other peptides in the product
known as Peptichemio[172]. Recently, a tripeptide (74) derived from the same
meta-L-PAM has been developed[173], which seems to show promising clinical
effectiveness[174].

74

A study effected in our laboratory supported the previous observation that the best therapeutic index of the three possible isomeric nitrogen mustards derived from aminobenzoic acid[1] belongs to the *meta* isomer, for which special transport properties were ascribed[175]. In an attempt to improve the antitumour properties, many derived nitrogen mustards have been prepared[176−178]. Among them 3-methyl-4-(N,N-bis-(chloroethyl)]amino-benzoic acid (**75**) (IOB-82) was in clinical use in Romania.

A large number of aromatic nitrogen mustards have also been prepared starting from aromatic polycyclic hydrocarbons, natural or synthetic oestro-gens[135,136], as well as from many aromatic heterocycles. Among these uracil mustard (Dopan) (**76**) (X = Cl) and its fluorinated analogue (**76**) (X = F) were of clinical interest[179, 180].

Aromatic nitrogen mustards linked to carrier molecules. Generally, *p*-substituted aromatic nitrogen mustards such as 4-[N,N-bis-(2-chloroethyl)-amino]phenylacetic or butyric acids are linked to potential carrier structures (especially steroid hormones) through ester or carbamoyl functions. A representative compound is Fenesterine (**42**) obtained by treating cholesterol with 4-[N,N-bis-(2-chloroethyl)amino]phenylacetic acid chloride (**77**)[136].

Using the same procedure, corresponding nitrogen mustards were prepared from pregnenolone, oestradiol, oestriol, testosterone, stilboestrol, etc., as well as from some steroidic lactames[138,181−183].

3.5.2.3 *Derivatisation of nitrogen mustards.* In order to obtain compounds with low chemical reactivity (according to the latent activity principle, see section 3.7 and Table 3.3), the nitrogen mustard moiety was modified.

Synthesis of cyclic phosphoramide nitrogen mustards. The synthesis of the first phosphoramide nitrogen mustard was described by Friedman and Seligman in 1954[184]. However, the preparation of cyclic nitrogen mustards and particularly that of CP was achieved by Arnold and Bourseaux[185].

Nor-HN$_2$ treated with POCl$_3$ in excess, under mild conditions, yields N,N-bis(2-chloroethyl)aminophosphorodichloride (78) which further undergoes condensation with 1-aminopropanol in inert solvents (e.g. 1,2-dichloroethane) at room temperature, and in the presence of an HCl acceptor, to afford CP (79) in good yield.

Deuterated CPs were also prepared (CP-4,6-d_4; CP-4,5,6-d_6 and CP-5-d_2) in the same way, using conveniently deuterated 1-aminopropanol. Thus 1,3-d_4-aminopropanol was, in turn, obtained by the reduction of cyanoacetate with LiAlD$_4$ and the 2-d_2 intermediate resulted by isotopic exchange (D$_2$O) using the same starting material, followed by reduction with LiAlH$_4$[186].

The synthesis of 3-(2-chloroethyl)-2-(2-chloroethylamino)perhydro-1,2,3-oxazaphosphorin-2-oxide, Iphosphamide (IF) (81), was reported around 1960. An efficient path for the synthesis of this compound is the following[187].

The synthesis of new IF derivatives of the type (83) were recently reported[108]. It was based on the rationale that any modification of the parent molecule which does not affect the activation process, but does change the alkylating activity of the ultimate alkylator, should affect the IF antitumour activity. The derivative (83) (X = Br) appears to be more effective with respect to IF or CP on animal tests[108]. On a similar basis, several 2-chloroethylnitrosoureas with an oxazaphosphorine ring (85) were obtained[188].

The CP activation (see section 3.6.1) results in enzymatic 4-hydroxylation of CP to the reactive 4-HO-CP. This is believed to exist both as the ring closed form and as the ring opened aldophosphamide (tautomeric equilibrium). 4-HO-CP and aldophosphamide are key intermediates because they seem to be responsible for the selectivity of CP against malignant cells. Therefore, the synthesis of these intermediates was intensely investigated, but with rather conflicting results[189,190]. A route to aldophosphamide ($n = 1$) (90) was described by Friedman et al.[191], using the Sarett oxidation of the precursor alcohol (89) under very mild conditions. The major product thus formed is extremely unstable. It was characterised by TLC, as well as by conversion to the corresponding semi-carbazone. If Sarett oxidation was performed under the more usual conditions, then 4-keto-CP (91) resulted directly, probably via aldophosphamide as an intermediate.

A more recent and elegant method for the preparation of 4-HO-CP (which also serves to obtain 4-HO-IF) uses 4-hydroperoxy-CP (IF), (94) as its precursor[192].

Both CP and IF are readily converted to their 4-hydroperoxides either by ozonolysis[195] or by Fenton oxidation[190,193,194,196]. The reduction of 4-

hydroperoxy-CP (**95**) to 4-HO-CP (**96**) can occur spontaneously[193] or in mild conditions using triethylphosphite[190,193,194].

4-HO-CP and their corresponding hydroperoxides may be readily converted into the corresponding 4-(S,R)-mercapto derivatives (**97**)[197,198]. Such a derivative of cysteine is probably involved in the delayed *in vivo* toxicity of CP (see section 3.6.1), because enzymatic or spontaneous hydrolysis would again release 4-HO-CP[197]. On this basis a new generation of CP and IF derivatives, with improved pharmacological properties, was obtained.

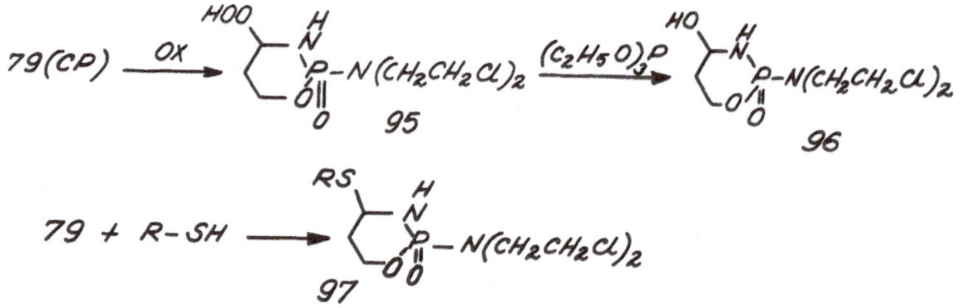

The ultimate alkylators arising from CP and IF, that is phosphoramide mustard (**98**) and isophosphoramide mustard (**99**), were also synthesised and their biological properties assayed[43,199].

$$(ClCH_2CH_2)_2NP(O) {<} {\overset{\displaystyle -OH}{\underset{\displaystyle -NH_2}{}}}$$

98

$$(ClCH_2CH_2NH)_2P(O)OH$$

99

The stereochemistry of oxazaphosphorines is unusually interesting; however, it does not influence decidely their biological properties. Optically active enantiomers appear both for CP and IF, due to the chirality of the phosphorus atom[200,201]. Recently, the separation of these enantiomers by preparative chromatography was claimed[202]. The biological tests performed on the racemates and enantiomers of CP, IF and compound (**83**) (X = Br)[108] did not indicate significant differences in activity among these forms.

The conformation of the 1,3,2-oxazaphosphorine ring was also investigated by ^{1}H-NMR and other spectral techniques. A chair-chair equilibrium could be demonstrated for the 5,5'-dimethyl substituted derivatives (**100**⇌**101**).

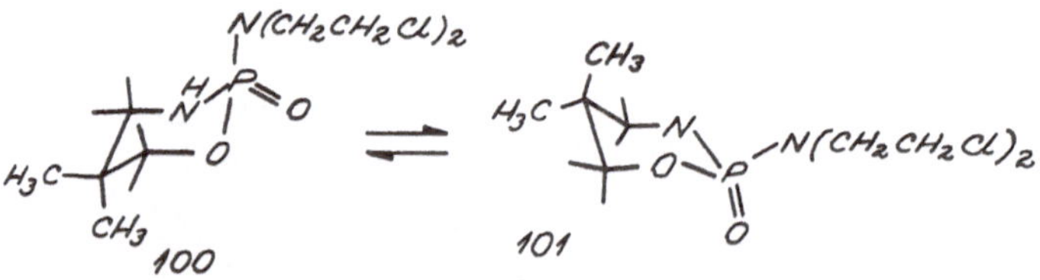

The nitrogen mustard moiety in CP preferentially adopts an equatorial conformation[204]. Substitution at C4 (4-HO-CP, 4-Ph-CP, etc.) leads to the

introduction of a second chiral centre[204]. The structures of the diastereoisomeric racemates were assigned by NMR and X-ray determinations[205].
There is an obvious difference in their biological activities for 4-Ph-CP: the
trans (RR/SS) diastereoisomeric racemates are effective inhibitors of tumour
growth, whereas the *cis (RS/SR)* are inactive[205]. No such phenomenon was
shown for 4-Me-CP. This difference might be related to the stereospecificity of
the enzymatic activation which occurs at the C_4 position, the bulky phenyl
substituent being able, in one of its conformations, to hinder the enzymatic
attack.

A large number of CP and IF derivatives substituted at positions 4, 5 and 6
have been synthesised. None of them, however, exhibited advantages over CP
or IF[43,206,207].

Urethane type nitrogen mustards. These compounds represent another
attempt to prepare latent nitrogen mustards, based on the possibility of
inducing by previous treatment (with non-toxic urethanes) the enzymes which
would selectively cleave the ester bond[1].

Besides the early attempts of Danielli[1], this approach was used mainly in
conjunction with steroid molecules. The synthesis of this category of
compounds actually raises the problem of coupling the N,N-bis(2-
chloroethyl)carbamoyl residue to the steroid molecules. The preparative
procedures vary with the nature of the hydroxylic groups involved in the
coupling[144]. If an aliphatic hydroxy group is available (as for instance the 17β-
OH in oestradiol), then the first step consists of the addition of phosgene
followed by coupling of the chloroformate thus formed with nor-HN_2. If the
hydroxyl group exhibits a phenolic character (e.g. the 3-OH group in natural
oestrogens or 4,4′-OH groups in stilboestrol, hexoestrol, etc.) its condensation
at room temperature with N,N-bis(2-chloroethyl)carbamoyl chloride in
pyridine affords the desired products[139,141]. An example using both synthetic
pathways is represented by the synthesis of 3- and 17β-[N-N-bis(2-chloro-
ethyl)carbamoyl]oestradiol derivatives.

4,4′-[Bis-(2-chloroethyl)carbamoyl]stilboestrol (**105**) (Stilbostat, ICI
85968) was prepared in the same manner. Some clinical data regarding its
activity in the treatment of prostatic carcinomas are available[208].

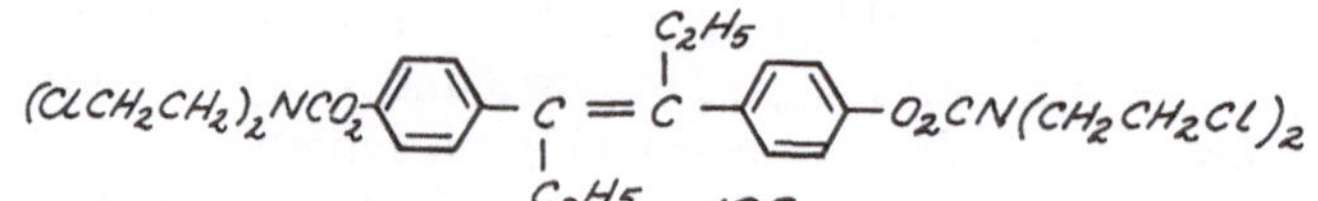

$$(ClCH_2CH_2)_2NCO_2 - \bigcirc - \underset{\underset{C_2H_5}{|}}{\overset{\overset{C_2H_5}{|}}{C}} = C - \bigcirc - O_2CN(CH_2CH_2Cl)_2$$

105

The key reagent in this synthesis, the bis (2-chloroethyl) carbamoyl chloride, is readily available by reaction of phosgene with nor-HN_2[140].

The syntheses of urethane type nitrogen mustards from cholesterol[209], cholic acid[209], natural[140,141,142] or synthetic oestrogens[39,109,210], androgens[139,141,142] and glucocorticoids[211,212] have been reported.

Triazene nitrogen mustards. These represent a more recent modality to synthesise latent nitrogen mustards. The triazene nitrogen mustards were readily obtained by coupling a diazonium salt with nor-HN_2. Some difficulties in separation and purification of the resulting products arise because of their instability, especially to light[213-215].

The nitrogen mustard analogue of DTIC (**106**) exhibited interesting tumour inhibitory properties[216]. Studies focused on its mechanism of action suggested that the reactive intermediate is MCTIC, 3-(2-chloroethyl)-1-triazeno-4(5)-imidazole-5(4)-carboxamide (**107**). This observation led to the design of a new latentiated alkylating agent, Mitozolamide, 8-carbamoyl-3-(2-chloroethyl)imidazo[5,1-*d*]-1,2,3,5-tetrazin-4-(3*H*)one, which proved to be clinically effective[217] (**108**) (see Chapter 6).

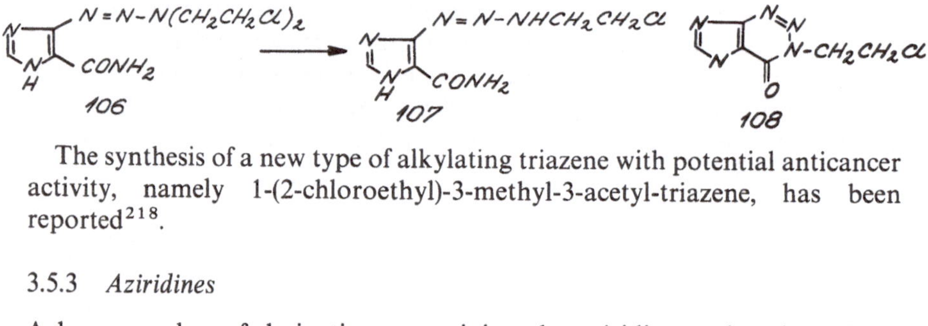

106　　　　**107**　　　　**108**

The synthesis of a new type of alkylating triazene with potential anticancer activity, namely 1-(2-chloroethyl)-3-methyl-3-acetyl-triazene, has been reported[218].

3.5.3　*Aziridines*

A large number of derivatives containing the aziridine moiety have been synthesised and tested in cancer chemotherapy. Several clinically useful drugs were found (Table 3.1).

3.5.3.1　*Phosphoramide derivatives.*

The first congeners belonging to this series were synthesised by Cyanamid Corporation, by treating halogenated or thiohalogenated derivatives of phosphorus with aziridine.

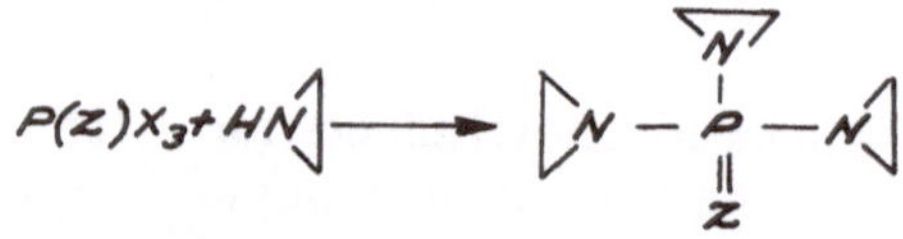

$X = Cl, Br;\ z = 0, s$　　　　**109**

The preparation of aziridine had been reported by Wenker in 1935[219], by the following route:

$$HOCH_2CH_2NH_2 + H_2SO_4 \longrightarrow {}^{\ominus}O_3SCH_2CH_2\overset{\oplus}{N}H_3 \xrightarrow{\text{NaOH}} HN\langle$$

Some improvements to this basic procedure were reported by Ullmann[219b].

The coupling of (thio)phosphoryl chloride with the amines could be performed step by step, by first introducing the desired amines and then coupling the intermediate dichloro (thio)phosphorylamide with aziridine. A large number of heteroamidophosphoryl aziridines were synthesised, in which, beside the aziridinyl moieties, the following amines were introduced: cyclohexylamine (Hexaphosphamide), thiazolidine, 1-phenylurea (Benzo TEP), diallylamine, piperidine (112)[130,220]. Some of them underwent clinical trials. For the study of their mechanism of action, thiophosphoric aziridines with spin markers (113) have been also synthesised[221].

$$X = Cl, Br \; ; \; Z = O, S$$

Melamine derivatives, which contain exclusively aziridine moieties (TEM), as well as those which have also other amines, besides aziridine, can also be mentioned (Chapter 11). Curious structures show the active aziridine derivatives built on an inorganic skeleton, for instance hexa-aziridinotriphosphazene (114) (Apholate or Fotrin)[222].

An interesting attempt to improve the cytotoxic effectiveness of these compounds is represented by the synthesis of 'dual antagonists', whose molecules contain both alkylating moieties and a pharmacophore group exhibiting cytotoxicity against cancer cells by a different mechanism of action (e.g. a urethane). Another point of interest was the idea that one can change the reaction mechanism from slow S_N2 to fast S_N1 by introducing gem-dialkyl substituents on the aziridine ring, as in compound (116) (prepared via (115)[23]).

$$P(Z)X_3 + H_2NCO_2R \longrightarrow X_2P(Z)NHCO_2R \xrightarrow{+HN\langle} (\langle N)_2P(Z)NHCO_2R$$
$$\textbf{115 } (Z = O, S) \qquad\qquad \textbf{116}$$

Recently, the antitumour activity of thiophosphoaziridines substituted with nitrophenyl and nitroimidazolyl moieties has been claimed[223].

3.5.3.2 *Aziridinylbenzoquinones.* The first impressive review of the quinones tested for antitumour activity at the National Cancer Institute shows that a number of active materials contain this structural feature[224]. Among these *p*-benzoquinones containing multiple aziridinyl groups were the only agents very active against murine leukaemia L 1210.

The first effective derivatives obtained in this series were 2,5-diaziridinyl-3, 6-di-*n*-propyloxy-*p*-benzoquinone (**117**) (E-39 Bayer) and 2,3,5-triaziridinyl-*p*-benzoquinone (**118**) (Trenimon)[225,226].

117   118

As part of a systematic search for new central nervous system (CNS) antitumour agents, it seemed reasonable to design aziridinylbenzoquinones, because they possess, beside a high cytotoxicity, several of the molecular characteristics necessary for CNS penetration, namely: (a) suitable lipophilicity and (b) low degree of ionisation. On this basis and taking the structure of the effective antitumour antibiotic, Mytomicin C(**119**), as a model, a number of *p*-benzoquinones have been synthesised[227-229]. Among these, as the result of the efforts of two different teams, 3,6-diaziridinyl-2,5-bis(carbethoxy)amino-1,4-benzoquinone (**123**) (Diazoquinone, AZQ) and 2,5-bis-(1-aziridinyl)-3-(2-carbamoyloxy)-1-methoxyethyl-6-methyl-1,4-benzoquinone (**120**) (Esquinone) have emerged as clinically useful derivatives.

119   120

While 2,5-diaziridinyl-1,4-benzoquinones could be prepared in a number of ways, substitution reactions, using alkoxy- or haloquinones, have been most often employed. Chloranil was generally used as starting material. However, the 2,5-diaziridinyl-3,6-dichloro-1,4-benzoquinone obtained in the first step of this synthetic pathway hardly undergoes further displacement of the remaining chlorine atoms. For this reason, improved synthetic methods were devised using fluoranil (**121**) as starting material[230].

When compound (123) (AZQ) is reduced *in vivo*, semiquinone free radicals also appear; they may also be involved in its antitumour effects[231].

3.5.4 *Methanesulphonates*

In an attempt to prepare drugs resembling the nitrogen mustards, Haddow and Timmis observed in 1951 that the introduction of sulphonic acid ester groups conferred biological alkylating properties to substituted molecules.

Primary and secondary alcohols react with mesyl chloride in dry pyridine, at room temperature, to give mesylates, usually with good yields. In turn, mesyl chloride is prepared from methylsulphonic acid and thionyl chloride. This method is, however, dependent on the availability of the appropriate alcohols and is limited to the preparation of those esters which are not sensitive to bases. Another method consists of heating a silver sulphonate with an alkyl iodide. This reaction was recently improved by using acetonitrile as solvent[232].

$$HO(CH_2)_nOH + CH_3SO_2Cl \longrightarrow CH_3SO_2O(CH_2)_nOSO_2CH_3$$
$$124 \ (n = 1\text{--}9)$$
$$CH_3SO_3Ag + RI \longrightarrow CH_3SO_3R + AgI$$

From the methanesulphonates thus synthesised, 1,4-dimethane-sulphonyloxybutane (124) $(n = 4)$[233] exhibits a remarkable activity in chronic leukaemias. Also, 3,3′-iminodi-1-propanoldimethanesulphonate (126) proved to be more effective than the corresponding nitrogen mustards[234,235]. Their synthesis was performed according to the scheme

$$Cl(CH_2)_3OH + H_2N(CH_2)_3OH \longrightarrow HN(CH_2CH_2CH_2OH)_2$$
$$125$$

$$\longrightarrow HN(CH_2CH_2CH_2OSO_2CH_3)_2$$
$$126$$

A large number of methanesulphonate derivatives have been synthesised and investigated[236]. The sugar methanesulphonates, namely, 1,2,5,6-tetramesyl-D-mannitol[236], 1,6-dimethanesulphonyl-1,6-dideoxy-D-manni-tol[237,238], 1,4-di-(2′-mesyloxyethylamino)-1,4-dideoxy-mesoerithritoldi-mesylate[239,240], have been extensively studied. Some are of clinical use. Recently, the synthesis of a 2-chloroethyl ester of mesyl-sulphonyl-methanesulphonic acid (2) (Clomesone) was reported. This compound exhibits very interesting antitumour properties like N-(2-chloroethyl)-N-nitrosoureas and does not give rise to cross-resistance with HN_2[241].

Another novel agent belonging to this class which proved to be an interesting DNA alkylator is 1,5;2,4-dioxathiepane-2,2,4,4-tetraoxide (127)[241c].

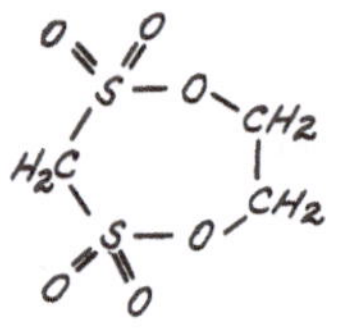

127

3.5.5 Oxiranes

The synthesis of the ethylene oxide function can occur in various ways. Current methods consist of the treatment of chlorohydrins with sodium or calcium hydroxide, or in the addition of one oxygen atom to an alkene bond with perbenzoic, peracetic, performic, or other peracids.

A correlation between the antitumour activity of diepoxides and their alkylating activity was reported by Ross[1]. They are moderately reactive biological alkylating agents which are transported as neutral molecules.

Among the first derivatives prepared in the class, 1,2; 5,6-dianhydrogalactitol (129) can be cited. It may be obtained readily by treating DBD (128) with alkali. A similar conversion also takes place in serum, being a more general reaction of halogenated hexitols[242].

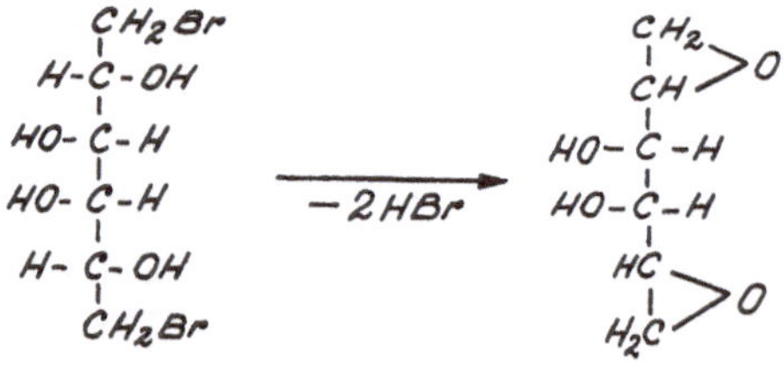

128   129

Thus synthesis and testing of 3,4-diacetyl-1,2;5,6-dianhydrogalactitol demonstrated that acetylation, in contrast with methylation, has a favourable effect on the antitumour properties of these products[243].

Another interesting oxirane derivative is 1,3,5-triglycidyl-s-triazenetrione (130) (α-TGT, Teroxinone). Its synthesis leads to a mixture of two diastereoisomers: α-TGT ($R,R,S/S,S,R$), more soluble and more active than β-TGT ($R,R,R/S,S,S$)[244,245]. A Teroxinone analogue namely α/β-triglycidyl-urazol (131) (TGU), is presently undergoing clinical trials[246].

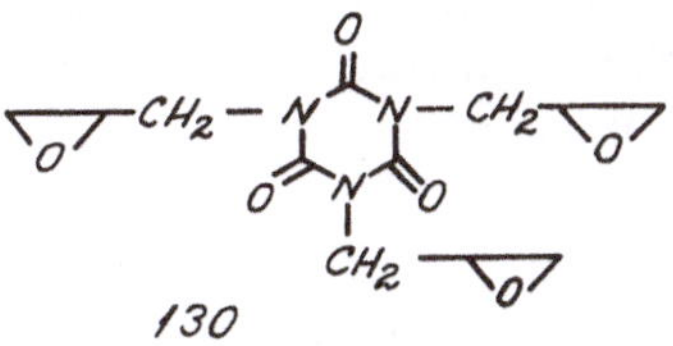

130

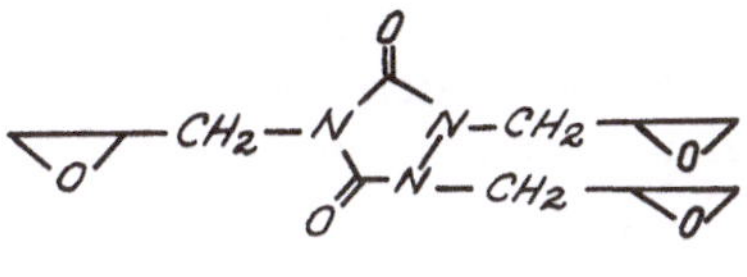

131

3.5.6 Benzo- and naphthoquinones

These are compounds activated by enzyme reduction. Their design, as well as that of aziridinylbenzoquinones (see section 3.5.3.2), was modelled on the Mytomicin C molecule (**119**). They require reduction by NADPH-dependent enzyme systems in order to exert their alkylating activity. Structure-activity studies on the relation between the redox potentials of benzo- and naphthoquinones and their antitumour efficacy showed that there is a positive correlation between the first half-wave reduction potential and antineoplastic activity, with agents having the lowest reduction potential being the most effective ($E_{1/2} = -0.40\,\mathrm{mV}$ for Mytomicin)[247]. It was suggested that derivatives such as (**132**) and (**133**) act as coenzyme Q electron competitors in the mitochondrial chain. They are reduced to dihydroquinone intermediates, spontaneously converted to reactive o-quinonemethides, which act as alkylating agents. There is chemical evidence which substantiates the formation of o-quinonemethide intermediates in the reductive amination of 2,3-dimethyl-5,6-bis-(acetoxymethyl)-1,4-benzoquinone ($R = CH_3$, $R_1 = COCH_3$) (**133**) by aniline and morpholine[248,249]. The reaction of 2- and 6-methylnaphthoquinones with thiols gives the following order of reactivity: GSH < cysteamine < dithiothreitol < cysteine[250]. The synthesis of compound (**133**) is a typical example of the synthetic paths employed in this class[248].

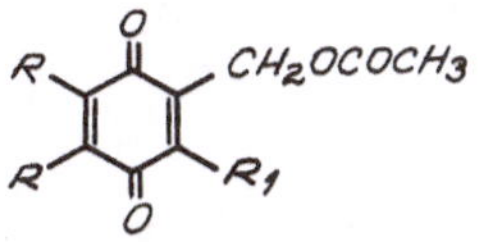

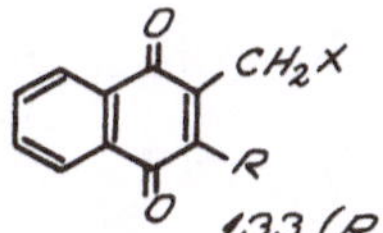

132 ($R = CH_3$, OCH_3; $R_1 = H$, $H_3C(CH_2)_{14}$-)

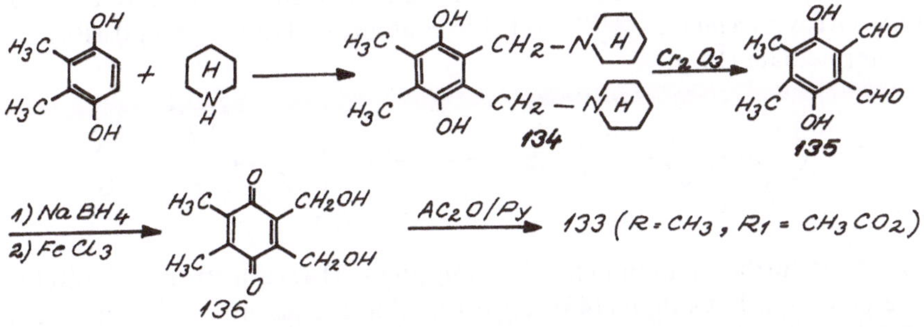

133 ($R = H$, $CH_3(CH_2)_{14}$; $X = Cl$, CH_3CO_2-)

134

135

136

In order to obtain derivatives with as low a reduction potential as possible, anthraquinone derivatives (137)[247], sulphone analogues (138)[251], and benzimidazol-4,7-diones (139)[252] were also prepared.

137 (R = Cl, Br, CH₃CO₂, etc.)   138 (R = Cl, CH₃CO₂)   139

There are also a large number of quinones and naphthoquinones of natural origin with antitumour properties, probably acting by a similar mechanism[253].

3.5.7 Piperazinediones

Piperazinediones are a class of alkylating agents whose mechanism of action has not yet been elucidated. There are data which claim they have alkylating properties[254,255] and thus inhibit DNA synthesis[256]. Their biological activity is similar to that of L-PAM[257].

The first derivatives belonging to this class were 1,2-bis-(3,5-dioxopiperazine-1-yl)-ethane (140) (ICRF 154), and (±)-1,2-bis-(3,5-dioxopiperazine-1-yl)-propane (141) (ICRF 159, Razoxane), the latter being clinically tested[258].

The synthesis of compound (141) (ICRF 154) starting from ethylenediaminotetraacetic acid is typical for this series[259].

Another piperazinedione derivative is 2,5-piperazinedione-3,6-bis-(5-chloro-2-piperidyl)-dihydrochloride (142) extracted from the culture broth of *Streptomyces griseoluteus*[260,261]. It behaves as an alkylating agent possibly via the intermediate (143).

However, further data indicate that compound (142) is a stronger inhibitor of the DNA synthesis than (143) suggesting that the parent compound is the true alkylator[262]. The influence of the introduction of similar aziridine structures into the CP molecule has also been investigated[263].

$$Br-\text{(ring)}P-N(CH_2CH_2Cl)_2 \longrightarrow \text{(ring)}P-N(CH_2CH_2Cl)_2$$

$$144 \qquad\qquad\qquad 145$$

Accordingly, compound (145) obtained by bromine elimination from 5-Br-CP (144) is highly toxic without microsomal activation and represents a new type of exploitable approach to 'pre-activation' of CP analogues[263].

3.5.8 Diazoalkanes

These are alkylating agents with an alternative type of antitumour activity, because the alkylation occurs by a different mechanism, which possibly involves carbenoid or radical intermediates.

The most important representatives of this class, extracted from *Streptomyces* cultures, are the analogues of glutamic acid, namely, azaserine (146), 6-diazo-5-oxo-norleucine (DON) (147) and azotomycin (148). The latter, a polymer consisting of two DON molecules and a glutamic acid residue, becomes active by *in vivo* conversion to DON[264].

$$N_2CHCO_2CH_2\underset{\underset{NH_2}{|}}{CH}CO_2H \qquad\qquad N_2CHCO(CH_2)_2\underset{\underset{NH_2}{|}}{CH}CO_2H$$

$$146 \qquad\qquad\qquad\qquad 147$$

$$N_2CHCO(CH_2)_2^-\underset{\underset{\underset{\underset{N_2CHCO(CH_2)_2-}{|}}{CO}}{NH}}{CH}-CO_2H$$

$$N_2CHCO(CH_2)_2-CH-NHCO(CH_2)_2\underset{\underset{NH_2}{|}}{CH}CO_2H$$

$$148$$

Although these derivatives inhibit the metabolism of purines, pyrimidines, phosphopyrimidine nucleotides and D-glucosamines, they are probably acting by a specific inhibition of the various L-glutamine-amino-transferases[265]. At high concentrations they also exhibit alkylating properties. However, it is not certain that their antitumour properties are due to this effect, because only the L-enantiomer is active[266,267].

The antitumour properties of 1,2-bis-diazoacylethane (149, Diazan)[268], diazo derivatives of other amino acids[269] or of synthetic oestrogens were reported (150)[270].

149   150

3.5.9 α-Methylene-γ-lactones

Sesquiterpene lactones such as vernolepin (**151**), elephantopin (**152**) helenalin (**153**) were found to possess significant *in vivo* antitumour activity[271].

151   152   153

The biological activity of these natural products appears to be associated with their ability to act as alkylating agents, by virtue of the conjugate addition of biological nucleophiles to the α-methylene-γ-lactone moiety[272]. These compounds alkylate the thiolic functions of the enzymes such as phos-phofructokinase, glucogen synthetase, as well as that of substrates like cysteine[273]. Their antitumour activity might be assigned to their reaction with nucleophilic centres of key regulatory enzymes involved in nucleic acid metabolism[274].

154   155

A general preparative procedure for obtaining α-methylene-γ-lactones is the Reformatsky reaction.

156   157

Various derivatives containing an α-methylene-γ-lactone structure have been reported, such as the steroid derivatives (**159**)[275]. The testosterone congener is remarkably cytotoxic, but its toxicity does not depend on the localisation of the α-methylene-γ-lactone in the A or D nucleus[275].

158   159

The synthesis of bifunctional derivatives was undertaken to increase the cytotoxicity of these structures (161)[276].

$$OHC(CH_2)_n CHO + Br CH_2 - \overset{\overset{CH_2}{\|}}{C} - CO_2 C_2 H_5 \longrightarrow$$

160 (n = 3-6)   161

α-Methylene-γ-lactones possessing a leaving group linked to the α-methylene undergo Michael-type nucleophilic addition elimination reactions much more readily than the simple enones and are in this respect more effective alkylators.

162     $\xrightarrow{NaH}$ 163   CHONa   CHOR   164 (R = SiMe₃, CH₃SO₂, CF₃SO₂) etc.

α-Methylenecyclopentenedione structures (165) were also assayed as anti-tumour agents[271]. There is evidence that their cytotoxic activity is probably due to their ability to inhibit DNA and RNA synthesis and not to their capacity to scavenge thiol groups[276].

165

3.5.10 Alkyl halides

Although alkyl halides are typical alkylating agents, they were not extensively used in cancer chemotherapy. Some of them are powerful carcinogens for animals[20]. However, some halogenated sugars were synthesised, some of which demonstrated clinical activity, such as 1,6-dibromo-1,6-dideoxy-D-mannitol (Mielobromol, DBD)[277].

3.6 The metabolism of alkylating agents

Because most of the alkylating agents react directly with DNA, their metabolism is not of paramount importance in determining their antitumour activity. However, alkylating agents, being generally recognised as xenobiotics, are modified by monooxygenase cytochrome P-450 dependent systems. They are usually transformed into more hydrophilic compounds. Thus, in the metabolisation of Chlorambucil, twelve metabolites were detected[278], resulting from sidechain oxidation, dechlorination, esterification and also mustard hydrolysis followed by cyclisation. L-PAM, besides its spontaneous hydrolysis, undergoes metabolic transformations, which involve primarily the phenylalanine portion of the molecule (decarboxylation, β-oxidation and deamination by monoamine oxidase, MAO)[13]. Metabolic changes may be noticed also in the 2-chloroethylnitrosourea series. Thus, the hydroxylation of the cyclohexane ring was demonstrated for CCNU, leading to at least equally cytotoxic metabolites[279,280].

However, for these alkylating agents the enzymatic metabolism influences the detoxication pathways, product distribution and toxicity, rather than their intrinsic cytotoxicity against cancer cells.

In contrast, for latent alkylating agents, the metabolism is of prime importance in determining the cytotoxicity and especially the specificity of the compound. Triazenes (see Chapter 5), hexamethylmelamine (see Chapter 11), benzo- and naphthoquinones (see section 3.5.6), cyclic phosphoramide nitrogen mustards (CP, IF, etc.) can be cited in this respect. In order to stress the importance of a knowledge of the metabolism for the design of congeners with improved effectiveness, we shall further discuss two relatively well understood cases.

3.6.1   Cyclophosphamide and iphosphamide metabolism

The metabolism of CP and IF has been studied intensively, both because of their clinical role and in an attempt to understand and explain their specificity against malignant cells. However, the elucidation of this last aspect still remains questionable.

It has been known since 1962 that CP activation takes place mainly in the liver and requires O_2[281]. This process is due to cytochrome P-450 dependent monooxygenases which convert the basic compound to 4-HO-CP (96) at equilibrium with its tautomeric form, aldophosphamide (90). This hydroxylation is probably stereoselective, the 4-cis-HO-CP isomer being preferentially formed (96a)[282]. However, trans-4-HO-CP was also detected, because of the proteolytic equilibrium existing between the two conformers. The ratio 4-HO-cis: 4-HO-trans: aldo-CP: aldo-CP·2H$_2$O was found at equilibrium to be 48:33:5:15 (Scheme 3.3).

4-HO-CP is actually the key product of CP metabolism, being the most selective among the CP metabolites. This activation step has a $t_{1/2} = 20$ min $(k_1 = 3.5 \times 10^{-2}$ min$)$[45,283].

Aldophosphamide further undergoes spontaneous or enzymatic toxication which occurs by acrolein elimination and release of a phosphoramide mustard (98a) (CP-mustard). It was recently demonstrated that his process is catalysed by 3′,5′-exonuclease[284,285]. The toxication step also exhibits quite slow kinetics, $t_{1/2} = 80$ min $(k_2 = 8.7 \times 10^{-3}$ min$^{-1})$[45]. CP-mustard may also

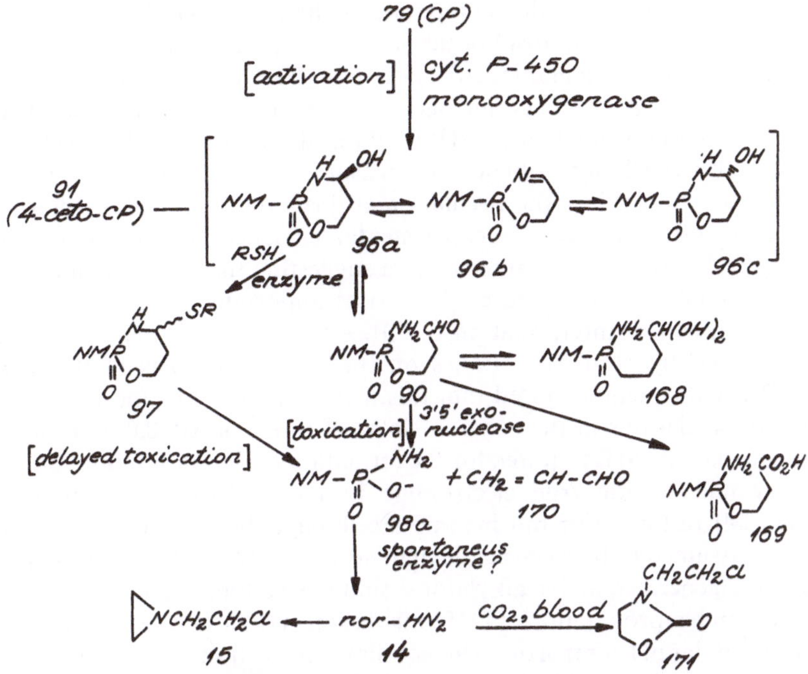

Scheme 3.3 Metabolism of cyclophosphamide.

undergo a dephosphorylation to HN_2[45, 286], or it could represent the ultimate alkylator.

The formation of both 4-keto-CP (91) and the acyclic acid (169) represent detoxication pathways, catalysed by aldehyde oxidases. Another detoxication pathway consists of the 4-HO-CP conversion to the corresponding thioether (97) by reaction with glutathione or cysteine. This last intermediate may be recycled in the metabolic turnover by the enzymatic cleavage of the thioether bond (delayed toxication)[103, 287]. A similar reaction occurs with protein thiol groups affording 4-(S-proteinyl)-CP[288]. The discovery of the delayed toxication process led to the design of a new generation of compounds with better pharmacological properties.

It was also suggested that the Fenton model systems for drug metabolising enzymes might represent an alternative pathway for the formation of 4-HO-CP or its anhydrodimer (172), a spontaneous conversion of the two intermediates into 4-HO-CP being observed[196] (see section 3.5.2.3).

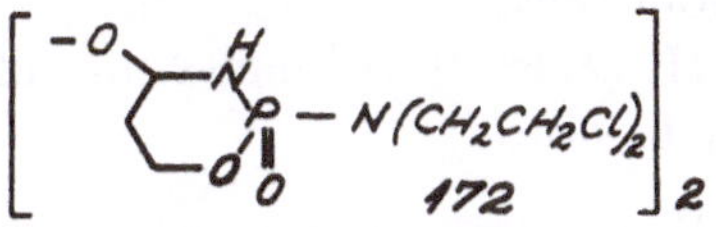

The most important question is to find the biochemical basis for the antitumour selectivity of this compound. A first explanation was that the selectivity of the oxazaphosphorines would be based mainly on the high specificity and the particular reactivity of the 4-HO-CP metabolite and not on the alkylating activity of the nitrogen mustard moiety. At this point it is important to establish whether 4-HO-CP (or aldophosphamide) reaches the tumour cells as such or just breaks down as it is formed in the liver, releasing the CP-mustard (98) into the circulation at this level. This aspect is not fully elucidated, although there is experimental evidence supporting the first hypothesis. However, there are also data suggesting that the contribution of the extra-cellular CP-mustard might also be important.

Another possible interpretation of CP selectivity acting via CP-mustard was provided by ^{1}H-NMR studies on the water solvolysis of this meta-bolite[289]. It was demonstrated that this metabolite exhibits mechanistically significant variations in the physiological pH range of 6.0–8.0 and that the conjugate base (98a) is required for the formation of the aziridinium ion (173) which would be the true electrophile in the reaction with DNA (see Scheme 3.4). Its formation rate in tumour cells might be 50-fold lower, because these are assumed to be more acidic than normal cells (pH 6.9 versus pH 7.4), so that the generated active alkylator would have a longer half-life. This fact leads to a higher probability for DNA alkylation in tumour cells[289]. Therefore, the intracellular pH control over this equilibrium reaction provides a chemical basis for rationalising at least a part of CP's specificity.

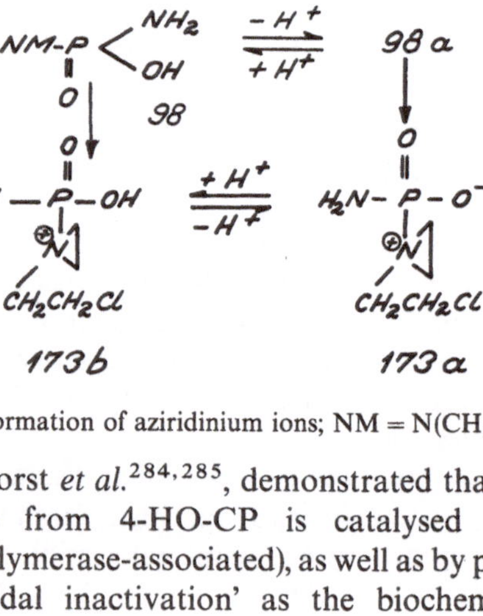

Scheme 3.4 Formation of aziridinium ions; NM = N(CH$_2$CH$_2$Cl)$_2$.

More recently, Hohorst et al.[284,285], demonstrated that the release of the alkylating metabolite from 4-HO-CP is catalysed by various 3′,5′-exonucleases (DNA polymerase-associated), as well as by phosphodiesterases. They suggested 'suicidal inactivation' as the biochemical basis of CP selectivity against tumour cells.

The metabolism of IF generally resembles that of CP[290-294]. However, several differences may be noted:

— the activation of IF proceeds more slowly than that of CP and yields isophosphamide mustard (99) (see Scheme 3.5).

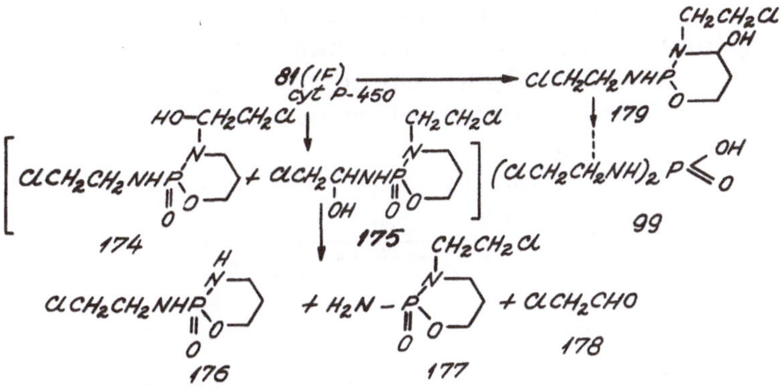

Scheme 3.5 Metabolism of iphosphamide.

—the lower hydroxylation rate makes other metabolic pathways more important, particularly dechloroethylation[295,296]. By this pathway chloroacetaldehyde (178) is released. The importance of this intermediate in the cytotoxic and immunosuppressive activity of IF has not yet been elucidated but it was recently demonstrated that chloroacetaldehyde may also form interstrand cross-links in DNA[296b].

The first impetus for the design of latent phosphoramide mustards was Gomori's observation that the level of phosphoamidases is increased in tumour cells. Although the reality did not correspond to the working hypothesis (the activation occurring under the effect of cytochrome P-450 dependent microsomal enzymes), this approach represented an extremely useful experiment which led to active compounds. Further elucidation of other metabolic aspects (e.g. delayed toxication of CP, IF) allowed the development of a second generation of more efficient products.

3.6.2   Metabolism of CB 1954

CB 1954, 5-(aziridin-1-yl)-2,4-dinitrobenzamide (179), exhibited an unusually high selectivity against the Walker carcinosarcoma[297]. This compound was considered as a monofunctional alkylating agent because it contains a single aziridine group. However, its activity was associated with possible purine antimetabolite behaviour[298,299].

Recently, it was demonstrated that the selective toxicity of this compound against the Walker 256 tumour could be correlated with its ability to cross-link the DNA of these cells, but not of other cell types[300], therefore the compound would need to be metabolically converted into a bifunctional agent. By studying the metabolism of this compound in the presence of cytosolic extracts from W 256 tumour cells, the conversion of the 4-NO_2 group to a hydroxylamine was demonstrated[301]. This could further interact with

DNA via a nitrenium ion intermediate, as was demonstrated for several similar hydroxamic acids. The enzyme which performs this activation was identified as a form of NAD(P)H dehydrogenase (quinone) (EC 1.6.99.2)[111]. It is specific for Walker 256 tumour cells and is absent from other tumours, as well as from normal cells.

This represents one of the very few examples of a compound showing a real tumour selectivity (although this concerns an animal tumour) and this finding will be of great interest for the design of effective antitumour compounds.

3.7 Quantitative structure-activity relationships of the alkylating agents

Quantitative structure-activity relationships (QSAR) are today extremely useful tools in the design of biologically effective compounds. Using large series of congeners tested in standard conditions, they allow us to outline the main physico-chemical and structural features which are responsible for the antitumour activity of a class of compounds. Thus it becames relatively easy to predict the most active compound in a series of congeners. On the other hand, such relationships can further suggest arguments supporting or rejecting a certain mechanism of action. Finally, such an approach allows the separation of structural features responsible for one type of biological property (e.g. toxicity, mutagenicity, etc.) from those responsible for other types (e.g. antitumour activity).

In the past 10 years the development of steric parameters (Sterimol[302], MTD[303]), as well as special computer algorithms, allowed a quantitative approach to the effector-reactor interactions. Nevertheless, the design of fundamentally new structures by such approaches remains an open question.

The alkylating agents are generally not well suited for such an analysis, because:

(a) for the direct alkylating agents, their cytotoxic properties depend mainly on their reactivity but not in a linear manner;
(b) receptor-effector interactions, fairly well described by the QSAR approach, are not involved in the case of direct alkylators, except for quite special cases (e.g. steroid hormones);
(c) for compounds with latent activity such an approach concerns the activation step rather than the interaction with the target.

However, significant QSAR studies have been developed for nitrogen mustards[59,60,304,305], triazenes[306-308], N-nitrosoureas[123,124,309-311], benzo- and naphthoquinones[312], oxiranes[44] and aziridines[313], affording very interesting conclusions. In several cases such relationships suggested new ways of development. Some examples are reported below.

3.7.1   Optimisation of lipophilicity by QSAR analysis

QSAR studies of alkylating agents enabled a rationalisation of the optimum lipophilicity. Thus, from the regression equations developed by Hansch *et al.*[60,305] for a series of nitrogen mustards tested on solid tumours (W 256 carcinosarcoma) (equation (3.16)), L 1210 (equation (3.17)) and P 388 leukaemias (equation (3.18)), it appears that the optimum lipophilicity for solid tumours is much higher $(-0.20 < \log P_0 < 1.34)$ than that for leukaemias $(\log P_0 < 0)$.

$$\log 1/C = -1.19\sigma^- + 0.75\,I - 1.00\pi - 0.53\pi^2 + 3.84 \qquad (3.16)$$

$$n = 14, \qquad r = 0.940, \qquad s = 0.291, \qquad \pi_0 = -0.95$$

$$\log 1/C = -0.96\sigma^- + 0.86\,I_0 - 0.31\pi + 4.07 \qquad (3.17)$$

$$n = 19, \qquad r = 0.926, \qquad s = 0.313$$

$$\log 1/C = -1.01\sigma + 0.19\,I_0 - 0.39\pi + 4.56 \qquad (3.18)$$

$$n = 19, \qquad r = 0.954 \qquad s = 0.236$$

where σ = Hammett constant; σ^- = electron withdrawing Hammett constant; π = lipophilicity constant; I = indicator variable, I = 0 for Cl, I = 1 for Br; I_0 = indicator variable for *ortho* effects. This finding, also confirmed by the QSAR analysis of many *N*-nitrosoureas[123, 314], has a practical value, for instance in the design of new central nervous system antitumour agents (e.g. aziridinylquinones, nitrosoureas, spirohydantoin mustards, etc.) able to cross the blood-brain barrier $(3.0 > \log P_0 > 0)$[46,227,228,314].

3.7.2   Separation of the structural features responsible for adverse effects from those responsible for activity

Separation of the structural features responsible for antitumour activity from those involved in carcinogenic or mutagenic activity was attempted for the triazenes[307, 308]. For instance, equation (3.19) describes the dependence of the mutagenicity (determined by the Ames test) on $\log P$ and σ^+ parameters.

$$\log 1/C = 1.09\log P - 1.63\sigma^+ + 5.58 \qquad (3.19)$$

$$n = 17, \qquad r = 0.974, \qquad s = 0.313$$

Comparing this equation with equation (3.20) resulting from the antitumour activity (L 1210, $T/C = 125$) of the triazenes

$$\log 1/C = 0.10\log P - 0.042\,(\log P)^2 - 0.316\sigma^+ + 0.18MR_{2,6} + 0.39\,E_s + 4.12 \qquad (3.20)$$

$$n = 61, \qquad r = 0.836, \qquad s = 0.191, \qquad \log P_0 = 1.18$$

where σ^+ = electron releasing Hammett constant; $MR_{2,6}$ = indicator variable featuring the steric effects at positions 2 and 6; E_s = Taft constant for the sidechain), shows that within this series of compounds the antitumour effect is much less dependent on the lipophilicity and on the electron releasing character of the substituents, than the mutagenic effect. Although Hansch concluded that for this series it is not possible to separate the two effects, this separation might be successful in other series of compounds. This might allow the design of better antitumour agents, while minimising the mutagenic properties.

3.7.3 Structural effects in the carrier molecules

Special emphasis was placed on the rationalisation by QSAR of the effects exerted by various substituents upon the reaction centre in the nucleophilic substitution reactions.

A good example is the finding that, for a series of aromatic nitrogen mustards[60], the *ortho* effects exert a favourable influence upon the antitumour properties. This fact was also confirmed by our own work[59,177]; this approach led to the development of the clinically used 4-methyl-3-[N,N-bis(2-chloroethyl)] aminobenzoic acid.

Another example is the determination in a natural or synthetic oestrogen molecule of the sites to which an alkylating moiety could be coupled in order to maintain a high affinity for oestrogen receptors (see section 3.4.2)[147].

3.8 Concluding remarks

Despite the enormous amount of work carried out in this area the alkylating agents remain an attractive class of compounds for several reasons:

(a) a large number of alkylating agents are still very important components of various cancer chemotherapeutic regimens;
(b) some new trends in the design of more specific alkylators have emerged during the past two decades. They open new ways especially toward more selective compounds (e.g. the use of monoclonal antibodies, the synthesis of specific DNA alkylators, etc.);
(c) the huge amount of data accumulated in this area allows for a more rational design of compounds to fulfill a given requirement.

Acknowledgements

We are deeply indebted to Dr. C. Truia, Dr. N. Voiculetz and T. Andrian for helpful discussions. We also thank to Miss Floria Lupu for her consistent help in preparing the English version of this manuscript.

References

1. W. C. J. Ross, in *Biological Alkylating Agents*, Butterworths, London (1962).
2. S. E. Jones, *Cancer Treat. Rep.* **66** (1982) 847.
3. D. L. Hill, in *A Review of Cyclophosphamide*, Charles C. Thomas, Springfield, IL (1975).
4. R. L. Comis, *Cancer Treat. Rep.* **60** (1976) 165.
5. Proceedings of the 6th New Drug Seminar: DTIC, *Cancer Treat. Rep.* **60** (1976) 123.
6. D. C. Tormey, G. Falkson, E. Perlin, J. Bull. J. Blom and M. E. Lippman, *Cancer Treat. Rep.* **60** (1976) 1593.
7. D. E. Johnson, W. W. Scott, R. P. Gibons, G. R. Prout, J. D. Schmidt, T. M. Chu, J. Gaeta, I. Saroff and G. P. Murphy, *Cancer Treat. Rep.* **61** (1977).319.
8. W. P. Brade, G. A. Nagel and S. Seeler, in *Ifosfamide in Tumor Therapy*, Karger, Basel (1987).
9. M. Zalupki and L. H. Baker, *J. Natl. Cancer Inst.* **80** (1988) 556.
10. J. W. Keller, W. H. Knospe and C. M. Huguley, *Proc. Am. Assoc. Cancer Res., ASCO* **17** (1976) 280.
11. (a) EORTC Brain Tumor Group, *Eur. J. Cancer* **12** (1976) 41; (b) T. H. Wasserman, M. Slavik and S. K. Carter, *Cancer Treat. Rev.* **1** (1974) 131.
12. (a) R. B. Livingston, L. Heilbrun, D. Lehane, *et al.*, *Cancer Treat. Rep.* **61** (1977) 1623; (b) E.C. Vonderheid, E. J. van Scott, P. E. Wallner and W. C. Johnson, *Cancer Treat. Rep.* **63** (1979) 681.
13. R. K. Furner and R. K. Brown, *Cancer Treat. Rep.* **64** (1980) 559.
14. Medical Research Council, *Br. Med. J.* **1** (1968) 201.
15. S. T. Crooke and W. T. Bradner, *Cancer Treat. Rev.* **3** (1976) 121.
16. B. L. Trantum, A. Haut and S. Rivkin, *Cancer* **35** (1975) 1153.
17. (a) National Bladder Cancer Collaborative Group, *Cancer Res.* **37** (1977) 2916; (b) R. D. Hard, M. Perloff and J. F. Holland, *Cancer* **48** (1981) 1522.
18. N. Brock, 'Experimental basis of cancer chemotherapy', in *Chemotherapy*, Vol. 7, Plenum Press, New York (1976).
19. U. S. Department of Health and Human Services, Fourth Annual Report on Carcinogens (NTP-85 002) (1985).
20. P. D. Lawley, 'Chemical carcinogens', *ACS Monograph* 182, Vol. I, ed. C. E. Searle, American Chemical Society, Washington, DC (1984).
21. S. A. Little and P. E. Mirkes, *Cancer Res.* **47** (1987) 5421.
22. W. C. J. Ross, *Adv. Cancer Res.* **1** (1953) 397.
23. T. J. Bardos, E. F. Chmielewicz and P. Hebborn, *Ann. N. Y. Acad. Sci.* **163** (1969) 1006.
24. K. W. Kohn, L. C. Erickson, G. Laurent, J. M. Ducore, N. A. Sharkey and R. A. G. Ewig, in *Nitrosoureas, Current Status and New Developments*, ed. W. Prestayo, S. T. Crooke, S. K. Karter and P. S. Schein, Academic Press, New York (1981) p. 69.
25. W. Davis and W. C. J. Ross, *J. Med. Chem.* **8** (1965) 787.
26. T. J. Bardos, N. Datta-Gupta, P. Hebborn and D. J. Triggle, *J. Med. Chem.* **8** (1965) 167.
27. C. C. Price, G. M. Gaucher, P. Koneru, R. Shibakawa, J. R. Sowa and M. Yamaguki, *Ann. N. Y. Acad. Sci.* **163** (1969) 593.
28. R. A. Sneen and J. W. J. Larson, *J. Am. Chem. Soc.* **91** (1969) 362.
29. G. H. Howie, P. E. Manni and J. M. Cassidy, *J. Med. Chem.* **17** (1974) 840.
30. P. J. Stang and W. L. Treptow, *J. Med. Chem.* **24** (1981) 468.
31. J. A. Haines, C. B. Reese and T. Lord, *J. Chem. Soc.* (1964) 1406.
32. J. A. Montgomery, R. James, G. S. McCaleb, M. C. Kirk and T. P. Johnstone, *J. Med. Chem.* **18** (1975) 568.
33. R. P. Bundrett, J. W. Cowens, M. Colvin and J. Jardine, *J. Med. Chem.* **19** (1976) 958.
34. A. G. Ogston, *Trans. Faraday Soc.* **44** (1948) 45.
35. A. Streitweiser, *Chem. Rev.* **56** (1956) 571.
36. E. E. Williamson and B. Witten, *Cancer Res.* **27** (1967) 33.
37. F. Badea, in *Reaction Mechanisms in Organic Chemistry*, ed. Stiintifică, Bucharest (1971) p. 132.
38. V. A. Levin, *J. Med. Chem.* **23** (1980) 682.
39. W. L. Petty, P. L. Nichols, Jr., *J. Am. Chem. Soc.* **76** (1954) 4385.
40. M. Yoshimoto, H. Miyazawa and H. Nakao, *J. Med. Chem.* **22** (1979) 491.
41. K. Sugiura, T. Kumura and M. Goto, *Mutat. Res.* **58** (1978) 159.

42. C. G. Swain and C. B. Scott, *J. Am. Chem. Soc.* **75** (1953) 141.
43. N. Brock, *Cancer Treat. Rep.* **60** (1976) 301.
44. D. Godeneche, J. C. Madelmont, M. F. Moreau, *et al., Cancer Chemother. Pharmacol.* **5** (1980) 1.
45. N. Brock and H. J. Hohorst, *Z. Krebsforsch.* **88** (1977) 185.
46. W. G. Peng, V. E. Marquez and J. S. Driscoll, *J. Med. Chem.* **18** (1975) 846.
47. B. L. van Duren and B. M. Goldschmidt, *J. Med. Chem.* **9** (1966) 77.
48. B. F. Cain and W. A. Denny, *J. Med. Chem.* **20** (1977) 515.
49. C. P. Spears, *Mol. Pharmacol.* **19** (1981) 496.
50. T. P. Johnstone and J. A. Montgomery, *Cancer Treat. Rep.* **70** (1986) 13.
51. I. Niculescu-Duvăz, G. Botez and A. Serban, *Neoplasma* **17** (1970) 143.
52. A. Cambanis, V. Dobre and I. Niculescu-Duvăz, *J. Med. Chem.* **12** (1969) 161.
53. I. Niculescu-Duvăz, M. Rotaru, V. Feyns and O. Maior, *Rev. Roum. Chim,* **18** (1973) 1617.
54. J. Epstein, R. W. Rosenthal and R. J. Ess., *Anal. Chem.* **27** (1955) 1435.
55. E. Sawicki and C. R. Sawicki, *Ann. N. Y. Acad. Sci.* **163** (1969) 895.
56. E. Sawicki, D. F. Bender, T. R. Hauser, R. M. Wilson and J. E. Meeker, *Anal. Chem.* **35** (1963) 1479.
57. G. P. Wheeler, B. J. Bowdon and J. A. Grimsley, *Cancer Res.* **34** (1974) 194.
58. O. M. Friedman and E. Borger, *Anal. Chem.* **33** (1961) 906.
59. I. Niculescu-Duvăz, G. Stihi, T. Craescu and Z. Simon, *Neoplasma* **27** (1980) 271.
60. A. Panthananikal, C. Hansch, A. Leo and F. R. Quin, *J. Med. Chem.* **21** (1978) 16.
61. C. G. Swain, R. B. Moseley and D. E. Brown, *J. Am. Chem. Soc.* **77** (1955) 3731.
62. R. G. Pearson, *J. Am. Chem. Soc.* **85** (1963) 3533.
63. R. G. Pearson and J. Songstadt, *J. Am. Chem. Soc.* **89** (1967) 1827.
64. R. G. Pearson, in *Hard and Soft Acids and Bases*, Dowden Hutchison and Ross, Strousburg, Pennsylvania (1973).
65. G. Klopman, *J. Am. Chem. Soc.* **90** (1968) 223.
66. T. L. Ho, in *Hard and Soft Acids and Bases Principle in Organic Chemistry*, Academic Press, New York, San Francisco, London (1977).
67. G. A. Olah, E. B. Baker and J. C. Evans, *J. Am. Chem. Soc.* **86** (1964) 1360.
68. A. Pullman and B. Pullman, *Int. J. Quant. Chem.* **7** (1980) 245.
69. C. C. Price, G. M. Gaucher and P. Konaru, *Biochim. Biophys. Acta* **166** (1968) 327.
70. R. O. Rahn, R. G. Schulman and J. W. Longworth, *J. Chem. Phys.* **45** (1965) 2955.
71. B. Singer, *Prog. Nucleic Acid. Res. Mol. Biol.* **15** (1975) 219.
72. R. Lavery, A. Pullman and B. Pullman, *Nucleic Acids Res.* **8** (1980) 1061.
73. B. Singer, *J. Toxicol. Environ. Health* **2** (1977) 1279.
74. W. J. Bodell, K. Tokuda and D. B. Ludlum, *Cancer Res.* **48** (1988) 4489.
75. M. R. Osborne, in 'Chemical carcinogens', 2nd edn., *ACS Monograph* 182, ed. C. E. Seale, American Chemical Society, Washington, DC (1984).
76. K. W. Kohn, in *Molecular Actions and Targets for Chemotherapeutic Agents*, eds. A. C. Sartorelli, J. S. Lazlo and J. R. Bertino, Academic Press, New York (1981) p. 3.
77. P. D. Lawley and P. Brookes, *Biochem. J.* **89** (1963) 127.
78. J. J. McCann, T. M. Lo and D. A. Webster, *Cancer Res.* **31** (1971) 1573.
79. (a) J. D. Gralla, S. Sasse-Dwight and L. G. Poljak, *Cancer Res.* **47** (1987) 5092; (b) L. A. Zwelling and K. W. Kohn, *Cancer Treat. Rep.* **63** (1979) 1439.
80. M. Thomasz, *Biochim. Biophys. Acta* **213** (1970) 299.
81. W. P. Tong and D. B. Ludlum, *Biochim. Biophys. Acta* **608** (1980) 174.
82. K. Yamaguki, T. Tanabe and M. Kinoshita, *J. Org. Chem.* **41** (1976) 3691.
83. E. Institoris, *Chem.-Biol. Interact.* **35** (1981) 207.
84. J. Doskocil and Z. Sormova, *Z. Collect. Czech. Chem. Commun.* **30** (1965) 481.
85. W. G. Verley and L. Brakier, *Biochim. Biophys. Acta* **174** (1969) 674.
86. E. Harbers, P. Warnecke, H. Hollandt and K. Kruze, *Z. Krebsforsch.* **88** (1977) 237.
87. K. D. Tew and D. M. Taylor, *J. Natl. Cancer Inst.* **58** (1977) 1413.
88. E. G. Trams, M. V. Nadkarni and P. K. Smith, *Cancer Res.* **21** (1961) 560.
89. J. Hansson, R. Lewensohn, U. Ringborg and B. Nilsson, *Cancer Res.* **47** (1987) 2631.
90. W. E. Ross, R. A. G. Ewig and K. W. Kohn, *Cancer Res.* **38** (1978) 1502.
91. E. H. L. Chun, L. Gonzales, I. S. Lewis, J. Jones and R. J. Rutman, *Cancer Res.* **29** (1969) 1184.

92. R. Chen, J. J. Mieyal and D. A. Goldthwait, *Carcinogenesis* **2** (1981) 73.
93. K. Hemminki, *Chem.-Biol. Interact.* **34** (1981) 323.
94. T. P. Brent, S. O. Lestrud, D. G. Smith and J. S. Remack, *Cancer Res.* **47** (1987) 3384.
95. J. A. Alexander, B. J. Bowdon and G. P. Wheeler, *Cancer Res.* **46** (1986) 6024.
96. W. P. Tong and D. B. Ludlum, *Cancer Res.* **41** (1981) 380.
97. W. P. Tong, M. C. Kirk and D. B. Ludlum, *Cancer Res.* **42** (1982) 3102.
98. J. L. Skibba and G. T. Bryan, *Toxicol. Appl. Pharmacol.* **18** (1971) 707.
99. W. P. Tong, M. C. Kirk and D. B. Ludlum, *Biochem. Pharmacol.* **32** (1983) 2011.
100. K. Hemminki and D. B. Ludlum, *J. Natl. Cancer. Inst.* **37** (1984) 1021.
101. J. J. Roberts, T. P. Brent and A. R. Crathorn, *Eur. J. Cancer* **7** (1971) 515.
102. Y. E. M. Chaw, L. E. Crane, P. Lange and R. Shapiro, *Biochemistry* **19** (1980) 5525.
103. (a) N. Brock, *J. Cancer Res. Clin. Oncol.* **111** (1986) 1; (b) D. Schmähl and H. Druckerey, *Naturwissenschaften* **43** (1956) 199.
104. G. Weber, *Cancer Res.* **43** (1983) 3466.
105. V. F. Shealey, C. A. Krauth and W. L. Laster, *J. Med. Chem.* **27** (1984) 664.
106. L. M. Cobb, T. A. Connors, L. A. Elson, A. H. Khan, B. C. V. Mitchley, W. C. J. Ross and M. E. Whissen, *Biochem. Pharmacol.* **18** (1969) 1519.
107. M. Rodriguez, J. L. Imach and J. Martinez, *J. Med. Chem.* **27** (1984) 1222.
108. K. K. Misiura, R. W. Kinas, V. J. Stec, H. Kusnierczyk, C. Radzikowski and A. Sanoda, *J. Med. Chem.* **31** (1988) 226.
109. H. Druckerey, *Dtsch. Med. Wochenschr.* **77** (1952) 1534.
110. A. M. Seligman, M. M. Nachlas and L. H. Manheimer, *Ann. Surg.* **130** (1948) 333.
111. R. J. Knox, M. P. Boland, F. Friedlos, B. Coles, C. Southan and J. J. Roberts, *Biochem. Pharmacol.* **37** (1988) 4671.
112. T. S. Lin, B. A. Teicher and A. C. Sartorelli, *J. Med. Chem.* **23** (1980) 1237.
113. K. Krowicki, J. Balzarini, E. de Clercq, R. A. Newman and J. W. Lown, *J. Med. Chem.* **31** (1988) 341.
114. P. N. Confalone, E. M. Huie, S. S. Ko and G. M. Cole, *J. Med. Chem.* **31** (1988) 482.
115. M. E. Dona, A. E. Pegg, N. K. Hora and L. C. Ericson, *Cancer Res.* **48** (1988) 3603.
116. N. Brock, J. Pohl, J. Stelar and W. Scheef, *Eur. J. Cancer Clin. Oncol.* **18** (1982) 1377.
117. W. Doppler, J. Hofmann, K. Maly and H. H. Grunicke, *Cancer Res.* **48** (1988) 2454.
118. N. Bodor and J. J. Kaminski, *J. Med. Chem.* **23** (1980) 566.
119. M. Szekerke, R. Wade and M. E. Whisson, *Neoplasma* **19** (1972) 199.
120. V. T. Oliverio, *Cancer Treat. Rep.* **60** (1960) 703.
121. A. Leo, C. Hansch and D. Elkins, *Chem. Rev.* **71** (1977) 929.
122. I. Niculescu-Duvăz and I. Baracu, *Cancer Treat. Rep.* **61** (1977) 929.
123. C. Hansch, R. N. Smith, R. Engle and H. Wood, *Cancer Chemother. Rep.* **56**, (1972) 443.
124. J. A. Montgomery, J. G. Mayo and C. Hansch, *J. Med. Chem.* **17** (1974) 477.
125. M. T. Hakala, in *Handbook of Experimental Pharmacology*, Vol. XXXVIII/1, New Series, eds. A. C. Sartorelli and D. C. Johns, Springer-Verlag, Brlin, New-York, (1974) p. 240.
126. U. Draeger and H. J. Hohorst, *Cancer Treat. Rep.* **60** (1976) 423.
127. A. Begleiter, H. Y. Lam, J. Grover, E. Froese and G. J. Goldenberg, *Cancer Res.* **39** (1979) 353.
128. A. Begleiter, E. Froese and G. J. Goldenberg, *Cancer Lett.* **10** (1980) 243.
129. M. Szekerke, *Cancer Treat. Rep.* **60** (1976) 347.
130. N N. Preobrazhenskaia, in *Method of Development of New Anticancer Drugs*, USA-USSR, Monography, ed. NCI, Bethesda, MD (1977) p. 195.
131. L. A. Elson, *Biochem. Pharmacol.* **5** (1960) 192.
132. L. H. Schmidt, R. Franklin, R. Sullivan and A. Flowers, *Cancer Chemother. Rep.* Suppl. 2, Parts I, II and III (1965).
133. T. S. Lin, P. H. Fisher and W. H. Prusoff, *J. Med. Chem.* **23** (1980) 1235.
134. A. Okruzek and J. G. Verkade, *J. Med. Chem.* **22** (1979) 882.
135. J. Raus, H. Martens and G. Leclercq, in *Cytotoxic Estrogen in Hormone Receptive Tumors*, Academic Press, London, New York (1980).
136. I. Niculescu-Duvăz, *Oncologia (Bucharest)* **18** (1979) 161.
137. R. G. Rao and C. C. Price, *J. Org. Chem.* **27** (1962) 205.
138. M. Wall, F. I. Abernathy, F. I. Caroll and D. J. Taylor, *J. Med. Chem.* **12** (1969) 810.
139. E. Ciorănescu and D. Răileanu, *Stud. Cercet. Chim.* **10** (1962) 295.
140. T. F. Nogrady, K. M. Vagi and V. M. Adamkiewicz, *Can. J. Chem.* **40** (1962) 2126.

141. I. Niculescu-Duvăz, A. Cambanis and E. Tărnăuceanu, *J. Med. Chem.* **10** .(1967) 172.
142. H. J. Fex, K. B. Hogberg and I. Konyves, Ger. Patent 2,001,305 (1970).
143. B. Lam, *J. Med. Chem.* **22** (1979) 200.
144. B. Forsgren, B. Hogberg, J. A. Gustawsson and A. Pousette, *Acta Pharm. Sued.* **15** (1978) 23.
145. E. P. Wollmer, D. J. Taylor, I. J. Masnik, *et al.*, *Cancer Chemother. Rep.* **4** (1973) 121.
146. L. Brandt, J. Konwes and T. R. Möller, *Acta Med. Scand.* **197** (1975) 317.
147. (a) I. Niculescu-Duvăz, I. Elian and T. Crăescu, *Oncologia (Bucharest)*, **18** (1979) 243; (b) R. Bucourt, M. Vignau, V. Torelli, *et al.*, *J. Biol. Chem.* **253** (1978) 8221.
148. H. Hamascher and E. Christ, *Arzneim-Forsch./Drug Res.* **33** (1983) 347.
149. (a) V. C. Jordan, L. Fenik, R. E. Allen, R. C. Cotton, D. Richardson, A. L. Walpole and J. Bowler, *Eur. J. Cancer.* **17** (1981) 193; (b) H. Hamacher and C. Mangold, *Arch. Pharm.* **316** (1983) 271.
150. (a) A. H. Soloway, I. Agranat. A. R. Chase, *et al.*, *J. Med. Chem.* **17** (1974) 918; (b) A. H. Soloway, J. E. Wright, V. Subranmanyani and J. J. Gozzo, *J. Med. Chem.* **20** (1977) 1357.
151. T. Ghose, A. H. Blair, P. Uadia, P. N. Kulkarni, A. Goundalkar, M. Mezei and S. Ferrone, *Ann. N. Y. Acad. Sci.* **446** (1985) 213.
152. W. C. J. Ross, *Chem.-Biol. Interact.* **10** (1975) 169.
153. A. H. Blair and T. Ghose, *J. Immunol. Methods* **59** (1983) 129.
154. I. Berenblum, *J. Pathol. Bacteriol.* **34** (1931) 731.
155. E. Gason, H. McI. Combie, A. H. Williams and F. N. Woodward, *J. Chem. Soc.* (1948) 44.
156. W. Davis, W. C. J. Ross, *J. Med. Chem.* **8** (1965) 757.
157. W. C. J. Ross, *J. Chem. Soc.* (1949) 183.
158. A. Gilman and R. S. Philips, *Science* **103** (1946) 409.
159. I. Vargha, L. Toldy, O. Feher and S. Lendvai, *J. Chem. Soc.* (1957) 805.
160. V. A. Mikhalev, *Khim.-Farm. Zh.* **6** (1972) 9.
161. V. S. Ghandi and E. Schwenk, *Indian J. Chem.* **39** (1962) 306.
162. S. Fabrician, M. de Nardo, C. Nisi, L. Morasca, E. Dolfini and G. Franchi, *J. Chem. Soc.* **19** (1976) 639.
163. M. A. Khaled, R. D. Morin, F. Benington and J. P. Daugherty, *Cancer Chemother. Pharmacol.* **13** (1984) 73.
164. J. J. Everett, J. J. Roberts and W. C. J. Ross, *J. Chem. Soc.* (1953) 2386.
165. M. Jarman, L. J. Griggs and M. J. Tisdale, *J. Med. Chem.* **17** (1974) 194.
166. F. Bergel and J. A. Stock, *Br. Emp. Cancer Comp. Annu. Rep.* **31** (1953) 6.
167. L. F. Larionov, E. N. Shkodinskaia, N. I. Troosheikina, A. S. Khokhlov, O. S. Vasina and M. A. Novikova, *Lancet* **ii** (1955) 169.
168. F. Fukuoka and Y. Shirasu, *Gann* **52** (1961) 191.
169. L. F. Larionov, *Cancer Res.* **21** (1961) 99.
170. S. L. Gurina, R. K. Batulina, L. V. Alekseeva and L. V. Pushkareva, *Khim. Geterotsikl. Soedin.* (1968) 431.
171. L. P. Rateikene, M. I. Dagene and I. L. Knuniants, *Usp. Khim.* **39** (1970) 1537.
172. (a) A. de Barbieri, P. Di Vittorio and M. Maugeri, Proceedings of the 6th International Congress of Chemotherapy (Tokyo (1970) vol. 2, p. 146; (b) W. E. Grose, M. Burgess and G. P. Bodey, *Cancer. Treat. Rep.* **63** (1979) 385.
173. De Barbieri, Eur. Pat. Appl. EP 56565 (1982).
174. M. J. Yagi, M. Zanjani, J. F. Holland and G. J. Bekesi, *Cancer Chemother. Pharmacol.* **12** (1984) 88.
175. I. Niculescu-Duvăz, V. Feyns, V. Savaniu, M. Ionescu and A. Cambanis, *Rev. Roum. Chim.* **14** (1969) 535.
176. I. Niculescu-Duvăz, M. Ionescu, A. Cambanis, M. Vitan and V. Feyns, *J. Med. Chem.* **11** (1968) 500.
177. I. Niculescu-Duvăz, V. Feyns and V. Dobre, *Cancer Treat. Rep.* **62** (1978) 2045.
178. I. Niculescu-Duvăz, A. Toma, M. Rotaru, D. Ionescu, I. Baracu, G. Botez, E. Tarnăuceanu and V. Dobre, *Rev. Roum. Chim.* **24** (1979) 5583.
179. J. A. Montgomery, *Handb. Exp. Pharmakol.* **38** (1974) 97.
180. L. F. Larionov and G. N. Platonova, *Vopr. Onkol.* **1** (1955) 36.
181. F. I. Caroll, A. Philip, T. J. Blackwell, D. J. Taylor and M. Wohl, *J. Med. Chem.* **15** (1972) 1158.
182. P. Dalmases, G. Gervantes, J. Quintana, J. J. Bonet, A. Giner-Sorola and F. A. Schimid, *Eur. J. Med. Chem. Chim. Ther.* **19** (1984) 465.

183. I. Elian, D. Ionescu, E. Tărnăuceanu, I. Niculescu-Duvăz, G. Leclercq and N. Dewleesh-ower, *Eur. J. Med. Chem. Ther.* **18** (1983) 385.
184. O. M. Friedman and A. M. Seligman, *J. Am. Chem. Soc.* **76** (1954) 655.
185. H. Arnold and F. Bourseaux, *Angew. Chem.* **70** (1958) 539.
186. P. J. Cox, P. B. Farmer, A. B. Foster, E. D. Gilby and M. Jarman, *Cancer Treat. Rep* **60** (1976) 483.
187. Sociedad Espaniola Especialidades Farmaco-Terapeutics Span. ES 526.194 (1985) (*Chemical Abstracts* **107** (1987) 7385).
188. G. Lavielle and C. Cudennec, Fr. Patent FR 2567129 (1986).
189. R. F. Struck, *Cancer. Treat. Rep* **60** (1976) 317.
190. A. Takamizawa, S. Matsumoto, T. Iwata, I. Tochino, K. Katagiri, K. Yamaguchi and O. Shiratori, *J. Med. Chem.* **18** (1975) 376.
191. O. M. Friedman, I. Wodinsky and A. Myles, *Cancer Treat. Rep.* **60** (1976) 337.
192. A. Takamizawa, S. Matsumoto, T. Iwata, I. Makino, K. Yamaguchi, N. Uchida, A. Kasai, O. Shiratori and S. Takase, *J. Med. Chem.* **21** (1978) 208.
193. J. Van Der Steen, E. C. Timmer, J. G. Westra and C. Benckhuysen, *J. Am. Chem. Soc.* **95** (1973) 7535.
194. A. Takamizawa, S. Matsumoto, T. Iwata, I. Makino, K. Yamaguchi, N. Uchida, H. Kasai, O. Shirartori and S. J. Takase, *J. Med. Chem.* **17** (1974) 1237.
195. A. Takamizawa, T. Iwata, K. Yamaguchi, O. Shiratori, M. Haraka, Y. Tochino and S. Matsumoto, *Cancer Treat. Rep.* **60** (1976) 361.
196. R. F. Struck, M. C. Thorpe, W. C. Coburn and W. R. Laster, *J. Am. Chem. Soc.* **96** (1974) 313.
197. G. Peter, T. Wagner and H. J. Hohorst, *Cancer Treat. Rep.* **60** (1976) 429.
198. G. Scheffer, U. Niameyer, N. Brock and R. Pohl, Ger. Offen. DE 3.220.432 (1983).
199. R. F. Struck, D. J. Dykes, T. H. Corbett, W. J. Suling and M. W. Trader, *Br. J. Cancer* **47** (1983) 15.
200. M. Jarman, R. A. V. Milsted, J. F. Smyth, R. W. Kinas, K. Pankiewicz and W. J. Stec, *Cancer Res.* **39** (1979) 2762.
201. K. Pankiewicz, R. W. Kinas, W. J. Stec, A. B. Foster, M. Jarman and J. M. S. Van Maanen, *J. Am. Chem. Soc.* **101** (1979), 7712.
202. G. Blaschke and P. Hilgard, *Arzneim.-Forsch.* **36** (1986) 1493.
203. H. Kusnierczyk, C. Radzikowski, M. Paprocka, W. Budzynski, R. Kinas, K. Misiura and W. Stec, *J. Immunopharmacol.* **8** (1986) 455.
204. W. N. Setzer, A. E. Sopchik and W. G. Bentrude, *J. Am. Chem. Soc.* **107** (1985) 2083.
205. V. L. Boyd, G. Zon, V. Hinus, J. K. Stalick, A. D. Mighell and H. V. Secor, *J. Med. Chem.* **23** (1980) 372.
206. S. M. Ludeman, G. Zon and W. Egon, *J. Med. Chem.* **22** (1979) 151.
207. A. B. Foster, M. Jarman, R. W. Kinas, J. M. S. Van Maanen and G. N. Taylor, *J. Med. Chem.* **24** (1981) 1399.
208. H. Zimmel and D. Bocancea, *Neoplasma* **21** (1974) 583.
209. I. Niculescu-Duvăz, I. Elian, M. Ionescu and E. Tărnăuceanu, *J. Prakt. Chem.* **321** (1979) 522.
210. (a) E. Grant and I. Niculescu-Duvăz, *Can. J. Chem.* **43** (1965) 1227; (b) I. Niculescu-Duvăz, V. Feyns, E. Grant and E. Tărnăuceanu, *Can. J. Chem.* **44** (1966) 1102.
211. D. Răileanu and E. Ciorănescu, *Stud. Cercet. Chim.* **11** (1965) 433.
212. A. H. El Masry, V. C. Braun, C. J. Niessen and W. B. Pratt, *J. Med. Chem.* **20** (1977) 1134.
213. Y. F. Shealey, C. A. Krauth and J. A. Montgomery, *J. Org. Chem.* **27** (1962) 2150.
214. Y. T. Ling, T. L. Loo, S. Vadlamudi and A. Goldin, *J. Med. Chem.* **15** (1972) 201.
215. T. Giraldi, T. Nisi, T. A. Connors and P. M. Goddard, *J. Med. Chem.* **20** (1977) 850.
216. Y. F. Shealy, *J. Pharm. Sci.* **59** (1970) 1533.
217. (a) M. F. G. Stevens, J. A. Hickman, R. Stone, N. W. Gibson, G. U. Baig, E. Lunt and C. G. Newton, *J. Med. Chem.* **27** (1984) 196; (b) C. J. Brindley, P. Antoniu and E. S. Newlands, *Br. J. Cancer* **53** (1986) 91.
218. R. H. Smith Jr., A. F. Mehl, D. L. Shantz, G. N. Chmurny and C. J. Michejda, *J. Org. Chem.* **53** (1988) 1467.
219. (a) J. Wenker, *J. Am. Chem. Soc.* **57** (1935) 2328; (b) *Ullmann's Encyklopädie der Techn. Chem* **3** (1953) 463.
220. R. C. Tsou, D. Bender, N. L. Santora, L. David and S. Damle, *J. Med. Chem.* **19** (1976) 806.
221. L. Gutierrez, M. Konieczny and G. Sosnovsky, *Z. Naturforsch. B* **36B** (1981) 1622.
222. T. T. S. Safonova, *Zh. Vses. Khim. Ora.* **18** (1973) 657.

223. T. J. Bardos and M. E. Perlman, US Patent 4617.398 (1986).
224. J. S. Driscoll, G. F. Hazard, H. B. Wood and A. Goldin, *Cancer Chemother. Rep.* Part 2 **4** (1974) 1.
225. W. Gaus, *Ber.* **91** (1958), 2216.
226. W. Gaus and S. Petersen, *Angew. Chem.* **69** (1957) 252.
227. A. H. Khan and J. S. Driscoll, *J. Med. Chem.* **19** (1976) 313.
228. F. Chou, A. H. Khan and J. S. Driscoll, *J. Med. Chem.* **19** (1976) 1302.
229. (a) H. Nakao and M. Arakawa, *Chem. Pharm. Bull.* **20** (1972) 1962; (b) H. Nakao, M. Arakawa and T. Nakamura, *Chem. Pharm. Bull.* **20** (1972) 1968.
230. (a) K. Wallenfals and Y. Draber, *Justus Liebigs Ann. Chem.* **667** (1963) 55; (b) A. N. Makarova, V. S. Martinov and A. YA. Berlin, *Zh. Obshch. Khim.* **33** (1963) 1643.
231. P. L. Gutierez, S. Biswall, R. Nardino and N. Biswall, *Cancer Res.* **46** (1986) 5779.
232. W. E. Emmons and A. V. Ferris, *J. Am. Chem. Soc.* **75** (1983) 2207.
233. G. M. Timmis, *Ann. N.Y. Acad. Sci.* **68** (1958) 727.
234. Y. Sakurai and M. M. El-Merzabani, *Chem. Pharm. Bull. (Tokyo),* **12** (1965) 954.
235. B. F. Cain and W. A. Denny, *J. Med. Chem.* **20** (1976) 515.
236. J. S. Sandberg, H. B. Wood, R. R. Engle, J. M. Venditti and A. Goldin, *Cancer Chemother. Rep.* Part 2 **30** (1972) 137.
237. A. Haddow, G. M. Timmis and B. B. Brown, *Nature* **182** (1958) 1164.
238. A. Vargha and J. Kuszmann, *Naturwissenschaften* **46** (1959) 84.
239. L. Vargha and T. Horvath, *Nature* **183** (1959) 394.
240. L. Vargha and T. Horvath, *Extract de Acta Union Internationale contra Cancer* **XX** (1964) 76.
241. (a) Y. F. Shealy, C. A. Krauth and W. R. Laster, *J. Med. Chem.* **27** (1984) 664; (b) N. W. Gibson, J. A. Hartley, J. M. Strong and K. W. Kohn, *Cancer Res.* **46** (1986) 553; (c) N. W. Gibson, J. A. Hartley and K. W. Kohn, *Cancer Res.* **46** (1986) 1679.
242. L. A. Elson, M. Jarman and W. C. J. Ross. *Eur. J. Cancer* **4** (1968) 617.
243. L. Vidra, L. Institoris and S. Somfal-Relle, *Proc. Int. Congr. Chemother.* **13** (1983) 284/106.
244. M. Budnowski. *Angew. Chem.* **7** (1968) 827.
245. G. Attassi, F. Spreafico, P. Dumont, P. Nayer and J. Klastersky, *Eur. J. Cancer* **16** (1980) 1561.
246. H. Wagner, K. Possinger, K. Bremer, R. Donhujsen-Ant, M. Peukert and W. Queisser, *Cancer Treat. Rep.* **71** (1987) 209.
247. T. S. Lin, B. A. Teicher, and A. C. Sartorelli, *J. Med. Chem.* **23** (1980) 1237.
248. A. J. Lin, C. W. Shanky and A. C. Sartorelli, *J. Med. Chem.* **17** (1974) 558.
249. A. J. Lin, R. S. Pardini, B. J. Lillis and A. C. Sartorelli, *J. Med. Chem.* **17** (1974) 668.
250. I. Wilson, P. Wardman, T. S. Lin and A. C. Sartorelli, *Chem.-Biol. Interact.* **61** (1987) 229.
251. M. H. Holshouser, L. J. Loeffer and I. H. Hall, *J. Med. Chem.* **24** (1981) 854.
252. I. Natonini, F. Claudi, G. Cristalli, P. Franchetti, M. Grifantini and S. Martelli, *J. Med. Chem.* **31** (1988) 260.
253. H. W. Moore, R. Czerniak and A. Hamdan, *Drug Exp. Clin. Res.* **12** (1986) 475.
254. S. E. Jones, W. G. Tucker, A. Haut, B. L. Tranum, C. Vaughn, E. M. Chase and B. G. M. Durie, *Cancer Treat. Rep.* **61** (1977) 1617.
255. A. Hagenbeek and A. C. M. Martens, *Cancer Treat. Rep.* **65** (1981) 975.
256. H. B. A. Shape, E. O. Field and K. Hellman, *Nature* **226** (1970) 524.
257. F. M. Schabel Jr., M. W. Trader, W. R. Laster, S. C. Shaddix and R. W. Brockman, *Cancer Treat. Rep.* **60** (1976) 1325.
258. J. F. Conroy, J. A. Blessing, A. Kessinger and H. D. Homesley, *Cancer Treat. Rep.* (1984) 439.
259. A. M. Greighton, K. Hellman and S. Whitecross, *Nature* **222** (1969) 384.
260. C. O. Gitterman, E. L. Rikes, D. E. Donald, J. Madas, S. B. Zimmerman, T. S. Stoudt and T. C. Denny, *J. Antibiot. (Tokyo)* **23** (1970) 305.
261. B. H. Arison and J. L. Beck, *Tetrahedron* **29** (1973) 2743.
262. R. W. Brockman, S. C. Shaddix, M. Williams and R. F. Struck, *Cancer Treat. Rep.* **60** (1976) 1317.
263. S.M. Ludeman, G. Zon and W. Egon, *J. Med. Chem.* **22** (1979) 151.
264. R. F. Pittillo, C. Wooley, L. S. Rice and R. W. Brockman, *Cancer Chemother. Rep.* **55** (1971) 47.
265. (a) R. V. Brockman, *Cancer Res.* **7** (1963) 129; (b) G. B. Linvingstone, J. M. Venditti and D. A. Cooney, *Adv. Pharmacol. Chemother.* **8** (1970) 57.

266. L. R. Duvall, *Cancer Chemother. Rep.* **7** (1960) 65.
267. H. S. Liija, E. T. Kulmann and J. D. Yager, *Proc. Am. Assoc. Cancer Res., ASCO* **19** (1978) 174.
268. N. M. Emanuel, N. P. Knovalova and R. F. Djachkowskala, *Cancer Treat. Rep.* **60** (1976) 1601.
269. C. C. Cheng and K. Y. Zee-Chang, *J. Pharm. Sci.* **61** (1972) 485.
270. J. A. Katzenellenbogen, R. J. McGorrin, T. Tatee R. J. Kempton, K. E. Carlson and D. H. Kinder, *J. Med. Chem.* **24** (1981) 435.
271. S. M. Kupchan, M. A. Eakin and A. M. Thomas, *J. Med. Chem.* **14** (1971) 1147.
272. S. M. Kupchan, *Trans. N.Y. Acad. Sci.* **32** (1970) 85.
273. J. M. Cassidy, S. R. Byrn and T. K. Stamos, *J. Med. Chem.* **21** (1978) 815.
274. K. H. Lee, T. Ibuka, O. Muraoka, H. Kiyokawa and I. H. Hall, *J. Med. Chem.* **24** (1981) 924.
275. S. S. Dehal, B. A. Marples, R. J. Stretton and J. R. Taynor, *J. Med. Chem.* **23** (1980) 90.
276. S. Inayama, K. Mamoto, T. Sahibata and T. Girose, *J. Med. Chem.* **19** (1976) 433.
277. B. Kellner, L. Nemeth, I. P. Hovarth and L. Institoris, *Nature* **213** (1967) 402.
278. C. Mitoma, T. Onodera, T. Takagoshi and W. Thomas, *Xenobiotica* **7** (1977) 201.
279. H. E. May, P. Boose and D. J. Reed, *Biochem. Biophys. Res. Commun.* **57** (1974) 426.
280. G. P. Wheeler, T. P. Jonstone, B. J. Bowden, G. S. McCaleb, D. L. Hill and J. A. Montgomery, *Biochem. Pharmacol.* **26** (1977) 2331.
281. N. Brock and H. J. Hohorst, *Naturwissenschaften* **49** (1962) 610.
282. G. H. Known, K. Maddism, L. Locastro and R. F. Borch, *Cancer Res.* **47** (1987) 1505.
283. H. J. Hohorst, V. Draeger, G. Peter and G. Voelcker, *Cancer Treat. Rep.* **60** (1976) 309.
284. L. Bielicki, G. Voelker and H.J. Hohorst, *J. Cancer Res. Clin Oncol.* **108** (1984) 195.
285. H. J. Hohorst, L. Bielicki and G. Voelker, in 'The enzymatic basis of cyclophosphamide specificity', *Advances in Enzyme Regulation*, ed. G. Weber, Pergamon Press, Oxford (1986).
286. R. F. Struck and D. S. Alberts, *Cancer Treat. Rep.* **68** (1984) 765.
287. U. Praeger, G. Peter and H. J. Hohorst, *Cancer Treat. Rep.* **60** (1976) 355.
288. G. Voelcker, T. Wagner and H. J. Hohorst, *Cancer Treat. Rep.* **60** (1976) 415.
289. T. Wangle, G. Zon and W. Egan, *J. Med. Chem.* **22** (1979) 897.
290. W. P. Brade, K. Herdrich and M. Varini, *Cancer Treat. Rev.* **12** (1985) 1.
291. S. M. Ludeman, V. L. Boyd, J. B. Regan, K. A. Gallo, B. Zon and K. Ishii, *J. Med. Chem.* **29** (1986) 716.
292. D. L. Hill, W. R. Laster, M. C. Kirk, S. El Dareer and R. F. Struck, *Cancer Res.* **33** (1973) 1016.
293. T. A. Connors, P. J. Cox, P. B. Farmer, A. B. Foster and M. Jarman, *Biochem. Pharmacol.* **23** (1974) 115.
294. B. M. Bryant, M. Jarman, H. M. Baker, I. E. Smith and J. E. Smith, *Cancer Res.* **10** (1980) 4734.
295. K. Norpoth, *Cancer Treat. Rep.* **60** (1976) 437.
296. (a) K. Norpoth, G. Mueller and H. Raid, *Arzneim.-Forsch.* **26** (1976) 1376; (b) S. J. Spengler and B. Singer, *Cancer Res.* **48** (1988) 4704.
297. L. M. Cobb, T. A. Connors, L. A. Elson, A. H. Khan, B. C. V. Mitchley, W. C. J. Ross and M. E. Whissen, *Biochem. Pharmacol.* **18** (1969) 1519.
298. T. A. Connors, H. G. Mandel and D. H. Melzack, *Int. J. Cancer* **9** (1972) 126.
299. M. J. Tisdale and A. D. Habberfield, *Biochem. Pharmacol.* **29** (1980) 2845.
300. J. J. Roberts, F. Friëdlos and R. J. Knox, *Biochem. Biophys. Res. Commun.* **140** (1986) 1073.
301. R. F. Knox, F. Friëdlos, M. Jarman and J. J. Roberts, *Biochem. Pharmacol.* **37** (1988) 4661.
302. A. Verloop, W. Hoogenstraten and J. Tipker, *Drug Design* **7** (1967) 165.
303. Z. Simon, A. Chiriac, S. Holban, D. Ciubotariu, G. I. Mihalas, in *Minimum Steric Differences. The MTD Method for QSAR Studies*, John Wiley, New York (1984).
304. E. J. Lien and G. L. Tong, *Cancer Chemother. Rep.* **57** (1973) 251.
305. A. Panthananickal, C. Hansch and A. Leo, *J. Med. Chem.* **19** (1976) 1267.
306. W. J. Dunn, M. J. Greenberg and S. S. Callejas, *J. Med. Chem.* **19** (1976) 1299.
307. G. J. Hatheway, C. Hansch, K. H. Kim, S. R. Mildstein, C. L. Schmidt, R. N. Smith and F. R. Quinn, *J. Med. Chem.* **21** (1978) 563.
308. G. Hansch, G. A. Hatheway, F. R. Quinn and N. Greenberg, *J. Med. Chem.* **21** (1978) 574.
309. C. Hansch, N. Smith and R. Engle, *Cancer Chemother. Rep.* **29** (1963) 91.

310. I. Baracu, G. Botez, R. Denes, V. Dobre, B. Ureche, T. Craescu, N. Voicule and I. Niculescu-Duvăz, *Rev. Roum. Biochem.* **18** (1981) 175.
311. C. Hansch, A. Leo, C. Schmidt and P. Y. C. Jow, *J. Med. Chem.* **23** (1980) 1095.
312. G. Prakash and E. M. Hodnett, *J. Med. Chem.* **21** (1978) 369.
313. M. Yoshimoto, M. Miyazawa, H. Nakao, K. Shinkai and M. Arakawa, *J. Med. Chem.* **22** (1979) 491.
314. T. P. Johnstone and J. A. Montgomery, *Cancer Treat.-Rep.* **70** (1986) 13.

4 Nitrosoureas

J. A. MONTGOMERY AND T. P. JOHNSTON

4.1 Introduction

The nitrosoureas were the first class of compounds derived from a lead compound identified by the random screening program of the National Cancer Institute to show useful activity against human cancer[1]. The activity of N-methyl-N-nitrosourea (1), not only against L 1210 leukaemia implanted intraperitoneally in the usual manner, but against the intracerebrally implanted disease, which contrasted with the inactivity of standard clinical agents such as methotrexate, 6-mercaptopurine, 5-fluorouracil, and cyclophosphamide[2], led to a drug development programme that resulted in the identification of the 2-chloro- and 2-fluoroethylnitrosoureas (2) as agents with a high level of activity against a variety of murine neoplasms followed by the establishment of the clinical activity of the 2-chloroethyl compounds[3-5].

The 2-chloroethylnitrosoureas possess a broad spectrum of activity against human neoplasms. The major indications include lymphoproliferative diseases, gliomas, melanomas, small-cell carcinomas of the lung and gastrointestinal cancers[6]. The lipophilicity of most nitrosoureas and their activity against intracerebrally implanted leukaemia L 1210[7], indicating their ability to cross the blood-brain barrier, caused them to be used in the treatment of human gliomas with reported response rates of 45% for BCNU (3), 37% for CCNU (4) and 23% for MeCCNU (5).

Five independent clinical trials of BCNU in Hodgkin's disease following relapse from previous therapy with other drugs have shown an overall objective response rate of 47% with a range of 34–55%[8,9]. In another study, CCNU proved superior to both BCNU and MeCCNU[8]. CCNU-containing combinations have produced complete remission rates that are either equivalent or superior to MOPP, the standard combination therapy for Hodgkin's disease. These studies included patients without prior therapy as well as those relapsed after primary radiation therapy. In addition, the CCNU-containing regimens resulted in a significantly longer duration of complete remission. Clearly the chloroethylnitrosoureas have an established role in the treatment of Hodgkin's disease.

Although the nitrosoureas have demonstrated only modest activity in advanced gastrointestinal cancer as single agents, with response rates of 8–18% in gastric cancer, 0–9% in pancreatic cancer, and 9–13% in advanced colorectal carcinoma[6], they have given better response rates in combination with 5-fluorouracil than can be obtained with either drug alone. Kovach *et al.*[10] reported a 40% objective response rate with BCNU plus 5-fluorouracil in advanced gastric cancer, with a significant improvement in survival. A study of MeCCNU plus 5-fluorouracil produced similar results[11]. A 33% response was achieved in patients with advanced pancreatic cancer treated with BCNU and 5-fluorouracil, but with little impact on survival[10], MeCCNU in combination with 5-fluorouracil and vincristine gave a 43% response rate in advanced colorectal cancer[12], and recent data support the use of MeCCNU with 5-fluorouracil in this disease[13,14].

The chemotherapeutic agent most widely used in the treatment of malignant islet cell tumours is streptozotocin (**6**), a methylnitrosourea with activity in Hodgkin's disease and reduced myelosuppression[15]. In a series of twelve patients with metastatic islet-cell carcinoma treated with streptozotocin, three achieved clinical disappearance of tumour and hormone production[16]. Other studies have confirmed the activity of this compound in pancreatic cancer[17,18].

Despite the excellent anticancer activity of the nitrosoureas against a variety of rodent neoplasms (leukaemias and solid tumours), they have not fulfilled their promise in humans, largely due to the problem of delayed and cumulative bone marrow toxicity, which has forced the use of intermittent schedules of treatment, to allow for full bone marrow recovery between each course. This type of scheduling, which prevents the use of intensive courses and complicates the design of combination chemotherapy trials, represents a real limitation of this class of agent. Unfortunately, attempts to date to produce new, highly active nitrosoureas with reduced marrow toxicity and a better therapeutic index have met with limited success[5].

4.2   Structure-activity relationships

4.2.1   *Leukaemia L 1210*

The most active nitrosoureas in the leukaemia L 1210 system contain either the *N*-methyl-*N*-nitrosoureido or the *N*-[2-chloro(fluoro)ethyl]-*N*-nitroso-ureido moieties*. It is clear, however, that the *N*-methyl compounds are less potent and less effective than the corresponding 2-haloethyl compounds (Table 4.1). They also have significantly longer half-lives and lower alkylating activity, but there is no clear relationship in the case of car-bamoylating activity except that this activity is more dependent on the nature of R than R'. These data also support the concept that alkylation is the most important factor for activity, with the 2-haloalkyl compounds being the best alkylating agents (in a series of (2-chloroethyl)nitrosoureas there is a good correlation between alkylating activity and half-life). The problem is more complicated than that, however, because the chemistry of the 2-haloalkyl compounds is more complex due to the lability of the carbon-halogen bond (Br > Cl > F) (see discussion below).

From a comprehensive study of over 100 2-haloethyl nitrosoureas Hansch *et al.*[19] derived equations relating activity, as judged by a 3-log kill of L 1210 cells *in vivo*, and toxicity, LD_{10}, to $\log P$. Although the r and s values are reasonably good for a study of this magnitude, no confidence limits could be placed on $\log P_0$ for either activity or toxicity and, thus, there is no clear separation of toxicity from activity as was hoped. The most active compounds are also the most toxic compounds, but at the same time they have the greatest therapeutic index. Furthermore, there is a $\log P_0$ range for maximum activity even though it is fairly broad. Data from approximately 300 compounds prepared at Southern Research Institute show that the ED_{50} (the dose required to cure 50% of a treated group of mice inoculated with leukaemia L 1210 (10^5 cells, i.p.)) varies 100-fold (from 0.024 to 2.3 mmol/kg), whereas the therapeutic index (TI), defined as the LD_{10}/ED_{50}, varies less than fourfold (from 1 to 3.9), again supporting the idea that the most potent compounds are also the most toxic. In fact, the compounds with a high LD_{10} (2.4–2.7 mmol/kg) are at the low end of the TI range.

---

*This discussion of the activity of the nitrosoureas in experimental animal systems is limited to data collected at Southern Research Institute, since it is not possible to make accurate comparisons of relative activities based on data collected in many different laboratories around the world.

Table 4.1 Comparisons of N-methyl- and N-(2-haloethyl)-N-nitrosoureas.

RNHCON(NO)R' R	R'	Schedule	LD_{10} or OD^a	% ILS	Cures^b	$t_{1/2}^c$ (min)	Alkylating activity^d (A_{540}^c)	Carbamoylating activity^e (dpm)^c
H	Me / CH_2CH_2Cl	Chronic	12	109	0	7.0	0.36	10890
	CH_2CH_2Cl	Chronic	0.90	147	5–10	1.3	3.0	8560
	Me / CH_2CH_2F	Chronic	150	46	0	250	0.02	8850
	CH_2CH_2F	Chronic	8.0	187	5–10^f	43	1.4	28700
X (chain)		Day 1	12	233	8	ND	1.3	19800
(branched)	Me / CH_2CH_2Cl	Day 1	300	27	0	490	0.02	19800
	CH_2CH_2Cl	Day 1	48	—	10	53	0.52	42000
	CH_2CH_2F	Day 1	34	—	5–10	68	0.64	49100
HO—(sugar)—OH, OH	Me	qd 1–9^g	50	31		48	0.19	1630
	CH_2CH_2F	Day 1	29	—	9	21	2.4	820

^{a}Optimal dose (mg/kg per dose).
^{b}Out of 10 treated animals.
^{c}Half-life.
^{d}A_{540} is a measure of concentration of alkylated 4-(p-nitrobenzyl)pyridine in an ethyl acetate extract of a mixture of the nitrosourea and the pyridine in acetate buffer (pH 6) incubated at 37°C for 2 h.
^{e}The dpm is a measure of radioactivity present in unidentified reaction products obtained upon incubation of the nitrosourea with [^{14}C] lysine in phosphate buffer (pH 7.4) at 37°C for 6 h.
^{f}Cures from single injection on day 1.
^{g}Optimal schedule.

Table 2 Activity[a] of the nitrosoureas versus solid tumours.

Compound	Advanced Lewis lung	Colon 6/A	Colon 7/A	Colon 8/A	Colon 26	Colon 38	Colon 51	C3H mammary	B16 mammary	M5 ovarian tumour
BCNU	–	+	ND	+	ND	ND	+	–	–	ND
CCNU	++	–	++	+	ND	–	ND	–	++	ND
MeCCNU	+++	+++	++	+++	++++	+++	++	++	+++	+++
PCNU	+	++++	++	ND	ND	ND	ND	–	+	ND
Chlorozotocin	–	ND	ND	–	–	ND	+	–	–	++

[a] Activity rating (approximate log cell kill): ++++, >2.6; +++, 1.6–2.6; ++, 0.9–1.5; +, 0.5–0.8; –, <0.5

4.2.2   Solid tumours

As indicated in the discussion above, many 2-chloro- or 2-fluoroethylnit-
rosoureas are highly active against the L 1210 leukaemia in mice and many
variations of this test system (i.p., i.v., i.c., 10^5–10^7 cells inoculated, various
treatment schedules) failed to differentiate between highly active structures
and point to a few structures for clinical investigation. This problem led to the
selection of the Lewis Lung carcinoma as a secondary screen for these
compounds. The activity of the nitrosoureas was determined against both the
early (before metastasis) and late (after metastasis) forms of the disease[20].
Although some exceptions were noted, compounds most active against the
early form of the disease were most active against the established tumour. A
differentiation in activity based on the Lewis Lung system was evident with
nitrosoureas equally active against leukaemia L 1210, with N-(2-chloroethyl)-
N'-(trans-4-methylcyclohexyl)-N-nitrosourea (MeCCNU) (5) and some of its
closely related congeners being the most active structures. A number of these
nitrosoureas have also been evaluated for activity against other solid tumours
in rodents, including several colon tumours, the B16 melanotic melanoma, the
C3H mammary adenocarcinoma and the M5 ovarian tumour (Table 4.2). In
general, the compounds most active against the Lewis Lung tumour were the
most active against these other solid tumours, indicating that the structural
characteristics necessary for activity against solid tumours are fairly general
(at least in the case of rodent neoplasms). The features do not ensure activity,
however, since, for example, none of the nitrosoureas tested is significantly
active against the Ridgeway osteogenic sarcoma, apparently because of poor
uptake by this tumour[21]. Further, the predictive value of rodent neoplasms for
human disease has yet to be established.

4.3   Toxicity

Most of the preclinical toxicity data on a vast majority of the nitrosoureas are
simply lethal toxicity, LD_{10} determinations[19] or in many cases, less precisely,
optimal dose value[22]. The myelosuppression observed clinically[23] led to a
detailed study of this specific toxicity in rodents and to attempts to develop less
myelosuppressive agents. Thus chlorozotocin (7)[24], its tetraacetate[25], and
other sugar-containing nitrosoureas[26-28] were shown to be less marrow toxic
than the clinically used BCNU and CCNU. Unfortunately, although curative
in the L 1210 leukaemia systems, chlorozotocin (7)[29], and presumably its
relatives, are much less active or inactive against a spectrum of solid tumours
(Table 4.2). The studies referred to above also showed that there is no
correlation of myelosuppression with carbamoylating activity[30-33], even
though carbamoylation was shown to interfere with the DNA repair
process[34,35]. At the same time the myelosuppressive nitrosoureas have a low
ratio of alkylation (L 1210:bone marrow DNA) (approx. 0.6) compared to the
marrow sparing compounds (1.3)[27,30]. Furthermore, the myelosuppressive

compounds alkylate nucleosome core particles, whereas the others alkylate internucleosomal linker regions of bone marrow chromatin[30].

A detailed study of the toxicity of N-(2-chloroethyl)-N'-(*trans*-4-methyl-cyclohexyl)-N-nitrosourea (MeCCNU) (**5**) in monkeys and dogs showed that fatal doses produced severe depression of haematopoietic activity and injury to the mucosa of the gastrointestinal tract in both species. Nephrotoxicity, stemming from primary injury to the proximal convoluted tubules, was a prominent feature of the fatal reactions of the rhesus monkey. Injury to the lung parenchyma of serious dimensions and a significant level of hepatotoxicity in the dog were produced by lethal doses of the drug[36]. Nephrotoxicity has also been observed in the rat, and further studies showed that MeCCNU metabolites and/or degradation products preferentially accumulated in the kidney, but the molecular basis for this induced nephropathy is not yet understood[37]. Chlorozotocin (**7**) was more nephrotoxic than MeCCNU, and the nephropathy produced by both drugs appears to be related to their alkylating activity[38].

Interest in the 2-fluoroethylnitrosoureas has been stimulated by the finding that, although the activity of these compounds is comparable to that of the 2-chloroethyl compounds (Table 4.1), the fluoroethyl compounds cross-link DNA poorly[39,40], a result of the greater stability of the carbon-fluorine bond. In addition, the $\log P$ values are about 0.6 lower, meaning they are more hydrophilic, and the fluoro analogue of CCNU, CFNU (**8**), has been reported to be less toxic to the bone marrow[41]. An examination of the chemical properties of the fluoro and the chloro compounds shows little difference in half-life, alkylating activity, or carbamoylating activity, the alkylating activity being a measure of the reactivity of the (2-haloethyl)diazotates and the stability of the parent nitrosourea[5].

There is, however, a significant difference in the toxicity of the fluoro and the chloro compounds, not in LD_{10} values, which are greater in some cases for the chloro compounds and in other cases for the fluoro compounds, but in median day of death, which varies from 6 to 27 days for the chloro compounds and from 2 to 5 days for the fluoro compounds. This difference suggested fluoroacetate toxicity, since a significant amount of the fluoro compounds decomposes to 2-fluoroethanol, which is readily oxidised via alcohol dehydrogenase to fluoroacetate. Sodium acetate in aqueous ethanol, which has been shown to reverse fluoroacetate toxicity in mice, given 10 min after a single dose of N-cyclohexyl-N'-(2-fluoroethyl)-N'-nitrosourea (CFNU) (**8**) raised its LD_{10} from 40 to 90 mg/kg, although it had no effect on myelosuppression[5]. Enhancement of its effectiveness against leukaemia L1210 could not be demonstrated because the maximum tolerated dose of CFNU alone cured 10/10 animals. Fluoroacetate toxicity reversal and activity enhancement was observed in the treatment of subcutaneously implanted B16 melanoma with the acetate of N-(2-fluoroethyl)-N'-[*trans*-4-(hydroxymethyl)cyclo-hexyl]-N-nitrosourea (**9**) and sodium acetate in aqueous ethanol[5].

4.4 Chemical basis of biologic activity

An understanding of the chemistry of the nitrosoureas is essential to an understanding of their biologic activity, although their chemistry alone certainly does not explain their apparent specificity for neoplastic cells and the resulting anticancer activity they display against both experimental animal neoplasms and human cancers.

Other similar N-nitroso compounds have shown no anticancer activity even though some are quite toxic, mutagenic, or carcinogenic, and therefore biologically active[3]. More recently, however, the curative action of certain 2-chloroethyl nitrosourethanes in the L 1210 leukaemia system has been demonstrated[42] and N-(2-chloroethyl)-N-nitroso-N'-nitroguanidine (10) has considerable activity in this system, albeit less than the corresponding urea[43]. At the same time, none of the precursors of these N-nitroso compounds, including the ureas, are very active biologically, clearly indicating the necessity for the N-nitroso group, which labilises the bond between the nitrosated nitrogen and the adjacent atom, so that when the adjacent atom is a carbon double bonded to a heteroatom, decomposition occurs spontaneously in aqueous media, the rate depending on the particular structure and the pH of

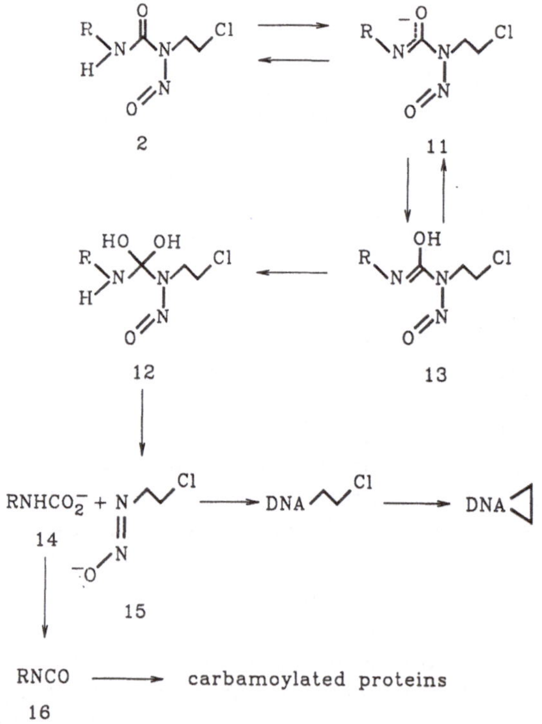

Figure 4.1 The important reactions of the 2-chloroethylnitrosoureas.

the medium[44,45]. Breakage of this bond, regardless of the exact mechanism, gives rise to a diazotate (Figure 4.1) capable of alkylating molecules such as DNA under physiologic conditions of pH, ionic strength and temperature. In the case of the N-nitrosoureas the reaction may be initiated by loss of a proton from the unnitrosated nitrogen. In support of this mechanism is the stability of the trisubstituted N-nitrosoureas, which are not toxic to cells in culture and must be metabolised—dealkylation in the liver—to show *in vivo* activity[46]. In the case of N-methyl-N-nitrosourea, the first nitrosourea found to have anticancer activity, this decomposition leads to methyldiazotate and isocyanic acid[44], compounds capable of methylating and carbamoylating biological macromolecules. It is now known that other N-nitroso compounds such as N-nitrosoamides and N-nitrosourethanes that produce alkylating but not carbamoylating moieties have anticancer activity, and at the same time, certain nitrosoureas that have little carbamoylating activity are quite active[5], whereas nitrosoureas with no detectable alkylating activity are very weakly active or inactive. Thus alkylating activity, but not carbamoylating activity, appears to be essential, although there is not a straight line correlation between alkylating and anticancer activity[47,48]. As mentioned in section 4.3, carbamoylation has been associated with the inhibition by the nitrosoureas of the repair of DNA damaged by ionising radiation or by alkylating agents[34,35] and thus could indirectly contribute to the cytotoxicity and anticancer activity of these compounds. At the same time, another type of chloroethylating agent with no carbamoylating activity, clomesone, has activity comparable to the chloroethylnitrosoureas and its fluoro analogue is similar[49].

Despite the large amount of experimental work on the complex organic chemistry of the chloroethylnitrosoureas, many aspects of their aqueous decomposition are still not understood. Of all the mechanisms proposed, formation of the *gem*-diol tetrahedral intermediate (12) (see Figure 4.1) by addition of water to the urea carbonyl is the most probable[50-52]. Because the N^3-monosubstituted CENUs have an acidic proton on N^3, hydrolysis could occur through the imino urea (13). The mechanism by which the carbamic acid anion (14) could be converted to the isocyanate (16) is unclear, although the carbamoylating activity of the nitrosoureas has been amply demonstrated. The important reactions of the CENUs are shown in Figure 4.1, but the decomposition is much more involved than this scheme indicates. In the case of BCNU, sixteen products have been identified. Furthermore, recent evidence has been presented to support the contention that the intact CENU molecule rather than the diazohydroxide (15) may attack DNA[52-54]. Such a mechanism could explain a number of baffling observations such as the predominant formation of 7-(2-hydroxyethyl)guanine. The exact nature of the intermediate formed by such an attack, which is consonant with all of the experimental data at hand, is yet to be defined.

In spite of the vagueness of the details of the chemistry involved, the involvement of interstrand cross-linking of DNA in the biologic activity of the

CENUs has long been appreciated. Indeed, the cross-linked dinucleoside, 1-[N^3-deoxycytidyl]-2-[N^1-deoxyguanosinyl]ethane (**17**), has been isolated

17

from DNA exposed to BCNU[55]. It is probable that this structure is responsible for the interstrand cross-linking observed previously by physical methods[39,40]. It has been proposed that initial attack occurs at O^6 of deoxyguanosine followed by an internal cyclisation with N^1 of deoxyguanosine before secondary reaction with N^3 of deoxycytidine of the other DNA chain. This hypothesis is supported by established chemistry[56,57] and by the observation that the addition of rat liver O^6-alkylguanine-DNA alkyltransferase prevents formation of the cross-link[58]. Further, DNA from resistant cells treated with N-(2-chloroethyl)-N-nitrosourea contains much less of the cross-link[59]. Also human tumour cells can be sensitised to the CENUs by inactivation of the transferase with O^6-methylguanine[60,61] or agents that alkylate the O^6 of guanine, although CCNU and MeCCNU were relatively ineffective[60].

This elegant work notwithstanding, the true role of interstrand cross-linking in the activity of the CENUs is not unequivocally established, since FENUs, which cross-link much less than the CENUs, are just as active against the rodent leukaemias, and a number of methylnitrosoureas, which cannot cross-link, are quite active albeit less than the haloethyl compounds[3-5]. Also, although the cross-linking of DNA to protein is unlikely to be the major cause of the cytotoxicity of the CENUs, it might be a contributing factor[62]. Since nitrosoureas are known to interact extensively with cellular proteins themselves by carbamoylation as well as alkylation, there is reason to consider these reactions as potential cytotoxic events. There is evidence that protein modifications *per se* can modulate cytotoxicity or, alternatively, interactions with histone or nonhistone proteins could induce toxicity indirectly in several ways[63].

The kinetics of cell killing by the nitrosoureas, like that of other alkylating agents, is pseudo first-order since the drug is always present in large excess. There is a small difference in the responsiveness of resting and dividing L 1210 cells, with the dividing cells being two or three times as sensitive. Cells are essentially equisensitive in all phases of the cell cycle and cell killing is rapid[5]. This behaviour is in contrast to that of the antimetabolites such as 1-β-D-arabinofuranosylcytosine (ara-C), the cell killing by which is, above a certain minimum concentration, concentration independent. Ara-C is an S-phase specific agent requiring long exposure relative to BCNU for maximum cell kill. These differences are the basis for the schedule dependency of these agents in the treatment of mice inoculated with L 1210 leukaemia[64]. The nitrosoureas are most effective when given as a single dose, whereas ara-C must be given every 3 h to realise its potential. Other antimetabolites, such as methotrexate, 6-mercaptopurine and 5-fluorouracil, are maximally effective on a daily or every other day schedule.

4.5 Structure and latentiation

4.5.1 *Enzymatic activation*

Structural modifications of the type of nitrosoureas most tested and judged most active in biological systems, that is the N-(2-chloroethyl)-N-nitrosoureas, can result in compounds of altered chemical and biological properties. Those modifications that result in compounds from which biologically active forms can be released under suitable conditions constitute latentiation, a concept originally demonstrated with other types of drugs[65]. Latent forms are rendered biologically active by either enzymatic or chemical activation, or both. Examples of both types have been recognised among the nitrosoureas. A demonstration of latentiation lay dormant for a number of years because of an infelicitous selection, for synthesis and evaluation, of the piperazinedicarboxamide (**18**), whose inactivity against L 1210 leukaemia[66] and aqueous stability under physiological conditions[67] did not engender further interest.

Full substitution of the ureido moiety, which was seen as blocking the course of decomposition necessary for biological activity, usually resulted in stabilisation, but not always in inactivity. An effective latentiation of antineoplastic nitrosoureas was achieved, for example, when the fourth substituent was a propyl group as in N-(2-chloroethyl)-N'-methyl-N-nitroso-N'-propylurea (**19**), which was relatively stable in aqueous solution and active against L 1210 leukaemia only *in vivo*, a finding indicative of enzymatic activation[46,68,69]. This compound had the highest therapeutic index in a series that included methyl, 2-chloroethyl and cyclohexyl groups as R. Similar effects including activity against L 5222 leukaemia in rats were noted with the morpholinyl derivative (**20**) which was described as remarkably non-toxic[70].

Latentiation was also demonstrated in a study of (2-chloroethyl)-nitrosoureas derived from amino acid amides as carriers[71] as in the sarcosinamide derivative (**21**). This series was characterised by high activity against L 1210 leukaemia and very slow decomposition in buffered aqueous solution, affording sarcosinamide as the sole ninhydrin-positive product. Nevertheless, the possibility exists that *in vivo* decomposition involves enzymatic demethylation. The prolinamide derivative (**22**) was less anti-leukaemic than the N-methyl congeners such as (**21**), but almost as stable.

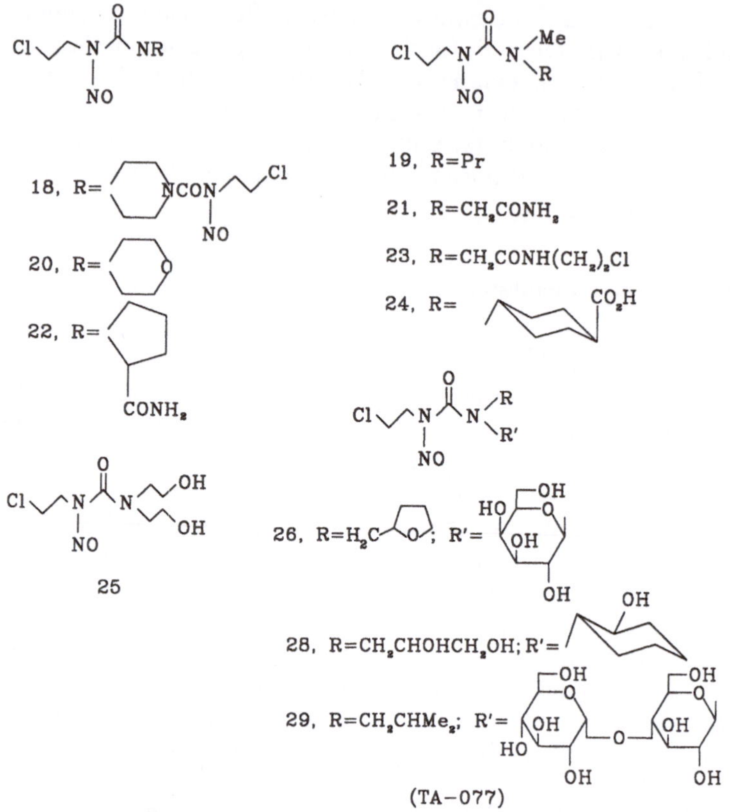

18, R=

19, R=Pr

20, R=

21, R=CH$_2$CONH$_2$

22, R=

23, R=CH$_2$CONH(CH$_2$)$_2$Cl

24, R=

25

26, R=H$_2$C–O ; R'=

28, R=CH$_2$CHOHCH$_2$OH; R'=

29, R=CH$_2$CHMe$_2$; R'=

(TA–077)

Similar results were observed in a series in which the amide nitrogen was substituted with a 2-chloroethyl group[72] as in the sarcosinamide derivative (23). The prospect of enhanced antileukaemic activity by introducing a 2-chloroethyl substituent in this series, however, is not great when seen in comparison with the inactivity of N-(2-chloroethyl)butaneamide against L 1210 leukaemia[73].

The effects of latentiation were extended to an experimental solid tumour when the N-methyl-N-cis-4-carboxylcyclohexyl derivative (24) afforded cures of murine Lewis Lung carcinoma although its in vitro half-life was several times that of the demethylated parent[67]. The reduced toxicity and enhanced stability of (24) made possible the use of the large doses required for maximal effectiveness, a characteristic already noted in some of the other enzymatically activated nitrosoureas mentioned.

4.5.2   Chemical activation

N-(2-Chloroethyl)-N',N'-bis(2-hydroxyethyl)-N-nitrosourea (25) and other congeners with suitably positioned hydroxyl groups were found to be unstable

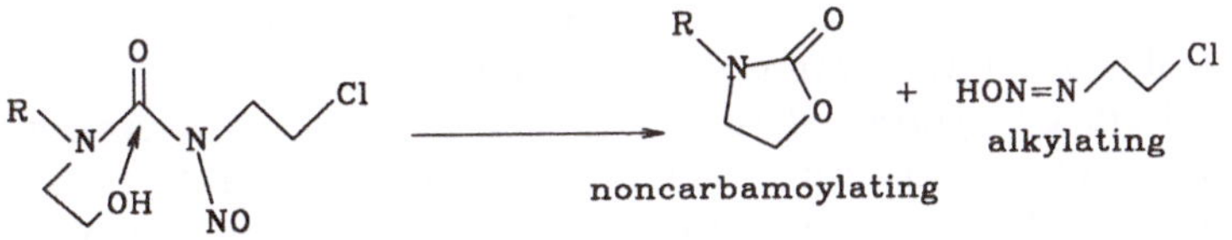

Figure 4.2 Chemical activation.

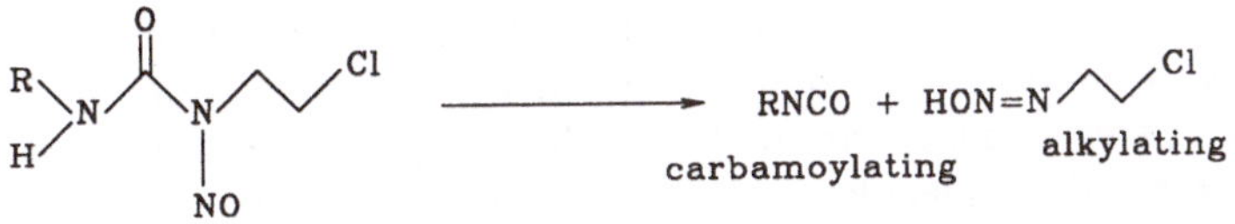

Figure 4.3 Normal decomposition.

in phosphate-buffered aqueous solution and quite active against L 1210 leukaemia without prior enzymatic activation[74]. These properties were said to be due to a chemical activation initiated by attack of a β-hydroxyl group on the carbonyl group, which results in N-substituted-oxazolidinone formation without the co-formation of a carbamoylating species (Figure 4.2). The many nitrosoureas that decompose in a normal fashion give a carbamoylating species in addition to the requisite alkylating species (Figure 4.3). Activity against Ehrlich ascites murine carcinoma distinguished the chemically activated compounds from those requiring enzymatic activation.

In subsequent extensions of the above finding[75,76], in search of reduced myelotoxicity and increased antitumour activity among water-soluble analogues related to chlorozotocin (7) and GANU (30), galacto- and arabinopyranosyl derivatives in general, and the galactopyranosyl derivative (26) and the arabinopyranosyl derivative (27) in particular, showed remarkable activity against L 1210 leukaemia. The activation of (27), for example, apparently involves the intracarbamoylation of the C-2 hydroxyl group (Figure 4.4).

Studies of positional isomers in the sugar series revealed that the rate of activation by a γ-hydroxyl group in C-6 isomers was significantly slower than by a β-hydroxyl group in C-2 and C-3 isomers[77-79]. All congeners in a series of

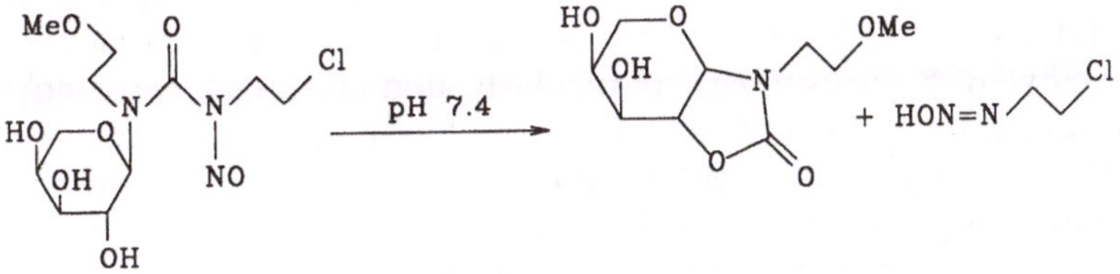

27

Figure 4.4 Activation of an arabinopyranosyl derivative.

derivatives of *trans*-2-hydroxycyclohexylamine were highly active against L 1210 leukaemia, but only those having two kinds of β-hydroxyl groups as in the N'-2,3-dihydroxypropyl derivative (28) were equally active against Ehrlich ascites carcinoma, suggesting differences in modes of activation[80]. Competition of R and R' hydroxyl groups in intracarbamoylation may be involved.

Numerous maltosyl congeners were curative of L 1210 leukaemia in the further modification of chemically activated nitrosoureas by disaccharide groups, whereas lactosyl and melbiosyl congeners were almost inactive[81]. In this series, the maltosyl derivative (29) emerged as the one of most interest on the basis of good activity against a spectrum of animal tumours including advanced Lewis Lung carcinoma[82,83]. This congener is described as a unique nitrosourea and is under clinical trials in Japan[84] as TA-077. A substantial loss of activity occurred when the hydroxyl groups of (29) were acetylated, thus supporting the requirement of a free hydroxyl group for activation. This disaccharide derivative was shown to be converted by maltase into the N'-(2-methylpropyl) derivative of GANU, which then can be non-enzymatically activated[85].

30, R=

R'=CH$_2$OH (GANU)
R'=H

31, R=

(ACNU)
•HCl

4.6 Structure and stability

Aqueous decomposition of many of the nitrosoureas that are known to be highly active against animal tumours gives an essential alkylating species and a non-essential carbamoylating species, possibly an isocyanate or a progenitor of it (see Figures 4.2 and 4.3). A nitrosourea can have antitumour activity in the absence or near absence of carbamoylating activity; for example: chlorozotocin (7), ACNU (31), and TA-077 (29), all involving intracarbamoylation. Carbamoylating nitrosoureas, particularly non-alkylating carbamoylating nitrosoureas, were found to be inhibitors of DNA repair[34] and may possibly be radiosynergists[35]. The essential requirement for such activity may be strong carbamoylating activity, which both alkylating and non-alkylating nitrosoureas may have, but the prototypical non-alkylating N,N'-dicyclohexyl-N-nitrosourea (DCyNU, 33) enhanced the cytotoxic effect of ionizing radiation on L 1210 leukaemia cells in culture to a greater extent than the alkylating nitrosoureas[35]. Compound (33) and its alkylating analogue CCNU (4) generate the same carbamoylating species, but differ widely in half-life under physiological conditions of pH and temperature (5.6 min, 53 min, respectively)[86]. In fact, a series of such congeners were characterised by instability, negligible alkylating activity, and good carbamoylating activity[87]. The *trans*-4-hydroxy derivative (39) of (33) provided some water solubility without significantly changing carbamoylating activity. Changes in ring size showed the cyclopropyl congener (40) had the shortest half-life (2.5 min) in phosphate buffer (pH 7.4) and 37°C and the highest carbamoylating activity (208% of that of (33)).

The effect of sterically hindered nitroso groups on stability as seen in the above-mentioned series of congeners is related to the rates of apparent first-order hydrolysis of a series of monosubstituted nitrosoureas (41) in phosphate buffer in which these rates decreased in the following order of R groups: cyclohexyl > benzyl > phenethyl, allyl > isobutyl > butyl, ethyl, methyl[88]. This order was viewed as reasonably consistent with the concept of inductive effects except for cyclohexyl and isobutyl, which can possibly assume more reactive configurations due to steric hindrance. Aromatic substituents were not included in the above comparison, but N-(2-chloroethyl)-N'-(4-chlorophenyl)-N-nitrosourea was too unstable for a determination of the octanol-water partition coefficient in the usual way[89]. The relative stabilities of DCyNU (33) and CCNU (4) were demonstrated in experiments in which equilibria favouring CCNU were observed when equimolar amounts of (33) and N-(2-chloroethyl)-N'-cyclohexylurea (32) or dicyclohexylurea (34) and CCNU were allowed to interact by transnitrosation in undiluted formic acid[66] (Figure 4.5).

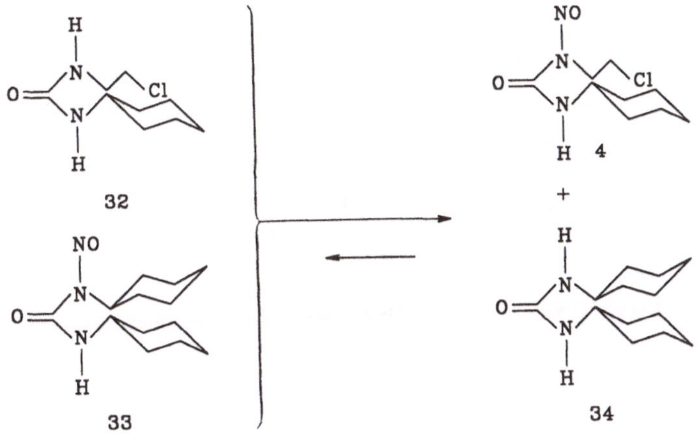

Figure 4.5 Transnitrosation.

Undiluted formic acid is a useful preparative medium for the selective formation of a nitrosourea in which the position of nitrosation can be controlled by steric hindrance; for example, the nitrosation of (32) in such a

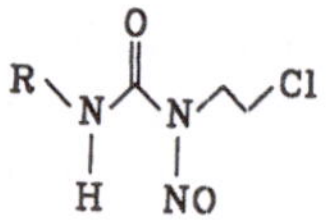

37, R=O$_2$N—⟨ ⟩—CO　(RFCNU)

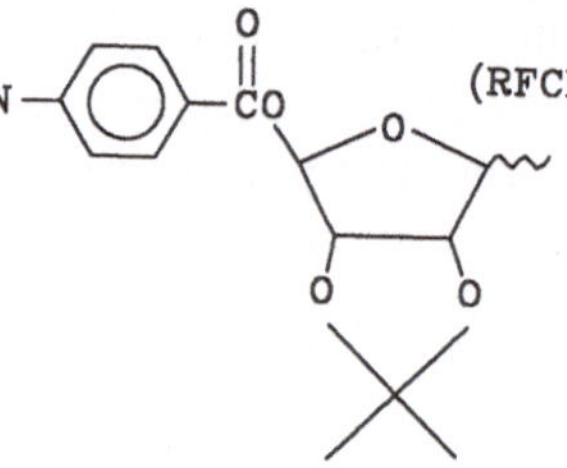

35, R=HO(CH$_2$)$_2$–(HECNU)

36, R= 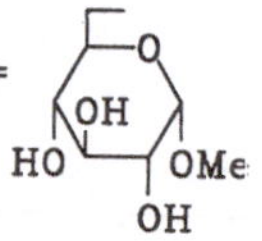　(MCNU)

38, R= 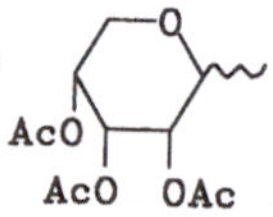　(RPCNU)

medium gives CCNU exclusively. Furthermore, the isomeric mixture obtained by nitrosation of (32) in aqueous formic acid converts to pure CCNU in undiluted formic acid as an example of irreversible transnitrosation[66]. Such transfer of the nitroso group as in iso-CCNU (42) could very well occur by abstraction as formyl nitrite (Figure 4.6).

The limitations of steric control were overcome in a number of one-step conversions of amines originally developed for the synthesis of streptozotocin (6) and its analogues[90] and (2-chloroethyl)nitrosoureas unattainable by last-

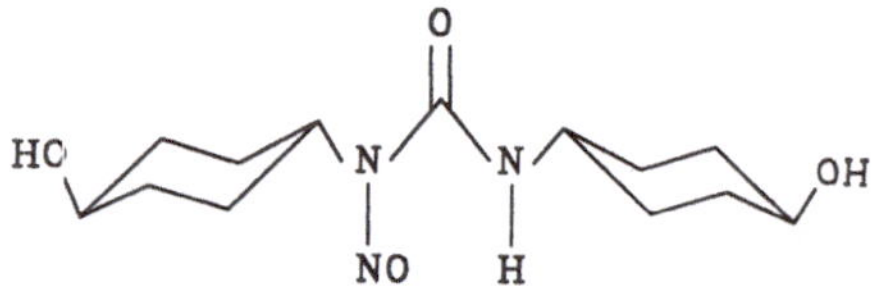

39

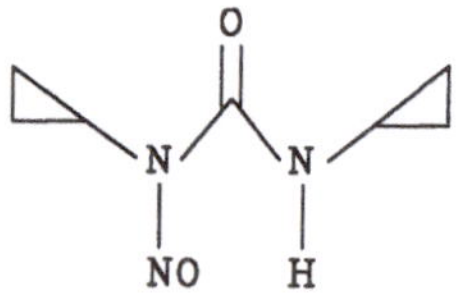

40

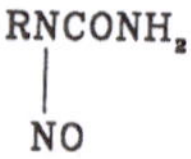

RNCONH$_2$
|
NO

41

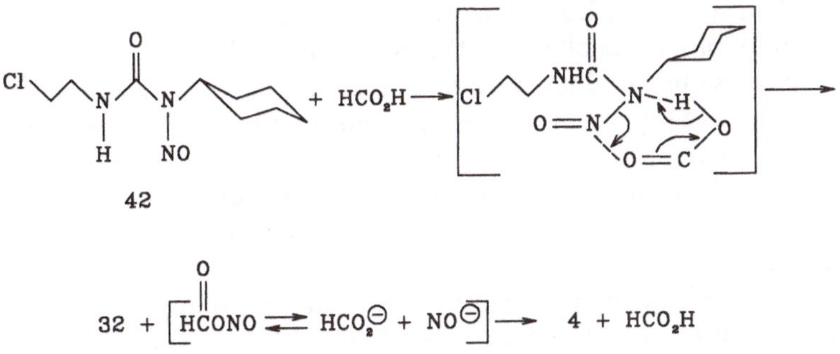

Figure 4.6 Irreversible transnitrosation via formyl nitrite.

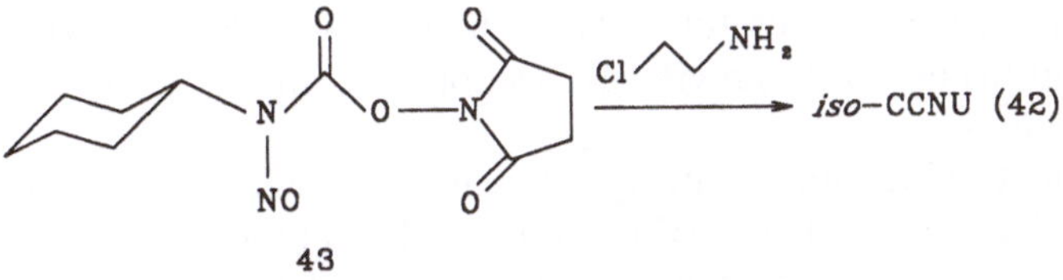

Figure 4.7 Synthesis of iso-CCNU.

step nitrosations[91]. These procedures involved displacements of azide ion from appropriate, potentially explosive nitrosocarbamic azides. Later variations involved the displacement of the corresponding 2-substituted phenol from 2-nitrophenyl or 2-cyanophenyl (2-chloroethyl)nitrosocarbamate by amino derivatives of glucose, which gave chlorozotocin and related compounds in high yield without the need of chromatographic purification[92]. Similar processes were used in the preparation of fluorozotocin (7, Cl = F) and other N-(2-fluoroethyl)-N-nitrosoureas[93]. The utility of this approach was further expanded by the use of other activated nitrosocarbamates including reagents derived from 2,4,5-trichlorophenol and N-hydroxy-succinimide[94]. The latter (43) enabled the synthesis of iso-CCNU (42), water-insoluble and inaccessible in pure form by direct nitrosation[95] (Figure 4.7). These methods maximise solubility differences between products and by-products in media that do not promote transnitrosation, thus negating the need for chromatographic purification.

4.6 Structure and metabolism

CCNU (4) is rapidly metabolised by microsomal material from rats and mice without loss of the cytotoxic nitrosoureido functions[96-98]. Five of the six possible monohydroxylated derivatives of CCNU were identified as metabolites in rats[96,99], whereas only two (trans-4 and cis-4) were identified in man[100]. All six were eventually synthesised[101,102] and their properties compared in a study whose data were consistent, for the most part, with attribution of biological effects to the major metabolites[101]. The therapeutic indexes of these metabolites were greater than that of CCNU itself against L 1210 leukaemia (i.p. and i.c.). It is apparent that hydroxylation can modify parameters such as protein binding, tissue distribution, and mode and rate of excretion[100]. Ring hydroxylation axial to the (2-chloroethyl)nitrosoureido group (*cis*-2, *trans*-3, *cis*-4) is favoured over equatorial hydroxylation (*trans*-2, *cis*-3, *trans*-4)[100].

In reversed-phase high-pressure liquid chromatography, elution of a synthetic mixture of monohydroxylated isomers occurred in the order of alternating diequatorial and axial-equatorial isomers; that is, in the order of *trans*-4, *cis*-4, *cis*-3, *trans*-3, *trans*-2 and *cis*-2[103]. Differences in retention times of these isomers reflect differences in log P values (P = octanol/water partition coefficients), which are useful in Hansch structure-activity correlations[104]. This relationship was previously established for a number of other nitrosoureas[89].

Attempts to block metabolic hydroxylation of CCNU for possible assessment of the role of hydroxylation by polydeuteration[105] and later by polyfluorination[106] fell somewhat short of expectations. Deuteration as in CCNU-d_{10} (**44**) did not have a significant effect on rate of hydroxylation, whereas the decafluoro–CCNU (**45**) was completely inactive against i.c.-implanted L 1210 leukaemia. The latter result was attributed to the observed rapid transformation to the imidazolidin-2-one (**47**) (Figure 4.8). It could also have been due to insolubility that placed its log P value outside the range calculated for biologic activity[104].

It is now apparent that certain intact nitrosoureas may undergo several rapid metabolic transformations that include cytochrome P-450-dependent

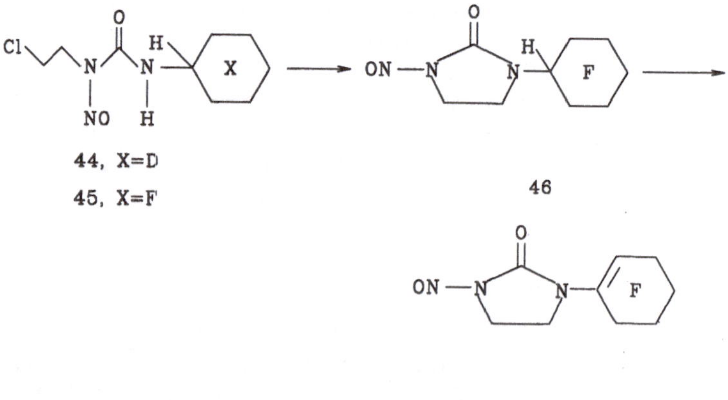

44, X=D

45, X=F

46

47

Figure 4.8 Transformation of CCNU-F_{10}.

monooxygenation to form monohydroxylated derivatives of the cyclohexyl ring of CCNU; the cyclohexyl ring, the methyl group, and the α-methylene group of MeCCNU[99]; and a denitrosation, particularly of BCNU[97] and MeCCNU[99]. The rapid *in vivo* transformations of nitrosoureas appear to depend on protein-mediated decomposition[107], cytochrome-*P*-450-dependent hydroxylation, and denitrosation in addition to chemical stability.

4.8   Water-soluble nitrosoureas and chlorozotocin analogues

A study of octanol water partition coefficients of selected nitrosoureas suggested that further study of hydrophilic nitrosoureas should lead to more potent and less toxic agents[104]. This and the observation that chlorozotocin was an anticancer agent of reduced bone marrow toxicity[109] accelerated the development of new synthetic methods and synthesis of water-soluble congeners of varied design[71,90-94]. Partition coefficients, however, have provided little utility in the design of nitrosoureas with greater selectivity because of the inseparability of antitumour activity and toxicity[108]. Partition coefficients, which sometimes are changed by metabolism[99], are but one factor that must be weighed against alkylating activity, biologic half-life, and perhaps carbamoylating activity[48].

The proposition that the marrow-sparing properties of chlorozotocin were related to water solubility and intracarbamoylation of hydroxyl groups was erased in a comparison of seven water-soluble nitrosoureas (**7, 30, 31, 48–50**) and lipophilic CCNU[110]. Only chlorozotocin (**7**) and GANU (**30**)[111] were not myelosuppressive in mice when given at an LD_{10} dose, but this property was clearly not mediated by water solubility alone. Congener (**48**) of *scyllo*-inositol and (**49**) of cyclopentanetetrol[112] and self-carbamoylating ACNU (**31**)[114], all water-soluble, exhibited marrow toxicity comparable to CCNU, as did the hydroxylated minor metabolites (**50**) and (**51**)[101]. Relative myelotoxicity did not correlate with alkylating activity, carbamoylating activity, and hydrophilicity, a conclusion later supported by a comparison that included the ethanolamine derivative HECNU (**35**)[114]. Decreased myelotoxicity appears to be mediated by the glucose carrier group in chlorozotocin and GANU and may be related to alkylation of sites on bone marrow chromatin different from those preferred by CCNU and ACNU[115]. In addition, in alkylations of DNA *in vitro*, chlorozotocin and GANU favoured alkylation of L 1210 leukaemia cells over murine bone marrow cells, whereas the opposite was true for myelotoxic CCNU and ACNU[116].

D-Glucose-mediated transport of chlorozotocin was considered a possibility because L-glucose is not actively transported in mammalian tissues, but a comparison of chlorozotocin and its L-isomer (**52**)[117] revealed no significant difference in activity against L 1210 leukaemia or toxicity. The isomers were identical in characterisation except for specific rotations, which varied with

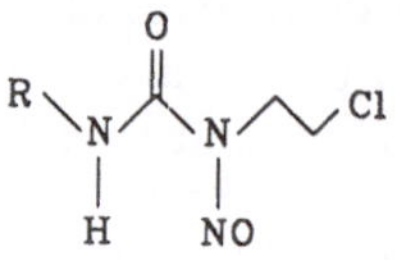

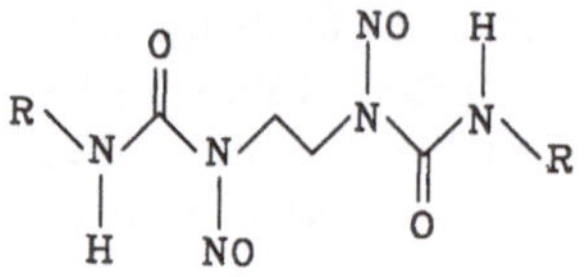

53, R=H (EBNU)

54, R=(CH$_2$)$_2$Cl

48, R=

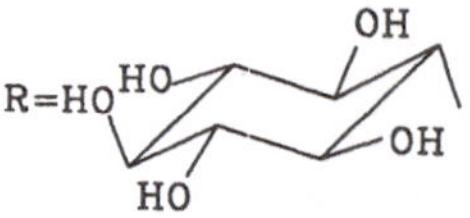

49, R=

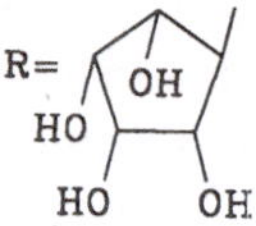

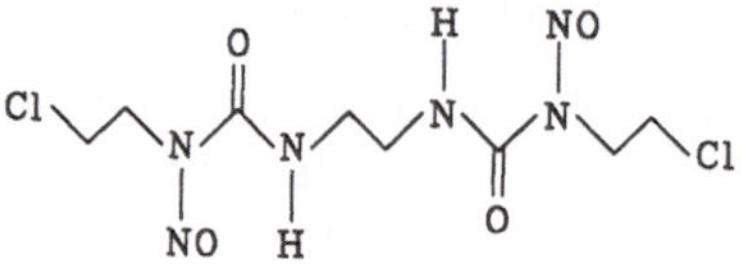

50, R=

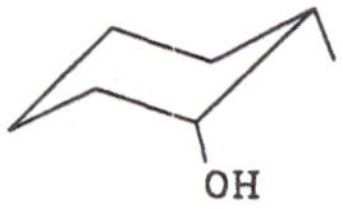

51, R=

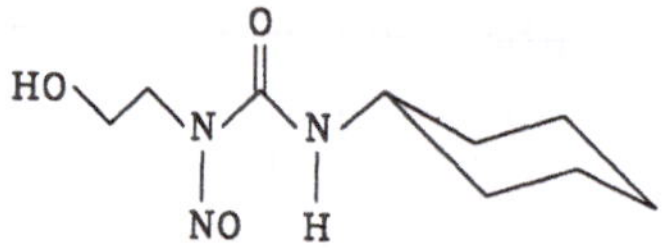

52, R=

55

56

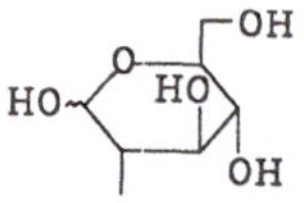

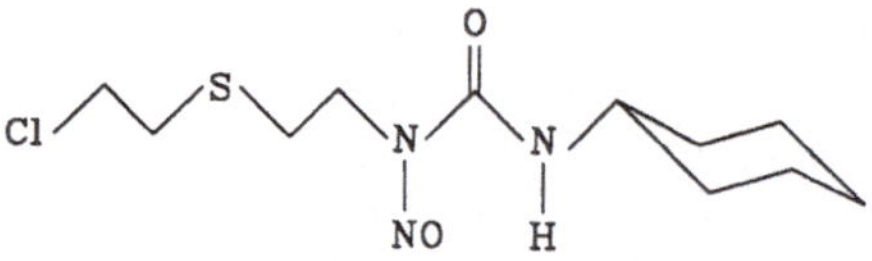

57

time due to mutarotation and aqueous decomposition, approaching a common value of opposite sign[117].

Four water-soluble nitrosoureas are in clinical trials in Japan[118]: ACNU (**31**), GANU (**30**), MCNU (**36**), and TA-077 (**29**). ACNU, the first of these in the clinic, being the hydrochloride of a weak base, is highly soluble at pH 3.3 but much less so at pH 6.8 and behaves as a lipid-soluble nitrosourea *in vivo*, showing activity against experimental brain tumours[119].

A series of ribose derivatives[120] produced two that have been used clinically in France: RFCNU (**37**) and RPCNU (**38**). Although rendered water-insoluble by O-substitution, they have been compared extensively with water-soluble chlorozotocin, HECNU (**35**), and the like. Against L 1210 leukaemia, RFCNU had the widest maximal effective dose range[121], showed low cumulative haematoxicity[122], and was judged not to be immunosuppressive[123], the role of the 4-nitrobenzoyl and isopropylidene protective groups being fortuitous. HECNU is about 30 times more water-soluble than BCNU, but retains lipophilic properties and crosses the blood-brain barrier effectively, and was as active as or more active than the most active nitrosoureas in nine of eleven rodent tumour models[124,125].

4.9 Structural modifications and special carrier groups: aspirations

A 2-chloroethyl or 2-fluoroethyl group on the nitroso nitrogen atom almost invariably characterises the most active nitrosoureas against experimental animal tumours[3,55]. One exception may be the ethylenebis(nitrosourea) EBNU (53), which effected cures of L 1210 leukaemia and surpassed by far other polymethylene series including guanidines and sulfonamides[126]. The terminal bis(2-chloroethyl) derivative (54) was inactive[127], but the congener (55) in which the nitroso and 2-chloroethyl groups are on the same nitrogen atom and whose structure suggests bifunctional carbamoylating activity was highly active against L 5222 leukaemia, as were polymethylene variants[91].

Further support for the superiority of the 2-chloroethyl group is found in the inactivities of the N-(2-hydroxy) analogue (56)[127] and of iso-CCNU (42)[95]. Crowding of the nitroso group in iso-CCNU imitates that in the non-alkylating carbamoylating DCyNU (36) of short half-life presented earlier (section 4.6). An attempt to increase the extent of DNA alkylation and cross-linking apparently succeeded in the case of the N-[[2-(2-chloroethyl)thio]ethyl] analogue (57) but activity against L 1210 leukaemia was only moderate[128]. The most effective variations have been those of the N'-substituent, which determines differences in solubility, stability, metabolism, transport, and the like. A distinctive variation of this type is the mixture (58, 59)

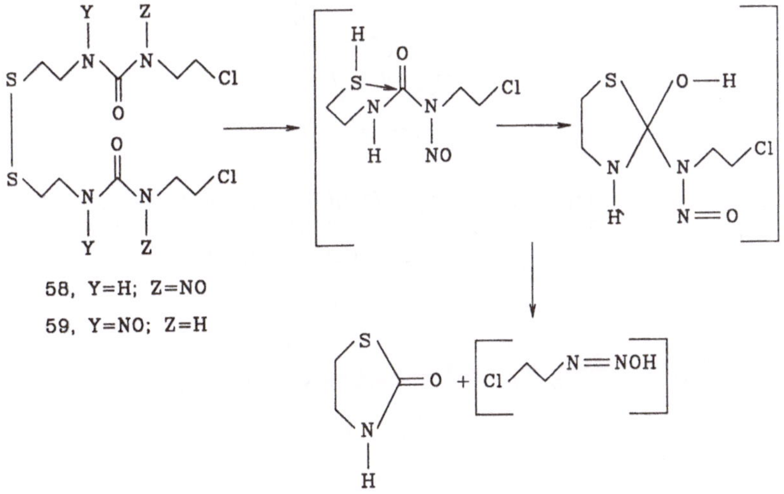

58, Y=H; Z=NO
59, Y=NO; Z=H

60

Figure 4.9 Proposed biotransformation of CNCC.

(CNCC) derived from cystamine, which was expected to undergo *in vitro* reduction and biotransformation to thiazolidin-2-one (**60**) and an alkylating species[121] (Figure 4.9). This variation, which precludes carbamoylation, afforded sufficient benefits in animal testing to warrant clinical trials[129], the mixture being more effective than pure (**59**).

Goals of better chemotherapeutic indexes have been pursued in the incorporation of N'-substituents derived from amino acids[130,131], amino acid esters[132], amino acid amides[71,72], and aminocarbohydrate-amino acid derivatives[133]. Nothing in the evaluation of, for example, the sarcosinamides (**21**, **23**), the serinamide (**61**) and the γ-aminobutyramide (**62**), all highly active against L 1210 leukaemia, suggests that biological properties associated with L-amino acids, such as active transport, played a special role in the observed results.

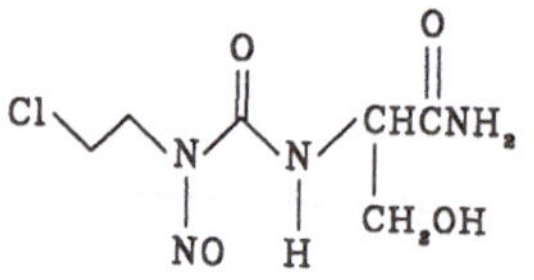

61

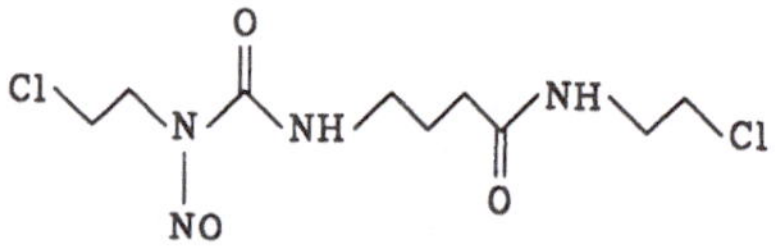

62

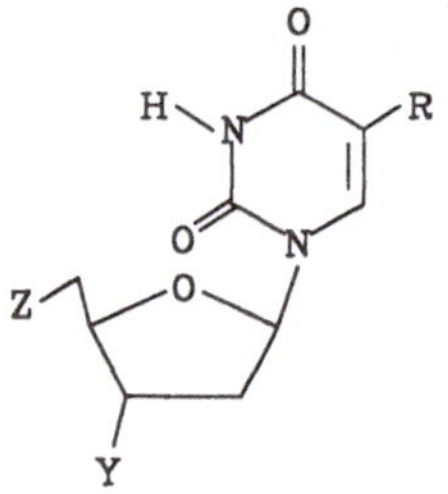

63, R=Me; Y=NHCON(NO)(CH$_2$)$_2$Cl; Z=OH
64, R=Me; Y=OH; Z=NHCON(NO)(CH$_2$)$_2$Cl
65, R=H; Y=OH; Z=NHCON(NO)(CH$_2$)$_2$Cl

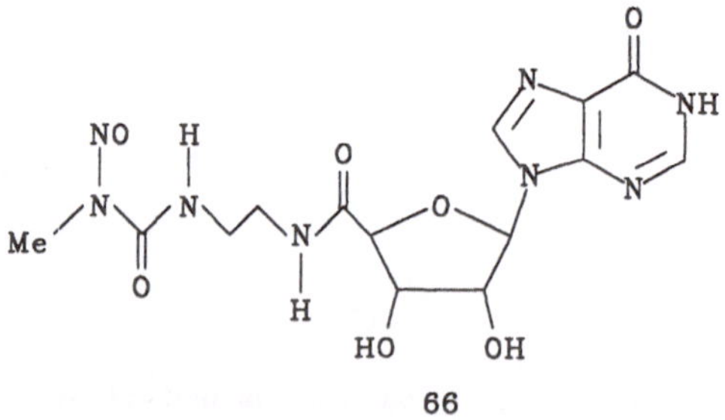

66

Combinations of nitrosoureas with nucleosides as in 3′-CTNU (63) and 5′-CTNU (64) have been studied in attempts to alter metabolism or uptake[134], cell growth inhibition caused by 3′-CTNU being reversible or preventable with thymidine, deoxyuridine, or deoxycytidine[135,136]. 5′-CTNU and its deoxyuridine counterpart (65) were also made in an effort to obtain active-site-directed enzyme inhibitors[137,138], some 30-day survivals being effected with (65). The similarity of the effects of (64) and (65) on DNA, RNA and protein synthesis in whole cells, which prevents cell division and leads to cell death, indicates that these compounds are active as CENUs rather than enzyme inactivators, a conclusion supported by the inactivity of the ethylnitroureido analogue of (64)[138]. A low level of biological activity attributable to unexpected stability was encountered with N-methyl-N-nitrosoureido derivatives like the inosine (66) and the adenine (67), with half-lives of hours instead of minutes[139,140].

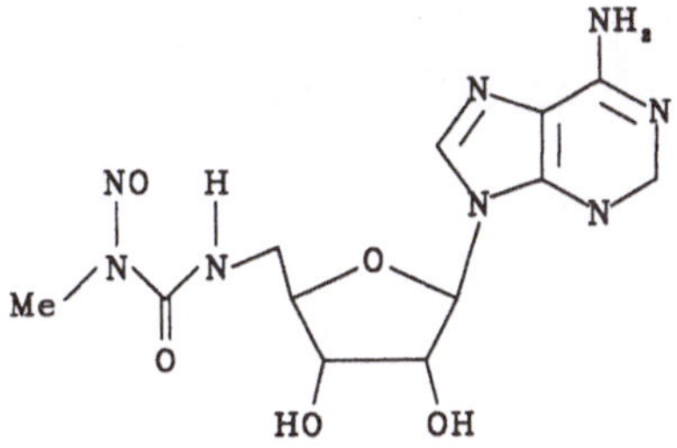

67

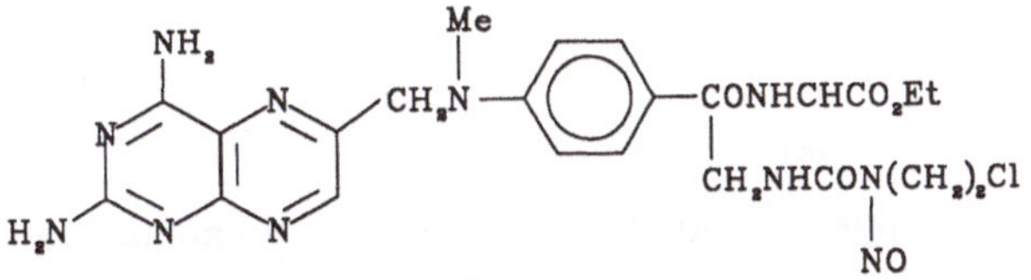

68

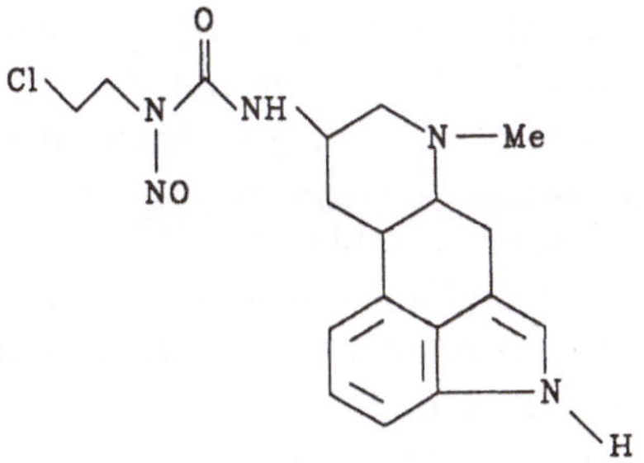

69

The appreciable activity against L 1210 leukaemia shown by the methotrexate analogue (68) did not appear to be attributable to the inhibition of dihydrofolate reductase[141].

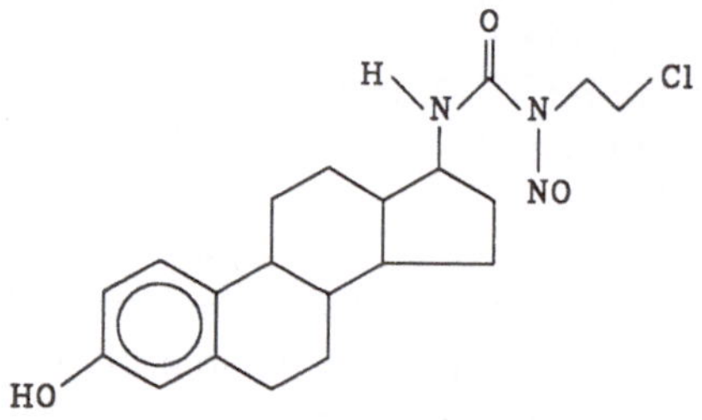

70

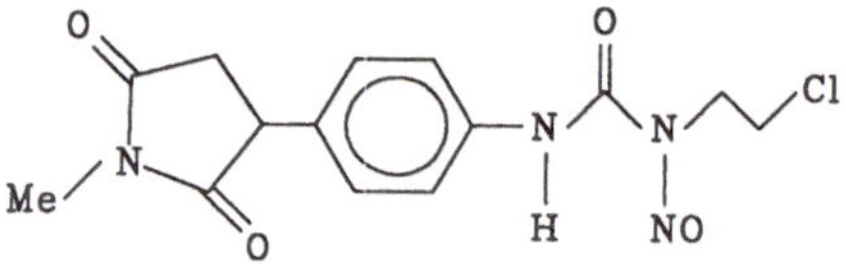

71

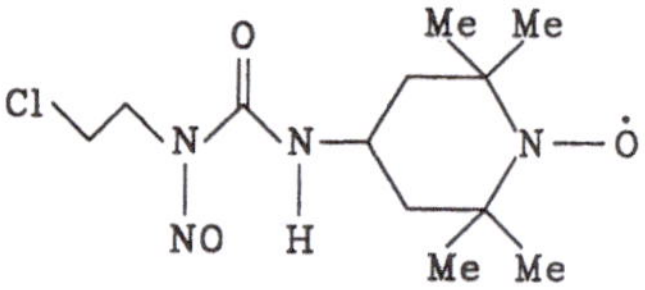

72

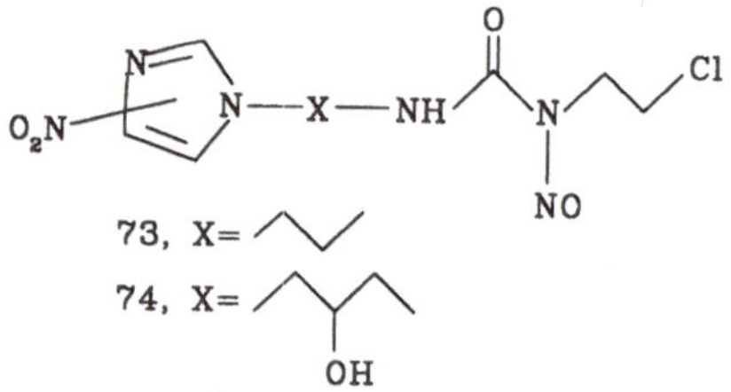

73, X=

74, X=

The nitrosoureas (69) and (70) derived from an ergot alkaloid and a steroidal hormone, respectively, were designed to reach tumours of specific organs[142,143]. The former which had only moderate prolactin-inhibiting activity might show specificity for a pituitary gland tumour. The steroidal derivative showed significant acitivity against a rat mammary tumour, but so did CCNU. Similar activity had been noted with the 3β-hydroxy-5α-androstan-17α-yl congener[144]. The phensuximide derivative (71) designed for transport to the central nervous system was not as active as MeCCNU or BCNU against an i.c. ependymoblastoma[145].

A spin-labelled nitrosourea (72) was inspired by observations of beneficial effects of such modifications on the toxicity and antitumour activity of other types of chemically reactive agents[146].

The chemosensitisation of hypoxic cells to CCNU by misonidazole and related radiation sensitisers[147,148], albeit only successful in cells deficient in

O^6-alkylguanine-DNA alkyltransferase, has led to the synthesis of CENUs tethered to a nitroimidazole radiosensitising unit (**73** and **74**)[149].

References

1. S. A. Schepartz, *Cancer Treat. Rep.* **60** (1976) 647.
2. H. E. Skipper, F. M. Schabel, Jr., M. W. Trader and J. R. Thomson, *Cancer Res.* **21** (1961) 1154.
3. J. A. Montgomery, *Cancer Treat. Rep.* **60** (1976) 651.
4. J. A. Montgomery, in *Chronicles of Drug Discovery*, Vol. 2, eds. J. Bindra and D. Lednicer, John Wiley, New York (1983) p. 189.
5. T. P. Johnston and J. A. Montgomery, *Cancer Treat. Rep.* **70** (1986) 13.
6. T. H. Wasserman, M. Slavik and S. K. Carter, *Cancer* **36** (1975) 1258.
7. F. M. Schabel, Jr., T. P. Johnston, G. S. McCaleb, J. A. Montgomery, W. R. Laster and H. E. Skipper, *Cancer Res.* **23** (1963) 725.
8. O. S. Selawry and H. H. Hansen, *Proc. Am. Assoc. Cancer Res.* **13** (1972) 46.
9. T. Anderson, V. T. DeVita and R. C. Young. *Cancer Treat. Rep.* **60** (1976) 761.
10. J. S. Kovach, C. G. Moertel, A. J. Schutt, R. G. Hahn and R. J. Reitemier, *Cancer* **33** (1974) 563.
11. H. O. Douglass, Jr., P. T. Lavin and C. G. Moertel, *Cancer Treat. Rep.* **60** (1976) 769.
12. C. G. Moertel, A. J. Schutt, R. G. Hahn and R. J. Reitemeier. *J. Natl. Cancer Inst.* **54** (1975) 69.
13. B. Fisher, N. Wolmark, H. Rockette, C. Redmond, M. Deutsch D. L. Wickerham, E. R. Fisher, R. Caplan, J. Jones, H. Lerner, P. Gordon, M. Feldman, A. Cruz, S. LeGault-Poisson, M. Wexler, W. Lawrence, A. Robidoux and other NSABP Investigators, *J. Natl. Cancer Inst.* **80** (1988) 21.
14. N. Wolmark, B. Fisher, H. Rockette, C. Redmond, D. L. Wickerham, E. R. Fisher, J. Jones, A. Glass, H. Lerner, W. Lawrence, D. Prager, M. Wexler, J. Evans, A. Cruz, N. Dimitrov, P. Jochimsen and other NSABP Investigators, *J. Natl. Cancer Inst.* **80** (1988) 30.
15. P. S. Schein, M. J. O'Connell, J. Blom, S. Hubbard, I. T. Magrath, P. A. Bergevin, P. H. Wiernik, J. L. Ziegler and V. T. DeVita, *Cancer* **34** (1974) 993.
16. P. Schein, R. Kahn, P. Gorden, S. Wells and V. T. DeVita, *Arch. Intern. Med.* **132** (1973) 555.
17. C. R. Kahn, A. G. Levy, J. D. Gardner, J. V. Miller, P. Gorden and P. S. Schein. *N. Engl. J. Med.* **292** (1975) 941.
18. L. E. Broder and S. K. Carter, *Ann. Intern. Med.* **79** (1973) 108.
19. C. Hansch, A. Leo, C. Schmidt, P. Y. C. Jow and J. A. Montgomery, *J. Med. Chem.* **23** (1980) 1095.
20. J. A. Montgomery, G. S. McCaleb, T. P. Johnston, J. G. Mayo and W. R. Laster, Jr., *J. Med. Chem.* **20** (1977) 291.
21. F. A. Schmid, G. M. Otter, G. C. Perri, D. J. Hutchison and F. S. Philips. *Cancer Res.* **43** (1983) 976.
22. T. P. Johnston, G. S. McCaleb and J. A. Montgomery. *J. Med. Chem.* **6** (1963) 669.
23. A. W. Prestayko, L. H. Baker, S. T. Crooke, S. K. Carter and P. S. Schein, eds., in *Nitrosoureas: Current Status and New Developments*, Academic Press, New York (1981).
24. T. Anderson, M.G. McMenamin and P. S. Schein, *Cancer Res.* **35** (1975) 761.
25. P. S. Schein, M. G. McMenamin and T. Anderson, *Cancer Res.* **33** (1973) 2005.
26. M. Aoshima and Y. Sakurai, *Gann* **68** (1977) 247.
27. L. C. Panasci, P. A. Fox and P. S. Schein, *Cancer Res.* **37** (1977) 3321.
28. P. A. Fox, L. C. Panasci and P. S. Schein, *Cancer Res.* **37** (1977) 783.
29. T. P. Johnston, G. S. McCaleb and J. A. Montgomery, *J. Med. Chem.* **18** (1975) 104.
30. D. Green, K. D. Tew, T. Hisamatsu and P. S. Schein, *Biochem. Pharmacol.* **31** (1982) 1671.
31. L. C. Panasci, D. Green and P. S. Schein, *J. Clin. Invest.* **64** (1979) 1103.
32. J. D. Ahlgren, D. C. Green, K. D. Tew and P. S. Schein, *Cancer Res.* **42** (1982) 2605.
33. A. Horikoshi, R. Smith, and M. J. Murphy, Jr., *Chemotherapy* **29** (1983) 135.
34. H. E. Kann, Jr., M. A. Schott and A. Petkas, *Cancer Res.* **40** (1980) 50.
35. H. E. Kann, Jr., B. A. Blumenstein, A. Petkas and M. A. Schott, *Cancer Res.* **40** (1980) 771.
36. L. H. Schmidt, Unpublished report to the National Cancer Institute, 1972.

37. R. A. Kramer and M. R. Boyd, *J. Pharm. Exp. Ther.* **227** (1983) 409.
38. R. A. Kramer and M. R. Boyd *Proc. Am. Assoc. Cancer Res.* **25** (1984) 380.
39. K. W. Kohn, *Cancer Res.* **37** (1977) 1450.
40. R. A. G. Ewig and K. W. Kohn, *Cancer Res.* **37** (1977) 2114.
41. R. K. Johnson, L. F. Faucette, I. Wodinsky and J. J. Clement *Proc. Am. Assoc. Cancer Res.* **23** (1982) 166.
42. J. A. Montgomery, *Prog. Cancer Res. Ther.* **28** (1983) 1.
43. K. A. Hyde, E. Acton, W. A. Skinner, L. Goodman, J. Greenberg and B. R. Baker, *J. Med Pharm. Chem.* **5** (1962) 1.
44. J. A. Montgomery, R. James, G. S. McCaleb and T. P. Johnston, *J. Med. Chem.* **10** (1967) 668.
45. J. A. Montgomery, R. James, G. S. McCaleb, M. C. Kirk and T. P. Johnston, *J. Med. Chem.* **18** (1975) 568.
46. M. Colvin, R. B. Brundrett, W. Cowens, I. Jardine and D. B. Ludlum, *Biochem. Pharmacol.* **25** (1976) 695.
47. G. P. Wheeler and S. Chumley, *J. Med. Chem.* **10** (1967) 259.
48. G. P. Wheeler, B. J. Bowdon, J. A. Grimsley and H. H. Lloyd, *Cancer Res.* **34** (1974) 194.
49. Y. F. Shealy, C. A. Krauth and W. R. Laster, Jr., *J. Med. Chem.* **27** (1984) 664.
50. J. K. Snyder and L. M. Stock, *J. Org. Chem.* **45** (1980) 1990.
51. J. W. Lown and S. M. S. Chauhan, *J. Org. Chem.* **46** (1981) 5309.
52. N. Buckley, *J. Org. Chem.* **52** (1987) 484.
53. W. P. Tong, K. W. Kohn and D. B. Ludlum, *Cancer Res.* **42** (1982) 4460.
54. S. Parker, M. C. Kirk, D. B. Ludlum, R. R. Koganty and J. W. Lown, *Biochem. Biophys. Res. Commun.* **139** (1986) 31.
55. W. P. Tong, M. C. Kirk and D. B. Ludlum, *Cancer Res.* **42** (1982) 3102.
56. R. W. Balsiger, A. L. Fikes, T. P. Johnston and J. A. Montgomery, *J. Org. Chem.* **26** (1961) 3446.
57. J. R. Piper, A. G. Laseter, T. P. Johnston and J. A. Montgomery, *J. Med. Chem.* **23** (1980) 1136.
58. D. B. Ludlum, J. R. Mehta and W. P Tong, *Cancer Res.* **46** (1986) 3353.
59. W. J. Bodell, K. Tokuda and D. B. Ludlum, *Cancer Res.* **48** (1988) 4489.
60. D. B. Yarosh, S. Hurst-Calderone, M. A. Babich and R. S. Day III, *Cancer Res.* **46** (1986) 1663.
61. M. E. Dolan, G. S. Young and A. E. Pegg, *Cancer Res.* **46** (1986) 4500.
62. G. P. Wheeler, Unpublished data.
63. K. D. Tew and P. S. Schein, *Handb. Exp. Pharmakol.* **72** (1984) 425.
64. H. E. Skipper, F. M. Schabel, Jr., L. B. Mellett, J. A. Montgomery, L. J. Wilkoff, H. H. Loyd and R. W. Brockman, *Cancer Chemother. Rep.* **54** (1970) 431.
65. N. J. Harper, *J. Med. Pharm. Chem.* **1** (1959) 467.
66. T. P. Johnston, G. S. McCaleb, P. S. Opliger and J. A. Montgomery, *J. Med. Chem.* **9** (1966) 892.
67. T. P. Johnston, G. S. McCaleb, W. C. Rose and J. A. Montgomery, *J. Med. Chem.* **27** (1984) 97.
68. W. Cowens, R. Brundrett and M. Colvin, *Proc. Am. Assoc. Cancer Res., ASCO* **16** (1975) 100.
69. R. B. Brundrett, J. W. Cowens and M. Colvin, *Proc. Am. Assoc. Cancer Res., ASCO* **17** (1976) 102.
70. M. Berger, W. J. Zeller, G. Eisenbrand, P. Z. Lin, M. Nakra and D. Schmahl, *Arzneim.-Forsch./Drug Res.* **32** (1982) 481.
71. T. Suami, T. Kato, H. Takino and T. Hisamatsu, *J. Med. Chem.* **25** (1982) 829.
72. M. Rodriguez, J.-L. Imbach and J. Martinez, *J. Med. Chem.* **27** (1984) 1222.
73. R. B. Brundrett, M. Colvin, E. H. White, J. McKee, P. E. Hartmann and D. L. Brown, *Cancer Res.* **39** (1979) 1328.
74. K. Tsujihara, M. Ozeki, T. Morikawa and Y. Arai, *Chem. Pharm. Bull.* **29** (1981) 2509.
75. K. Tsujihara, M. Ozeki, T. Morikawa, N. Taga, M. Miyazaki, M. Kawamori and Y. Arai. *Chem. Pharm. Bull.* **29** (1981) 3262.
76. T. Morikawa, M. Ozeki, N. Umino, M. Kawamori, Y. Arai and K. Tsujihara, *Chem. Pharm. Bull.* **30** (1982) 534.
77. T. Morikawa, M. Takeda, Y. Arai and K. Tsujihara, *Chem Pharm. Bull.* **30** (1982) 2386.
78. T. Morikawa, K. Tsujihara, M. Takeda and Y. Arai, *Chem. Pharm. Bull.* **30** (1982) 4365.

79. T. Morikawa, K. Tsujihara, M. Takeda and Y. Arai, *Chem. Pharm. Bull.* **31** (1983) 3924.
80. T. Morikawa, K. Tsujihara, M. Takeda and Y. Arai, *Chem. Pharm. Bull.* **31** (1983) 1646.
81. K. Tsujihara, M. Ozeki, T. Morikawa, M. Kawamori, Y. Akaike and Y. Arai, *J. Med. Chem.* **25** (1982) 441.
82. Y. Akaike, Y. Arai, H. Taguchi and H. Satoh, *Gann* **73** (1982) 480.
83. S. Fujimoto and M. Ogawa, *Jpn. J. Cancer Res.* (*Gann*) **76** (1985) 651.
84. B. Serrou, P. S. Schein and J. L. Imbach, eds., in *Nitrosoureas in Cancer Treatment.* Elsevier/North-Holland, Amsterdam (1981) p. 249.
85. S. Fujimoto and M. Ogawa, *Cancer Chemother. Pharmacol.* **9** (1982) 134.
86. G. P. Wheeler, In *Cancer Chemotherapy* (ACS Symposium Series No. 3), ed. A. C. Sartorelli, American Chemical Society, Washington, DC (1976) p. 87.
87. T. P. Johnston, G. P. Wheeler, G. S. McCaleb, B. J. Bowdon and J. A. Montgomery, in *Nitrosoureas in Cancer Treatment*, eds. B. Serrou, P. S. Schein and J.-L. Imbach, Elsevier, New York (1981) p. 139.
88. E. R. Garrett, S. Goto and J. F. Stubbins. *J. Pharm. Sci.* **54** (1965) 119.
89. W. J. Haggerty, Jr. and E. A. Murrell, *Res./ Dev.* **25** (1974) 30.
90. A. Meier, F. Stoos, D. Martin, G. Buyuk and E. Hardegger *Helv. Chim. Acta* **57** (1974) 2622.
91. G. Eisenbrand, H. H. Fiebig and W.J. Zeller, *Z. Krebsforsch Klin. Onkol.* **86** (1976) 279.
92. G. Kimura, U.S. Patent 4, 156,777 (1979).
93. T. P. Johnston, C. L. Kussner, R. L. Carter, J. L. Frye, N. R. Lomax, J. Plowman and V. L. Narayanan, *J. Med. Chem.* **27** (1984) 1422.
94. J. Martinez, J. Oiry, J.-L Imbach and F. Winternitz, *J. Med. Chem.* **25** (1982) 178.
95. J. Oiry, J. Martinez, J.-L. Imbach and F. Winternitz, *Eur. J. Med. Chem, Ther.* **16** (1981) 539.
96. H. E. May, R. Boose and D. J. Reed, *Biochem. Biophys. Res. Commun.* **57** (1974) 426.
97. D. L. Hill, M. C. Kirk and R. F. Struck, *Cancer Res.* **35** (1975) 296.
98. J. Hilton and M. D. Walker, *Biochem. Pharmacol.* **24** (1975) 2153.
99. D. J. Reed, in *Nitrosoureas: Current Status and New Developments*, eds. A.W. Prestayko, S. T. Crooke, L. H. Baker, S. K. Carter and P. S. Schein, Academic Press. New York (1981) p. 51.
100. J. Hilton and M. D. Walker, *Proc. Am. Assoc. Cancer Res.* **16** (1975) 103.
101. G. P. Wheeler, T. P. Johnston, B. J. Bowdon, G. S. McCaleb, D. L. Hill and J. A. Montgomery, *Biochem. Pharmacol.* **26** (1977) 2331.
102. T. P. Johnston, G. S. McCaleb and J. A. Montgomery, *J. Med. Chem.* **18** (1975) 634.
103. J. A. Montgomery, T. P. Johnston, H. J. Thomas, J. R. Piper and C. Temple, Jr. in *Advances in Chromatogrpahy*, Vol. 15, eds. J. C. Giddings, E. Gweshka, J. Cazes and P. R. Brown, Marcel Dekker, New York and Boston (1977) p. 169.
104. C. Hansch, N. Smith, R. Engle and H. Wood, *Cancer Chemother. Rep.* (Part I) **56** (1972) 443.
105. P. B. Farmer, A. B. Foster, M. Jarman, M. R. Oddy and D. J. Reed, *J. Med. Chem.* **21** (1978) 514.
106. A. B. Foster, M. Jarman, P. L. Coe, J. Sleigh and J. C. Tatlow, *J. Med. Chem.* **23** (1980) 1226.
107. R. J. Weinkam, T.-Y. J. Liu and H.-S. Lin, *Chem.-Biol. Interact.* **31** (1980) 167.
108. J. A. Montgomery, *Cancer Treat. Rep.* **60** (1976) 651.
109. T. Anderson, M. G. McMenamin and P. S. Schein, *Cancer Res.* **35** (1975) 761.
110. J. M. Heal, P. Fox and P. S. Schein, *Biochem. Pharmacol.* **28** (1979) 1301.
111. T. Machinami, S. Nishiyama, K. Kikuchi and T. Suami, *Bull. Chem. Soc. Japan* **48** (1975) 3763.
112. T. Suami, K. Tadano and W. T. Bradner, *J. Med. Chem.* **22** (1979) 314.
113. F. Shimuzu and M. Arakawa, *Gann* **66** (1975) 149.
114. R. Osieka, P. Glatte, R. Pannenbacker and C. G. Schmidt, *Cancer Chemother. Pharmacol.* **11** (1983) 147.
115. D. Green, K. D. Tew, T. Hisamatsu and P. S. Schein, *Biochem. Pharmacol.* **31** (1982) 1671.
116. D. Green, J. D. Ahlgren, and P. S. Schein, *Nitrosoureas in Cancer Treatment*, eds. B. Serrou, P.S. Schein and J.-L. Imbach, Elsevier North-Holland, Amsterdam (1981) p. 45.
117. T. P. Johnston, G. S. McCaleb, T. Anderson and D. S. Murinson, *J. Med. Chem.* **22** (1979) 597.
118. M. Ogawa, in *Nitrosoureas in Cancer Treatment*, eds. B. Serrou, P. S. Schein and J.-L. Imbach. Elsevier North-Holland, Amsterdam (1981) p. 249.
119. H. Hasegawa and W. R. Shapiro, *Proc. Am. Assoc. Cancer Res., ASCO* **19** (1978) 153.
120. J.-L. Montero and J.-L. Imbach, *C.R. Acad. Sci. Ser. C* **279** (1974) 809.
121. J.-L. Imbach, J. Martinez, J. Oiry, C. Bournt, E. Chenn, R. Maral and C. Mathe, *Nitrosoureas*

in Cancer Treatment eds. B. Serrou, P. S. Schein and J.-L. Imbach, Elsevier North-Holland, Amsterdam (1981) p. 123.

122. J. L. Rebischung, M. D. Gougeon, K. J. Mori, G. Lemaigre, G. Mathe and C. Jasmin, *Cancer Res.* **44** (1984) 503.

123. J.-L. Imbach, J.-L. Montero, A. Moruzzi, B. Serrou, E. Chenn, M. Hayat and G. Mathe, *Biomedicine* **23** (1975) 410.

124. G. Eisenbrand, M. Habs, W. J. Zeller, H. Fiebig, M. Berger, O. Zelesny and D. Schmahl, *Nitrosoureas in Cancer Treatment*, eds. B. Serrou, P. S. Schein and J.-L. Imbach, Elsevier North-Holland, Amsterdam (1981) p. 175.

125. F. Spreafico, S. Fillippeschi, P. Falantano, G. Eisenbrand, H. H. Fiebig, M. Habs, V. Zeller, M. Berger, D. Schmahl, L. M. VanPutten, T. Smink, E. Csqnyi and S. Somfai-Relle, in *Nitrosoureas: Current Status and New Development*, eds. A. W. Prestayko, L. H. Baker, S. T. Crooke, S. K. Carter and P. S. Schein, Academic Press, New York (1981) p. 27.

126. M. Miyahara, M. Nakadate, M. Miyahara, I. Suzuki, M. Ishidate, Jr. and S. Odashima, *Gann* **68** (1977) 573.

127. I. Baracu, G. Botez, R. Denes, V. Dobre, B. Ureche, T. Craescu, N. Voiculet and I. Niculescu-Duvăz, *Rev. Roum. Biochim.* **18** (1981) 175.

128. J. W. Lown, A. V. Joshua and L. W. McLaughlin, *J. Med. Chem.* **23** (1980) 798.

129. J. L. Imbach, Personal communication.

130. T. P. Johnston and G. S. McCaleb, Unpublished results.

131. W. J. Zeller, M. Berger, G. Eisenbrand, W. Tang and D. Schmahl, *Arzrneim-Forsch/Drug Res.* **32** (1982) 484.

132. J.-L. Montero, A. Leydet, G. Dewynter, A. Seguin, J.-M. Favre and J.-L. Imbach, *Eur. J. Med. Chem. Chim. Ther.* **17** (1982) 257.

133. T. Suami, T. Kato, and T. Hisamatsu, in *Nitrosoureas in Cancer Treatment*, eds. B. Serrou, P. S. Schein and J.-L. Imbach, Elsevier North-Holland, Amsterdam (1981) p. 97.

134. T.-S. Lin, P. H. Fischer, G. T. Shiau and W. H. Prusoff, *J. Med. Chem.* **21**, (1978) 130.

135. P. H. Fischer, T.-S. Lin, M. S. Chen and W. H. Prusoff, *Biochem. Pharmacol.* **28** (1979) 2973.

136. T.-S. Lin, P. H. Fischer, J. C. Marsh and W. H. Prusoff, *Cancer Res.* **42** (1982) 1624.

137. R. D. Elliott, R. W. Brockman and J. A. Montgomery, *J. Med. Chem.* **24** (1981) 350.

138. R. D. Elliott, H. J. Thomas, S. C. Shaddix, D. J. Adamson, R. W. Brockman, J. M. Riordan and J. A. Montgomery, *J. Med. Chem.* **31** (1988) 250.

139. J. A. Montgomery and H. J. Thomas, *J. Med. Chem.* **22** (1979) 1109.

140. J. A. Montgomery, H. J. Thomas, R. W. Brockman and G. P. Wheeler, *J. Med. Chem.* **24** (1981) 184.

141. J. R. Piper, G. S. McCaleb. J. A. Montgomery, F. A. Schmid and F. M. Sirotnak, *J. Med. Chem.* **28** (1985) 1016.

142. A. M. Crider, C. K. L. Lu, H. G. Floss, J. M. Cassady and J. A. Clemens, *J. Med. Chem.* **22** (1979) 32.

143. H.-Y. P. Lam, A. Begleiter, G. J. Goldenberg and C.-M. Wong, *J. Med. Chem.* **22** (1979) 200.

144. F. I. Carroll, A. Philip, J. T. Blackwell, D. J. Taylor and M. E. Wall, *J. Med. Chem.* **15** (1972) 1158.

145. A. M. Crider, T. M. Kolczynski and K. M. Yates, *J. Med. Chem.* **23** (1980) 324.

146. G. Sosnovsky and S. W. Li, *Life Sci.* **36** (1985) 1479.

147. P. Workman, *Cancer Treat. Rep.* **70** (1986) 1139.

148. R. T. Mulcahy, A. Carminati, J.-L. Barascut and J.-L. Imbach, *Cancer Res.* **48** (1988) 798.

149. A. Carminati, J.-L. Barascut, E. Chenut, C. Bourut, G. Mathe and J.-L. Imbach, *Anticancer Drug Design* **3** (1988) 57.

5 Triazenes

K. VAUGHAN

5.1 Introduction

This chapter reviews the chemistry of the antitumour triazenes, which are related by having the common feature of three contiguous nitrogen atoms in a non-cyclic situation with one double bond between N1 and N2, i.e. $-\underset{1}{N}=\underset{2}{N}-\underset{3}{N}\!\!<$

5-(3,3-Dimethyl-1-triazeno)imidazole-4-carboxamide(**1a**, DTIC, Dacarbazine, NSC 45388) is a drug used in the treatment of malignant melanoma, and is the single most active agent available for the control of this disease[1]. DTIC is not unique among dimethyltriazenes in having antitumour activity; the inhibition of the growth of the mouse sarcoma 180 by 1-aryl-3,3-dimethyltriazenes (**2a**) has been long established[2] and these aryldialkyltriazenes have been examined extensively in a continuing search for second-generation analogues of DTIC[3]. Proposed models for the mechanism of action of the dimethyltriazenes suggest that the drugs undergo metabolic *N*-demethylation to the monomethyltriazenes (**1b** and **2b**), via the *N*-hydroxymethyltriazenes (**1c** and **2c**)[4]. The cytotoxic action of the drugs is thought to arise from DNA methylation by the monomethyltriazene metabolite.

A recent development in this field of cancer drug design is the discovery of the imidazolotetrazinone (**3**) (Mitozolomide, NSC 353451), which is a novel broad spectrum antitumour agent[5,6]. Although mitozolomide itself is not an acyclic triazene, its mode of action has been shown to involve hydrolysis of the tetrazinone ring to afford the monochloroethyltriazeno-imidazolecarboxamide (**1d**, MCTIC).

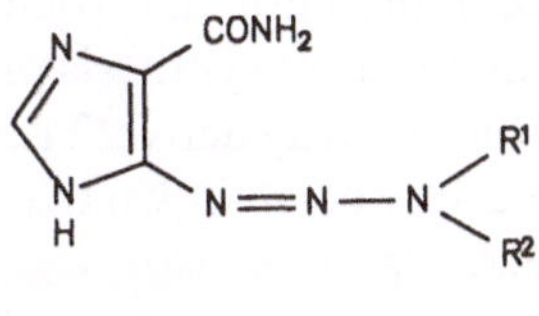

1

a) $R^1 = R^2 = Me$
b) $R^1 = Me$, $R^2 = H$
c) $R^1 = Me$, $R^2 = CH_2OH$
d) $R^1 = H$, $R^2 = CH_2CH_2Cl$

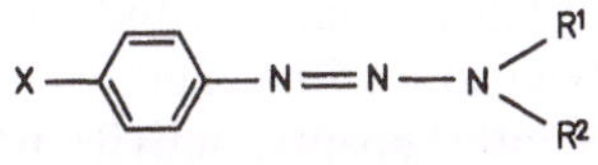

2

a) $R^1 = R^2 = Me$
b) $R^1 = Me$, $R^2 = H$
c) $R^1 = Me$, $R^2 = CH_2OH$

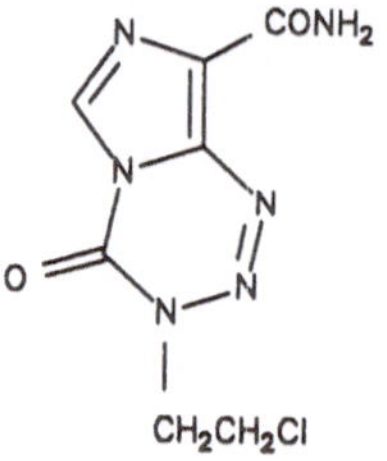

3

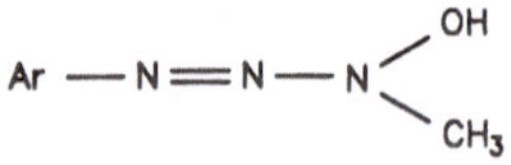

4

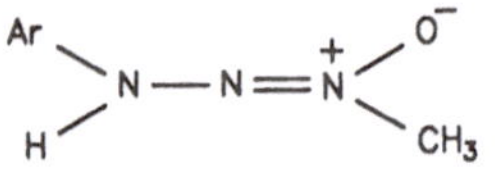

5

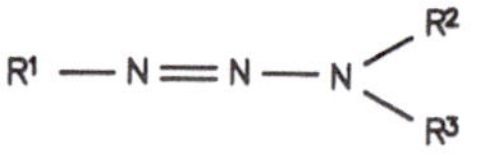

6

a) $R^1 = R^2 = Me$, $R^3 = H$
b) $R^1 = R^2 = R^3 = Me$
c) $R^1 = R^2 = Me$, $R^3 = Ac$
d) $R^1 = R^2 = Me$, $R^3 = CO_2 Et$
e) $R^1 = R^2 = Me$, $R^3 = PO(OEt)_2$
f) $R^1 = R^2 = Me$, $R^3 = CONHMe$
g) $R^1 = COOR$, $R^2 = R^3 = Me$
h) $R^1 = CONHR$, $R^2 = R^3 = Me$

Antitumour activity has also been associated with the 3-hydroxytriazene structure (4)[4], although an adequate explanation of this activity is still lacking. The recognition that (4) exists predominantly as the tautomeric N-oxide form (5) by Wilman[7] has opened up renewed interest in the structure and chemistry of these novel compounds and other triazene N-oxides.

Interest has also been re-awakened in the chemistry of 1,3-dialkyltriazenes and of the 1,3,3-trialkytriazenes (6), due to the work of Smith and Michejda[8]. These workers have developed much improved methods of synthesis of these truly 'alkyl' triazenes and have found them to be biologically active. Triazenes of type (6) are direct acting mutagens[9] and have recently been shown to possess chemotherapeutic activity by Smith and Michejda[10].

Thus, the thrust of this chapter is to review the chemistry of the di-methyltriazenes (1a and 2a), the proposed metabolites (1b, 1c, 2b and 2c), the triazene N-oxides and the triazenes of type (6). The chemistry of Mitozolomide (3) and of MCTIC (1d) are discussed in detail in Chapter 6; the chemistry of MCTIC will only be considered here when it is relevant to the behaviour of other open-chain triazenes.

5.2   1-Aryl- and 1-hetaryl-3,3-dialkyltriazenes

5.2.1   Synthesis and structure

1-Aryl-3,3-dialkyltriazenes are readily synthesised by the coupling of the appropriate diazonium ion (7) with a secondary amine[11].

$$\text{Ar–N}_2^+ + \text{R}^1\text{NHR}^2 \xrightarrow{\text{base}} \text{ArN=N–N}\begin{smallmatrix}\nearrow \text{R}^1 \\ \searrow \text{R}^2\end{smallmatrix} \qquad (5.1)$$

$$\quad\; 7 \qquad\quad 8$$

A large number of 3,3-dialkyltriazenes have been prepared by this general method; the scope of the method is limited only by the availability of aryl and heteroaryl diazonium salts and of the secondary amine (8). The diazonium salt can be prepared in solution and reacted *in situ* with the secondary amine; Kolar[12] has shown that the diazonium fluoroborate can be isolated and then reacted with the amine in a suitable solvent. A ^{13}C-NMR study by Axenrod et al.[13] of some 1-aryl-3,3-dimethyltriazenes has established a rotational barrier around the N2–N3 bond consistent with some considerable double bond character between N2 and N3. The results suggest that there is an important contribution to the ground state of the triazene molecule from the 1,3-dipolar resonance form shown in the following structure:

$$\text{Ar}^- - \overset{\cdot\cdot}{\text{N}} - \overset{\cdot\cdot}{\text{N}} = \overset{+}{\text{N}}\begin{smallmatrix}\nearrow \text{Me} \\ \searrow \text{Me}\end{smallmatrix}$$

5.2.2   Antitumour activity

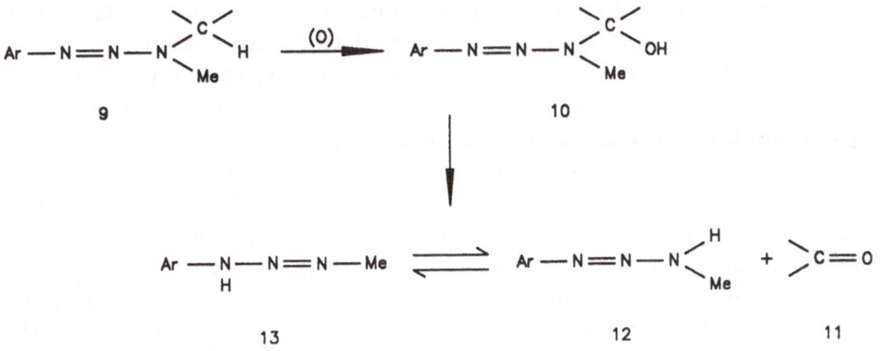

The antitumour activity of 1-aryl-3,3-dimethyltriazenes was first observed as inhibition of the growth of mouse sarcoma 180 by Clarke et al.[2]. Later experiments on the same tumour model showed that the presence of at least one methyl group at N-3 is essential for activity[14]. A more rigorous structure-activity correlation by Connors et al.[4] showed that only dialkyltriazenes that could undergo *N*-dealkylation to a monomethyltriazene were capable of inhibition of tumour growth, leading to the substructure (9) as the essential structural requirement for activity. The aryl group which may be a heterocycle, as in DTIC, is considered to be simply a carrying group for the triazene moiety.

These structure-activity correlations have led to a generally accepted hypothesis for the metabolic activation of the 3,3-dialkyltriazenes. Oxidative metabolism in the liver of one of the alkyl groups at N-3 produces the α-hydroxyalkyltriazene (10), followed by chemical loss of the α-hydroxyalkyl group as the appropriate carbonyl compound (11). The monoalkyltriazene (12) so formed tautomerises to the 3-aryl-1-alkyltriazene form (13)[15], which is inherently unstable and is capable of alkylation of DNA[16]. Indeed, it has recently been demonstrated that bioactivation of DTIC yields a methylating intermediate in rats. Intraperitoneal injection of [methyl-^{14}C] DTIC resulted in formation of 7-[methyl-^{14}C]guanine and O^6-[methyl-^{14}C]guanine in the DNA of selected tissues[17]. The acceptance of this hypothesis has led to considerable interest in the chemistry and pharmacology of the proposed triazene metabolites (10) and (12), which form a major segment of this chapter.

5.2.3 Metabolism

Although it is clear that the monomethyltriazene is the most likely active metabolite of a dialkyltriazene, an adequate explanation of this structural requirement is still lacking. The apparent uniqueness of a monomethyltriazene over other monoalkyltriazenes remains a mystery. All attempts to unravel this puzzle have invariably terminated in question marks rather than definitive answers.

An attempt to relate the conformation and electronic structure of 3,3-dimethyltriazenes to biological activity by Ramos and Pereira[18] led to the suggestion that antitumour activity correlates with the LUMO orbital energy, suggesting a charge-transfer mechanism during metabolism. Furthermore, it was suggested that metabolic demethylation should occur more readily with a planar conformation. This suggestion was corroborated by a study of the pH dependence of rotational barriers about the N(2)–N(3) bond in DTIC (1a); the rotational barrier at physiological pH is sufficient to prevent equilibration of the diastereotopic methyl groups[19]. One diastereotopic methyl group is preferentially oxidised; however, this study also suggested that the rotational barrier in the hydroxymethyl metabolite (1c) would not be sufficient to prevent equilibration between isomers of the metabolite.

Several groups have attempted to settle the monomethyltriazene controversy with in vitro metabolism experiments. The kinetics of the enzyme-catalysed N-demethylation of 3,3-dimethyltriazenes has been followed by measurement of the rate of formaldehyde release by Godin et al.[20]. The electronic character of the substituent in the 1-aryl-group has no effect at all on the rate of formaldehyde release, suggesting that electron density in the triazene moiety is not important in metabolic oxidation. A further surprising aspect of this study was the result obtained when a monomethyltriazene was incubated with liver microsomes; no formaldehyde was observed in the incubation mixture. However, it has been shown by Farina et al.[21] that a monomethyltriazene rapidly disappears during microsomal incubation; the identity of the product of monomethyltriazene metabolism was not established. The work of Farina et al.[21] also showed that biotransformation of a 3,3-dimethyltriazene gave products with selective toxicity towards TLX5 lymphoma cells. HPLC analysis of the products of this in vitro metabolism showed the presence of the monomethyltriazene but in an amount insufficient to account for the observed cytotoxicity.

The selectivity of monomethyltriazenes has been explored in experiments with two mouse tumours, the TLX5(S) lymphoma which is sensitive to dimethyltriazenes in vivo and the TLX5(R) lymphoma which has been made resistant in vitro[22]. The monomethyltriazenes are directly cytotoxic and, moreover, they are non-selectively cytotoxic, being approximately equipotent towards sensitive and resistant cells. An unselective cytotoxic species can also be generated by metabolic activation of a 3,3-diethyltriazene, which is surprising in view of the complete inactivity of diethyltriazenes in vivo[4]. These results[22] suggested that metabolism of dimethyltriazenes generates a mixture of selective and non-selective metabolites and that the selective species may not be monomethyltriazenes. This view is supported by a recent investigation of the effect of an inducer and an inhibitor of hepatic metabolism on the antitumour action of dimethyltriazenes; although treatment with the inducer or inhibitor did modify the N-demethylation of the dimethyltriazenes, there was no effect of either agent on their observed antitumour and antimetastatic activity[103].

A plausible explanation for the difference in the antitumour activity of 3,3-dimethyl- and 3,3-diethyltriazenes has arisen from a study by Farina et al.[23] of the metabolism of these two types of triazene. This study confirmed the inactivity of the 3,3-diethyltriazene on tumours against which the analogous dimethyltriazene is active. However, the diethyltriazene is much more toxic than the dimethyltriazene suggesting that the diethyltriazene is more prone to metabolism to a toxic intermediate. These results are in contradiction to those of Wilman et al.[24], who showed that a dimethyl- and a diethyltriazene undergo metabolism at equal rates.

Metabolic dealkylation has been used by Wilman et al.[24] to assess the best combination of N-3 methyl and N-3 alkyl groups in the homologous series $(NH_2CO \cdot C_6H_4 \cdot N=N-NMe-R)$ in relation to antitumour activity. It has been shown that the extent of oxidative metabolism of the dialkyltriazenes in vitro and their antitumour activity in vivo are dependent upon the hydrophobicity of the alkyl group. These studies have led to the selection of a second-generation triazene for clinical investigation, namely 1-(4-carboxyphenyl)-3,3-dimethyltriazene (CB 10–277) which can be readily metabolised to a cytotoxic species and can be readily formulated for i.v. administration. Clinical evaluation of CB 10-277 commenced under the auspices of the Cancer Research Campaign (U.K.) Phase I Committee in 1987[101].

5.2.4　Chemistry

Many of the side effects of DTIC when used as a drug in humans have been attributed to the photoinstability of the drug; it has been claimed by Baird and Willoughby[25] that troublesome venous pain and other side effects can be reduced by scrupulously shielding the drug from light at all times. The photochemistry of DTIC has been elucidated by Horton and Stevens[26], and it has been shown that the outcome of the photolytic degradation is pH dependent.

In the pH range 2–5.2, the product is a stable imidazole derivative, characterised as 4-carbamoylimidazolium-5-olate (16). When the drug is photodecomposed at clinically realistic concentrations (1 mg ml^{-1}) at pH 3–4, the solution goes pink and a dark maroon precipitate of the imidazolium olate (17) rapidly forms; this coupling product arises by interaction between the imidazolium olate (16) and diazo-imidazole carboxamide (14). Diazo-IC (14) is reported as one of the photoproducts of DTIC[27], and can undergo cyclisation to 2-azahypoxanthine (15), a secondary photoproduct, Photolysis of DTIC in ethanol gives a variety of products derived from C–H insertion reactions of the triplet carbene generated by loss of N_2 from diazo-IC (14) according to Kang and Schecter[28].

3,3-Dialkyltriazenes have some novel chemistry which has been applied in organic synthesis. For example, the reaction of a dialkyltriazene with hydrogen fluoride has been developed by Rosenfeld and Widdowson[29] as a mild and efficient method for aromatic fluorination

$$ArN{=}N{-}NR_2 \xrightarrow{\text{HF}} Ar{-}F + R_2NH + N_2 \qquad (5.2)$$

$$(97\%)$$

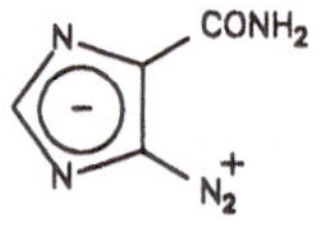

14

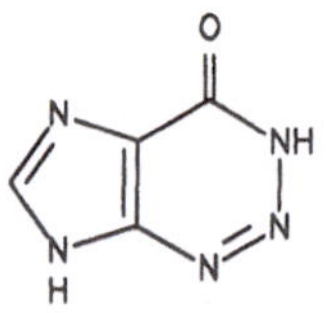

15

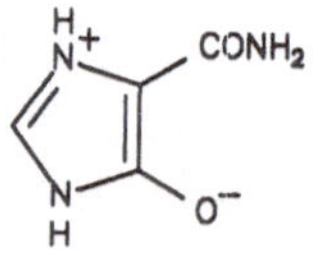

16

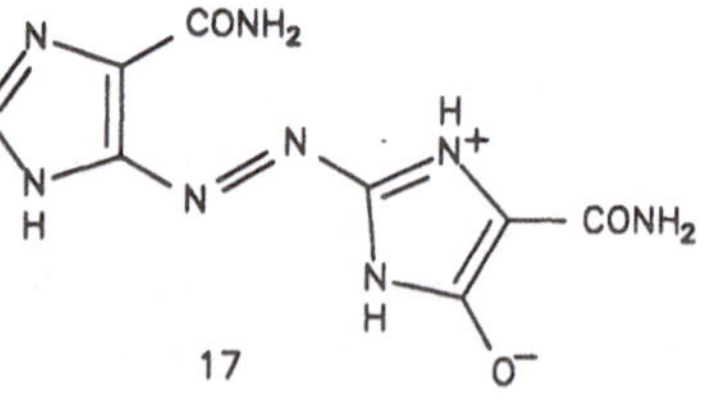

17

Extension of this approach has been applied to the iodination of aromatic systems; the reaction is catalysed by a cation exchange resin (hydrogen form)[30]. Because of its regiospecificity, this reaction is claimed to be particularly applicable to the preparation of aryl radioiodides for use in medicinal diagnosis.

$$ArN=N-NEt_2 \xrightarrow[\text{resin(H}^+\text{)}]{\text{*I}^-} Ar-\text{*I} + N_2 + HNEt_2 \qquad (5.3)$$

A novel reaction of dialkyltriazenes, which has been used as a probe for estimating electron density in the triazene moiety, is methylation by trimethyl-oxonium tetrafluoroborate affording the novel triazenium salt (18). Intra-molecular modification of this reaction gives the cyclic triazenium salt (19)[31].

18   19

Oxidation of 1-aryl-3,3-dimethyltriazenes with Bu^{t}OOH in the presence of V_2O_5 gives the 1-aryl-3-formyl-3-methyltriazenes, ArN=N-NMe-CHO, which have been considered as potential metabolites of the 1-aryl-3,3-dimethyltriazenes[32]. Although there is as yet no metabolic evidence for the formation of a formyltriazene from a dimethyltriazene *in vivo*, it has recently been shown[102] that biomimetic oxidation of dialkyltriazenes by the cytochrome P-450 model oxygenase system, tetraphenyl-porphyrinatomanganese(III) chloride-iodosobenzene, affords a synthetic route to N-alkyl-N-formyltriazenes. Formation of the N-formyl-triazene was preceded by formation of the N-hydroxymethyl derivative (2c), which has also been shown to yield the N-alkyl-N-formyltriazene by biomimetic oxidation. Furthermore, the N-alkyl-N-formyltriazenes are stable to further oxidation using this oxidising system, which also failed to oxidise the unsubstituted N-formyltriazene (ArN=N-NHCHO).

Reaction of the N-alkyl-N-formyltriazenes with nitrogen nucleophiles has been shown[102] to afford the monomethyltriazene, viz.

$$ArN=N-NMeCHO + R_2NH \rightarrow ArN=N-NHMe + R_2NCHO$$

and thus it appears quite likely that the N-alkyl-N-formyltriazenes may play an important role in the cytotoxicity of dimethyltriazenes.

The unsubstituted N-formyltriazenes have been the centre of a recent controversy concerning the possible existence of 'diazoisocyanides', $Ar-N=N-\overset{+}{N}\equiv\overline{C}$: Ignasiak *et al.*[33] reported that formamide couples in ethereal solution with diazonium ions to give stable 1-aryl-3-formyltriazenes

$$Ar-N_2^+ + NH_2CHO \rightarrow Ar-N=N-NH-CHO \qquad (5.4)$$

Other workers[34] have had difficulty reproducing these results. Further-more, Ignasiak and co-workers claimed in the same paper that dehydration of the 3-formyltriazene by thionyl chloride in pyridine affords the novel diazoisocyanide. However, recent work by Butler et al.[35] has shown that the reaction of the 3-formyltriazene with $SOCl_2$ in pyridine affords the aryl azide, $Ar-N=\overset{+}{N}=\bar{N}$. They conclude that diazoisocyanides are presently unknown and that they may not exist at all!

5.3　3-Alkyl-1-aryltriazenes

5.3.1　Synthesis

Because of their central role in the presumed metabolism of the 1-aryl-3,3-dialkyltriazenes, the 3-alkyl-1-aryltriazenes ('monoalkyltriazenes') have at-tracted much attention[36]. Synthesis of a monoalkyltriazene by diazonium coupling with a primary alkylamine (equation (5.5)) is not as straightforward as it would appear

$$ArN_2^+ + RNH_2 \rightarrow ArN=N-NHR \xrightarrow{ArN_2^+} ArN=N-\overset{\overset{\displaystyle R}{|}}{N}-N=N-Ar \quad (5.5)$$

20

Formation of the monoalkyltriazene is frequently accompanied by formation of a pentaazadiene (**20**), the product of further diazonium coupling. Penta-azadiene formation can be minimised or eliminated by the incorporation of strongly electron withdrawing groups in the aryl moiety[37]. An alternative general synthesis developed by Dimroth[38] is the reaction of an aryl azide with a Grignard reagent

$$Ar-N_3 + R-MgBr \rightarrow \xrightarrow{H^+} Ar-N=N-NHR \quad (5.6)$$

Of only specific application to 1-aryl-3-trifluoromethyltriazenes is the reaction of an arylhydrazine with nitrosotrifluoromethane[39]

$$ArNHNH_2 + CF_3NO \rightarrow ArNH-N=N-CF_3 \rightleftharpoons ArN=N-NHCF_3$$

5.3.2　Structure

The presence of a potentially mobile hydrogen atom at N-3 gives the monoalkyltriazene the potential for tautomerisation (**12**⇌**13**) not possible for 3,3-dialkyltriazenes. The tautomers (**12** and **13**) can be detected by NMR spectroscopy of solutions of the monoalkyltriazene[15]. Low temperature NMR spectra clearly show separate signals for the N−Me protons in the tautomers, a low field singlet arising from the N−Me group in the unconjugated tautomer (**13**) and a high field doublet from the N−Me group of (**12**). The relative intensity of the two N−Me signals provides a measure of the proportions of

(12) and (13) in the tautomeric mixture and reveals the effect of substituents in the aryl group on the tautomerism. Triazenes with electron donating groups in Ar favour the conjugated tautomer, whereas those with electron withdrawing groups show equal parts of both tautomers. Those triazenes with an *ortho* substituent capable of hydrogen bonding exhibit complete preference for the intramolecularly hydrogen-bonded form (e.g. 21).

21

A further study of the effect of solvents on the tautomerism in monoalkyl-triazenes by Hooper and Vaughan[40] corroborates these conclusions. An X-ray crystal structure determination of 1-*p*-tolyl-3-methyltriazene by Randall *et al.*[41] shows that tautomer (12) is the preferred form in the solid state, showing a similarity of structural behaviour of this compound in solution and in the solid state.

5.3.3 Chemistry

Thermal decomposition of 3-methyl-1-*p*-tolyltriazene in tetrachloroethene results in the formation of products rationalised in terms of homolytic breakdown of both tautomers[42]

$$\text{Ar-NH-N=N-CH}_3 \rightarrow \text{Ar}\dot{\text{N}}\text{H} \quad \dot{\text{N}}\text{=N-CH}_3 \rightarrow \text{ArNHCH}_3 + \text{N}_2 \quad (5.7)$$
$$5\%$$
$$\downarrow$$
$$\text{ArNH}_2 + \text{N}_2 + \dot{\text{C}}\text{H}_3$$
$$42\%$$

$$\text{ArN=N-NHCH}_3 \rightarrow \text{ArN=}\dot{\text{N}} \quad \dot{\text{N}}\text{HCH}_3 \rightarrow \text{Ar}\dot{\text{·}} + \text{N}_2 + \dot{\text{N}}\text{HCH}_3 \quad (5.8)$$
$$\downarrow$$
$$\text{ArCl}\,(10\%)$$

Significantly, the major products of thermolysis come from fragmentation of the unconjugated tautomer, which appears to be the more reactive form. Similar reactivity is shown in the reaction of monoalkyltriazenes with positive halogen compounds[43]. (equation (5.9)).

Of obvious interest to the pharmacology of monoalkyltriazenes is their stability in aqueous media and several accounts of the kinetics of decompo-

$$\text{Ar} - \underset{\text{H}}{\text{N}} - \text{N} \equiv \text{N} - \text{R} \; + \; \underset{}{>}\text{N} - \text{X} \longrightarrow \text{Ar} - \underset{\underset{\text{X}}{|}}{\text{N}} - \text{N} \equiv \text{N} - \text{R} \; + \; \underset{}{>}\text{N} - \text{H}$$

$$\text{ArNH}_2 \longleftarrow \text{Ar} - \ddot{\text{N}}: \xrightarrow{\times 2} \text{Ar} - \text{N} \equiv \text{N} - \text{Ar}$$

36% 62%

(5.9)

sition have appeared. In general, these reports support a protolysis mechanism resulting in fragmentation to the arylamine and a carbocation

$$\text{ArNH-N=N-R} \xrightarrow{\text{H}^+} \text{ArNH}_2 + \text{N}_2 + \text{R}^+ \tag{5.10}$$

More than one investigation[44] has shown that electron withdrawing groups in the aromatic ring slow down the hydrolysis. The presence of surfactants, which might mimic the micellar environment of cellular lipids, has a mixed effect on the rate of hydrolysis. A cationic surfactant (e.g. $\text{RNMe}_3^+ \text{Br}^-$) causes a rate decrease at all pH values studied, whereas an anionic surfactant ($\text{ROSO}_3^- \text{Na}^+$) enhances acid-catalysed hydrolysis but decreases the observed rate constants in the pH-independent region[45].

An intriguing application of this reaction is to use the triazene as a percursor of a 'hot' carbocation which can cause active-site directed irreversible inhibition of certain enzymes in *Escherichia coli*[46,47]. This is achieved by incorporating a β-D-galactopyranosylmethyl group as the R group; the β-D-galactopyranosylmethyl cation so generated specifically inhibits the enzyme β-galactosidase and it has been shown[48] that alkylation of methionine-500 of the lac Z β-galactosidase is the cause of the inhibition. The mechanism of decomposition of these galactopyranosylmethyltriazenes has been studied and reported to involve unimolecular heterolysis of the unconjugated tautomer (ArNH–N = N–R) (equation (5.10))[49]. It is concluded that gly-cosylmethyl (aryl) triazenes are active-site directed irreversible inhibitors of glycosidases, for which both substrate and product stereochemistries at the anomeric centre are the same as that of the reagent.

5.3.4 *Monoalkyltriazenes as heterocyclic synthons*

1-Aryl-3-alkyltriazenes with appropriate substituents in the aryl and alkyl groups (X or Y) have been used as precursors for a variety of heterocyclic systems (Scheme 5.1). Interaction of a reactive *ortho*-substituent (X) in the aryl group with N-3 of the triazene chain, path (a), leads to 1,2,3-triazine derivatives such as, in the case of X = acyl, benzotriazin-4-ols[50], which in turn can undergo dehydration to 4-methlyenebenzo-1,2,3-triazines[51,52]. Alternatively

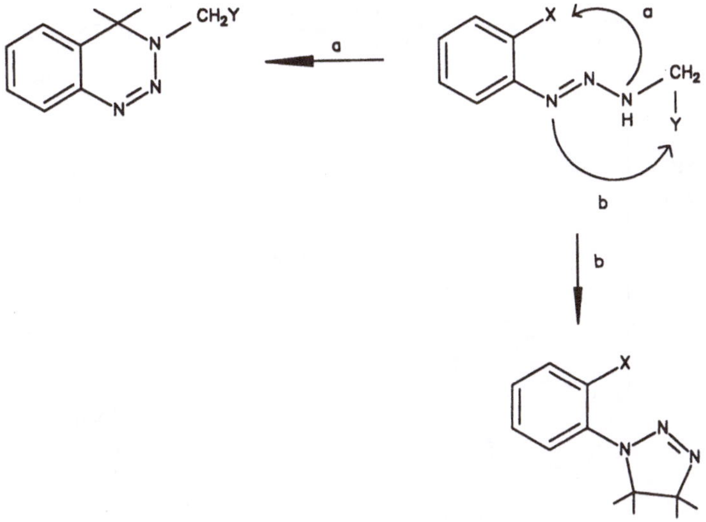

Scheme 5.1

if the alkyl group at N-3 has a reactive substituent, Y, then interaction with N-1 of the triazene chain, path (b), leads to 1,2,3-triazole derivatives. Thus 3-cyanomethyltriazenes (**22a** ⇌ **22b**) cyclise readily to give the 5-aminotriazoles (**23**) which undergo facile Dimroth-type rearrangement to the 5-arylamino-triazoles (**24**)[53].

An analogous reaction takes place when the *N*-arylazoglycinate (**25**) is treated with alcoholic potassium hydroxide[54]. Deprotonation of the triazene affords the triazenido anion (**26**) which cyclises to give the potassium triazolate (**27**). Careful acidification of solutions of (**27**) with acetic acid affords the free 5-hydroxy-1-aryltriazole (**28**). The acetate derivatives (**30**) of (**28**) are prepared

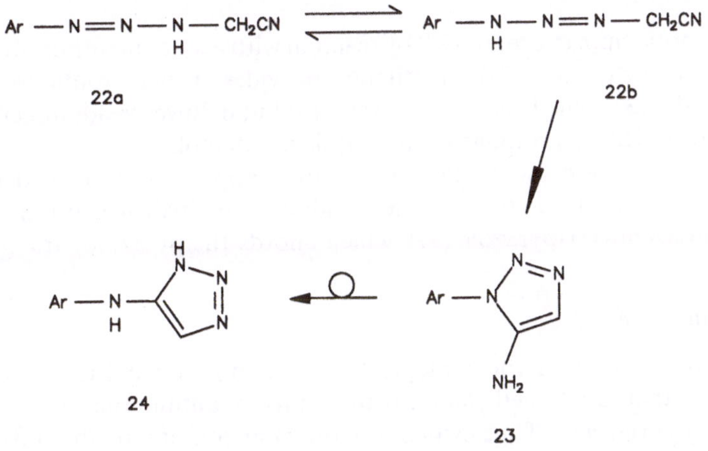

(R¹ = glycosyl)

from the potassium triazolate (**27**) by reaction with acetic anhydride. Refluxing the 5-hydroxytriazole (**28**) in ethanol provides a new route to the α-diazoamides (**29**), which can also be obtained but in lower yields directly from the triazene (**25**) by prolonged thermolysis in ethanol.

A novel variation on the triazene synthon approach, developed by Ege *et al.*[55], is the lead tetraacetate oxidation (dehydrogenation) of the glycosyltriazeno-1H-pyrazole (**31**), which affords the pyrazolotetrazole (**32**).

5.3.5 *Biochemistry*

A central question in the biological role of monomethyltriazenes is the hypothesis that DNA methylation is the source of antitumour activity of the 3,3-dimethyltriazenes. Thus evidence of direct action of a methyltriazene on

DNA is crucial to this hypothesis. One indication that monomethyltriazenes interact with DNA is the observation that they are direct-acting mutagens on *Salmonella typhimurium* bacteria[56]. A more direct experiment is to incubate a monomethyltriazene, with DNA *in vitro*; MTIC (**1b**) gives for instance several methylated heterocyclic bases[16]. The major methylated product was 7-methylguanine, with minor amounts of 3-methyladenine and methylphosphotriesters. Nevertheless, most (99.5%) of the available methyl cation is found as methanol. MTIC induces repair synthesis of DNA but inhibits semi-conservative replication.

In vivo studies of the epigenetic effect of MTIC on human melanoma cells with different DNA-repair characteristics have had mixed results. MTIC increases the thymidine and deoxycytidine pools but not the deoxyguanosine pool in MM253c1 melanoma cells[57], but it was not clear from this study whether DNA methylation or epigenetic effects such as RNA methylation are involved in the activity of MTIC. A clear result was obtained in a study of the effects of MTIC and related triazenes, on the differential cytotoxicity towards human cells of the Mer$^+$ and Mer$^-$ phenotypes by Gibson *et al.*[58] MTIC is preferentially cytotoxic towards the BE cell line (Mer$^-$) compared to the HT-29 cell line (Mer$^+$). A similar preferential cytotoxicity was found with the analogous triazenes in the phenyl series (**2b** and **2c**)[59]. The BE cell line is deficient in the repair of O^6-methylguanine lesions, whereas the HT-29 cell line is proficient in the repair of guanine alkylation. Thus preferential cytotoxicity towards the BE cell line is an indication of DNA alkylation, specifically at the O^6-guanine position. However, a striking result of this study was the observation that the monomethylimidazole triazene (**1b**) shows the same differential cytotoxicity as the monochloroethyltriazene MCTIC (**1d**). Neither the formation of DNA single-strand breaks nor DNA-protein cross-links could account for the differential cytotoxicity observed in the Mer$^+$ and Mer$^-$ cells. These results tend to suggest that lesions other than DNA interstrand cross-links may be responsible for the mechanisms of cell killing by chloroethylating agents. More relevant perhaps to the theme of this chapter is the observation that a monomethyltriazene is cytotoxic to cells which cannot repair the O^6-methylation lesion, whereas a monoethyltriazene is not cytotoxic to the same cell line. Therefore, the inactivity of the diethyltriazenes towards *in vivo* tumour models may be explained by the apparent inability of the monoethyltriazene metabolite to cause a DNA-damaging lesion. Alternatively, DNA ethylation may be more easily repaired than methylation.

In order to elaborate on these significant results Hartley *et al.*[60] examined the base sequence selectivity for reaction at the guanine-N^7 position for a series of structurally related triazenes using a modification of a standard DNA sequencing method. The monomethyl and monochloroethyl triazenes alkylate guanines extensively at the N^7 position with a general preference for runs of contiguous guanines, similar to, but not as striking as, that observed with chloroethylnitrosoureas. By contrast the monoethylating analogues alkylated weakly with little sequence preference. It may be significant that certain oncogenes are high in content of triplets of GGG and could thus constitute regions of enhanced reactivity with the triazene-type of alkylating agents.

Another striking difference between monomethyl- and monoethyltriazenes was found by Tisdale[61] in a study of induction of haemoglobin synthesis in the human leukaemia cell line K562. The monomethyltriazene (4-CN·C$_6$H$_4$N = N·NHMe) can induce haemoglobin synthesis in these cells, but the monoethyltriazene analogue does not. Furthermore, K562 cells, unlike the TLX5 cells used in earlier studies by Gescher et al.[22], are capable of differentiating the toxicity of methyl- and ethyltriazenes in vitro. The reason for the difference between the K562 and TLX5 cell lines is not known, but there may be a connection here to the fact that methyl- and ethyltriazenes can be differentiated in their toxicity towards the BE (Mer⁻) cell line, which is deficient in the repair of O^6-alkylguanine lesions[58,59]. Nevertheless, these results suggest that methylating agents are more effective than ethylating agents in the alteration of gene expression in certain cell lines.

Other experiments with 1-aryl-3-(2-haloethyl)triazenes (ArN=N–NHCH$_2$CH$_2$Cl) have shown that these triazenes react with and cause strand breakage of DNA[62,63]. In contrast to the 2-haloethylnitrosoureas, the 2-haloethyltriazenes appear to react via type II single-strand scission of DNA by base alkylation followed by depurination or depyrimidination and subsequent hydrolysis of the apurinic site. Overall, these triazenes show a preference for reaction at the more acidic phosphate sites in DNA owing to their unique acid-promoted decomposition, which may account for the lack of detection of DNA interstrand cross-links. There appears to be a fundamentally different mechanism of action of 3-(2-haloethyl)triazenes compared to the 2-haloethylnitrosoureas.

5.4 Hydroxymethyltriazenes

As discussed earlier in this chapter, the accepted metabolic activation pathway of a dimethyltriazene involves hydroxylation to give an 'unstable' α-hydroxyalkyltriazene (9→10). The presumed instability of the α-hydroxyalkyltriazene apparently inhibited a search for such derivatives until comparatively recently. Metabolic evidence for the possible stability of a derivative of type (10) was obtained in experiments in which 1-(2,4,6,-

33

trichlorophenyl)-3,3-dimethyltriazene was fed to rats and the urine of the animals was analysed for triazene derivatives[64]. The evidence from mass spectrometry pointed to a urinary metabolite with an intact diazoamino structure and it was suggested that this metabolite is the glucuronide derivative of the α-hydroxymethyltriazene (33).

Further studies by these workers showed that a metabolite of DTIC could be detected by thin-layer chromatography; this novel metabolite has lower mobility than DTIC and is suggested to be the hydroxymethyltriazenoimid-azole (1c), which has been independently synthesised by reaction of MTIC (1b) with formaldehyde by Kolar et al.[65]. Similar studies by Rutty et al.[66] have confirmed that oxidative N-demethylation of DTIC occurs in vivo by HPLC identification of the hydroxymethyltriazene (1c) and the monomethyltriazene (1b) in the plasma of treated mice, rats and patients. This study also revealed species differences between the extent of metabolism of DTIC compared with aryldimethyltriazenes. Overall the aryltriazene yields much higher levels of its cytotoxic monomethyl metabolite than does DTIC, especially in the rat.

A significant breakthrough in the study of hydroxymethyltriazenes occurr-ed with the discovery that stable, crystalline 3-hydroxymethyltriazenes (2c) could be isolated from the reaction of aryl diazonium salts with formaldehyde/methylamine mixtures[67].

$$Ar-N_2^+ + MeNH_2/CH_2O \longrightarrow Ar-N=N-N\begin{smallmatrix} CH_3 \\ CH_2OH \end{smallmatrix} \qquad (5.11)$$
$$(2c)$$

Formation of the hydroxymethyltriazene (2c) is favoured by electron with-drawing groups (X) in the aryl group. These hydroxymethyltriazenes lose formaldehyde readily during electron-impact mass spectrometry and in solutions with mixed aqueous-organic solvent systems. They behave chemi-cally and biochemically as 'masked' monomethyltriazenes. Hydroxymethyl-triazenes have half-lives in buffer solution very close to the half-life of the analogous monomethyltriazene and have a similar level of antitumour activity[68]. Hydroxymethyltriazenes are cytotoxic to tumour cells in culture without metabolic activation[69] and display the same preferential cytotoxicity as monomethyltriazenes towards human tumour cells of the Mer$^+$ and Mer$^-$ phenotypes[58,59]. Thus, it appears that hydroxymethyltriazenes are potentially good prodrugs for the monomethyltriazene, but their apparent instability may not lend itself to prodrug use. This does not preclude the possibility that a more stable derivative of a hydroxymethyltriazene might have a long enough half-life to be considered as a clinical prodrug, and the search for such derivatives is continuing.

For example, several crystalline derivatives of the hydroxymethyltriazenes have been reported by Hemens et al.[70]. Acetylation by acetic anhydride in pyridine affords the 3-acetoxymethyltriazenes (34); benzoylation yields the analogous benzoate (35). The acetoxymethyltriazenes (34) fragment spontaneously, presumably to give the novel iminium (36); in methanol, the iminium ion is trapped to give the methyl ether (37). Iley et al. [71] have shown recently that the methyl ethers (37) can be obtained directly from the hydroxymethyltriazene by reaction with the anhydrous alcohol in the presence of hydrogen chloride. Reaction of the acetate (34) with sodium azide in aqueous acetone affords the α-azidomethyltriazene (38)[72]. These studies have also shown that a monomethyltriazene (2b, X = CN) can react with formaldehyde to give the hydroxymethyltriazene (2c, X = CN)[70]. Direct formation of an ether of type (37) was observed when a diazonium fluoroborate was treated with a mixture of benzylamine and formaldehyde in alcoholic solution (equation (5.12)), but the generality of this reaction has not been established[70].

$$Ar-N_2^+ + PhCH_2NH_2/CH_2O \xrightarrow{\text{ROH}} Ar-N=N-N\underset{CH_2OR}{\overset{CH_2Ph}{<}} \quad (5.12)$$

The involvement of iminium ions (36) in the mechanism of decomposition of the acetoxymethyltriazenes (34) has been established by a kinetic study of the methanolysis and solvolysis reaction of (34)[72]. The rate of solvolysis in various solvents correlates with the Grunwald–Winstein parameter for solvent ionising power, supporting the hypothesis of an S_N1 mechanism involving an iminium ion intermediate. The hypothesis is further supported by a non-common ion effect; the presence of lithium chloride in the solvent greatly increases the rate of reaction, whereas lithium acetate causes a slight decrease in rate, attributable to a common ion effect. Furthermore, it is apparent that hydroxymethyltriazenes do not react via iminium ions and that functionalisation to a derivative such as the acetate is necessary for iminium ion generation.

A thorough kinetic study of the solvolysis reactions of 1-aryl-3-benzoyloxymethyl-3-methyltriazenes (35) by Iley et al.[73] has confirmed beyond any reasonable doubt that acyloxymethyltriazenes in general undergo solvolysis via iminium ion formation. Whereas in the study of Hemens and Vaughan[72], it was not possible to clearly detect a difference between the rates of solvolysis of the hydroxymethyltriazene and the acetoxymethyltriazene, Iley et al.[73] were able to follow the liberation of benzoate ion from the triazene (35) by ultraviolet spectroscopy. It has also been claimed[73] that it is possible to observe the decomposition of the hydroxymethyltriazene (2c) to the monomethyltriazene prior to the decomposition of the latter to the corresponding arylamine.

These results may have important biological consequences. Hydroxymethyltriazenes, such as (2c), do not form iminium ions directly. However, it is possible that in vivo metabolic conjugation of a hydroxymethyltriazene could afford a derivative capable of iminium ion formation in the same manner as observed in vitro by the acetoxymethyltriazenes (34). 'However, the reactivity of these compounds is such that unless stabilised in some as yet unknown way, they are unlikely to be anything but transitory intermediates in vivo'[73].

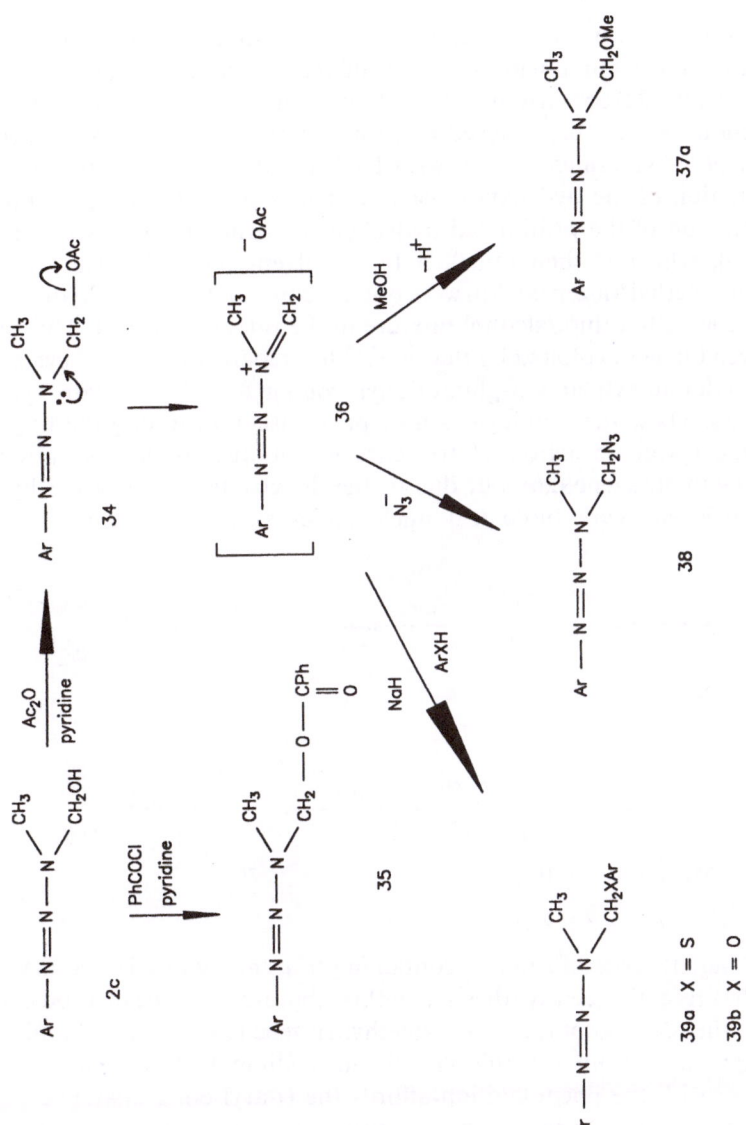

Nevertheless, the triazene system represents an excellent model for studying the metabolism of xenobiotic N-alkyl compounds generally; this aspect of the toxicology of N-alkyl compounds has recently been reviewed by Overton et al.[74]

Thus it is clear from the foregoing discussion that hydroxymethyltriazenes (2c) do not form iminium ions directly and that activation by conversion of (2c) to the acetate (34) or benzoate (35) facilitates iminium ion formation. That such activation can also be achieved by protonation has recently been shown by Iley et al.[71] Treatment of (2c) with HCl in anhydrous methanol results in protonation of the hydroxyl oxygen, rather than on the triazene nitrogen. Dissociation of the protonated hydroxymethyltriazene affords the iminium ion (36), which is then attacked by a solvent molecule. The analogous alkylthiomethyltriazenes (37b) were obtained by reaction of the hydroxymethyltriazene with a thiol/alcohol mixture in the presence of HCl. This method has been further exploited by Iley et al.[75] to prepare a series of S-cysteinyl, S-(N-acetylcysteinyl) and S-glutathionyl conjugates of the hydroxymethyltriazenes. These new conjugates may prove useful in testing the hypothesis that the cytotoxic action of triazenes is not due to the formation of a monomethyltriazene (2b), but due to the direct interaction of the hydroxymethyltriazene with biological nucleophiles such as protein and nucleic acid[76].

A different series of sulphur-containing triazenes with CH_2–S–aryl group (39a) has recently been synthesised in this laboratory[77]. The method of choice here is the reaction of the acetoxymethyltriazene (34) with the thiophenolate ion, generated from the thiophenol and sodium hydride. The analogous reaction with the phenolate ion affords the O-aryl ether analogues (39b)[78]. These S-aryl and O-aryl derivatives display a wide divergence in stability at physiological pH. The S-aryl compounds decompose quite rapidly in pH 7.5 buffer, and have half-lives comparable to the hydroxymethyltriazene. On the other hand, the stability of the O-aryl derivatives can be modulated by the introduction of a para substituent in the O-aryl group. Half-lives in excess of 60 min have been observed for triazenes of type (39b); these compounds represent the most likely candidate for a clinically useful prodrug form of the hydroxymethyltriazene and could be a viable alternative to the experimentally successful, but clinically toxic drug, Mitozolomide.

The decomposition of hydroxymethyltriazene by loss of formaldehyde has been exploited as an excellent method for preparation of a monomethyltriazene free from contamination by arylamine or pentaazadiene; the reported synthetic procedure of Cheng and Iley[79] employs methylamine as the base catalyst

$$ArN=N-N\begin{array}{c}CH_3\\CH_2OH\end{array} \xrightarrow{\text{MeNH}_2} ArN=N-N\begin{array}{c}CH_3\\H\end{array} + CH_2O \quad (5.13)$$

Further examination of this reaction, using a large number of amines as base catalysts, reveals a specific structural requirement in the catalyst[80]. The base requires both a lone pair of electrons and an X–H bond (X = O or N), suggesting the concerted mechanism for formaldehyde elimination. Electron density at N-3 of the triazene evidently plays a part in the reaction since electron withdrawing groups in Ar stabilise the hydroxymethyltriazene. One of the most stable of the hydroxymethyltriazenes used in this work is the 1-(3-pyridyl)triazene, which survives HPLC analysis; conversion to the monomethyltriazene can be followed chromatographically. The elimination of formaldehyde from a hydroxymethyltriazene is also catalysed by the presence of Lewis acids[81], e.g. Fe^{3+}, Fe^{2+}, Zn^{2+}, Cu^{2+}.

$$Ar-N=N-N\cdots \xrightarrow{} Ar-N=N-N\begin{array}{c}H\\CH_3\end{array} + HCHO + HX$$

$$(5.14)$$

(Ar = 3–pyridyl)

Several of the derivatives of hydroxymethyltriazenes have shown promise for application as prodrugs in cancer treatment. The acetoxymethyltriazenes (**34**) display broad spectrum activity against the TLX5, P388 and PC6 tumours[82] and against the L 1210 lymphoid leukaemia, the LX-1 lung xenograft and the M5076 ovarian carcinoma in mice (Vaughan, unpublished results, NCI tumour screening panel). The methoxymethyltriazenes (**37**) are active on the TLX5 and PC6 tumours but not on the P388 tumour. The inactivity of these compounds against the P388 tumour is not readily explained.

5.5 Bis-triazenes: *N,N*-bis-(1-aryl-3-methyltriazen-3-ylmethyl)-methylamines (40)

The coupling reaction of a diazonium salt with formaldehyde/methylamine mixtures is not always as simple as suggested by equation (5.11). The hydroxymethyltriazene is certainly the major, often the only product, when the arenediazonium ion carries an electron withdrawing substituent in Ar. In other cases a second type of triazene-containing product is formed. This realisation was made simultaneously by two groups, Cheng et al.[83] and LaFrance et al.[84] and the second product was characterised unequivocally as the 'bis-triazene' (40). Formation of the bis-triazene is favoured by (a) the absence of $-M$ groups in Ar, (b) a low formaldehyde/methylamine ratio. A high $CH_2O/MeNH_2$ ratio favours formation of the hydroxymethyltriazene[85]. These results suggest that a complex set of equilibria exist in the formaldehyde/methylamine mixture and that the major product arises from diazonium coupling to the predominant or most reactive species in the mixture (41 or 42).

The bis-triazenes have been shown to possess antitumour activity comparable to analogous triazenes. Furthermore, the bis-triazenes are cytotoxic to tumour cells *in vitro* implying that they can break down spontaneously to generate the monomethyltriazene. Indeed, kinetic studies show that a bis-triazene hydrolyses in aqueous buffer with almost exactly the same half-life as the corresponding acetoxymethyltriazene with the same aryl-substituent[86].

Diazonium coupling with a methylamine/formaldehyde mixture can result in the formation of a non-triazene product if the coupling is carried out at low pH. The product is a 3,7-bisaryl-1,5,3,7-dioxadiazocine (43)[87]. The most likely explanation for this observation is that the initially formed triazene (2c or 40) is broken down rapidly under acidic conditions to give the arylamine, which then condenses with formaldehyde, albeit in an unusual manner, to give the observed product.

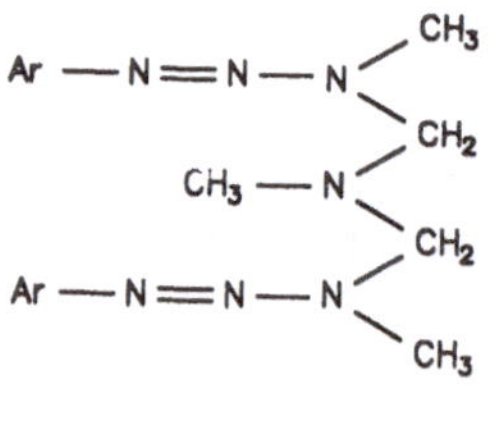

40

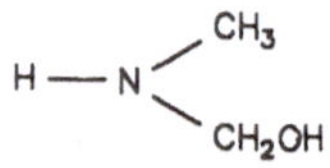

41

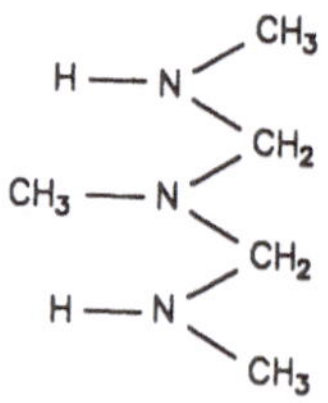

42

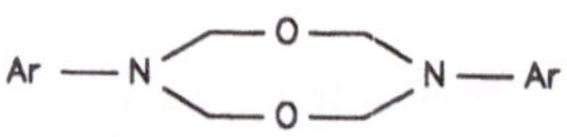

43

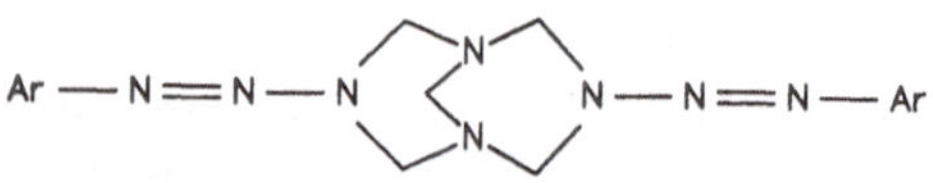

44

A logical extension of the hydroxymethyltriazene synthesis is to couple a diazonium salt with a formaldehyde/ammonia mixture, the expectation being that an appropriate recipe might afford the elusive 'monomethyloltriazene', $ArN = N-NHCH_2OH$. This species is a possible candidate as the unidentified metabolite of the monomethyltriazene alluded to in section 5.2.3. However, diazonium coupling with ammonia/formaldehyde mixtures in a variety of proportions affords only the novel 3,7-bisarylazo-1,3,5,7-tetraazabicyclo[3, 31]nonanes (44)[88]. These bicyclic compounds are also obtained by diazonium coupling to hexamethylenetetramine (HMT), which is presumably formed as the major equilibrium component in NH_3/CH_2O mixtures. The bis-arylazobicyclononanes (44) are the cyclic analogues of the bis-triazene (40), but do not undergo fragmentation under physiological conditions. Hydrolysis of (44) is only possible at low pH.

5.6   1,3-Dialkyl- and 1,3,3-trialkyltriazenes

5.6.1   *Synthesis*

The starting point for the synthesis of triazenes of type (6) is the reaction of an alkyl azide with an organometallic reagent, either a Grignard reagent[89], or an alkyllithium reagent[8] (Scheme 5.2), which affords the 1,3-dialkyltriazene (e.g. 6a). Further alkylation of the 1,3-dialkyltriazene is achieved by treatment with potassium *t*-butoxide, followed by methyl iodide, which affords the trialkyltriazene (6b). An alternative base used to deprotonate the dialkyltriazene is potassium hydride with 18-crown-6. A variable temperature NMR study of the trialkyltriazenes shows that the rotational barrier about the N2–N3 bond is $11 \, kcal \, mol^{-1}$, about $3 \, kcal \, mol^{-1}$ lower than in the 1-aryl-3,3-dialkyltriazenes[8].

Acetylation of a 1,3-dialkyltriazene by acetyl chloride occurs regioselectively at N-3 to give 3-acetyl-1,3-dialkyltriazenes (e.g. 6c). Analogous acylation of the anion (45) with ethyl chloroformate or diethyl chlorophosphite affords the ester (6d) and phosphite (6e) derivatives. The carbamoyltriazene (6f) is synthesised directly from the dialkyltriazene by reaction with methyl isocyanate[90].

A different route has been found to prepare the structurally related 1-acyl-3,3-dialkyltriazenes (6g and 6h). Nucleophilic attack by a secondary amine at an electrophilic nitrogen of the azo-diester or -diamide (equation (5.15)) affords the 3,3-dialkyltriazene-1,2-dicarboxylic derivatives (46). Oxidative hydroacyl-elimination of (46) by lead tetraacetate affords the 3, 3-dialkyltriazenecarboxyl derivatives[91].

5.6.2   *Reactions*

1,3-Dimethyltriazene reacts with the *t*-butoxy radical to give the relatively stable, 1,3-dimethyltriazene radical (47) which has been detected by ESR

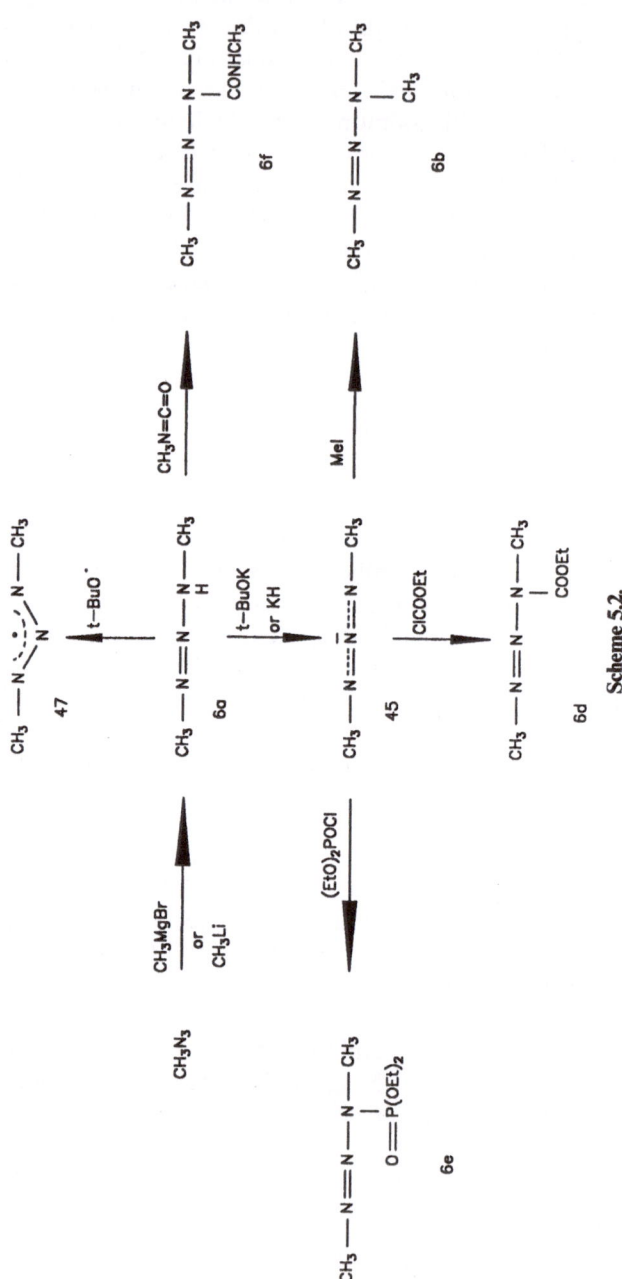

Scheme 5.2.

$$(5.15)$$

$$(X = OPr, NHPr)$$
$$(R = Me, Et)$$

spectroscopy and shown to be a σ-radical[92,93]. 1, 3-Dialkyltriazenes are stable as pure liquids and in aprotic solutions, but decompose in buffered, aqueous solution over the pH range of 9–12. The reaction is acid catalysed since the rate is inversely proportional to pH. The catalytic mechanism is somewhat dependent on the buffer system, varying from specific acid catalysis in one buffer to specific acid, followed by general base, in another. The dependence of the rate of decomposition on buffer concentrations of dimethyl- and diethyltriazene is explained in terms of nucleophilic attack of the buffer anions on N-2 of the protonated triazenes. Di-isopropyltriazene, on the other hand, is protonated and dissociates directly to the isopropyl carbonium ion[94].

Similar kinetic studies have been carried out with the trialkyltriazenes by Sieh and Michejda[95]. The products of acid-catalysed decomposition of trialkyltriazenes are consistent with protonation and formation of alkyl-diazonium cations. The rate of hydrolysis of the tri-substituted triazenes is affected by the basicity of N3, since the presence of an acyl group at N3 slows down the hydrolysis considerably; the order of hydrolysis rates of the various tri-substituted triazenes[96] is **6e** > **6b** > **6d** > **6c** > **6f**. The common mechanism of decomposition appears to be:

$$(5.16)$$

5.6.3 *Biological activity*

In view of the demonstrated proclivity of trialkyltriazenes to generate alkyldiazonium ions in aqueous solution, it is not surprising that these and other trisubstituted triazenes are direct-acting mutagens.[9] Trialkyltriazenes have a completely different spectrum of mutagenic potency than aryldialkyltriazenes. An interesting comparison within the range of trialkyltriazenes is seen in the mutagenic potency of the isomeric benzyldimethyltriazenes (**48** and **49**).

$$PhCH_2-N=N-N \begin{matrix} Me \\ Me \end{matrix} \qquad Me-N=N-N \begin{matrix} CH_2Ph \\ Me \end{matrix}$$
$$\mathbf{48} \qquad\qquad\qquad \mathbf{49}$$

Compound (**49**) is much more potent in its action on the TA 1535 strain of *Salmonella typhimurium* than (**48**), since (**49**) decomposes to the extremely reactive methyl diazonium ion, while (**48**) leads to the relatively less reactive benzyl diazonium ion.

Similar results are obtained with the 1,3-dimethyl-3-acyltriazenes; (**6c**), (**6d**) and (**6e**) are direct-acting mutagens in strains of *Salmonella typhimurium* that require a base substitution to revert to the wild type, i.e. TA 1535 and TA 100[90]. The mutagenicity of the ester (**6d**) was enhanced by incubation with esterase, whereas these derivatives did not cause mutation in frameshift strains. The *N*-methylcarbamoyltriazene (**6f**) was found to be non-mutagenic on all strains tested.

Preliminary antitumour evaluation of triazenes of type (**6**) is very promising[10]. Compounds (**6d**) and (**6f**) are active in the P388 leukaemia NCI antitumour agent screen and have been selected for additional testing in the human tumour xenograft panel. Both compounds are active against several tumours. The corresponding 3-acyl-1-(2-chloroethyl)-3-methyltriazenes have also been prepared and they are significantly more toxic than the 1-methyl analogues[97].

5.7 Triazene *N*-oxides

In establishing the structural requirements for antitumour activity of aryldialkyltriazenes, Connors *et al.*[4] investigated the properties of 3-(4-carbamoylphenyl)-1-methyltriazene 1-oxide (**51**); later, Wilman[7] showed it to be active against the TLX5 lymphoma and the AdjPC6/A plasmacytoma. Compounds of this type are commonly synthesised by the reaction of an aryldiazonium salt with *N*-methylhydroxylamine, and were originally described as 1-aryl-3-hydroxy-3-methyltriazenes (**50**). However, infrared studies by Mitsuhashi *et al.*[98] and 1 H- and ^{13}C-NMR data from Guimanini *et al.*[99] tended to suggest that the structure is better described as the *N*-oxide form (**51**). Crystallographic data from Kuroda and Wilman[100] have established unequivocally that these compounds exist in the solid state as the N-oxide form.

As regards the observed antitumour activity of this N-oxide, it is tempting to envisage the metabolic reduction of the N-oxide to the analogous mono-methyltriazene. As indicated earlier, in order for triazenes to show antitumour activity, they must be capable of chemical or biological degradation to a monomethyltriazene. Supporting evidence from Wilman[7] for this suggestion is the lack of activity of the O-methyl derivative (52) against the TLX5 tumour which can only N-demethylate to give an inactive N-methoxytriazene.

50 (R = H)

52 (R = CH₃)

51

5.8 Concluding remarks

Undoubtedly, interest in the chemistry of triazenes has been stimulated to a large degree by the antitumour activity of some triazene derivatives, and considerable chemical information has been generated by the search for a better antitumour analogue of DTIC. Many of the new triazenes synthesised have proven useful, deliberately or otherwise, in experiments designed to elucidate the metabolic changes that triazenes undergo *in vivo*, and to determine the exact identity of the 'active' metabolite of DTIC. Nevertheless, a definitive answer to the triazene metabolism question remains elusive. Further studies of the chemistry of the trialkyltriazenes, of the triazene-N-oxides and of the hydroxymethyltriazenes, and the derivatives thereof, may well lead to some new proposals for triazene metabolism. Whether or not the metabolic puzzle is solved, the new chemistry revealed by these investigations is certain to be rewarding.

References

1. V. S. Lucas, Jr. and A. T. Huang, 'Chemotherapy of Melanoma', in *Developments in Oncology*, Vol. 5, *Clinical Management of Melanoma*, ed. H. F. Siegler, Martinus Nijhoff, The Hague, The Netherlands (1982) p. 382.
2. D. A. Clarke, R. K. Barclay, C. C. Stock and C. S. Rondesvedt, *Proc. Soc. Exp. Biol. Med.* **90** (1955) 484.
3. D. E. V. Wilman and P. B. Farmer, 'Alkylating Agents', in *Progressive Stages of Neoplastic Growth*, ed. H. E. Kaiser, Martinus Nijhoff, Dordrecht, The Netherlands (1989).
4. T. A. Connors, P. M. Goddard, K. Merai, W. C. J. Ross and D. E. V. Wilman, *Biochem. Pharmacol.* **25** (1976) 241.
5. M. F. G. Stevens, J. A. Hickman, R. Stone, N. W. Gibson, G. U. Baig, E. Lunt and C. G. Newton, *J. Med. Chem.* **27** (1984) 196.

6. E. S. Newlands, G. Blackledge, J. A. Slack, C. Goddard, C. J. Brindley, L. Holden and M F. G. Stevens *Cancer Treat. Rep.* **69** (1985) 801.
7. D. E. V. Wilman, Triazene *N*-oxides, in *Biological Oxidation of Nitrogen in Organic Molecules, Chemistry, Toxicology and Pharmacology*, eds. J. W. Gorrod and L. A. Damani, Ellis Horwood, Chichester (1985) pp. 297–302.
8. R. J. Smith Jr. and C. J. Michejda, *Synthesis* (1983) 476.
9. D. H. Sieh, A. W. Andrews and C. J. Michejda, *Mutat. Res.* **73** (1980) 227.
10. R. J. Smith Jr. and C. J Michejda, *Proc. Am. Assoc. Cancer Res.* **28** (1987) 1210.
11. J. Elks and D. H. Hey, *J. Chem. Soc.* (1943) 441.
12. G. F. Kolar, *Z. Naturforschung.* **27b** (1972) 1183.
13. T. Axenrod, P. Mangiaracina and P. S. Pregosin, *Helv. Chim. Acta* **59** (1976) 1655.
14. C. S. Rondesvedt Jr. and S. J. Davis, *J. Org. Chem.* **22** (1957) 200.
15. K. Vaughan, *J. Chem. Soc., Perkin Trans. 2* (1977) 17.
16. N. S. Mizuno and R. W. Decker, *Biochem. Pharmacol.* **25** (1976) 2643.
17. L. Meer, R. C. Janzer, P. Kleihues and G. F. Kolar, *Biochem. Pharmacol.* **35** (1986) 3243.
18. M. N. Ramos and S. R. Pereira, *J. Chem. Soc., Perkin Trans. 2* (1986) 131.
19. B. Golding, T. J. Kemp, R. Narayanaswamy and B. W. Waters, *J. Chem. Res.* (S) (1984) 130.
20. J. R. P. Godin, K. Vaughan and K. W. Renton, *Can. J. Physiol. Pharmacol.* **59** (1981) 1234.
21. P. Farina, A. Gescher, J. A. Hickman, J. K. Horton, M. D'Incalci, D. Ross, M. F. G. Stevens and L. Torti, *Biochem. Pharmacol.* **31** (1982) 1887.
22. A. Gescher, J. A. Hickman, R. J. Simmonds, M. F. G. Stevens and K. Vaughan, *Biochem. Pharmacol.* **30** (1981) 89.
23. P. Farina, L. Torti, R. Usro, J. K. Horton, A. Gescher and M. D'Incalci, *Biochem. Pharmacol.* **35** (1986) 209.
24. D. E. V. Wilman, P. J. Cox, P. M. Goddard, L. I. Hart, K. Merai and D. R. Newell, *J. Med. Chem.* **27** (1984) 870.
25. G. M. Baird and M. L. N. Willoughby, *Lancet* **ii** (1978) 681.
26. J. K. Horton and M. F. G. Stevens, *J. Pharm. Pharmacol.* **33** (1981) 808.
27. Y. F. Shealy, C. A. Krauth and J. A. Montgomery, *J. Org. Chem.* **27** (1962) 2150.
28. U. G. Kang and H. Schecter, *J. Am. Chem. Soc.* **100** (1978) 651.
29. M. N. Rosenfeld and D. A. Widdowson, *J. Chem. Soc. Chem. Commun.* (1979) 914.
30. N. Satyamurthy and J. R. Barrio, *J. Org. Chem.* **48** (1983) 4394.
31. H. Hansen, S. Hunig and K-I Kishi, *Chem. Ber.* **112** (1979) 445.
32. L. Lassiani, C. Nisi, F. Sigon, G. Sava and T. Giraldi, *J. Pharm. Sci.* **69** (1980) 1098.
33. T. Ignasiak, J. Suszko and B. Ignasiak, *J. Chem. Soc. Perkin Trans. 1* (1975) 2122.
34. S. Treppendahl, P. Jacobsen and J. Wieczorkowski, *Acta Chem. Scand.* Ser. B. **37** (1983) 155; M. D. Threadgill and A. P. Gledhill *J. Chem. Perkin Trans. 1* (1986) 873.
35. R. N. Butler, P.D. O'Shea and L. A. Burke, *J. Chem. Soc. Chem. Commun.* (1987) 1210.
36. K. Vaughan and M. F. G. Stevens, *Chem. Soc. Rev.* **7** (1978) 377.
37. T. P. Ahern, H. Fong and K. Vaughan, *Can. J. Chem.* **55** (1977) 1701.
38. O. Dimroth, *Chem. Ber.* **38** (1905) 670.
39. S. P. Makarov, A. Ya. Yakubovich, A. S. Filatov, N. A. Englin and T. Ya. Nikiforova, *Zh. Obshch. Khim.* **38** (1968) 709 (*Chemical Abstracts* **69** (1968) 18506).
40. D. L. Hooper and K. Vaughan, *J. Chem. Soc. Perkin Trans. 2* (1981) 1161.
41. A. J. Randall, C. H. Schwalbe and K. Vaughan, *J. Chem. Soc. Perkin Trans. 2* (1984) 251.
42. K. Vaughan and M. T. H. Liu, *Can. J. Chem.* **59** (1981) 923.
43. O. O. Orazi, R. A. Corral, J. Zinczuk and H. Schuttenberg, *Tetrahedron Lett.* **23** (1982) 293.
44. F. Delben, S. Paoletti, G. Manzini and C. Nisi, *J. Pharm. Sci.* **70** (1981) 892; V. Zverina, M. Remes, J. Divis, J. Marhold and M. Matrka, *Collect. Czech. Chem. Commun.* **38** (1973) 251.
45. C. Ebert, L. Lassiani, P. Linda, M. Lovrecich, C. Nisi and F. Rubessa, *J. Pharm. Sci.* **73** (1984) 1691.
46. M. L. Sinnott and P. J. Smith, *J. Chem. Soc. Chem. Commun.* (1976) 223.
47. P. J. Marshall, M. L. Sinnott, P. J. Smith and D. Widdows, *J. Chem. Soc. Perkin Trans. 1* (1981) 366.
48. M. L. Sinnott and P. J. Smith, *Biochem. J.* **175** (1978) 525.
49. C. C. Jones, M. A. Kelly, M. L. Sinnott and P. J. Smith, *J. Chem. Soc. Chem. Commun.* (1980) 322.
50. P. L. Faye, K. Vaughan and D. L. Hooper, *Can. J. Chem.* **61** (1983) 179.

51. H. Fong and K. Vaughan, *Can. J. Chem.* **53** (1975) 3714.
52. K. Vaughan, R. J. LaFrance, Y. Tang and D. L. Hooper, *Can. J. Chem.* **63** (1985) 2455.
53. K. M. Baines, T. W. Rourke, K. Vaughan and D. L. Hooper, *J. Org. Chem.* **46** (1981) 856.
54. K. M. Baines, K. Vaughan, D. L. Hooper and L. F. Leveck, *Can. J. Chem.* **61** (1983) 1549.
55. G. Ege, K. Gilbert and R. Heck, *Angew. Chem. Int. Ed.* **21** (1982) 6984.
56. H.F. Thomas, D. L. Brown, P. E. Hartman, E. H. White and Z. Hartman, *Mutat. Res.* **60** (1979) 25.
57. I. P. Hayward and P. G. Parsons, *Austr. J. Exp. Biol. Med. Sci.* **62** (1984) 597.
58. N. W. Gibson, J. Hartley, R. J. LaFrance and K. Vaughan, *Carcinogenesis* **7** (1986) 259.
59. N. W. Gibson, J. A. Hartley, R. J. LaFrance and K. Vaughan, *Cancer Res.* (1986) 4999.
60. J. A. Hartley, W. B. Mattes, K. Vaughan and N. W. Gibson, *Carcinogenesis* **9** (1988) 669.
61. M. J. Tisdale, *Biochem. Pharmacol.* **34** (1985) 2077.
62. J. W. Lown and R. Singh, *Can. J. Chem.* **59** (1981) 1347.
63. J. W. Lown and R. Singh, *Biochem. Pharmacol.* **31** (1982) 1257.
64. G. F. Kolar and R. Carubelli, *Cancer Lett.* **7** (1979) 209.
65. G. F. Kolar, M. Maurer and M. Wildschutte, *Cancer Lett.* **10** (1980) 235.
66. C. J. Rutty, D. R. Newell, R. B. Vincent, G. Abel, P. M. Goddard, S. J. Harland and A. H. Calvert, *Br. J. Cancer* **48** (1983) 140.
67. A. Gescher, J. A. Hickman, R. J. Simmonds, M. F. G. Stevens and K. Vaughan, *Tetrahedron Lett.* (1978) 5041.
68. K. Vaughan, Y. Tang, G. Llanos, J. K. Horton, R. J. Simmonds, J. A. Hickman and M. F.G. Stevens, *J. Med. Chem.* **27** (1984) 357.
69. D. J. Kohlsmith, K. Vaughan and S. J. Luner, *Can. J. Physiol. Pharmacol.* **62** (1984) 396.
70. C. M. Hemens, H. W. Manning, K. Vaughan, R. J. LaFrance and Y. Tang, *Can. J. Chem.* **62** (1984) 741.
71. J. Iley, E. Rosa and L. Fernandes, *J. Chem. Res.* (S) (1987) 264.
72. C. M. Hemens and K. Vaughan, *J. Chem. Soc. Perkin Trans.* 2 (1986) 11.
73. J. Iley, R. Moreira and E. Rosa, *J. Chem. Soc. Perkin Trans.* 2 (1987) 1503.
74. M. Overton, J. A. Hickman, M. D. Threadgill, K. Vaughan and A. Gescher, *Biochem. Pharmacol.* **34** (1985) 2055.
75. J. Iley, R. Moreira, G. Ruecroft and E. Rosa, *Tetrahedron Lett.* **29** (1988) 2707.
76. A. H. Soloway, R. J. Brumbaugh and D. T. Witiak, *J. Theor. Biol.* **102** (1983) 361.
77. K. Vaughan, H. W. Manning, M. P. Merrin and D. L. Hooper, *Can. J. Chem.* **66** (1988) 2487.
78. R. Snooks, R. J. LaFrance, M. P. Merrin and K. Vaughan, Unpublished results.
79. S. C. Cheng and J. Iley, *J. Chem. Res.* (S) (1983) 320.
80. S. C. Cheng, L. Fernandes, J. Iley and E. Rosa, *Tetrahedron Lett.* (1985) 1557.
81. S. C. Cheng, L. Fernandes, J. Iley and E. Rosa, *Proceedings of the Conference on the Organic and Biological Chemistry of Carcinogenic and Carcinostatic Agents Containing Nitrogen–Nitrogen Bonds*, Harpers Ferry, West Virginia (1985).
82. L. M. Cameron, R. J. LaFrance, C. M. Hemens, K. Vaughan, R. Rajaraman, D. C. Chubb and P. M. Goddard, *Anti-Cancer Drug Design* **1** (1985) 27.
83. S. C. Cheng, L. Fernandes, J. Iley and E. Rosa, *J. Chem. Res.* (S) (1983) 108.
84. R. J. LaFrance, Y. Tang, K. Vaughan and D. L. Hooper, *J. Chem. Soc. Chem. Commun.* (1983) 721.
85. H. W. Manning, C. M. Hemens, R. J. LaFrance, Y. Tang and K. Vaughan, *Can. J. Chem.* **62** (1984) 749.
86. H. W. Manning, L. M. Cameron, R. J. LaFrance, K. Vaughan and R. Rajaraman, *Anti-Cancer Drug Design* **1** (1985) 37.
87. G. F. Kolar and M. Schendzielorz, *Tetrahedron Lett.* **26** (1985) 1043.
88. R. D. Singer, K. Vaughan and D. L. Hooper, *Can. J. Chem.* **64** (1986) 1567.
89. D. H. Sieh, D. J. Wilbur and C. J. Michejda, *J. Am. Chem. Soc.* **102** (1980) 3883.
90. R. J. Smith Jr., A. F. Mehl, A. Hicks, C. L. Denlinger, L. Kratz, A. W. Andrews and C. J. Michejda, *J. Org. Chem.* **51** (1986) 3751.
91. N. Von Egger, L. Hoesch and A. S. Dreiding, *Helv. Chim. Acta* **66** (1983) 1416.
92. F. Bernardi, M. Guerra, L. Lunazzi, G. Panciera and G. Placucci, *J. Am. Chem. Soc.* **100** (1978) 1607.
93. J. C. Brand and B. P. Roberts, *J. Chem. Soc. Chem. Commun.* (1981) 748.
94. R. J. Smith Jr., C. L. Denlinger, R. Kupper, A. F. Mehl and C. J. Michejda *J. Am. Chem. Soc.* **108** (1986) 3726.

95. D. H. Sieh and C. J. Michejda, *J. Am. Chem. Soc.* **103** (1981) 442.
96. R. J. Smith Jr, C. L. Denlinger, R. Kupper, S. R. Koepke and C. J. Michejda, *J. Am. Chem. Soc.* **106** (1984) 1056.
97. R. H. Smith Jr., A. F. Mehl, D. L. Shantz Jr., G. N. Chmurny and C. J. Michejda, *J. Org. Chem.* **53** (1988) 1467.
98. T. Mitsuhasha, Y. Osamura and O. Simamura, *Tetrahedron Lett.* **30** (1965) 2593.
99. A. G. Giumanini, L. Lassiani, C. Nisi, A. Petrie and B. Stanovnik, *Bull. Chem. Soc. Jpn.* **56** (1983) 1887.
100. R. Kuroda and D. E. V. Wilman, *Acta Crystallogr.* **C41** (1985) 1543.
101. D. E.V. Wilman, *Cancer Treat. Rev.* **15** (1988) 69.
102. J. Iley and G. Ruecroft, *J. Chem. Res.* (S) (1988) 24.
103. G. Sava, S. Zorzet, L. Perissin, T. Giraldi and L. Lassiani, *Cancer Chemother. Pharmacol.* **21** (1988) 241.

6 The chemistry of azolotetrazinones

M. D. THREADGILL

6.1 Introduction

In 1978, it was believed[1,2] that the 1,2,3,5-tetrazine ring was inherently unstable, despite claims[3] of the preparation of this system. Indeed, there is only one confident report[4] of a [2,5-dihydro]-1,2,3,5-tetrazine without fusion to an azole. In 1979, Ege[5] published the first report of pyrazolo[5,1-d]-1,2,3,5-tetrazin-4(3H)-ones. There followed comprehensive reports[6-8] by the Aston group of the synthesis and antitumour activity of 3-alkyl and 3-aryl-8-substitutedimidazo[5,1-d]-1,2,3,5-tetrazin-4(3H)-ones and an indazolotetrazinone. 6-Alkylimidazotetrazinones and pyrazolotetrazinones have also been prepared by this group[9]. A group from the University of Kansas and from Warner Lambert/Parke-Davis have described[10] a further two pyrazolo[5,1-d]-1,2,3,5-tetrazin-4(3H)-ones closely analogous to the active lead compounds of Stevens et al.[7] The Heidelberg group followed up their first report of the ring system with an extensive study[11] of the synthesis of some forty pyrazolo[5,1-d]-1,2,3,5-tetrazin-4(3H)-ones, together with some 3-aryltriazolotetrazinones, by three distinct routes. Pyrrolo- and tetrazolo-tetrazinones remain unknown. The ring structures and numbering system (according to the IUPAC convention) of the two series of azolotetrazinones (1, 2) are shown in Figure 6.1.

Extensive studies have been carried out on the antitumour and other biological activities of the imidazotetrazinones. 8-Carbamoyl-3-(2-chloroethyl)imidazo-[5,1d]-1,2,3,5-tetrazin-4(3H)-one (3, mitozolomide, CCRG 81010, M & B 39565, NSC 353451) was the first azolotetrazinone to show good antitumour activity[12,13] in a murine screen, although results against cultures of human cancer cells have been mixed[14]. It has been the subject of Phase 1[15] and Phase 2 clinical trials and, despite displaying some activity, the further use of Mitozolomide may be limited by severe delayed toxicity[16]. It shows curative activity[17] towards the L1210 and P388 murine leukaemias and is highly potent against the TLX5 lymphoma, B16 melanoma, M5076 sarcoma, ADJ/PC6A plasmacytoma and the Colon 26 and Colon 38 tumours. Subsequent structure-activity studies[8,9] have shown a requirement for a 3-(2-chloroethyl) group for curative activity in murine tumour screeens. A 3-methyl group, as in temozolomide (4, CCRG 81045), confers less but definite activity, including the differentiation of K562 cells to a less malignant phenotype[18]. Other 3-substituents lead to the abolition of antitumour activity

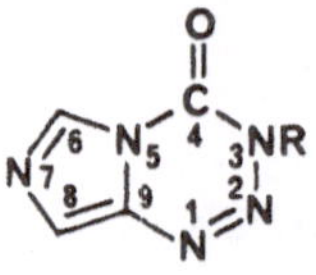

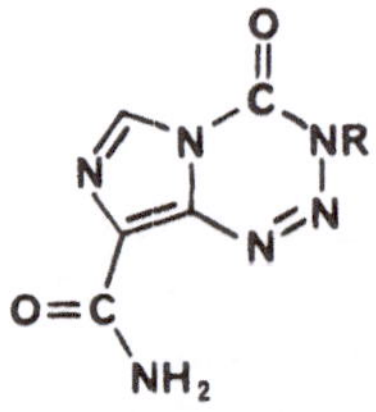

3: R = CH$_2$CH$_2$Cl
4: R = CH$_3$

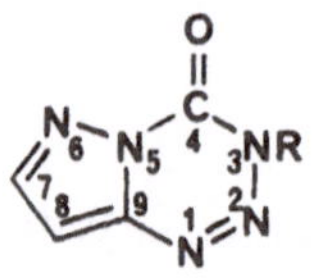

Figure 6.1 Numbering schemes of imidazotetrazinones and pyrazolotetrazinones. Structures of mitozolomide (3) and temozolomide (4).

in vivo. A small alkyl group is tolerated at the 6-position, whereas there is a requirement for the 8-substituent to be electron-withdrawing (carbonyl or sulphonyl) and, preferably, to bear a hydrogen-bonding N–H moiety. In the pyrazolo series, only 3-(2-chloroethyl)pyrazolotetrazinone-8-carboxamide inhibited the growth of the P388, L1210 and M5076 tumours[10].

6.2 Synthesis

Most published syntheses of pyrazolo[5,1-*d*]-1,2,3,5-tetrazin-4(3*H*)-ones[5,9] and all of imidazo[5,1-*d*]-1,2,3,5-tetrazin-4(3*H*)-ones[6,7,9] have involved treatment of the corresponding diazoazole (5, Scheme 6.1) with an alkyl or aryl isocyanate (6) at ambient temperature in a non-hydroxylic organic solvent, often for prolonged reaction times[9]. Systems with two liquid phases can be used for the more water-soluble diazopyrazoles and diazotriazoles[11]. Higher temperatures result in degradation of the diazoazole, particularly in the case of 5-diazoimidazole-4-carboxamide (7) which gives 2-azahypoxanthine (8) (Scheme 6.2). The cycloaddition of 5-diazoimidazole-4-carboxamide proceeds

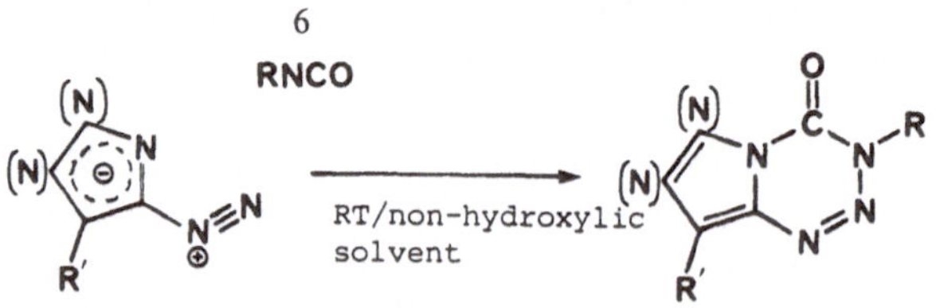

Scheme 6.1 Synthesis of azolotetrazinones from diazoazoles and isocyanates.

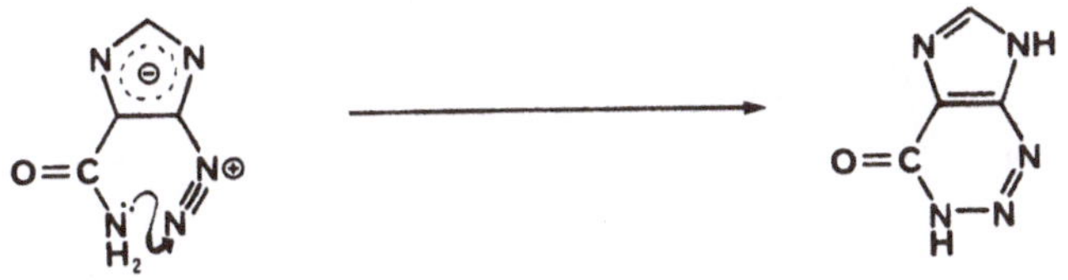

Scheme 6.2 Side reactions of 4-diazoimidazole-5-carboxamide.

smoothly in heterogeneous reactions with a wide range of isocyanates but fails with cyclohexyl isocyanate and with *tert*-butyl isocyanate[19], presumably owing to steric interactions. The corresponding bis(pyrazolotetrazinones) are formed if α, ω-diisocyanates are used[11]. Radiolabelled isotopomers of some imidazotetrazinones have been prepared with ^{14}C in the 3-substituent[20] or at C-4[21]. Synthesis of 3-unsubstituted azolotetrazinones (**1, 2**: R = H) by direct or indirect routes remains to be achieved. Treatment of 5-diazoimidazole-4-carboxamide with other heterocumulenes (isothiocyanates and carbodiimides) also failed[19] to furnish bicycles.

1,3-Dipolar cycloadditions have been studied extensively as routes to heterocycles[22–25]. The mechanism of the cycloaddition between diazoazoles and isocyanates to form azolotetrazinones has yet to be determined satisfactorily. Both concerted and stepwise pathways are conceivable. The fully concerted [4 + 2] route[11], shown as A in Scheme 6.3A, is unlikely owing to the considerable increase in reaction rate in the presence of hexamethylphosphoric triamide[26]. Gilchrist and Storr[27] have also generally discounted concerted mechanisms where heterocumulenes are involved as dipolarophiles. In the stepwise ionic pathway (Scheme 6.3B), there is an initial nucleophilic attack at the isocyanate carbon, followed by ring closure. More attractive is mechanism C in which an initial [3 + 2] concerted cycloaddition is followed by a [1,5] sigmatropic rearrangement; this compares closely with the mechanism[28] of the reaction between diazopyrazole (**9**) and 1,1-dimethoxyethene (**10**) to give the pyrazolo-1,2,4-triazine (**11**) (Scheme 6.3D).

Stepwise synthetic assembly of the tetrazinone ring in two distinct senses has been reported[11] for the pyrazolotetrazinones. In the first, involving insertion of N-2 (Scheme 6.4A), diazotisation/ring closure of carbamoylpyrazole (**12**) is effected by a nitrosating agent to give the pyrazolotetrazinone (**13**). Insertion of the carbonyl group by phosgene or its equivalent is also possible[11], converting the pyrazolyltriazene (**14**) to the bicycle (**13**) (Scheme 6.4B).

The infrared spectra[5,7,9,11] of the azolotetrazinones show that the tetrazinone carbonyl is in a relatively electron deficient environment (v_{max} 1725–1790 cm^{-1}). The electronic absorption maximum of the 8-carbamoylimidazo[5,1-*d*]-1,2,3,5-tetrazin-4(3*H*)-ones occurs in the ultraviolet region between 325 nm and 338 nm and is largely independent of the nature of

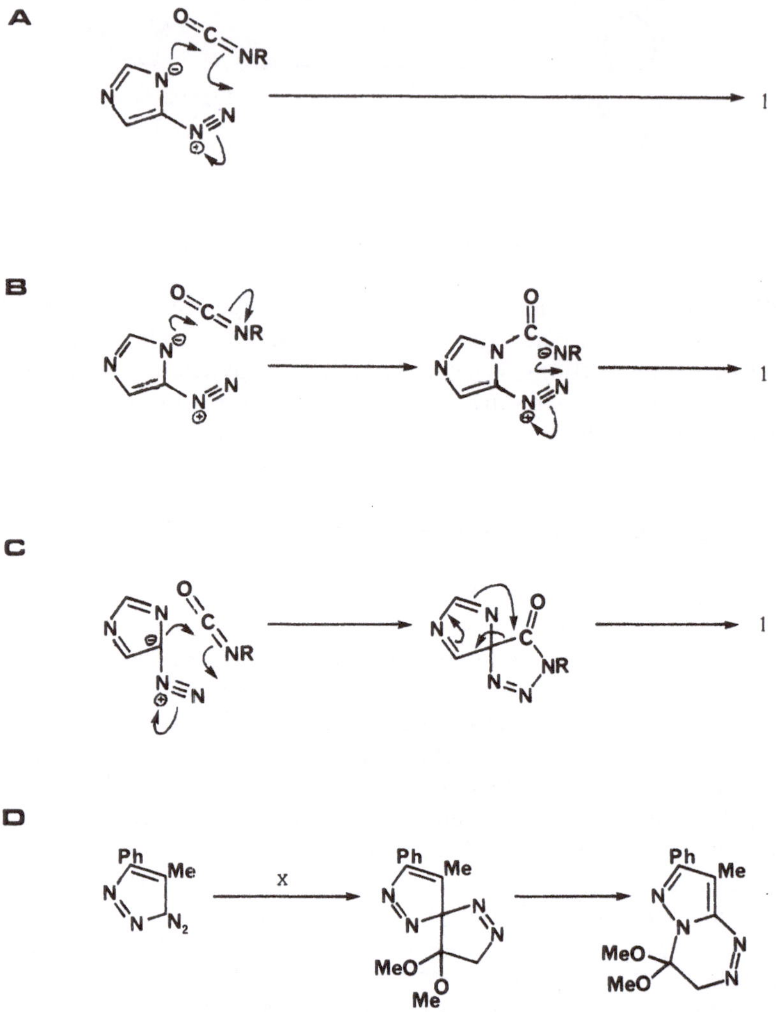

9   11

Scheme 6.3 Proposed mechanisms for the reaction of diazoimidazoles with isocyanates. (A) Concerted mechanism; (B) stepwise ionic mechanism; (C) stepwise 1,3-dipolar cycloaddition followed by 1,5 shift.

the 3-substituent (aryl or alkyl). The corresponding 8-carbamoylpyrazolote-trazinones absorb[10] at shorter wavelengths (310–314 nm). In the NMR spectra of the 6-unsubstituted imidazotetrazinones, the principal character-istic signal arises from the 6-proton which resonates[7,9,19,29] between δ 8.60 and δ 9.05 ppm. The 3-CH_3 protons of temozolomide (**4**) give rise to a sharp singlet at δ 3.90; this strong deshielding again reflects the electron–deficient and electrophilic nature of the 1,2,3,5-tetrazin-4-one ring. In the mass spectra,

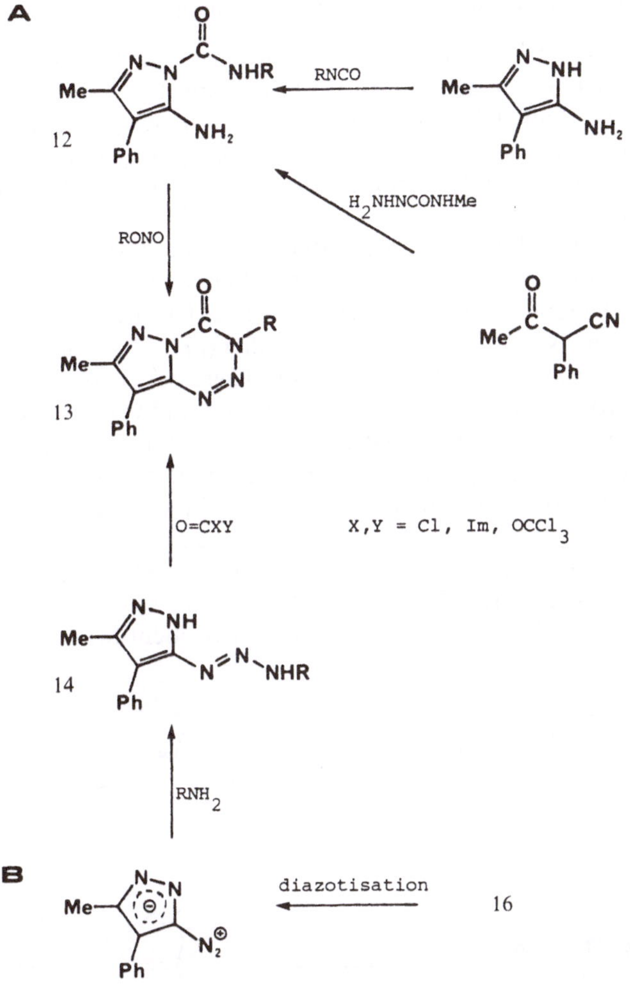

Scheme 6.4 Alternative synthetic approaches to pyrazolotetrazinones.

the factor determining the fragmentation pathway is the nature of the 3-substituent. 3-Arylimidazotetrazinones do not give a molecular ion[19] but only ions corresponding to the products of retrocyclisation. In contrast, the 3-alkyl imidazotetrazinones and pyrazolotetrazinones show abundant molecular ions[10,19]. Interestingly, a major ion (often the base peak) in the mass spectra of 3-(2-chloroethyl)azolotetrazinones is the 3-methyleneiminium ion[10,21,30]. X-ray crystal structure determinations have been reported for mitozolomide (3)[30], for the analogous 3-(2-methoxyethyl) compound[31] and for temozolomide (4)[31]. The two rings of each compound lie effectively in one plane which also contains the atoms of the carboxamide moiety.

6.3   Reactions involving opening of the tetrazinone ring

Comprehensive studies have been carried out on the reactivity of the imidazotetrazinones. However, study of the chemistry of the pyrazolotetrazinones has been limited. Pyrazolotetrazinones are generally[10,19] stable to light; whereas the imidazo compounds are moderately photolabile[19], giving rise to red pigments, presumably azoimidazoles. Thermolysis of azolotetrazinones leads to retrocyclisation[7].

Treatment of azolotetrazinones with nucleophiles results in opening of the tetrazinone ring, whereas acidic or electrophilic reagents usually permit this structure to be retained. Stone[19] determined the half-lives of 8-carbamoylimidazotetrazinones in aqueous phosphate buffer (pH 7.4) in the dark and found them to be in the range 2.9 min to 155 min. These data compare well with the rates of hydrolysis of mitozolomide in plasma[32] and with the observed great sensitivity to pH of the rate of hydrolysis of mitozolomide[33] ($t_{1/2} = 2.1$ h at pH 7.0 and 0.9 h at pH 7.5). Baig and Stevens[34] reported that mitozolomide (3) and temozolomide (4) both decomposed rapidly in boiling water to give 5-aminoimidazole-4-carboxamide (15, Scheme 6.5) as the major product, together with azahypoxanthine (8) and the imidazolylazoimidazole (16). The minor products arise from initial thermal 6π electrocyclic ring opening giving 5-diazoimidazole-4-carboxamide. The tentative identification of the inter-

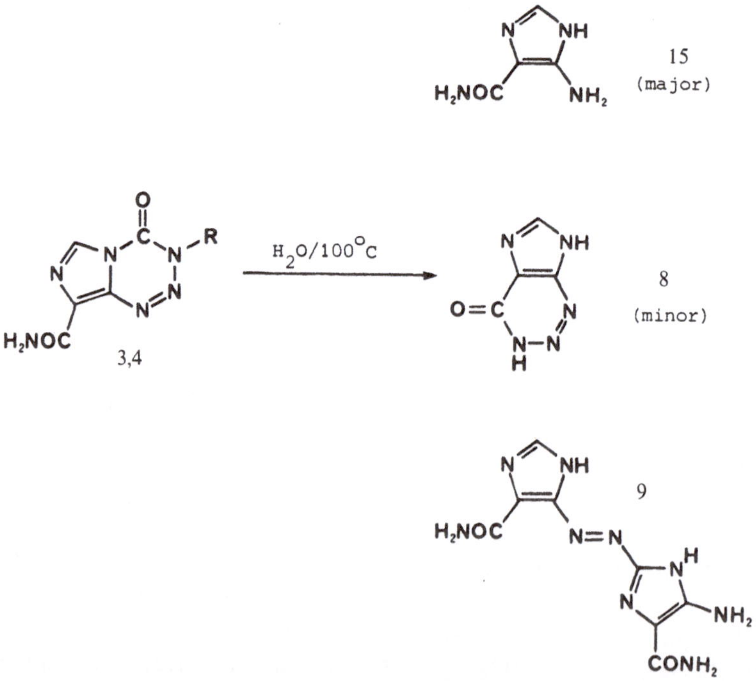

Scheme 6.5 Decomposition of mitozolomide (3) and temozolomide (4) in water.

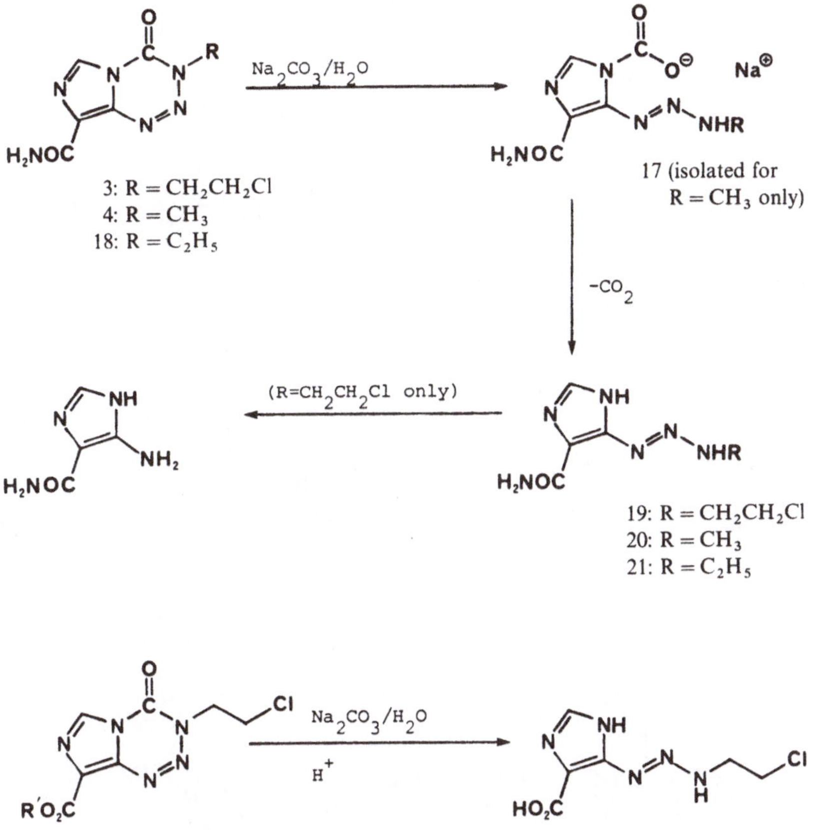

3: R = CH$_2$CH$_2$Cl
4: R = CH$_3$
18: R = C$_2$H$_5$

17 (isolated for
R = CH$_3$ only)

19: R = CH$_2$CH$_2$Cl
20: R = CH$_3$
21: R = C$_2$H$_5$

22: R' = C$_2$H$_5$
23: R' = H

24

Scheme 6.6 Hydrolysis of imidazotetrazinones in aqueous base.

mediate carbamate salt (**17**) in the hydrolysis of temozolomide in cold aqueous sodium carbonate shows that the formation of triazene (**20**) involves initial nucleophilic attack of water (or hydroxide) at the carbonyl group at the 4-position of the tetrazinone ring (Scheme 6.5). The main products of such base-catalysed hydrolysis[34] of the 3-alkyl-8-carbamoylimidazotetrazinones (**3, 4, 18**) were the corresponding triazenes (**19–21**). The ester (**22**) and the carboxylic acid (**23**) both gave the triazene carboxylic acid (**24**). Base-catalysed hydrolysis of pyrazolotetrazinones proceeds similarly[11]. The 2-chloroethyltriazene (MCTIC, **19**) arising from hydrolysis of mitozolomide is itself labile[33] to hydrolysis at pH 7.4 giving a mixture of 2-chloroethanol and (**15**) with the evolution of nitrogen[35].

The imidazotetrazinones react in a similar manner with alcohols, although the reported rates of reaction vary from worker to worker (Scheme 6.7). The

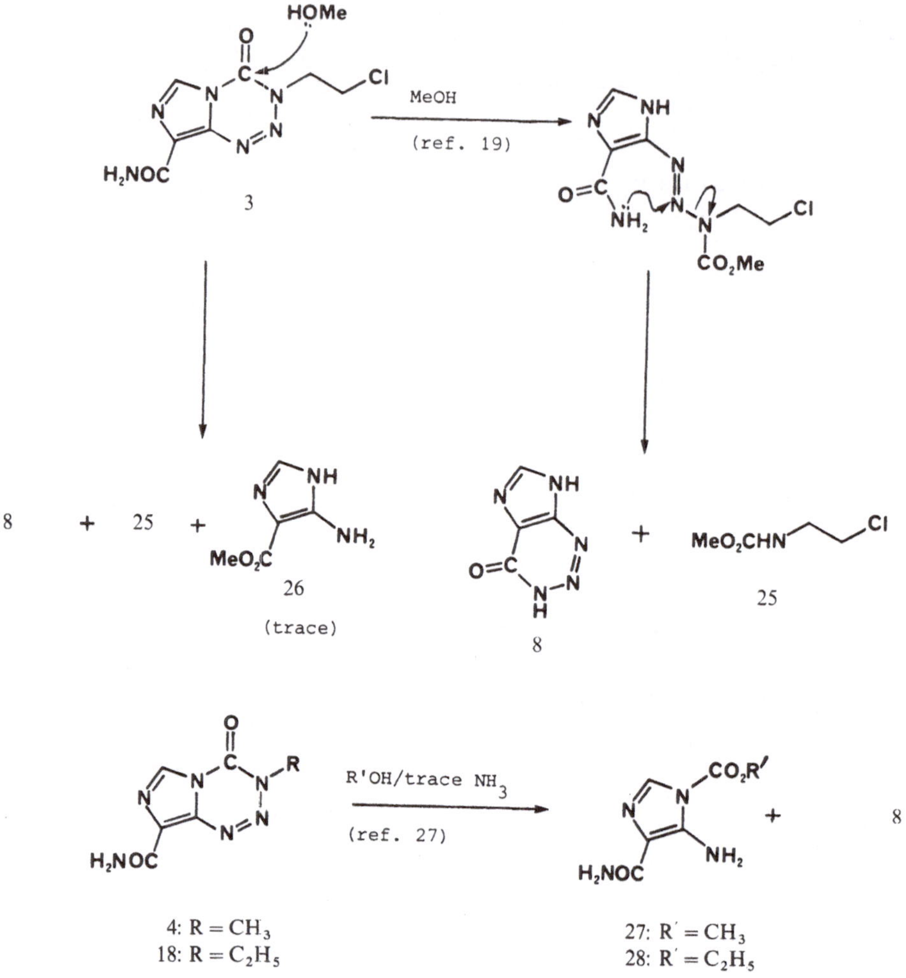

Scheme 6.7 Alcoholysis of imidazotetrazinones.

products from the treatment of mitozolomide with cold methanol were reported by Stone[19] to be 2-azahypoxanthine (**8**) and methyl N-(2-chloroethyl) carbamate (**25**); in boiling methanol, a small trace of methyl 5-aminoimidazole-4-carboxylate (**26**) was observed[7]. However, Baig[34] reported the 3-methyl and 3-ethyl imidazotetrazinones (**4**, **18**) to be very stable to boiling methanol. The rate of reaction with alcohols is greatly enhanced by addition of a trace of ammonia[19,29]. Under these base-catalysed conditions, the main product from the 3-methyl and 3-ethyl imidazotetrazinones was still 2-azahypoxanthine but the alkoxycarbonylimidazoles (**27**, **28**) were also formed[34]. The characterisation of these products of alcoholysis was complicated owing to conflicting identifications[36,37] of the products of alkoxycarbonylation of 5-aminoimidazole-4-carboxamide (**15**). Mitozolomide gave exclusively 2-azahypoxanthine, the product of cleavage of the 4,5 bond of the tetrahedral intermediate. It is interesting to note the lengths[30,31] of the 3, 4 (1.374 ± 0.003 Å) and 4,5 (1.392 ± 0.003 Å) bonds of the imidazotetrazinones in this context as indicators of bond order and bond energy and that corresponding lengths are not appreciably different between analogues.

The reactivity of the 3-(halo)alkylimidazotetrazinones with a representative nitrogen nucleophile, hydrazine[34], parallels that with the oxygen nucleophiles detailed above. The four 3-alkyl compounds (**4,18,29,30**) react slowly to form the carbohydrazide (**31**) through cleavage of the 3,4 bond (Scheme 6.8). In contrast, hydrazine caused fission of the 4,5 bond of mitozolomide (**3**), the final isolated product being 4-azidoimidazole-5-carboxamide (**32**), presumably via one of the tetrazenes (**33**, **34**). This mechanistic proposal is supported by the

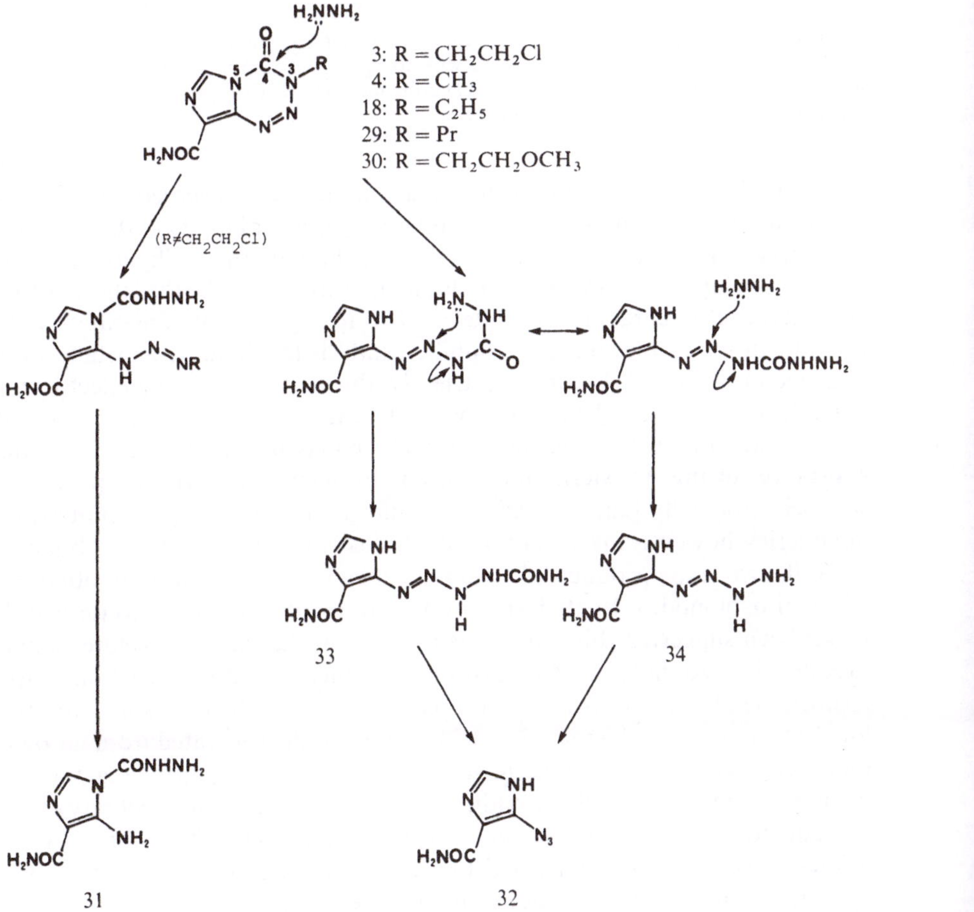
H2NNH2
3: R = CH2CH2Cl
4: R = CH3
18: R = C2H5
29: R = Pr
30: R = CH2CH2OCH3
(R≠CH2CH2Cl)
CONHNH2
NHCONHNH2
NHCONH2
H2NOC
31
32
33
34

Scheme 6.8 Hydrazinolysis of imidazotetrazinones.

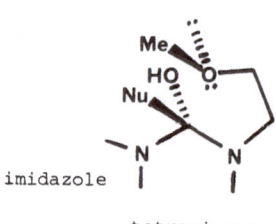

imidazole

tetrazinone

imidazole

tetrazinone

Figure 6.2 Stereoelectronic requirements for intramolecular nucleophilic catalysis in the reaction of 3-(2-chloroethyl) imidazotetrazinones with oxygen and nitrogen nucleophiles. Steric interaction between the methyl group (Me) and the nucleophiles (Nu) preclude this anchimeric assistance in 3-(2-methoxyethyl) analogues.

ready synthesis of this azide from 5-diazoimidazole-4-carboxamide (**7**) and hydrazine[38]. An explanation for the apparently greater lability of the 4,5 bond in mitozolomide, when compared with the 4,5 bond in the 3-alkylimidazotra-zinones, during reaction with alcohols and hydrazine may be that the chlorine atom in the tetrahedral intermediate can, owing to geometric constraints, only provide an intramolecular nucleophilic catalysis (anchimeric assistance) for cleavage of this bond (Figure 6.2). Clearly, this bond-weakening effect is not possible when the 3-substituent does not bear a heteroatom. The ethereal oxygen atom in the 2-methoxyethyl sidechain may be unable to perform this assistance owing to steric interactions from the methyl group in the stereoelectronically required reacting conformation. In the pyrazolotetrazi-none series, however, the 4,5 bond is intrinsically weaker than the 3,4 bond[11].

Sandmeyer-type products (4-haloimidazole-5-carboxamides) are obtained when mitozolomide is treated with bromide or iodide ions in hot acetic acid[34]. It has been suggested this represents reaction of the thermal cycloreversion product, 5-diazoimidazole-4-carboxamide, rather than direct reaction of the tetrazinone although this is not completely clear. Products arising from the formal trapping of 5-diazoimidazole-4-carboxamide, generated from mitozo-lomide, by carbon nucleophiles have also been observed[34] during reactions in hot acetic acid and in boiling pyridine (Scheme 6.9). The regioselectivity of the subsequent cyclisations to form imidazotriazines (**35–41**) is determined by the acidic or basic catalysis provided by the solvents and by the nature of the substituents attached to the active methylene function.

The effect of powerful oxidants on azolotetrazinone rings has not been

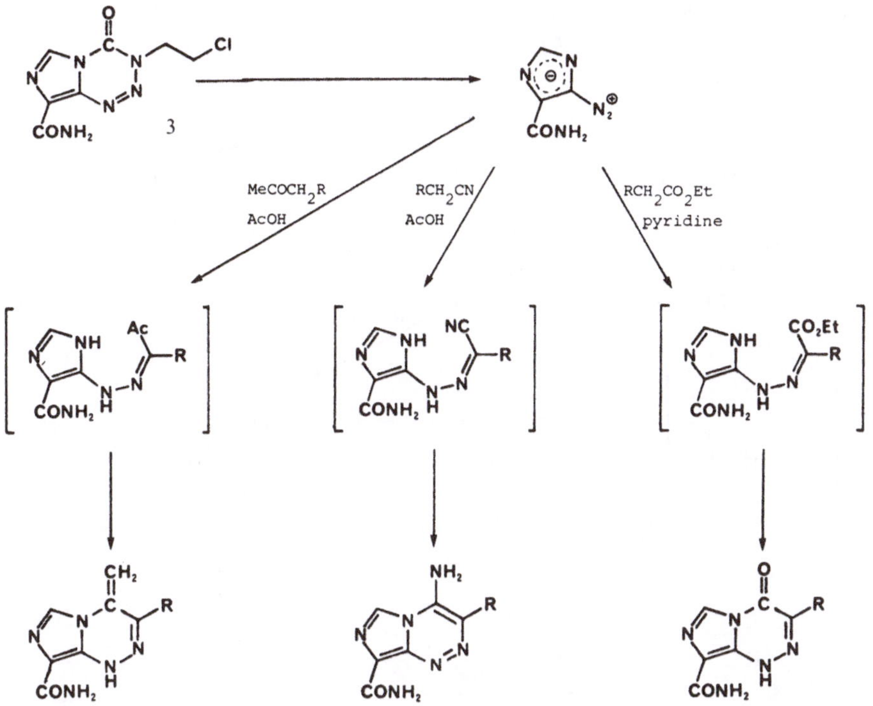

35: R = CO_2Me
36: R = CO_2Et
37: R = Ac

38: R = CN
39: R = $C0_2Et$

40: R = CN
41: R = Ac

Scheme 6.9 Reactions of mitozolomide (**3**) with carbon nucleophiles.

reported. Similarly, the effect of reducing agents is unknown but simple triazenes are unaffected by complex hydrides[39].

In summary, the chemistry of the imidazotetrazinones with nucleophiles is characterised by opening of the tetrazinone ring. The immediate products are frequently subject to further transformation under the reaction conditions.

6.4 Reactions involving retention of the tetrazinone ring

Acidic or electrophilic reagents usually permit the imidazotetrazinone ring structure to be retained. Indeed, mitozolomide (**3**) has been recovered unchanged from warm concentrated sulphuric acid, in striking contrast to the considerable acid-lability of the open-chain triazenes[40]. Presumably, this reflects the basicity of the N = N–N group in the different systems.

Some useful chemical transformations of functional groups attached to azolotetrazinones have, however, been accomplished using acidic and electrophilic reagents[29]. Treatment of mitozolomide with concentrated 'nit-

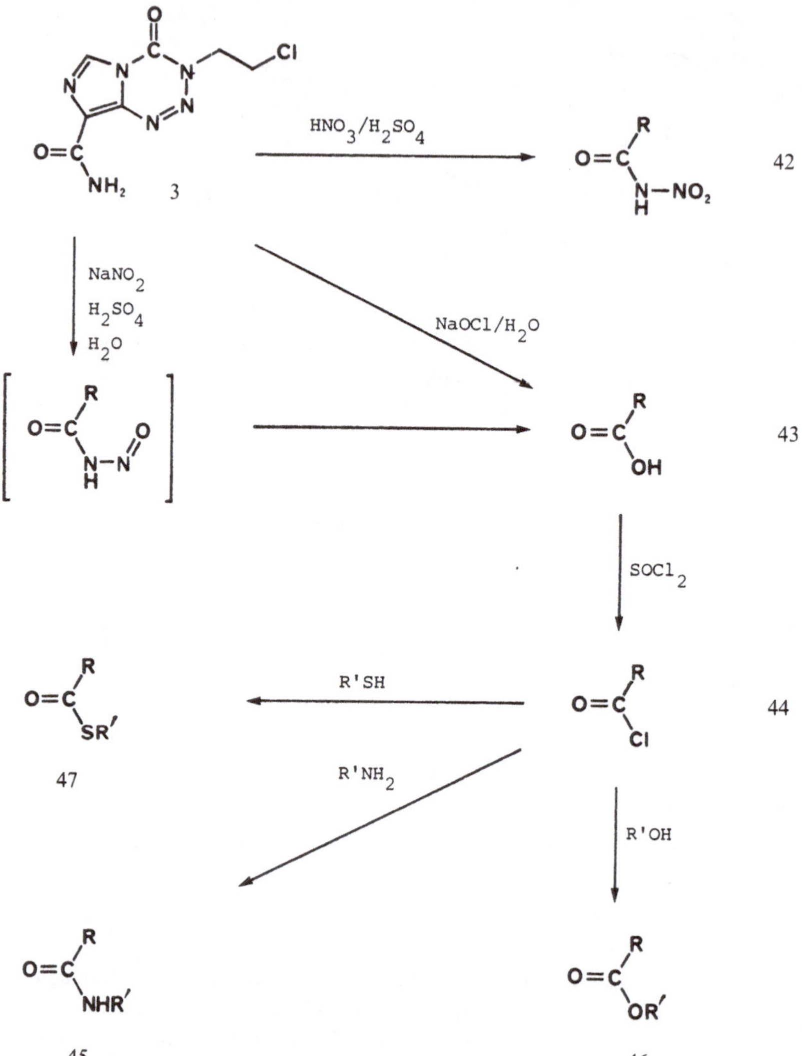

Scheme 6.10 Reactions of mitozolomide (**3**) with electrophiles. R = 3-(2-chloroethyl)-4-oxoimidazo[5, 1-*d*]-1, 2, 3, 5-tetrazin-8-yl.

rating mixture' gave only the *N*-nitroamide (**42**) (Scheme 6.10). This compound was found to be stable to mild hydrolytic conditions and to attempted rearrangement[32]. Nitrosation at carboxamide nitrogen affords a convenient synthesis of the carboxylic acid (**43**); this hydrolysis can also be effected by aqueous sodium hypochlorite. The chemistry of this carboxylic acid is straightforward, provided that aqueous basic conditions are avoided. The acid chloride (**44**), formed by treatment with thionyl chloride, reacts vigorously[42] with the appropriate nucleophiles to give amides, esters and thioesters (**45–47**).

Metabolic oxidation of the *N*-methyl groups of the *N,N*-dimethylcarboxamide analogue of mitozolomide by cytochrome P_{450} requiring enzymes of rodent systems has been described recently by Horspool[40]. The characterised products are the *N*-methylamide and mitozolomide presumably arising via *N*-(hydroxymethyl) intermediates in analogy with other aromatic amides[41].

Efficient deprotection of *N*-(4-methoxybenzyl) protected imidazotetrazinone-8-carboxamides and -8-sulphonamides by a standard[43] method of acid-catalysed nucleophilic substitution has been reported[29]. When the protecting group is attached directly to the tetrazinone ring, as in 3-(4-methoxybenzyl)imidazotetrazinones, this method appears to fail and the 3-unsubstituted imidazotetrazinone is not formed[29].

In summary, the chemistry of the azolotetrazinones with electrophiles and with acids is confined to interactions with peripheral functional groups.

6.5 Relationship of the chemistry of the azolotetrazinones to their biochemistry

The potent cytotoxicity of the lead compound, mitozolomide (**3**), has been attributed to interstrand cross-linking of DNA[44,45]. Tisdale[46] has recently extended this study to a wider range of cell lines and to the active antitumour agent temozolomide (**4**), and to the 3-ethyl analogue (**18**) which is cytotoxic *in vitro* but is not an antitumour agent *in vivo*. Discussions of the relationship of the chemistry to the (bio)chemical modes of action of the imidazotetrazinones must therefore take account of these observations and of the structure/antitumour activity relationship.

Of the two types of electrophilic biochemical reactivity (carbamoylation and alkylation) considered by Horgan and Tisdale[47], the former was discounted on the grounds that mitozolomide does not inhibit glutathione reductase, chymotrypsin and γ-glutamyltranspeptidase under conditions where these enzymes are inhibited by carbamoylating agents. The partial inhibition of an esterase of EMT6 cells by mitozolomide may, however, indicate some carbamoylation by the reagent[48]. The most likely reactive antitumour metabolite or metabonate of mitozolomide (**3**) is the *N*-(2-chloroethyl)triazene (MCTIC, **29**), the known product of aqueous hydrolysis. The cytotoxic activity of mitozolomide has then been explained[47] in terms of the reaction of this potent alkylating electrophile with nucleophilic sites on DNA, in particular with O^6 of guanine bases, with subsequent cross-linking of the two strands of the macromolecule. The differential sensitivity of the cell lines VA13 (sensitive, Mer⁻) and IMR90 (more resistant, Mer⁺) to mitozolomide is then consistent with the ability of the latter to repair the O^6-(2-chloroethyl) lesion prior to the cross-linking reaction. Temozolomide (**4**) is thought to exert its cytotoxicity through a similar mechanism involving the monomethyltriazene (**20**); although cross-linking is precluded. The 3-ethyl analogue (**18**), is an alkylating agent towards RNA, DNA and proteins and is a cytotoxin *in vitro*; however, alkylation of the O^6 position of guanine is not involved[20,46].

True enzymic metabolism of the imidazotetrazinones, as distinct from chemical change in physiological media, has been reported to result either in activation of the substrate as an antitumour agent or in deactivation, depending on the precise nature of the peripheral functional groups[49-51].

6.6 Concluding remarks

The azolotetrazinones represent a highly reactive series of heterocycles, both chemically and biologically. Suitably substituted, they show both electrophilic and nucleophilic character, with opening or retention of the ring system, respectively. Both the antitumour efficacy of some members of the class and their toxicity towards cells and organisms result from potently alkylating electrophiles generated by hydrolytic chemical reactions *in vivo*. Further detailed analysis of the interactions of azolotetrazinones and their metabonates with DNA may be necessary to explain satisfactorily the observed structure-activity relationship at a molecular level and to permit the design of a more selective and, therefore, more therapeutically useful agent.

References

1. P. F. Wiley, in *Chemistry of 1,2,3-Triazines and 1,2,4-Triazines, Tetrazines and Pentazines* Wiley Interscience, New York (1978) p. 1296.
2. M. H. Palmer, A. J. Gaskell and R. H. Findlay, *J. Chem. Soc. Perkin Trans. 2* (1974) 778.
3. K. Kubo, T. Nonaka and K. Odo, *Bull. Chem. Soc. Japan* **49** (1976) 1339.
4. R. N. Butler, D. Cunningham, P. McArdle and G. A. O'Halloran, *J. Chem. Soc. Chem. Commun.* (1988) 232.
5. G. Ege and K. Gilbert, *Tetrahedron Lett.* (1979) 4253.
6. E. Lunt, M. F. G. Stevens, R. Stone and K. R. H. Wooldridge, British Patent 2104522A.
7. M. F. G. Stevens, J. A. Hickman, R. Stone, N. W. Gibson, G. U. Baig, E. Lunt and C. G. Newton, *J. Med. Chem.* **27** (1984) 196.
8. M. F. G. Stevens, in *New Avenues in Developmental Cancer Chemotherapy*, Bristol-Myers Cancer Symposium Vol. 8, ed. K. R. Harrap and T. A. Connors, Academic Press, Orlando, Florida, USA.
9. E. Lunt, C. G. Newton, C. Smith, G. P. Stevens, M. F. G. Stevens, C. G. Straw, R. J. A. Walsh, P. J. Warren, C. Fizames, S. P. Langdon and L. M. Vickers, *J. Med. Chem.* **30** (1987) 357.
10. C. C. Cheng, E. F. Elslager, L. M. Werbel, S. R. Priebe and W. R. Leopold III, *J. Med. Chem.* **29** (1986) 1544.
11. G. Ege, K. Gilbert and K. Maurer, *Chem. Ber.* **120** (1987) 1375.
12. J. A. Hickman, N. W. Gibson, R. Stone, M. F. G. Stevens, F. Lavelle and C. Fizames, *Proc. 13th Int. Cancer Congress*, Geneva, UICC (1982) p. 1544.
13. J. A. Hickman, M. F. G. Stevens, N. W. Gibson, S. P. Langdon, C. Fizames, F. Lavelle, G. Atassi, E. Lunt and R. M. Tilson, *Cancer Res.* **45** (1985) 3008.
14. E. Erba, S. Pepe, P. Ubezio, A. Lorico, L. Morasca, C. Mangioni, F. Landoni and M. D'Incalci, *Br. J. Cancer* **54** (1986) 925.
15. E. S. Newlands, G. Blackledge, J. A. Slack, C. Goddard, C. J. Brindley, L. Holden and M. F. G. Stevens, *Cancer Treat. Rep.* **69** (1985) 801.

16. M. Harding, D. Northcott, J. Smyth, N.S.A. Stuart, J. A. Green and E. Newlands, *Br. J. Cancer* **57** (1988) 113.
17. J. A. Hickman, M. F. G. Stevens, N. W. Gibson, S. P. Langdon, C. Fizames, F. Lavelle, G. Atassi, E. Lunt and R. M. Tilson, *Cancer Res.* **45** (1985) 3008.
18. M. J. Tisdale, *Biochem. Pharmacol.* **34** (1985) 2077.
19. R. Stone, Ph.D. Thesis, Aston University, Birmingham (1981).
20. V. L. Bull and M. J. Tisdale, *Biochem. Pharmacol.* **36** (1987) 3215.
21. J. C. Madelmont, M. F. Moreau, D. Godeneche, P. Labarre and A. Veyre, *J. Labelled Compd. Radiopharm.* **25** (1988) 1135.
22. R. Huisgen, *Angew. Chem. Int. Ed. Engl.* **2** (1963) 565.
23. R. Huisgen, R. Grshey and J. Sauer, in *The Chemistry of Alkenes*, ed. S. Patai, Interscience, London (1964) p. 806.
24. R. Huisgen, *J. Org. Chem.* **41** (1976) 403.
25. G. W. Cowell and A. Ledwith, *Q. Rev. Chem. Soc.* **24** (1970) 119.
26. M. D. Threadgill, Unpublished observation.
27. T. L. Gilchrist and R. C. Storr, in *Organic Reactions and Orbital Symmetry*, 2nd edn., Cambridge University Press, Cambridge (1979) p. 201.
28. A. Padwa and T. Kumagai, *Tetrahedron Lett.* (1981) 1199.
29. G. U. Baig, Ph.D. Thesis, Aston University, Birmingham (1986).
30. P. R. Lowe, C. H. Schwalbe and M. F. G. Stevens, *J. Chem. Soc. Perkin Trans. 2* (1985) 357.
31. P. R. Lowe, Ph.D. Thesis, Aston University, Birmingham (1985).
32. C. Goddard, Ph.D. Thesis, Aston University, Birmingham (1985).
33. C. M. T. Horgan, Ph.D. Thesis, Aston University, Birmingham, (1985).
34. G. U. Baig and M. F. G. Stevens, *J. Chem. Soc. Perkin Trans. 1* (1987) 665.
35. Y. F. Shealy, C. A. O'Dell and C. A. Krauth, *J. Pharm. Sci.* **64** (1975) 177.
36. E. Shaw, *J. Biol. Chem.* **185** (1950) 439.
37. T. A. Pospelova, V. S. Mokrushin and Z. V. Pushkareva, USSR Patent 891622 (*Chemical Abstracts* **96** (1982) 199689s).
38. Y. F. Shealy and C. A. O'Dell, *J. Heterocycl. Chem.* **10** (1973) 839.
39. M. D. Threadgill and M. F. G. Stevens, *Synthesis* (1983) 289.
40. K. Vaughan and M. F. G. Stevens, *Chem. Soc. Rev.* **7** (1978) 377.
41. K. Horspool, A. Gescher, C. P. Quarterman, M. F. G. Stevens and E. Lunt, *Br. J. Cancer* **56** (1987) 221.
42. D. Ross, P. B. Farmer, A. Gescher, J. A. Hickman and M. D. Threadgill, *Biochem. Pharmacol.* **32** (1983) 1773.
43. M. I. Jones, C. Froussios and D. A. Evans, *J. Chem. Soc. Chem. Commun.* (1976) 472.
44. N. W. Gibson, L. C. Erikson and J. A. Hickman, *Cancer Res.* **44** (1984) 1767.
45. N. W. Gibson, J. A. Hickman and L. C. Erikson, *Cancer Res.* **44** (1984) 1772.
46. M. J. Tisdale, *Biochem. Pharmacol.* **36** (1987) 457.
47. C. M. T. Horgan and M. J. Tisdale, *Biochem. Pharmacol.* **33** (1984) 2185.
48. C. Dive, P. Workman and J. V. Watson, *Cancer Chemother. Pharmacol.* (1989) in press.
49. P. Workman and F. Y. F. Lee, *Br. J. Cancer* **50** (1984) 251.
50. C. J. Brindley, P. Antoniw and E. S. Newlands, *Br. J. Cancer* **53** (1986) 91.
51. A. Gescher and M. D. Threadgill, *Pharmacol. Ther.* **32** (1987) 191.

7 Chemistry of antifolates

M. G. NAIR

7.1 Introduction

Cancer can be defined as a disease that is characterised by a failure of the control mechanisms that are involved in cell division. Rapidly proliferating cells have a higher demand for DNA, and consequently require large quantities of purine and pyrimidine nucleotide substrates. Folic acid (1) plays a central role in the *de novo* biosynthesis of DNA. The coenzyme N^5, N^{10}-methylenetetrahydrofolate provides the one carbon unit necessary for the conversion of deoxyuridine monophosphate (dUMP) to thymidylic acid (dTMP). Thymidylate synthase catalyses the formation of dTMP, which is an obligatory intermediate in DNA biosynthesis (Figure 7.1). Tetrahydrofolate derivatives are also important cofactors in the biosynthesis of the purine ring system. The enzymes GAR-formyltransferase and AICAR-formyltransferase require N^{10}-formyltetrahydrofolate (2) as the one carbon donor for the construction of the purine ring. Nucleotides derived from the pyrimidine base thymine and the purine bases adenine and guanine are essential intermediates for DNA biosynthesis. In addition to these very important and crucial roles of folate coenzymes in cell division and growth they also function as carbon

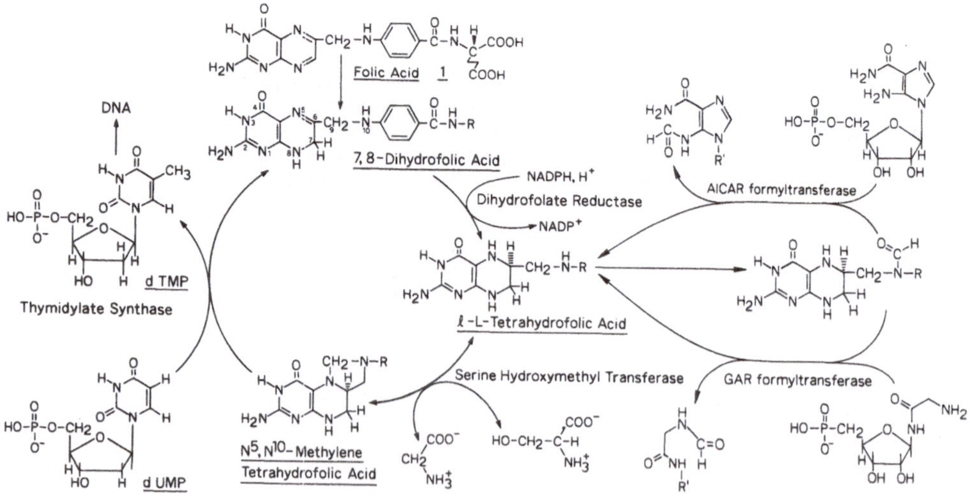

Figure 7.1

Aminopterin, R=—H
Methotrexate, R=—CH$_3$

donors in various biochemical reactions involved in one carbon metabolism. Prominent examples are the reactions catalysed by the folate based enzymes, serinehydroxymethyl transferase, N^5,N^{10}-methylenetetrahydrofolate dehydrogenase, tetrahydrofolate formiminotransferase, N^5,N^{10}-methylenetetrahydrofolate reductase, N^5-formiminotetrahydrofolate cyclodeaminase, C_1-tetrahydrofolate synthase and methionine synthase.

Antifolates are antimetabolites that interfere at various stages of folate metabolism. Antifolate antimetabolites are potentially capable of blocking the biosynthesis of purine and pyrimidine precursors of DNA, and many of them selectively inhibit the division and growth of certain neoplastic cells. Therefore several antifolate antimetabolites are known to be useful anticancer agents. Although a number of folate based enzymes are good targets for cancer chemotherapy, selective inhibition of only a few has been clinically successful thus far. In this chapter we focus on the chemistry of antifolate anticancer drugs that are currently under use as well as promising compounds presently undergoing clinical trials or commercial development.

7.2 Inhibitors of dihydrofolate reductase (EC 1.5.1.3)

Drugs such as methotrexate and aminopterin that are used for antifolate therapy of human cancers are directed toward dihydrofolate reductase (DHFR, EC 1.5.1.3) as the target enzyme[1–6]. The stoichiometric inhibition of DHFR by methotrexate[7–10] results in a complete blockade of one carbon metabolism, which is accompanied by a high level of toxicity to rapidly proliferating normal tissues such as bone marrow and gastrointestinal mucosa[11]. In order to minimise the toxicity of this drug, two types of rescue techniques have been advanced in recent years with some success. These are the high dose methotrexate/citrovorum rescue[12–15] and the thymidine rescue[16–18]. The second problem associated with the clinical use of methotrexate is resistance to the drug. Both intrinsic and acquired resistance to methotrexate are known[19]. Such resistance is accompanied by either an elevation of levels of the target enzyme, DHFR[20–23], or decreased transport of the drug[24-27]. A combination of these two characteristics, as well as changes in the enzyme itself, has also been noted[19-28]. In spite of the problems associated with toxicity and resistance, methotrexate continues to be a useful drug.

The main reason for the high toxicity of methotrexate is the inhibition of DHFR[7-10,29]. This enzyme converts dihydrofolate to tetrahydrofolate. All biologically relevant coenzymatic forms of the vitamin folic acid have a fully reduced pteridine ring. The source of folic acid in man is exogenous and the presence of 7,8-dihydrofolic acid in the cell is a result of its formation as a co-product in the reaction catalysed by thymidylate synthase. Dihydrofolate thus generated can become active only if it is reduced to the tetrahydro derivative. Since methotrexate inhibits DHFR, dihydrofolic acid is accumulated in the form of a dead-end pool and consequently all the coenzymatic forms of tetrahydrofolate are depleted. The consequence of this complete blockade of one carbon metabolism is very high toxicity to the cells. These observations have been documented. The incorporation of [^{14}C] formate into the purines is blocked up to 90% in animals receiving methotrexate[30,31]. Methotrexate causes marked inhibition of *de novo* purine biosynthesis in Ehrlich ascites cells *in vivo*[32] and the incorporation of labelled amino acids into protein in *E. coli* extracts[33]. Methotrexate has also been shown to block histidine degradation at the formimino transferase reaction, and patients receiving the drug excrete large amounts of formiminoglutamic acid[34]. These data taken together provide evidence that methotrexate blocks all the major routes of one carbon metabolism and resultant toxicity.

Since methotrexate (MTX) and aminopterin (AMT) are close analogues of the vitamin folic acid, the chemical syntheses of these compounds closely parallel the synthesis of the vitamin itself. The various strategies that are

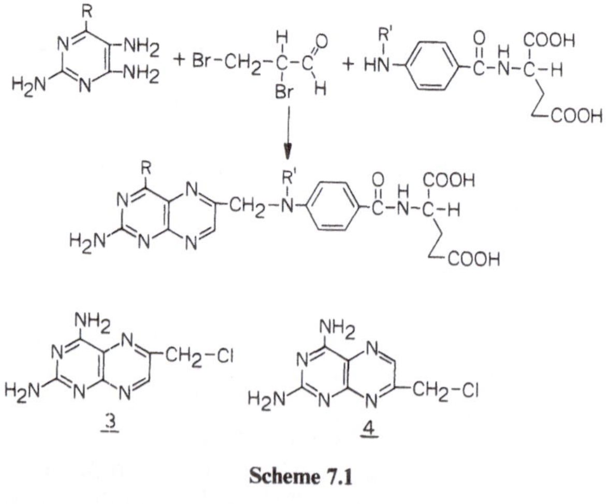

Scheme 7.1

available for the construction of 6-substituted pteridines have been applied to the synthesis of AMT and MTX. The earliest synthesis reported for MTX was based on the Waller reaction[35]. In this method a mixture of an appropriately 4-substituted 2,5,6-triaminopyrimidine, a *p*-aminobenzoyl-L-glutamate derivative and a three carbon reactant such as 2,3-dibromopropionaldehyde is reacted to obtain the corresponding 6-substituted pteridine in low yield (Scheme 7.1). Methotrexate was synthesised according to this procedure by the reaction of 2,4,5,6-tetraaminopyrimidine, 2,3-dibromopropionaldehyde and *p*-methylaminobenzoyl-L-glutamate. When *p*-aminobenzoyl-L-glutamate was substituted for the *N*-methyl derivative in the above reaction, the 4-amino analogue of folic acid, aminopterin, was obtained in acceptable, but low yield. When 2,2,3-trichlorobutyraldehyde was reacted with 2,5,6-triamino-4-pyrimidinol and *p*-aminobenzoyl-L-glutamic acid, 9-methylpteroylglutamic acid was obtained. Although many types of three carbon reactants have been used 2,3-dibromopropionaldehyde has been the choice of many investigators[36–38]. Suster and coworkers reported the use of 1,1,3-trichloroacetone as the three carbon reactant in the Waller type reaction in a successful synthesis of MTX and its 7 isomer. 6-Chloromethyl-2,4-diaminopteridine (3) and its 7 isomer (4) were postulated as intermediates in this reaction[37,38].

A detailed mechanistic investigation of the Waller reaction was carried out by Temple *et al.*[36]. Their results suggested that dehydrobromination of 2,3-dibromopropionaldehyde by the *p*-aminobenzoyl sidechain gives rise

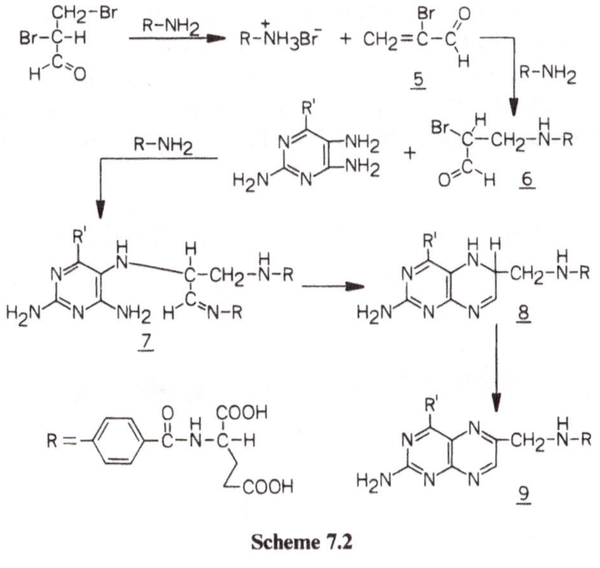

Scheme 7.2

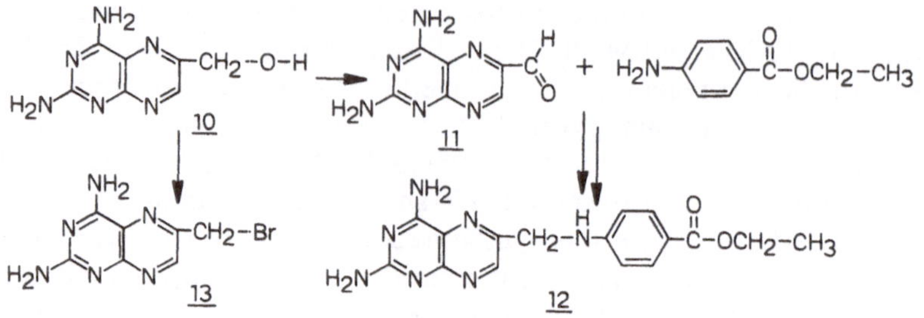

Scheme 7.3

to 2-bromopropenal (**5**, Scheme 7.2) which on reaction with the sidechain produces an intermediate (**6**). Alkylation of the 5-amino group of the pyrimidine with (**6**) yields (**7**) which by way of a transamination reaction produces the dihydropteridine (**8**). Air oxidation of (**8**) gives rise to the 6-substituted Waller product (**9**).

Montgometry et al. reported[39] a general and excellent synthesis of various aminopterin analogues via the crucial intermediate, 6-(hydroxymethyl)-2,4-diaminopteridine (**10**, Scheme 7.3) that was prepared by the method of Baugh and Shaw[40]. Conversion of (**10**) to the corresponding aldehyde (**11**) was accomplished by a Pitzner–Moffat oxidation procedure. Reaction of aldehyde (**11**) with ethyl-p-aminobenzoate gave the corresponding Schiff's base which on NaBH$_4$ reduction and subsequent oxidation gave ethyl-4-amino-4-deoxypteroate (**12**) albeit in low yields. However, treatment of (**10**) with triphenyl phosphonium dibromide and subsequent workup gave (**13**), which on reaction in DMAC with various amines yielded the desired 6-substituted pteridines. By this method Montgomery et al.[41] prepared a series of AMT analogues including MTX in good yield and purity.

Taylor et al.[42] carried out several pteridine syntheses taking advantage of the chemical reactivity of a key pyrazine intermediate, 2-amino-3-cyano-5-chloromethyl pyrazine (**14**, Scheme 7.4). This intermediate was prepared by

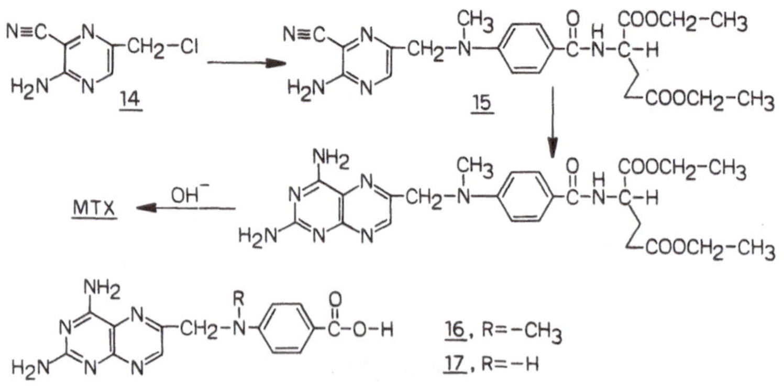

Scheme 7.4

PCl_3 deoxygenation of the N-oxide of (14) which was prepared by the condensation of aminomalononitrile with β-chloropyruvaldoxime. Reaction of (14) with diethyl p-methylaminobenzoyl-L-glutamate gave the pyrazine intermediate (15) which on cyclisation with guanidine gave the diethyl ester of methotrexate from which racemic MTX was prepared by careful alkaline hydrolysis[43]. Several analogues of MTX in which the L-glutamate moiety has been replaced with various other amino acids have been prepared using 4-amino-4-deoxy-N^{10}-methylpteroic acid (16) as a crucial intermediate[44].

4-Amino-4-deoxy-N^{10}-methylpteroic acid (16) can be conveniently prepared by the bacterial conversion of MTX using the organism *Pseudomonad* sp.[45,46] in an appropriate medium. It can also be prepared by the reaction of (13) with p-methylaminobenzoic acid in DMAC[41]. A synthesis of 4-amino-4-deoxypteroate (17) that is useful for the synthesis of several aminopterin analogues has been reported from this laboratory according to Scheme 7.5[47]. Alternatively this compound can be prepared either by the reaction of p-aminobenzoic acid with (13) as described for (16)[41], or by the reaction of (14) with p-aminobenzoic acid and cyclisation of the resulting product with guanidine[43].

Another interesting series of compounds that are powerful inhibitors of DHFR are the 10-deazaaminopterins[48]. These compounds exhibit a wide spectrum of antitumour activity and they are preferentially accumulated in tumour compared to normal tissues[49,50]. Consequently, 10-deazaaminopterins have distinctly superior therapeutic advantage over methotrexate. The first synthesis of 10-deazaaminopterin (10-DAAM) (24) was reported by DeGraw *et al.* 1974[48]. It was soon recognised that the 10-ethyl analogue of 10-deazaaminopterin (25) was equally active as the parent compound and a number of syntheses for the facile preparation of 10-DAAM and its 10-alkyl analogues soon emerged[51].

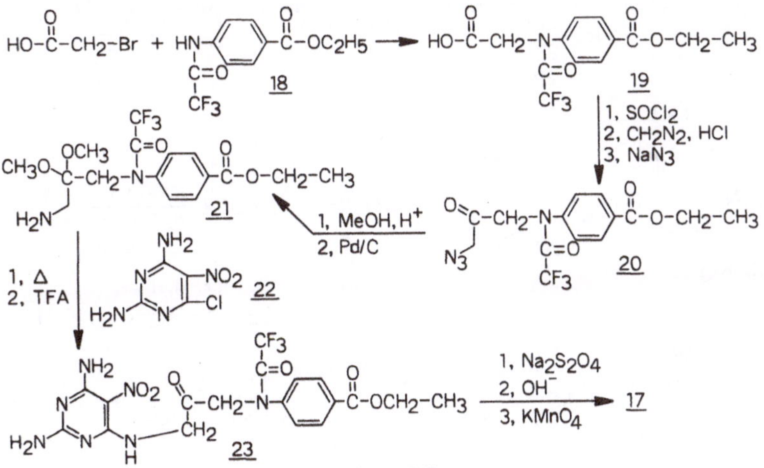

Scheme 7.5

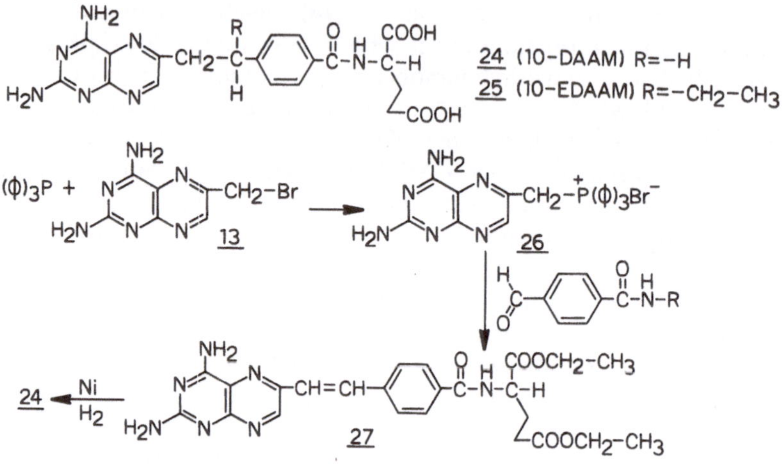

Scheme 7.6a

Piper and Montgomery[52] reacted the bromomethyl pteridine (13) with triphenylphosphine to obtain the corresponding Wittig salt (26) which on treatment with diethyl-p-formylbenzoylglutamate and base gave the olefin (27). Hydrogenation of (27) with Raney nickel followed by oxidation and base hydrolysis of the hydrogenation product gave 10-deazaaminopterin (24) in moderate yield[52] (Scheme 7.6a). DeGraw et al.[51] developed an alternative strategy for the synthesis of 10-alkyl-10-deazaaminopterins as summarised in Scheme 7.6b. The synthesis starts with the alkylation of the dianion derived

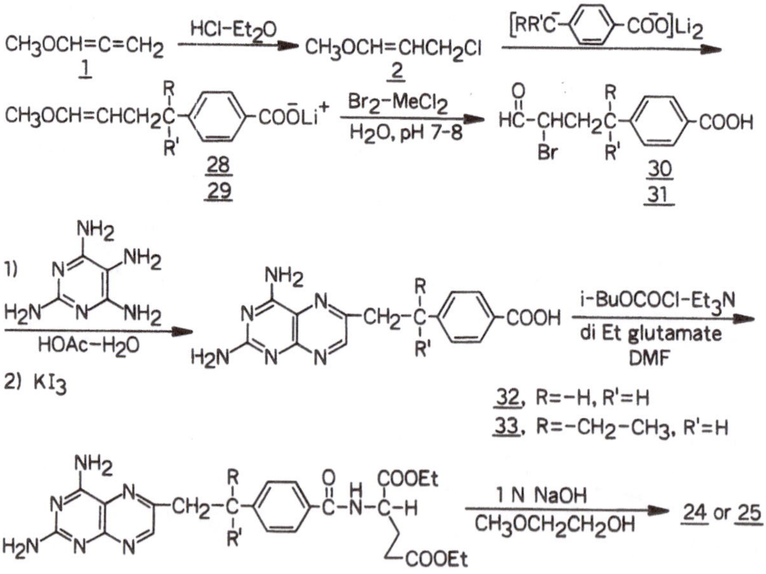

Scheme 7.6b

from 4-alkylbenzoic acids with methoxyallylchloride, followed by bromination and selective hydrolysis of the resulting product (28) or (29) to the bromoaldehyde (30) or (31). Reaction of (30) or (31) with 2,4,5,6-tetraaminopyrimidine and oxidation of the product with KI_3 gave the corresponding 4-amino-4-deoxy-10-substituted-10-deazapteroid acid (32) or (33) in acceptable yield. Subsequent elaboration of the pteroic acid analogues to (24) and (25) was carried out by standard procedures.

Another convenient and general synthesis of 10-alkyl-10-deazaaminopterins was reported from this laboratory[53] according to the procedure outlined in Scheme 7.7. The Wittig reagent (34) was prepared from N-(3-bromo-2-oxopropyl)phthalimide. Reaction of (34) with p-formylmethylbenzoate gave the enone (35) which on reduction with Zn/HOAc gave the saturated ketone (36). Alternatively (35) can be converted to (37) or (38) by the conjugate addition of organometallic reagents in the presence of Cu^+ ions. The carbonyl groups of (36), (37) and (38) were protected as the oxime, and subjected to hydrazinolysis to give the corresponding masked α-aminoketones (39), (40) and (41). Reaction of (39) or (41) with 6-chloro-2,4-diamino-5-nitropyrimidine, gave the pyrimidine intermediate (42) or (43). Treatment of (42) or (43) with a mixture of TFA and HCl followed by reduction with $Na_2S_2O_4$ and oxidation with $KMnO_4$ gave (32) or (33) in good yield. This procedure has the added advantage of preparing a large number of 10-substituted-10-deazaaminopterins due to the versatility of the conjugate

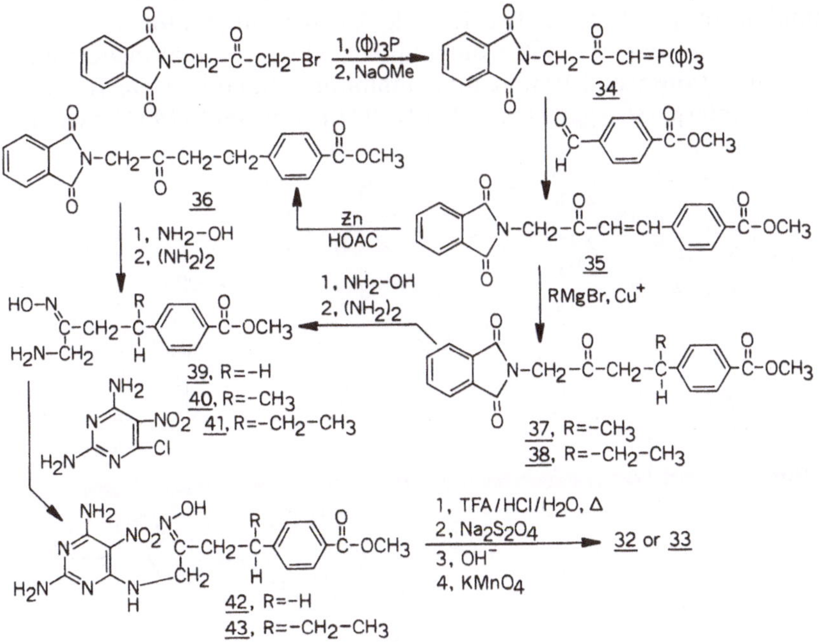

Scheme 7.7

addition reaction. For example, 10-deaza-10-methylaminopterin could be prepared from (40) by the above procedure.

Classical antifolates such as aminopterin and MTX enter the cell by a carrier-mediated route[54]. The effectiveness of these antifolates is however limited by the development of resistance. The most common cause of acquired resistance is a defect in the membrane transport of MTX[55,56]. A number of compounds in the 4-aminoantifolate series have been developed in an attempt to overcome this type of resistance. These antifolates do not possess the free carboxyl groups of the glutamate moiety. They are envisioned to be more lipophilic and capable of entering the cell by passive diffusion. Although many of these non-classical 4-aminoantifolates exhibit interesting biological properties, the most interesting compound in this series appears to be the pyridopyrimidinediamine (44), code-named BW301U, which is a potential candidate for the treatment of brain tumours[57]. The chemical synthesis of BW301U was reported by Grivsky et al.[58] Treatment of 2,5-dimethoxy-benzaldehyde with ethylacetoacetate gave the corresponding olefin[45] which was hydrogenated with Pd/C to give (46) (Scheme 7.8). Reaction of 2,4,6-pyrimidinetriamine with (46) gave the pyrimidinone (47). Chlorination of (47) to (48) followed by hydrogenolysis gave the dimethoxypyridopyrimidine (44).

Another class of compounds that are potent inhibitors of DHFR are the 6-substituted 2,4-diaminoquinazolines. Hynes et al.[59] synthesised a series of quinazolines that can be viewed as the 5,8-dideaza analogues of aminopterin and isoaminopterin[60]. A number of these compounds exhibited very strong inhibition of rat liver and L1210 leukaemia dihydrofolate reductases comparable to that of MTX. Compounds having a normal folate configuration at positions 9 and 10 were more inhibitory than their isomeric reversed bridge counterparts. Against an L1210 cell line resistant to MTX, by virtue of

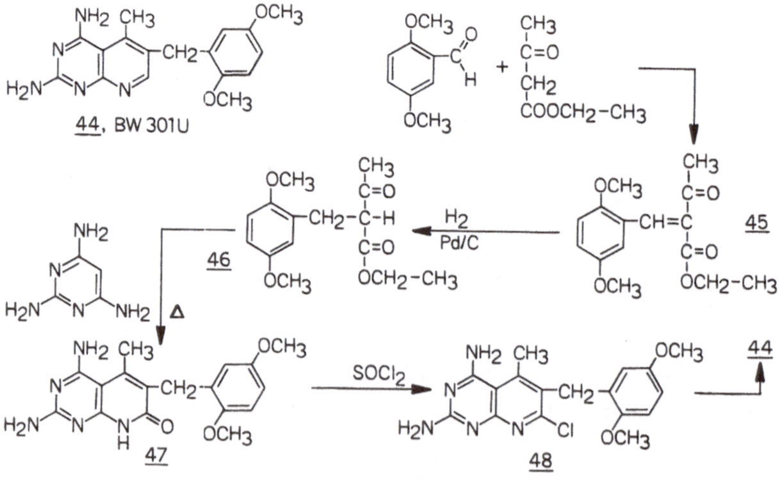

Scheme 7.8

altered transport and overproduction of DHFR, partial but not complete cross-resistance was observed for certain analogues. The syntheses of 5,8-dideazaaminopterin analogues were carried out starting from 2-amino-5-nitrobenzonitrile (**49**). Reaction of (**49**) with guanidine followed by stannous chloride reduction gave the triaminoquinazoline (**50**). By modification of the procedure of Davoll and Johnson[61], (**50**) was converted to 6-cyano-2,4-diaminoquinazoline (**51**) which on reductive cyclisation with di-*t*-butyl-*p*-aminobenzoylglutamate gave the di-*t*-butyl ester of 5,8-dideazaaminopterin (**52**). Treatment of (**52**) with TFA gave 5,8-dideazaaminopterin (**53**) which on reductive alkylation with formaldehyde and NaBH$_3$CN gave 5,8-dideaza-methotrexate (**54**) (Scheme 7.9).

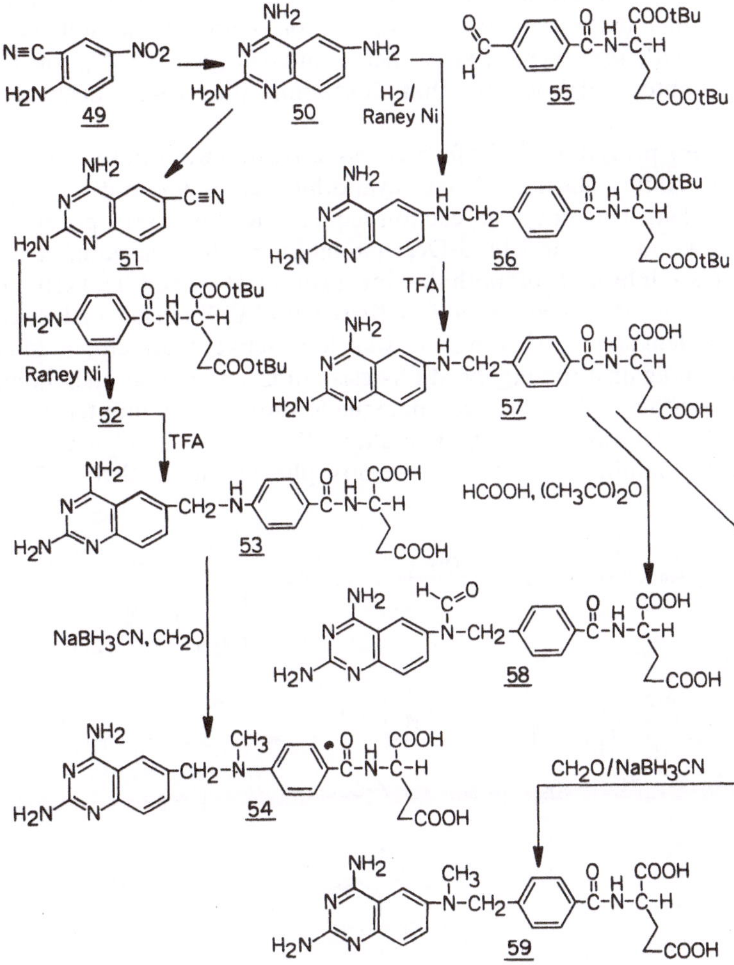

Scheme 7.9

The triaminoquinazoline (50) was condensed reductively with di-*t*-butyl-*p*-formylbenzoyl-L-glutamate (55) to the diester (56) which on treatment with TFA gave 5,8-dideazaisoaminopterin (57). Formylation of (57) with formic acid and acetic anhydride gave the N^9-formylisoaminopterin analogue (58) while the N^9-methyl analogue (59) was prepared by the reductive alkylation of (57) with formaldehyde and $NaBH_3CN$.

Two other interesting antifolates in the deazaaminopterin series are 5-deazaaminopterin (60) and 5,10-dideazaaminopterin (61). Temple *et al.*[62] reported the synthesis of a number of pyrido[2,3-*d*]pyrimidines including 5-deazaaminopterin according to Scheme 7.10. Subsequently, Taylor reported the synthesis of (61) by an alternative and more elaborate route[63] that is depicted in Scheme 7.11. In Temple's procedure[62], 2,4,6-triaminopyrimidine was reacted with triformylmethane to obtain 2,4-diamino-6-formyl-5-deazapteridine (62). Reductive alkylation of diethyl (*p*-aminobenzoyl)-L-glutamate with (62) using Raney nickel and H_2 gave a 32% yield of 5-deazaaminopterin diethyl ester, that on saponification gave 5-deazaaminopterin (60).

In Taylor's procedure[63], the formyl derivative (62) was obtained by a more elaborate route starting from cyanothioacetamide and 2-methyl-3-ethoxyacrolein. The nine step reaction sequence leading to the production of (60) is shown in Scheme 7.11. 5-Deazaaminopterin (60) was equipotent with MTX as an inhibitor of both bovine liver DHFR and L 1210 murine leukaemia cells. It had comparable activity with MTX *in vivo* against L 1210 and P388 leukaemia. However it was less active than either MTX or aminopterin against the highly MTX-resistant G 361 human melanoma cell line. The synthesis of the related analogue 5,10-dideazaaminopterin (61) was reported by DeGraw *et al.*[64] by two alternative routes. Condensation of the piperidine enamine of 4-*p*-carbomethoxyphenylbuteraldehyde (70) with

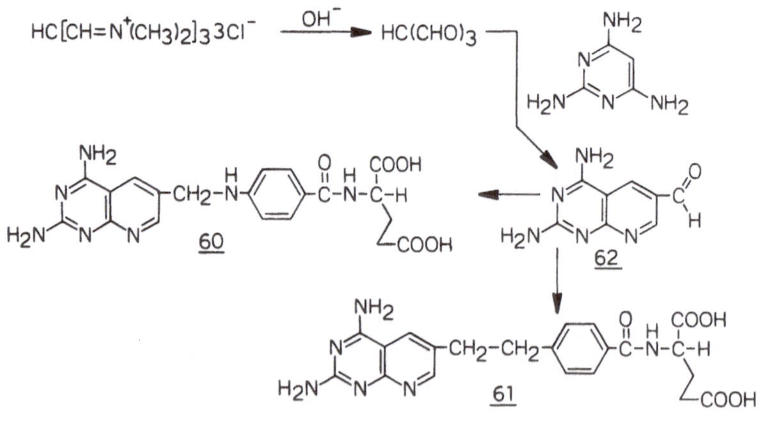

Scheme 7.10

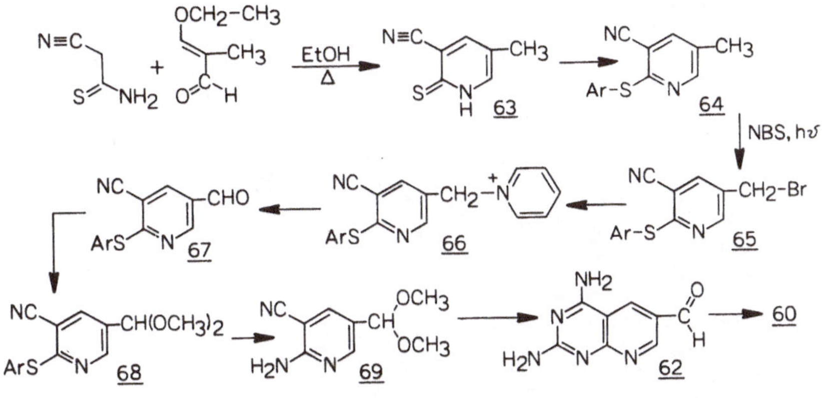

Scheme 7.11

ethoxymethylenemalononitrile followed by treatment of the resultant aryl-ethyleneaminomalononitrile (**71**) with methanolic ammonia produced the aminocyanopyridine (**72**). Cyclisation of (**72**) with guanidine followed by base hydrolysis afforded 4-amino-4-deoxy-5,10-dideazapteroic acid (**73**). Coupling of (**73**) with glutamate yielded 5,10-dideazaaminopterin (**61**) (Scheme 7.12).

Alternatively reduction of 2,4-diamino-6-formyl-5-deazapteridine (**62**) with NaBH$_4$ gave the corresponding hydroxymethyl compound (**74**). Conversion of (**74**) to the bromide (**75**) was followed by alkylation of dimethyl homopter-ephthalate (**76**) to afford methyl 4-amino-4-deoxy-10-carbomethoxy-5,10-dideazapteroate (**77**). Decarboxylation of (**77**) with ester cleavage gave 4-amino-4-deoxy-5,10-dideazapteroic acid (**73**) which was elaborated to (**61**) (Scheme 7.13). Details of the chemical synthesis of 5,10-dideazaaminopterin

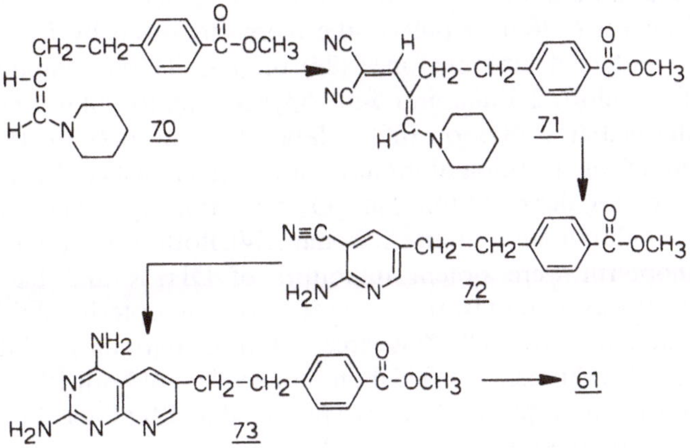

Scheme 7.12

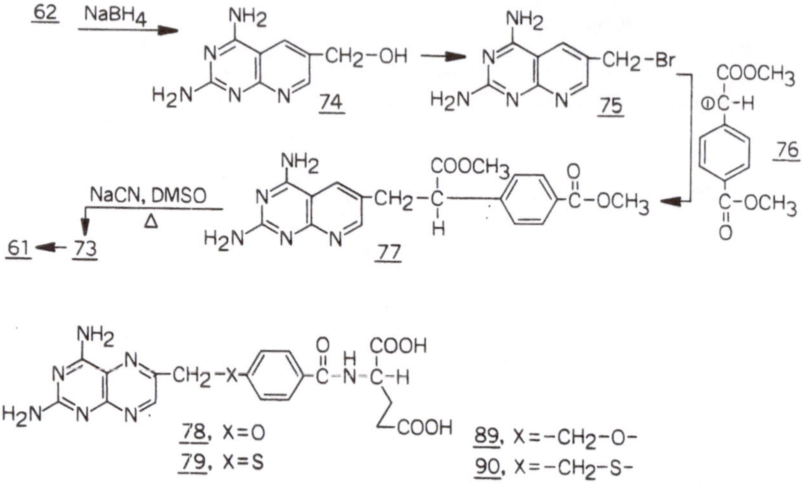

Scheme 7.13

(61) developed by Taylor et al.[65] are discussed in a later section (7.4) of this chapter which deals with inhibitors of GAR-formyltransferase. Compound (61) was an effective growth inhibitor of folate dependent bacteria, *Streptococcus faecium* and *Lactobacillus casei*. 5,10-Dideazaaminopterin and its, 5,6,7,8-tetrahydro-derivative exhibited significant *in vivo* activity against L 1210 leukaemia which was comparable to that observed with methotrexate.

The amino group at the tenth position of aminopterin was replaced with a heteroatom such as oxygen[66] or sulphur[67]. The synthetic strategies for the construction of 10-oxaaminopterin (78) and 10-thioaminopterin (79) were developed in this laboratory[66,37] and elsewhere[68,39]. Thus reaction of N-3-(bromo-2-oxopropyl)phthalimide (80) with either (81) or (82) gave the corresponding phthalimide intermediates (83) or (84). Protection of the carbonyl function of (83) or (84) as the oxime followed by hydrazinolysis yielded the protected aminoketones (85) or (86). These compounds were reacted with 6-chloro-2,4-diamino-5-nitropyrimidine to obtain the key pyrimidine intermediates (87) or (88). Elaboration of (87) or (88) to 10-oxaaminopterin or 10-thioaminopterin was accomplished by a series of reactions involving deprotection, NaS_2O_4 reduction, cyclisation, oxidation, hydrolysis and glutamate coupling (Scheme 7.14). Both 10-oxaminopterin and 10-thioaminopterin were potent inhibitors of DHFR and bacterial cell growth[66-68]. It was of interest to note that 10-oxaaminopterin inhibited folate and MTX influx in HeLa cells[69]; whereas 10-thioaminopterin inhibited only MTX transport in this cell line. The homologues[70,71] of both 10-oxaaminopterin and 10-thioaminopterin were also prepared and biologically evaluated. Both 10-oxahomoaminopterin (89), and 10-thiohomoaminopterin (90) had significantly lower activity compared to their predecessors, indicating the

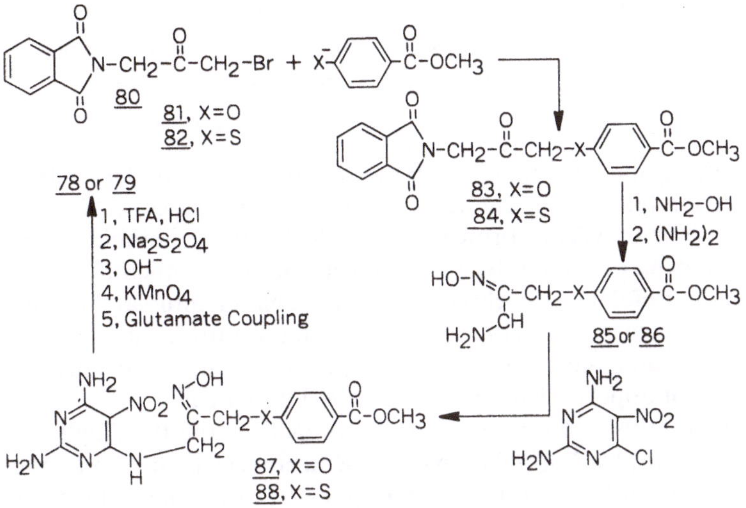

Scheme 7.14

biological sensitivity of classical 4-aminoantifolates to structural changes at the C^9, N^{10}-bridge region.

Although a large number of classical analogues of aminopterin such as 10-oxaaminopterin[66], 10-thioaminopterin[67], 11-thiohomoaminopterin[70], iso-aminopterin[60] and 9-nor-10-deazaaminopterin[72] were prepared and evaluated as potential anticancer agents, only those compounds belonging to the 10-deazaaminopterin series exhibited biological activity superior to methotrexate. The 10-deazaaminopterin compounds were shown to be active against a wide variety of experimental tumours in mice and they are presently undergoing advanced clinical trials[73,74] and commercial development in Europe. Other examples of classical analogues of aminopterin that have exhibited significant antitumour activity are 1-deazaaminopterin, 3-deaza-aminopterin, 3-deazaamethopterin[75] and the 2,6-diaminopurine analogues of MTX and AMT[76]. The activity exhibited by these analogues was not interesting from a therapeutic point of view.

Certain non-classical inhibitors of DHFR such as trimethoprim[77], BW 310U[58], and trimetrexate[78] show promise in the treatment of opportunistic infections in immunocompromised patients. Pulmonary infection is common-ly recognised as a life-threatening disorder in AIDS patients and Pneumocyst-is Carinii Pneumonia is the major opportunistic infection associated with AIDS[78-80]. The degree of synergism of antifolates with NADPH is reflected in the type of enzyme inhibition, and provides the basis of their selectivity as antibacterial agents. Subtle differences in the topology of the binding sites of homologous enzymes can manifest themselves as large differences in ligand affinity[81]. Antifolates such as trimethoprim (**91**) and trimetrexate (**92**) do not

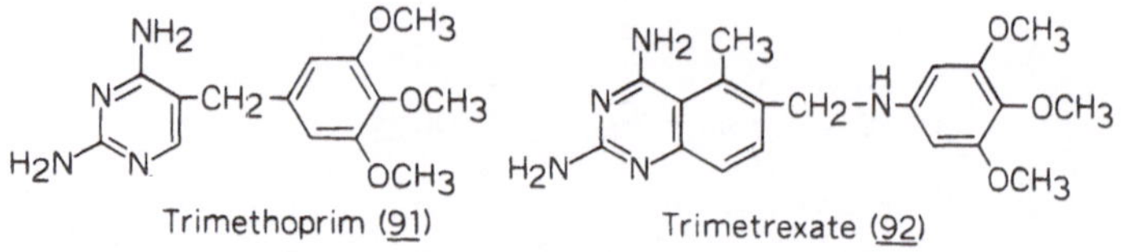

Trimethoprim (91)   Trimetrexate (92)

effectively inhibit mammalian DHFR, whereas they are very selective and powerful inhibitors of the bacterial enzyme. Such differences in binding affinity towards human versus protozoal DHFR by these antifolates are being currently exploited for selective antiprotozoal chemotherapy. Since 5-formyltetrahydrofolate is a well known rescue agent of antifolate toxicity, which cannot be transported efficiently in certain protozoa, a combination of trimetrexate and 5-formyltetrahydrofolate shows considerable promise in the treatment of opportunistic infections in AIDS patients.

Trimethoprim[77] can be conveniently prepared from the dihydrocinnamic acid derivative (93). Formylation of (93) with ethyl formate to (94) followed by condensation with guanidine gave the hydroxy pyrimidine (95). Conversion of the hydroxyl group of (95) to the desired amino group was accomplished by successive treatment of (95) with $POCl_3$ and ammonia (Scheme 7.15).

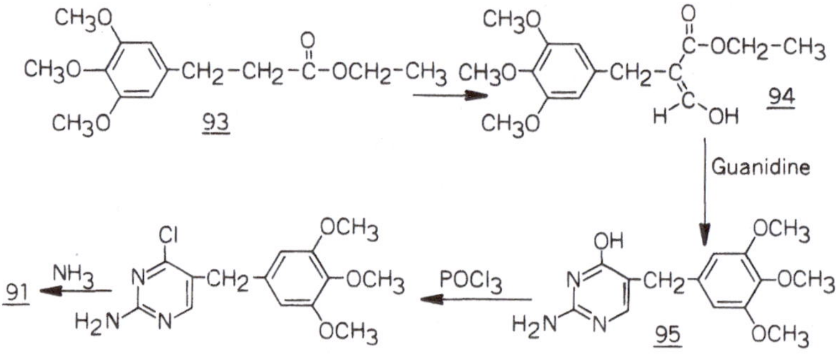

Scheme 7.15

7.3   Inhibitors of thymidylate synthase (EC 2.1.1.45)

Thymidylate synthase (TS) catalyses the reductive methylation of deoxyuridine monophosphate (dUMP) to thymidylate (dTMP), a reaction in which N^5, N^{10}-methylene-tetrahydrofolate participates as a cofactor[82] (Figure 7.1). Methylation of dUMP is the final step in the formation of dTMP that is required exclusively for DNA biosynthesis. This unique feature of the enzyme makes it a very attractive target for cancer chemotherapy[83,84]. Two series of TS inhibitors are well documented. Those belonging to the substrate class are analogues of dUMP and are well known inhibitors of TS; discussions of such compounds are beyond the scope of this chapter. TS inhibitors of the second series are analogues of folate coenzymes. One of the earliest known compounds in this series is tetrahydrohomofolic acid **(96)**[85]. *d,l*-Tetrahydrohomofolate was active against a methotrexate resistant strain of L 1210 leukaemia[86]. 7,8-Dihydrohomofolic acid was a substrate of *L. casei* DHFR, and the resulting *l*-L-tetrahydrohomofolate has been shown to be a powerful inhibitor of *E. coli* thymidylate synthase. *d,l*-Tetrahydrohomofolate was prepared by catalytic reduction of homofolic acid **(97)** which was synthesised according to Scheme 7.16. The alanine derivative **(98)** that was

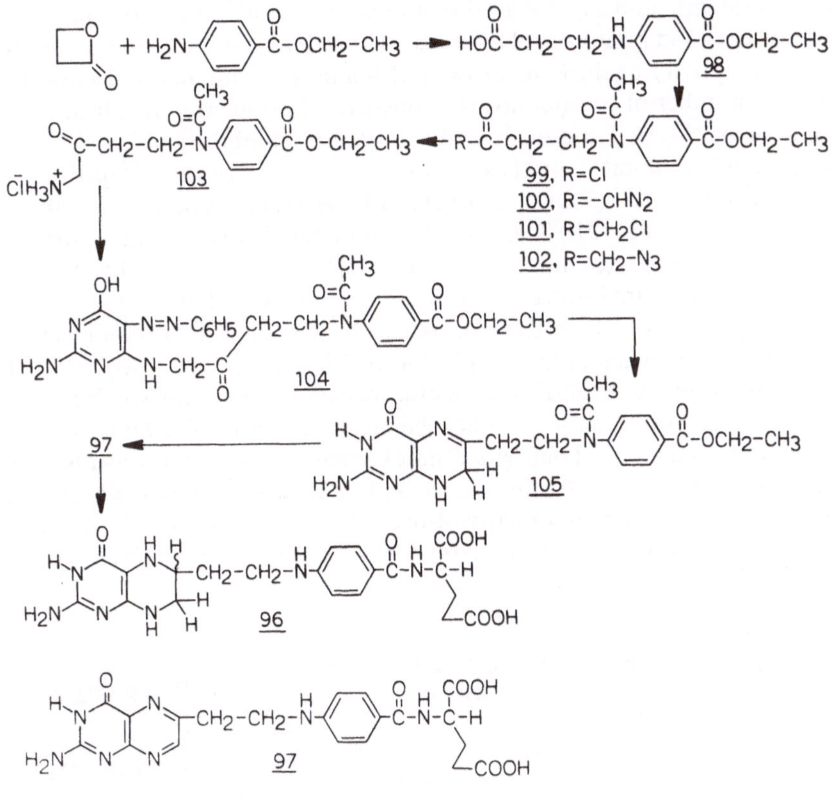

Scheme 7.16

obtained by reaction of β-propiolactone with ethyl-*p*-aminobenzoate was converted to the aminoketone **(103)** via **(99)**, **(100)**, **(101)** and **(102)** by standard procedures. After conversion of **(103)** to the semicarbazone, it was condensed with 2-amino-4-hydroxy-5-phenylazo-6-chloropyrimidine. Deprotection of the carbonyl group to **(104)** followed by hydrogenation gave **(105)**, which was oxidised and hydrolysed to homopteroic acid. After protection of the 2 and 10 amino groups of homopteroic acid by trifluoroacetylation, it was coupled to diethyl glutamate by the mixed anhydride method and hydrolysed to homofolic acid (**97**). Catalytic hydrogenation of (**97**) gave *d,l*-L-tetrahydrohomofolic acid (**96**) in acceptable yield.

Recent evidence suggests that the antitumour activity exhibited by tetrahydrohomofolate is a result of its polyglutamylation *in vivo*, and the ability of these metabolites to interfere with tetrahydrofolate utilisation at the level of GAR-formyltransferase[87,88]. Several homofolate analogues and their reduced derivatives, which were altered at the bridge region, have been subsequently synthesised and evaluated as TS inhibitors. These include 11-oxahomofolate[71,89], 11-thiohomofolate[70], 1′,2′,3′,4′,5′,6′-hexahydrohomofolate[90], 11-deazahomofolate and 11-deaza-10-methylhomofolate[91]. None of these compounds were good inhibitors of thymidylate synthase, although they exhibited significant activity against the growth of folate-requiring microorganisms.

A large number of compounds that possess a 5,8-dideazafolate framework have been synthesised and evaluated as potential inhibitors of TS[92-97]. The most potent TS inhibitor synthesised thus far in this series is N^{10}-propargyl-5,8-dideazafolic acid[96] (PDDF; CB 3717). PDDF (**108**) has been shown to be a specific inhibitor of L 1210[96], *L. casei*[97] and human[98] thymidylate synthases. The polyglutamyl metabolites [99,100] of PDDF were shown to be the most potent antifolate inhibitors of *L.casei*[101] and L 1210[100] thymidylate synthases. In spite of its excellent *in vivo* antitumour activity against L 1210 leukaemia and inhibition of TS ($K_i = 1 \times 10^{-9}$ M)[96,100], the therapeutic utility of PDDF has suffered because of its poor solubility and consequent nephrotoxicity[102]. The chemical synthesis of PDDF is summarised in Scheme 7.17. Diethyl *p*-aminobenzoyl-L-glutamate was alkylated with propargyl bromide to the monoalkylation product (**106**). Reaction of (**106**) with 2-amino-6-bromomethyl-4-hydroxyquinazoline (**107**)[103] gave the corresponding alkylation product that was hydrolysed to PDDF (**108**) in excellent yield[96].

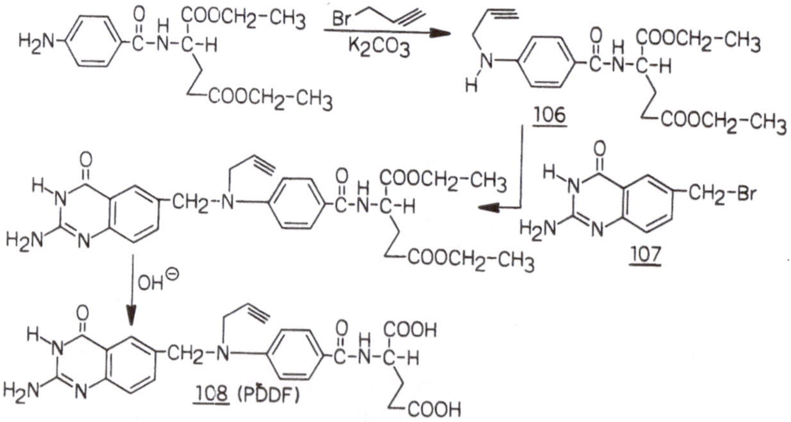

Scheme 7.17

The remarkable potency of PDDF as a specific inhibitor of TS, coupled with its excellent *in vivo* activity as an antileukaemic agent in murine tumour models, has prompted many investigators to synthesise several analogues of this lead compound. Noteworthy among these are N^{10}-propargylfolic acid[104], N^{10}-propargyl-5,8-dideaza-5,6,7,8-tetrahydrofolic acid[105] and several analogues that are altered at the N^{10}-region of the PDDF molecule[92]. None of these modifications resulted in a better inhibitor of TS than PDDF. Recently, two groups independently synthesised[106,107] several analogues of PDDF which were modified at the pyrimidine ring of PDDF for two different reasons. While Hughes[106] replaced the 2-amino group of PDDF with a hydrogen and several alkyl groups to disrupt hydrogen bonding and thereby increase solubility, we have carried out essentially the same modifications[107] to increase the lipophilicity of PDDF, so that tissue uptake of the drug could be enhanced relative to PDDF. Of the several compounds synthesised in this series 2-desamino-2-methyl-N^{10}-propargyl-5,8-dideazafolate (DMPDDF) was the most interesting. DMPDDF was a strong inhibitor of human and *L. casei* TS, and it exhibited excellent growth inhibition of Manca human lymphoid leukaemia and H35 hepatoma cells in culture. The inhibitory activities of DMPDDF were 43- and 65-fold greater than PDDF, respectively, in these cell lines. H35R cells that were resistant to MTX by virtue of a transport defect were cross-resistant to DMPDDF but not to PDDF. Transport studies *in vitro* established that DMPDDF effectively inhibits MTX influx to H35 hepatoma cells, whereas PDDF has no effect on MTX transport in this cell line. The enhanced activity of DMPDDF, in spite of its relatively weak TS inhibition, suggested that DMPDDF is taken up more effectively by tumour cells and it might be a better substrate of folylpolyglutamate synthetase than PDDF. DMPDDF has been synthesised by two independent routes, 2,6-Dimethylquinazolinone (109) was converted to the corresponding bromide (110) by allylic bromination and reacted with diethyl-*N*-propargylaminobenzoyl-L-glutamate. The alkylation product (111)[106] thus obtained was saponified to DMPDDF (Scheme 7.18). In the second

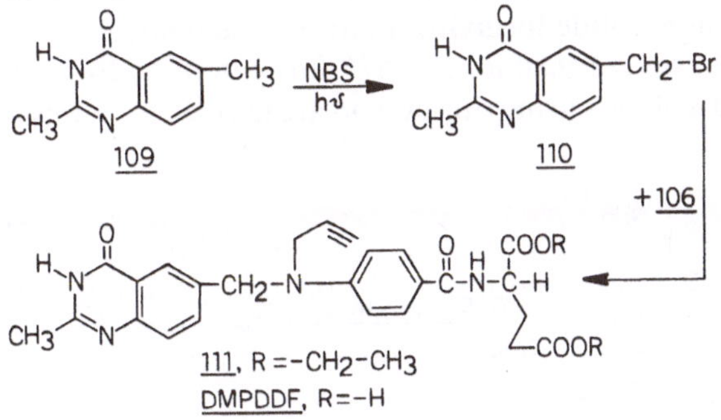

Scheme 7.18

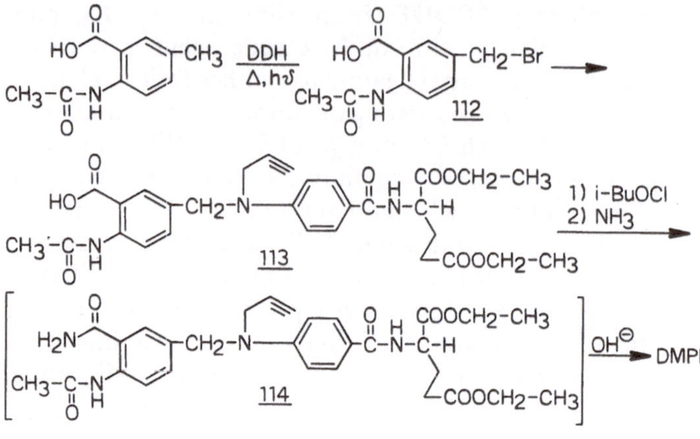

Scheme 7.19

procedure[107], which was developed in this laboratory (Scheme 7.19), 2-acetamido-5-methylbenzoic acid was allylically brominated to (112) and reacted with (106) to obtain the alkylation product (113). Activation of the carboxyl group of (113) by isobutylchloroformate followed by treatment with NH$_3$ gave the intermediate (114) which on saponification gave DMPDDF. Since details of the pre-clinical pharmacology of DMPDDF are not available at the present time, it is premature to predict the therapeutic potential of this compound. It should be pointed out, however, that development of PDDF analogues that are capable of altered transport and metabolisms as anti-tumour agents remains very appealing. The 3-methyl derivative of DMPDDF[107] was 1000 times weaker as an inhibitor of TS, indicating that the presence of the N^3 proton is necessary for enzyme binding presumably via hydrogen bonding.

7.4 Inhibitors of purine biosynthesis

Two important enzymes that take part in *de novo* purine biosynthesis are glycinamide ribonucleotide formyltransferase and aminoimidazole car-boxamide ribonucleotide formyltransferase. The folate cofactor required for both formyltransferase reactions is N^{10}-formyltetrahydrofolate (Figure 7.2). Carbons 2 and 8 of the purine ring system are derived from this folate cofactor.

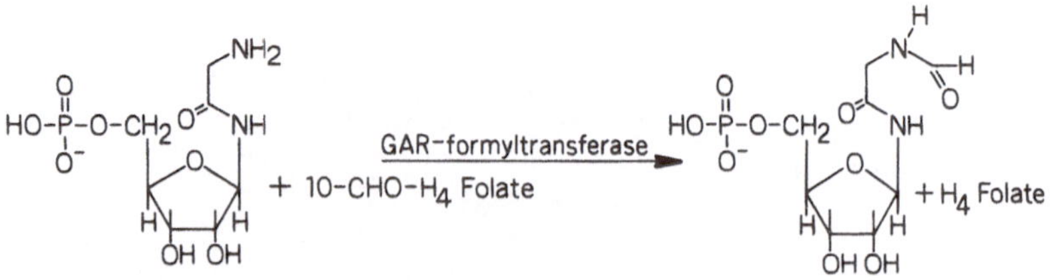

Figure 7.2

Analogues of tetrahydrofolate that are capable of inhibiting these formylation steps are potential anticancer agents. A specific blockade of *de novo* purine biosynthesis should result in the depletion of purine nucleotides required for DNA biosynthesis and hence cell division. The therapeutic utility and host toxicity of such analogues might depend on the relative importance of the *de novo* versus salvage purine biosynthetic pathway between tumour and normal proliferative tissues for growth and cell division.

The first indication that a specific inhibition of GAR-formyltransferase may contribute to antitumour activity came from the work of Hakala[87]. In 1971, she demonstrated the antipurine effect of tetrahydrohomofolate. This was later corroborated by other workers[108], who established that polyglutamyl derivatives of tetrahydrohomofolate are inhibitors of GAR-formyltransferase. The first example of a powerful and specific inhibitor of GAR-formyltransferase was reported by Taylor in 1985[65]. Replacement of the 5- and 10-amino groups of tetrahydrofolic acid resulted in a compound that was found to have a wide spectrum of antitumour activity. 5,10-Dideaza-5,6,7,8-tetrahydrofolic acid (DDATHF) was synthesised according to Scheme 7.20. The bromomethyl pyridine derivative (**65**) was converted to the corresponding phosphonium bromide (**115**), which was reacted with *t*-butyl-*p*-formyl-benzoate to yield the Wittig product (**116**). Reaction of (**116**) with ammonia gave (**117**) which on cyclisation with guanidine gave (**118**). Removal of the carboxyl protective group to (**119**) by HCl, followed by saponification, gave 5-deaza-9,10-dehydropteroic acid (**120**). After acetylation, (**120**) was coupled with diethyl-L-glutamate, catalytically reduced to the tetrahydrofolate

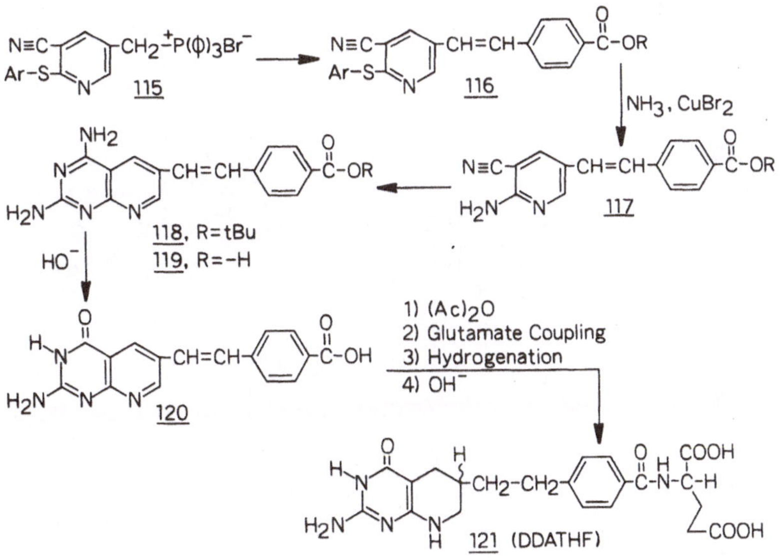

Scheme 7.20

analogue and deprotected by saponification to **DDATHF (121)**[65]. The preparation of the 10-methyl analogue of DDATHF has also been described by Taylor *et al.*[109].

A related analogue, 11-deazatetrahydrohomofolate **(130)**, was synthesised in this laboratory and was found to be potent inhibitor of *L. casei* GAR-formyltransferase. However, it was not an effective inhibitor of the mammalian enzyme[110]. 11-Deazahomofolate was synthesised using the general procedure developed for the construction of a number of 10-deazaanalogues of folic acid[111]. Reaction of **(34)** (Scheme 7.7) with the aldehyde **(122)** gave the phthalimide intermediate **(123)**(Scheme 7.21). Catalytic reduction of **(123)** gave the saturated ketone **(124)**, which was converted to the oxime **(125)** and subjected to hydrazinolysis to obtain the masked alpha-aminoketone **(126)**.

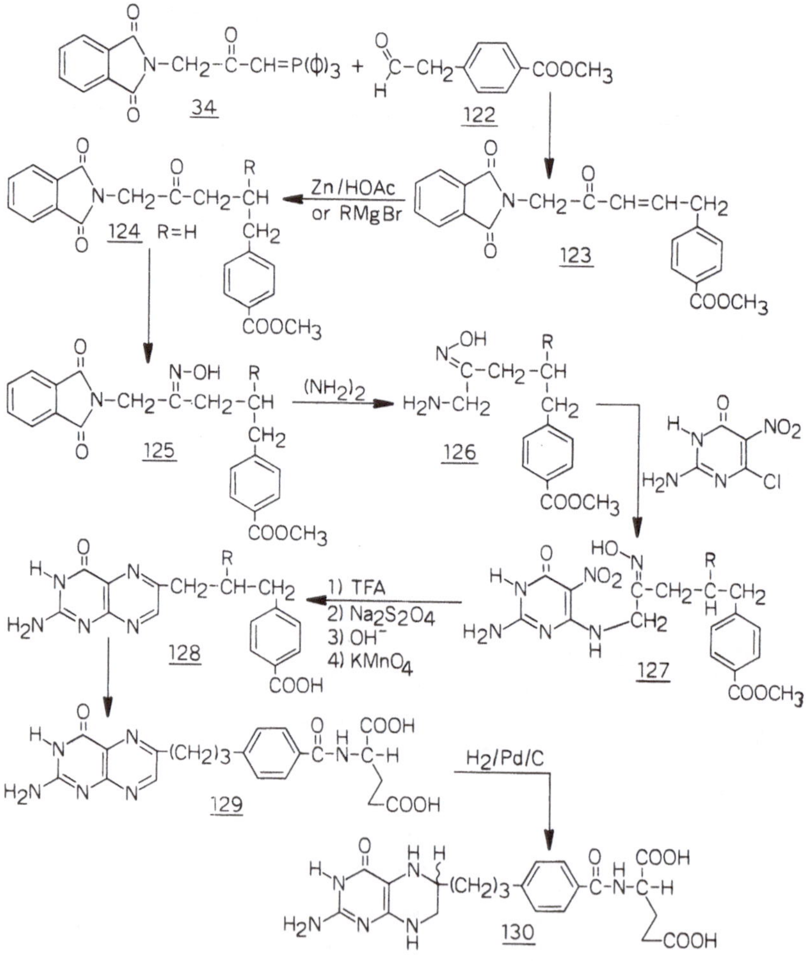

Scheme 7.21

Reaction of (126) with 2-amino-6-chloro-4-hydroxy-5-nitropyrimidine gave the pyrimidine intermediate (127). Treatment of (127) with TFA followed by $Na_2S_2O_4$ reduction, base treatment and $KMnO_4$ oxidation gave 11-deaza-homopteroic acid (128). Conversion of (128) to 11-deazahomofolic acid (129) was accomplished by coupling (128) with diethyl-L-glutamate by the mixed anhydride method followed by saponification. 11-Deazatetrahydrohomofolic acid (130) was prepared from (129) by catalytic hydrogenation. Treatment of (123) with CH_3MgBr in the presence of cuprous ions gave the conjugation addition product, which permitted elaboration of (123) to 10-methyl-11-deazahomofolic acid. We have also examined the effect of replacing the 8 amino group of DDATHF with a methylene group. Thus 5,8,10-trideazatetrahydrofolic acid (134) was synthesised[105,112] according to Scheme 7.22, but it was found to be inactive either as a GAR-formyltransferase inhibitor or as a cytotoxic agent, indicating the profound influence of N^8 on enzyme binding. Rosowsky et al.[113] also synthesised (134) and they have confirmed these observations. It is noteworthy that an acyclic analogue[114] of DDATHF (135) synthesised at Burroughs Wellcome Co. had very similar biological activity, indicating the possibility of developing a second generation of antitumour agents belonging to this series. An alternative synthesis of DDATHF was recently accomplished by Boschelli et al.[135] in seven steps from readily available reagents. Wittig condensation of 2-acetamido-6-formyl-4-(3H)-pyrido[2,3-b]pyrimidone (142) and the triphenylphosphonium bromide (143) gave the 9,10-dehydroderivative (144). Catalytic hydrogenation of (144)

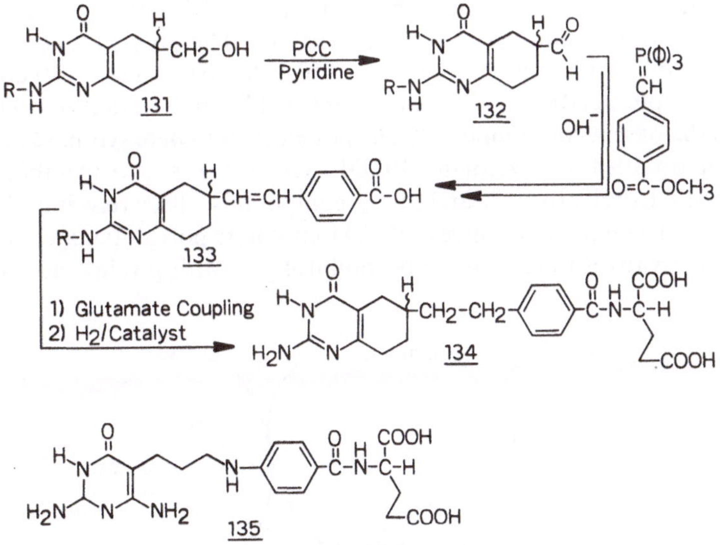

Scheme 7.22

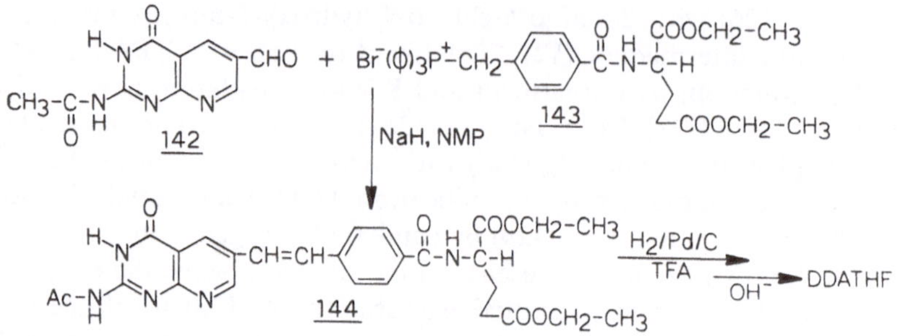

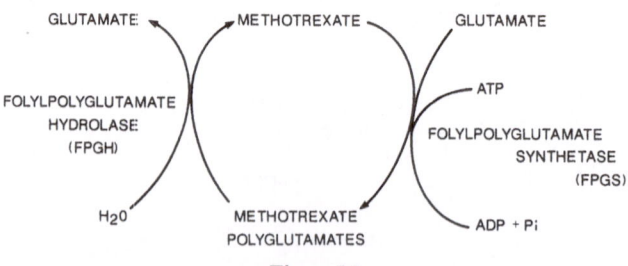

Scheme 7.23

followed by base hydrolysis gave DDATHF in 20% yield based on (**142**) (Scheme 7.23).

7.5   The polyglutamyl metabolites of antifolates

The role of poly-γ-glutamyl metabolites of antifolates as determinants of antifolate cytotoxicity has been established[115,116]. Like folate coenzymes, several antifolates including methotrexate (MTX) are metabolised to their poly-γ-glutamyl derivatives (Figure 7.3) of varying chain length in tumour cells[115–119], murine tissues[45,120] and in man[120]. Kisliuk et al.[121,122] found that polyglutamyl metabolites of MTX are potent inhibitors of thymidylate synthase (TS, EC 2.1.1.45) and dihydrofolate reductase (EC 1.5.1.3) derived from several species. The effects of various folate and MTX polyglutamates on human thymidylate synthase activity was investigated by Cheng et al.[123] and they found that increasing the number of glutamyl residues increases their respective binding affinities toward the enzyme. The poly-γ-glutamyl metabolites of 10-deazaaminopterin (10-DAAM) and 10-ethyl-10-deazaaminopterin (10-EDAAM) were remarkably more inhibitory to human TS than the parent compounds[124]. The potent thymidylate synthase inhibitor N^{10}-propargyl-5,8-dideazafolate (PDDF, CB 3717) is also metabolised to polyglutamyl derivatives[99], and these compounds effectively inhibited TS derived from a number of sources[98,100]. Homofolate polyglutamates and their reduced derivatives were found to be inhibitory towards glycinamideribonuc-

Figure 7.3

leotide formyl transferase[125], which is a newly discovered target for the design of novel antitumour agents. Investigations by Jolivet *et al.* established that the conversion of MTX to poly-γ-glutamates in human MCF 7 and ZR 75 B cells was dose and time dependent[118]. The rate of disappearance of the metabolites from these cells decreased with increasing chain length. Preferential polyglutamylation of 4-aminoantifolates in tumour versus normal proliferative tissues in tumour-bearing animals has also been documented[126]. All currently available biochemical and pharmacological data indicate that polyglutamylation of antifolates enhances their cytotoxicity, and these metabolites interfere more effectively with tetrahydrofolate utilisation at various stages of folate metabolism[115,116]. Folylpolyglutamate synthetase (FPGS), the enzyme responsible for polyglutamylation of folates and antifolates, is widely distributed[127,128], and is present in normal as well as neoplastic tissues. Invariably, almost all classical antifolates that possess an L-glutamate moiety behave as synthetic substrates for this enzyme. Methods are now available to screen antifolates *in vitro* as substrates of mammalian FPGS[129]. The interesting TS inhibitor DMPDDF and GAR-formyltransferase inhibitor DDATHF are substrates of FPGS, and indeed polyglutamylation of these antifolates appears to be a determinant of their cytotoxicity. The first chemical synthesis of MTX-polyglutamates was accomplished by modifications of the Merrifield solid phase method[45] (see Figure 7.4). Subsequently, the polyglutamyl metabolites of PDDF (CB 3717), 10-deazaaminopterins and homofolate[101,130,125] were also prepared in this laboratory by similar procedures.

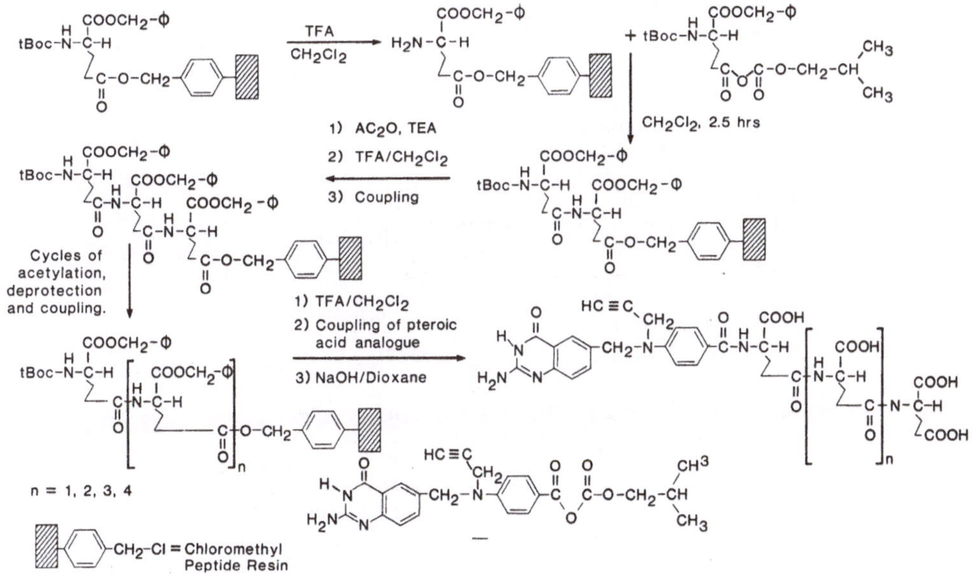

Figure 7.4 General method for the synthesis of antifolate polyglutamates by the Merrifield solid phase chemistry.

The synthesis of MTX and PDDF polyglutamates by classical peptide synthesis in solution was also reported by other workers[100,131]. At any rate, the current availability of these antifolate metabolites is scarce, and a convenient method was desired to make them in large quantities for use as analytical standards, inhibitors and substrates of various folate dependent enzymes. This was accomplished very recently in this laboratory using the new polyamide KH resin and using the Fmoc chemistry. α-t-Butyl-Fmoc-L-glutamate (136) was converted to its symmetrical anhydride (137) and coupled to the pepsin KH resin via an ester linkage. Deprotection of the Fmoc protective group was accomplished by piperidine. The deprotected resin was again reacted with (137) and the process of deprotection and coupling was repeated till the desired glutamate chain length was reached. After final deprotection, the resin was treated with a solution of the mixed anhydride derived from the pteroic acid analogue and isobutylchloroformate. The resin-bound product was cleaved from the resin by 1% TFA in CH_2Cl_2 and final deprotection was accomplished with 0.1 N HCl. A series of folate and several antifolate polyglutamates were synthesised by the Fmoc method and yields and purity of the products were consistently better than the Merrifield procedure in all cases (Schemes 7.24a, 7.24b, 7.24c).

Since folate coenzymes exist as poly-γ-glutamyl derivatives of varying chain length, it is conceivable that inhibitors of folate polyglutamylation may function as potential antifolates. More importantly, development of inhibitors of folate dependent enzymes that are also inhibitors of FPGS is appealing from

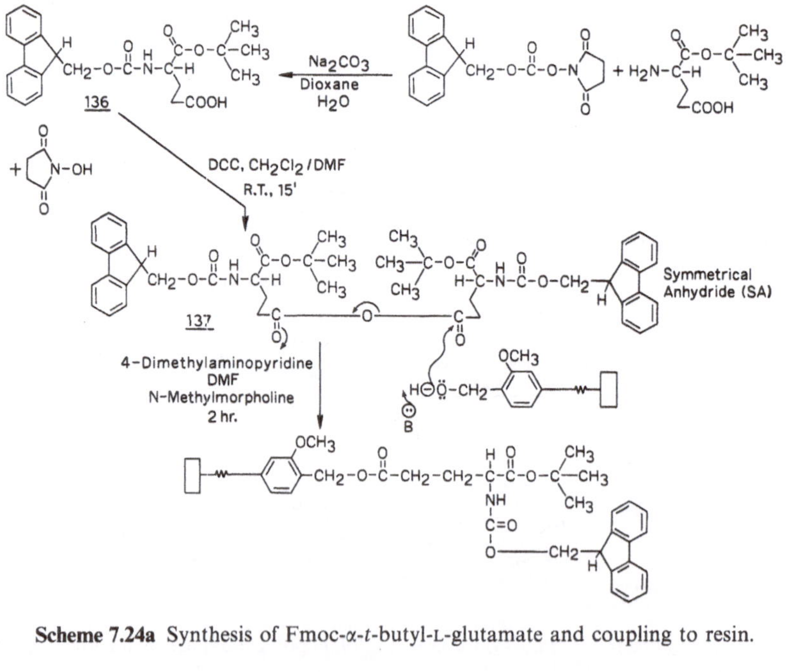

Scheme 7.24a Synthesis of Fmoc-α-t-butyl-L-glutamate and coupling to resin.

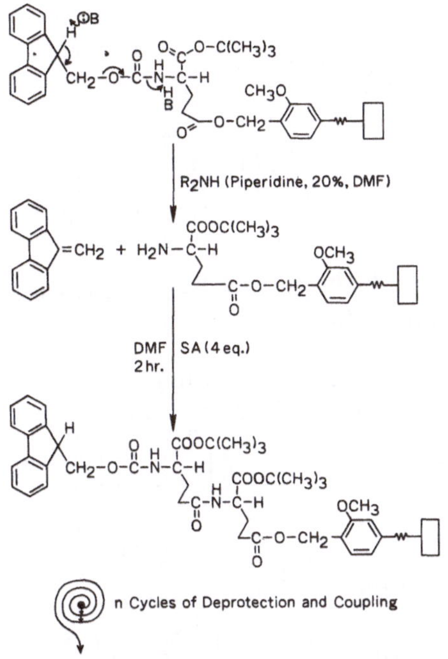

Scheme 7.24b Deprotection and coupling.

a therapeutic point of view due to their self-potentiating property. It has been postulated that inhibition of FPGS would result in preferential cell kill in rapidly proliferating cells[132]. Elevated levels of FPGS in organs such as liver and kidney that are sensitive to antifolate cytotoxicity have been observed[133]. Although there have been several attempts to synthesise specific inhibitors of FPGS, so far only very few compounds have shown any significant inhibition of this enzyme. The most interesting compound in this series appears to be 4-amino-4-deoxypteroyl-L-ornithine (**138**), which has been reported to be a potent inhibitor of DHFR and FPGS, but is a relatively weak inhibitor of cell

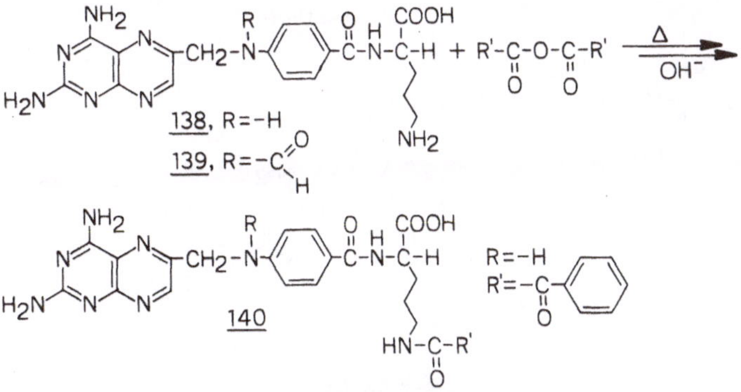

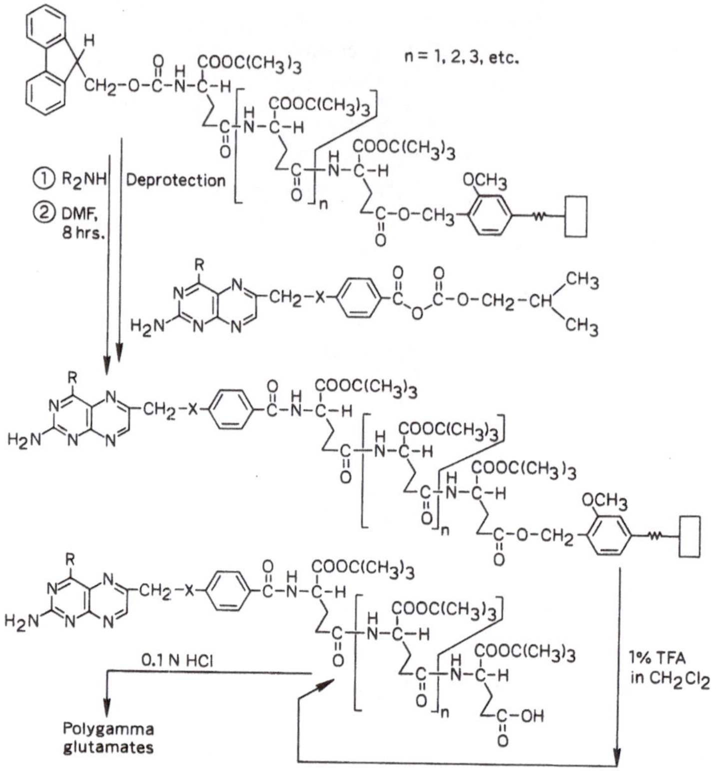

Scheme 7.24c

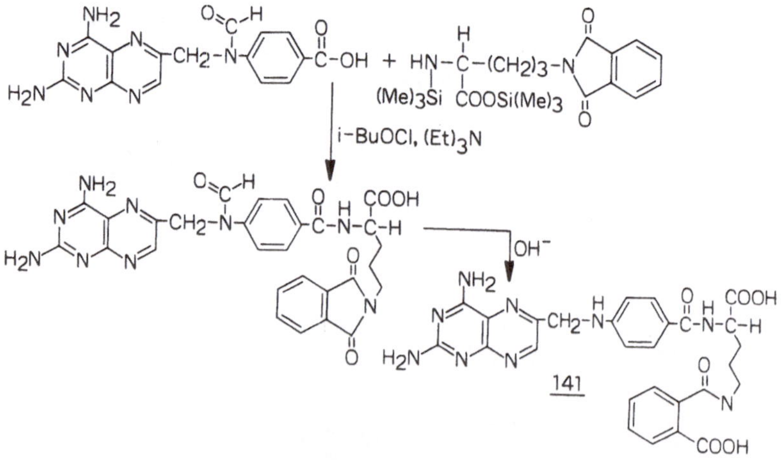

Scheme 7.25

growth in culture presumably due to impaired influx to tumour cells[134]. Several analogues of (138) in which the terminal group has been modified to potentially cleavable N-acyl groups have been synthesised by Rosowsky et al. and evaluated as antitumour agents in cell culture. Two compounds, the N-benzoyl (140) and N-hemiphthaloyl (141) derivatives of (138), with IC_{50} values against L1210 cells lower than MTX, appear to be transported more efficiently and cleaved within the cell to (138). N-Acyl-N-(4-amino-4-deoxypteroyl)-L-ornithines were prepared by reacting 4-amino-4-deoxy-N^{10}-formylpteroyl-L-ornithine (139) with acid anhydrides and subsequent removal of the N^{10}-formyl group by alkaline hydrolysis. The hemiphthaloyl derivative (141) was prepared according to Scheme 7.25.

From these results[134] it appears that antifolates can effectively enter mammalian cells in the absence of the γ-glutamate moiety and a variety of potentially cleavable functional groups can be substituted at this position. Such masked inhibitors of FPGS with appropriate structural modifications may find therapeutic utility as antifolate anticancer agents.

7.6   Concluding remarks

Specific inhibitors of three folate based enzymes, dihydrofolate reductase, thymidylate synthase and GAR-formyltransferase, have shown impressive anticancer activity. The successful clinical use of a folate analogue inhibitor of TS, or a GAR-formyltransferase inhibitor has not yet been realised. Impressive preclinical pharmacological results of DMPDDF and DDATHF suggest that these compounds, or their close analogues, will be more specific and less toxic anticancer drugs for the immediate future. Of particular importance is the more favourable therapeutic indices exhibited by 10-deazaaminopterins, not because they are better inhibitors of DHFR, but because they are transported more efficiently and metabolised differently in tumour to normal proliferative tissues. Although DMPDDF is a weaker inhibitor of TS than PDDF, it is a better antitumour agent because it is transported differently and accumulated more efficiently in tissues due to polyglutamylation. Metabolism to polygluta-mates appears to be a prerequisite for the in vivo antitumour activity of DDATHF. These observations suggest that better anticancer drugs of the antifolate series need not necessarily be better inhibitors of target enzymes. The optimal inhibitory potencies toward three folate based enzymes have already been achieved with MTX for DHFR, PDDF for TS and DDATHF for GAR-formyltransferase. Attention should now be focused on developing inhibitors of these enzymes that have more favourable transport character-istics and tissue accumulation. Development of inhibitors of folate dependent enzymes that are better substrates of FPGS are clearly warranted. In order to understand the role of polyglutamylation as a determinant of antifolate cytotoxicity, we also need to develop powerful and specific inhibitors of FPGS as well as antifolates that are incapable of polyglutamylation. We should be searching for antifolate drugs that might penetrate the cell by multiple transport pathways, including the reduced folate pathway, and undergo differential metabolism in tumour versus normal proliferative tissues. Development of folate analogues targeted towards serine hydroxymethyltransferase and AICAR formyltransferase should be pursued. A better understanding of the relative activities of folate based enzymes in normal versus neoplastic tissues would aid the medicinal chemist in designing antifolate drugs that might be preferentially accumulated or generated in tumour tissues. A coordinated multidisciplinary effort along these lines should pave the way for development of useful antifolate anticancer drugs for the treatment of human cancers in the near future.

Acknowledgement

The work described from the author's laboratory was supported by NIH Grants CA 27101 and CA 32687 from the National Cancer Institute, DHHS. The contributions made by several coworkers and collaborators are gratefully acknowledged.

References

1. J. R. Bertino, in *Antineoplastic and Immunosuppressive Agents*, eds. A. C. Sartorelli and D. G. Johns, Springer-Verlag, Berlin (1975) Part II, pp. 468–483.
2. W. B. Pratt, *Chemotherapy of Infection*, Oxford University Press, New York (1977) pp. 189–198.
3. D. G. Johns and J. R. Bertino, in *Cancer Medicine*, eds. J. F. Holland and E. Frie III, Lea and Febiger, Philadelphia (1973) p. 739–754.
4. W. B. Pratt and R. W. Ruddon, *The Anticancer Drugs*, Oxford University Press (1979) pp. 98–147.
5. L. E. Broder and S. K. Carter, *Meningeal Leukemia*, Plenum Press, New York (1972).
6. M. Tattersal, N. Jaffe and E. Frie, III, in *Pharmacological Basis of Cancer Chemotherapy*, Williams and Wilkins, Baltimore (1975) pp. 105–107.
7. W. C. Werkheiser, *J. Biol. Chem.* **236** (1961) 888.
8. C. C. Fau, K. S. Vitols and F. M. Hunnekens, *Adv. Enzyme Regul.* **18** (1980) 41.
9. J. C. White, *J. Biol. Chem.* **254** (1979) 889.
10. J. W. Williams, R. G. Duggleby, R. Cutler and J. F. Morrison, *Biochem. Pharmacol.* **29** (1980) 589.
11. J. R. Bertino and C. Lindquist, in *Advances in Chemotherapy*. eds. S. K. Carter, A. Goldin, K. Kurethani, G. Mathe, Y. Sakurai, S. Tsukagoshi and B. Umegara, Japan Scientific Press. Tokyo, University Park Press, Baltimore (1978) pp. 155–164
12. F. M. Sirotnak, R. C. Donsbach, D. M. Dorick and D. M. Moccio, *Cancer Treat. Rep.* **61** (1977) 565.
13. I. Djerassi, *Cancer Chemother. Rep.* **6** (1975) 3.
14. H. Bruckner, M. Rubinoff and S. Waxman, *Eur. J. Cancer* **16** (1980) 1057.
15. L. L. Samuels, *Proc. Am. Assoc. Cancer Res.* **21** (1980) 273.
16. R. G. Moran, M. Mulkins and C. Heidelberger, *Proc. Natl. Acad. Sci. (USA)* **76** (1979) 5924.
17. R. C. Jackson, *Mol. Pharmacol.* **18** (1980) 281.
18. S. B. Howell and R. K. Tamerius, *Eur. J. Cancer* **16** (1980) 1427.
19. K. R. Harrap, B. T. Hill, M. E. Furness and L. I. Hart, *Ann. N. Y. Acad. Sci.* **186** (1971) 312.
20. D. M. Misra, S. R. Humphreys, M. Friedkin, A. Goldin and E. J. Crawford, *Nature* **189** (1961) 39.
21. R. J. Kaufman, P. C. Brown and R. T. Schimke, *Proc. Natl. Acad. Sci. (USA)* **76** (1979) 5669.
22. F. W. Alt, R. E. Kellems and R. T. Schmike, *J. Biol. Chem.* **251** (1976) 3063.
23. J. L. Biedler, P. W. Malera and B. A. Spengler, *Cancer Genet. Cytogenet.* **2** (1980) 47.
24. F. M. Sirotnak, S. Kurita and D. J. Hutchison, *Cancer Res.* **28** (1968) 75.

25. H. J. Ryser and W. C. Shen, *Cancer* **45** (1980) 1207.
26. M. R. Hamrell, D. Sedwick and J. Laszlo, *Proc. Am. Assoc. Cancer Res.* **21** (1980) 3.
27. P. W. Rosson, L. M. Mangini, C. A. Cucchi and E. J. Modest, *Proc. Am. Assoc. Cancer Res.* **21** (1980) 264.
28. D. W. Kufe, M. N. Wick and H. T. Ableson, *J. Invest. Dermatol.* **75** (1980) 357.
29. P. T. Condit, *Cancer* **13** (1960) 222.
30. H. E. Skipper and L. L. Bennet, *Cancer Res.* **12** (1952) 677.
31. H. E. Skipper, J. H. Mitchell and L. L. Bennet, *Cancer Res.* **10** (1950) 510.
32. E. M. Hersch, V. G. Wong, E. S. Henderson and E. J. Freiriech, *Cancer* **19** (1958) 600.
33. J. Eisenstadt and P. Lengyel, *Science* **154** (1966) 524.
34. H. P. Broquist, *J. Am. Chem. Soc.* **78** (1956) 6205.
35. D. R. Seeger, D. B. Cosulich, J. M. Smith and M. E. Hultquist, *J. Am. Chem. Soc.* **71** (1949) 1753.
36. C. Temple, Jr., J. D. Rose and J. A. Montgomery, *J. Heterocycl. Chem.* **13** (1976) 567.
37. D. C. Suster, G. Ciustea, A. Dumirescu, L. V. Feyns, E. Tarnauceanu, G. Botez, S. Angelescu, V. Dobre and I. Niculescu-Duvăz, *Rev. Roum. Chim.* **22** (8) (1977) 1195.
38. D. C. Suster, L. V. Feyns, G. Ciustea, G. Botez, V. Dobre, R. Bick and I. Niculescu-Duvăz, *J. Med. Chem.* **17** (1974) 758.
39. J. A. Montgomery, J. D. Rose, C. Temple and J. R. Piper, in *Chemistry and Biology of Pteridines*, ed. W. Pfleiderer, Walter de Gruyter, Berlin (1975) p. 486.
40. C. M. Baugh and E. Shaw, *J. Org. Chem.* **29** (1964) 3610.
41. R. D. Elliot, C. Temple and J. A. Montgomery, *J. Org. Chem.* **35** (1970) 1976.
42. E. C. Taylor and T. Kobayashi, *J. Org. Chem.* **38** (1973) 2817.
43. E. C. Taylor, in *Chemistry and Biology of Pteridines*, ed. W. Pfleiderer, Walter de Gruyter, Berlin (1975) p. 543.
44. A. Rosowsky, R. Forsch, J. Uren, M. Wick, A. A. Kumar and J. H. Freischeim, *J. Med. Chem.* **26** (1983) 1719.
45. M. G. Nair and C. M. Baugh, *Biochemistry* **12** (1973) 3923.
46. C. C. Levy and P. Goldman, *J. Biol. Chem.* **242** (1967) 2933.
47. M. G. Nair, S. R. Adapa and T. Bridges, *J. Org. Chem.* **46** (1981) 3152.
48. J. I. DeGraw, R. L. Kisliuk, Y. Gaumont, C. M. Baugh and M. G. Nair, *J. Med. Chem.* **17** (1974) 552.
49. F. M. Sirotnak, J. I. DeGraw, D. M. Moccio, L. L. Samuels and L. J. Goutas, *Cancer Chemother. Pharmacol.* **12** (1984) 18.
50. F. M. Sirotnak, J. I. DeGraw, F. A. Schmid, L. J. Goutas and D. M. Moccio, *Cancer Chemother. Pharmacol.* **12** (1984) 26.
51. J. I. DeGraw, V. H. Brown, H. Tagawa, R. L. Kisliuk, Y. Gaumont and F. M. Sirotnak, *J. Med. Chem.* **25** (1982) 1227.
52. J. R. Piper and J. A. Montgomery, U. S. Patent (1979) 4172200.
53. M. G. Nair, *J. Org. Chem.* **50** (1985) 1879.
54. H. Diddens, D. Neithammer and R. C. Jackson, in *Chemistry and Biology of Pteridines*, ed. J. A. Blair, Walter de Gruyter, Berlin (1983) p. 954.
55. J. Galivan, *Cancer Res.* **41** (1981) 1757.
56. M. T. Hakala, *Biochim. Biophys. Acta* **102** (1965) 198.
57. D. S. Duch, M. P. Edelstein, S. W. Bowers and C. A. Nicol, *Mol. Pharmacol.* **18** (1979) 100; *Cancer Res.* **42** (1982) 3987.
58. E. M. Grivsky, S. Lee, C. W. Sigel, D. S. Duch and C. A. Nicol, *J. Med. Chem.* **23** (1980) 327.
59. J. B. Hynes, S. J. Harmon, G. G. Floyd, M. Farrington, L. D. Hart, G. R. Gale, W. L. Washtien, S. S. Susten and J. H. Freisheim, *J. Med. Chem.* **28** (1985) 209.
60. M. G. Nair, L. P. Mercer and C. M. Baugh, *J. Med. Chem.* **17** (1974) 1268.
61. J. Davoll and A. M. Johnson, *J. Chem. Soc.* C (1970) 997.
62. C. Temple, Jr., R. D. Elliot and J. A. Montgomery, *J. Org. Chem.* **47** (1982) 761.
63. E. C. Taylor, D. C. Palmer, T. J. George, S. R. Fletcher, C. P. Tseng and P. J. Harrington, *J. Org. Chem.* **48** (1983) 4852.
64. J. I. DeGraw, H. Tagawa, P. H. Christie, J. A. Lawson, E. G. Brown, R. L. Kisliuk and Y. Gaumont, *J. Heterocycl. Chem.* **23** (1986) 1.
65. E. C. Taylor, P. J. Harrington, S. R. Fletcher, G. P. Beardsley and R. G. Moran, *J. Med. Chem.* **28** (1985) 914.
66. M. G. Nair and P. T. Campbel, *J. Med. Chem.* **19** (1976) 825.

67. M. G. Nair, P. T. Campbel, E. Braverman and C. M. Baugh, *Tetrahedron Lett.* **31** (1975) 2745.
68. H. G. Mautner, Y. H. Kim, Y. Gaumont and R. L. Kisliuk, in *Chemistry and Biology of Pteridines*, ed. Walter Pfleiderer, Walter de Gruyter, Berlin (1975) p. 515.
69. R. H. Hornbeak and M. G. Nair, *Mol. Pharmacol.* **14** (1978) 299.
70. M. G. Nair, S. Y. Chen, R. L. Kisliuk, Y. Gaumont and D. Strumpf, *J. Med. Chem.* **22** (1979) 850.
71. M. G. Nair, C. Saunders, S. Y. Chen, R. L. Kisliuk and Y. Gaumont, *J. Med. Chem.* **23** (1980) 59.
72. M. G. Nair, M. K. Rozmyslovicz, R. L. Kisliuk, Y. Gaumont and F. M. Sirotnak, in *Chemistry and Biology of Pteridines*, ed. J. A. Blair, Walter de Gruyter, Berlin, New York (1983) p. 375.
73. E. W. Cheng, V. E. Currie and A. Yagoda, *Am. J. Clin. Oncol.* **6** (1983) 469.
74. J. I. De Graw, Personal communication.
75. R. D. Eliott, C. Temple, J. L. Frye and J. A. Montgomery, *J. Org. Chem.* **36** (1971) 2818.
76. L. T. Weinstock, B. F. Grabowski and C. C. Cheng, *J. Med. Chem.* **13** (1970) 995.
77. P. Stenbuck and H. M. Hood, U. S. Patent 3,049,544 (1961).
78. C. J. Allegra, B. A. Chabner, C. U. Tuazon, *N. Engl. J. Med.* **317** (1987) 978.
79. C. J. Allegra, J. A. Kovacs, J. C. Drake, *J. Exp. Med.* **165** (1987) 926.
80. J. A. Kovacs and H. Masur, in *AIDS*, eds. V. T. DeVita, Jr., S. Hellman and S. A. Rosenberg, J. B. Lippincott, New York (1988) p. 199.
81. S. R. Stone and J. F. Morrison, *Biochim. Biophys. Acta* **869** (1986) 275.
82. M. Friedkin, *Adv. Enzymol.* **38** (1973) 235.
83. C. Heidelberger P. V. Danenberg and R. G. Moran, *Adv. Enzymol.* **54** (1983) 57.
84. D. V. Santi, *J. Med. Chem.* **23** (1980) 103.
85. J. I. DeGraw, J. P. Marsh, Jr., E. M. Acton, O. P. Crews, C. W. Mosher, A. N. Fujiwara and L. Goodman, *J. Am. Chem. Soc.* **30** (1965) 3404.
86. J. A. R. Mead, A. Goldin, R. L. Kisliuk, M. Friedkin, L. Plante, E. J. Crawford and G. Kwok, *Cancer Res.* **26** (1966) 2374.
87. M. Hakala, *Cancer Res.* **31** (1971) 813.
88. J. Thorndike, Y. Gaumont, R. L. Kisliuk, F. M. Sirotnak, B. R. Murthy and M. G. Nair, *Proc. Am. Assoc. Cancer Res.* **29** (1988) 285.
89. B. A. Domin, Y. C. Cheng and M. G. Nair, *Biochem. Pharmacol.* **31** (1982) 255.
90. M. G. Nair, E. B. Otis, R. L. Kisliuk and Y. Gaumont, *J. Med. Chem.* **28** (1983) 135.
91. M. G. Nair, T. R. Toghiyani, B. R. Murthy, R. L. Kisliuk and Y. Gaumont, in *Pteridines and Folic Acid Derivatives*, eds. B. A. Cooper and V. M. Whitehead, Walter de Gruyter, Berlin (1986) p. 45.
92. T. R. Jones, A. H. Calvert, A. L. Jackman, M. A. Eakin, M. J. Smithers, R. F. Betteridge, D. R. Newell, A. J. Hayter, A. Stocker, S. J. Harland, L. C. Davies and K. R. Harrap, *J. Med. Chem.* **28** (1985) 1468.
93. J. E. Oatis and J. B. Hynes, *J. Med. Chem.* **20** (1977) 1393.
94. K. J. Scanlon, W. Rode and J. B. Hynes, *Proc. Am. Assoc. Cancer Res.* **19** (1978) 136.
95. A. H. Calvert, T. R. Jones, P. J. Dady, B. Grzelakowska-Sztabert, R. M. Paine, G. A. Taylor and K. R. Harrap, *Eur. J. Cancer* **16** (1980) 713.
96. T. R. Jones, A. H. Calvert, A. L. Jackman, S. J. Brown, M. Jones and K. R. Harrap, *Eur. J. Cancer* **17** (1981) 11.
97. M. G. Nair, D. C. Salter, R. L. Kisliuk, Y. Gaumont, G. North and F. M. Sirotnak, *J. Med. Chem.* **26** (1983) 605.
98. Y. C. Cheng, G. E. Dutschman, M. C. Starnes, M. H. Fisher, N. T. Nanavati and M. G. Nair, *Cancer Res.* **45** (1985) 598.
99. M. G. Nair, A. P. Mehtha and I. G. Nair, *Fed Proc., Fed. Am. Soc. Exp. Biol.* **45** (1986) 821.
100. E. Sikora, A. L. Jackman, D. R. Newell, K. R. Harrap, A. H. Calvert, T. R. Jones, K. Pawclczdak and B. Rzeszotarska, in *Pteridines and Folic Acid Derivatives*, eds, B. A. Cooper and V. M. Whitehead, Walter de Gruyter, Berlin (1986) p. 675.
101. M. G. Nair, N. T. Nanavati, I. G. Nair, R. L. Kisliuk, Y. Gaumont, M. C. Hsiao and T. I. Kalman, *J. Med. Chem.* **29** (1986) 1754.
102. D. R. Newell, Z. H. Siddik, A. H. Calvert, A. L. Jackman, K. G. McGhee and K. R. Harrap, *Proc. Am. Assoc. Cancer Res.* **23** (1982) 181.
103. S. P. Acharya and J. B. Hynes, *J. Heterocycl. Chem.* **12** (1975) 1283.

104. M. Ghazala, M. G. Nair, T. R. Toghiyani, R. L. Kisliuk, Y. Gaumont and T. I. Kalman, *J. Med. Chem.* **29** (1986) 1263.
105. M. G. Nair, R. Dhawan, M. Ghazala, T. I. Kalman, R. Ferone, Y. Gaumont and R. L. Kisliuk, *J. Med. Chem.* **30** (1987) 1256.
106. L. R. Hughes, U. K. Patent, GB2 188319 A (1987)
107. S. D. Patil, C. Jones, M. G. Nair, J. Galivan, F. Maley, R. L. Kisliuk, Y. Gaumont, J. Thorndike, D. Duch and R. Ferone, *J. Med. Chem.* **32** (1989) 1284.
108. J. Thorndike, Y. Gaumont, R. L. Kisliuk, F. M. Sirotnak, B. R. Murthy, M. G. Nair and J. R. Piper, *Cancer Res.* **49** (1989) 158.
109. E. C. Taylor, G. S. K. Wong, S. R. Fletcher, P. J. Harrington, G. P. Beardsley and C. J. Shih, in *Chemistry and Biology of Pteridines*, eds. B. A. Cooper and V. M. Whitehead, Walter de Gruyter, Berlin (1986) p. 61.
110. M. G. Nair, B. R. Murthy, S. D. Patil, R. L. Kisliuk, J. Thorndike, Y. Gaumont, R. Ferone, D. S. Duch and M. P. Edelstein, *J. Med. Chem.* **32** (1989) 1277.
111. M. G. Nair, T. R. Toghiyani, B. R. Murthy, R. L. Kisliuk and Y. Gaumont, in *Chemistry and Biology of Pteridines*, eds. B. A. Cooper and V. M. Whitehead, Walter de Gruyter, Berlin (1986) p. 45.
112. M. G. Nair, S. D. Patil and R. Ferone, University of South Alabama, and Burroughs Wellcome Co., unpublished results (1988).
113. A. Rosowsky, R. Forsch and R. G. Moran, *J. Med. Chem.* **32** (1989) 709.
114. R. Ferone, Personal communication; Burroughs Wellcome Co., N. C. USA.
115. I. D. Goldman and B. Chabner, eds., *Folyl and Antifolyl Polyglutamates*, Plenum Press, New York (1981).
116. I. D. Goldman, ed., *Proceedings of the Second Workshop on Folyl and Antifolylpolygluta-mates*, Praeger Scientific, New York (1984).
117. J. Galivan, *Cancer Res.* **39** (1979) 735.
118. J. Jolivet, R. Schilsky, B. Bailey, J. C. Drake and B. A. Chabner, *J. Clin Invest.* **70** (1982) 351.
119. T. B. Johnson, M. G. Nair and J. Galivan, *Cancer Res.* **48** (1988) 2426.
120. C. M. Baugh, C. L. Krumdieck and M. G. Nair, *Biochem. Biophys. Res. Commun.* **52** (1973) 27.
121. R. L. Kisliuk, Y. Gaumont, C. M. Baugh, J. Galivan, G. F. Maley and F. Maley, in *Chemistry and Biology of Pteridines*, ed. R. L. Kisliuk and G. M. Brown, Walter de Gruyter, New York (1979) p. 431.
122. P. Kumar, R. L. Kisliuk, Y. Gaumont, M. G. Nair, C. M. Baugh and B. T. Kaufman, *Cancer Res.* **46** (1986) 5020.
123. D. W. Szeto, Y. C. Cheng, A. Rosowsky, C. S. Yu, E. J. Modest, J. R. Piper, C. Temple, R. D. Elliott, J. D. Rose and J. A. Montgomery, *Biochem. Pharmacol.* **28** (1979) 2633.
124. T. Ueda, G. E. Dutschman, M. G. Nair, J. I. DeGraw, F. M. Sirotnak and Y. C. Cheng, *Mol. Pharmacol.* **30** (1986) 149.
125. J. Thorndike, Y. Gaumont, R. L. Kisliuk, F. M. Sirotnak, B. R. Murthy and M. G. Nair, *Cancer Res.* **49** (1989) 158.
126. F. M. Sirotnak, J. I. DeGraw, F. A. Schmid, L. J. Goutas and D. M. Moccio, *Cancer Chemother. Pharmacol.* **12** (1984) 26.
127. R. Ferone, S. C. Singer, M. H. Hanlon and S. Roland, in *Chemistry and Biology of Pteridines*, ed. J. A. Blair, Walter de Gruyter, Berlin (1983) p. 585.
128. A. L. Bognar and B. Shane, *J. Biol. Chem.* **258** (1983) 12581.
129. R. G. Moran, A. Rosowsky, P. Colman, R. Forsch, V. C. Solan, H. Bader, P. Harvison and T. I. Kalman, in *Proceedings of the Second Workshop on Folyl and Antifolyl Polyglutamates*, ed. I. D. Goldman, Praeger Scientific, New York (1985) p. 51.
130. M. G. Nair, N. T. Nanavati, P. Kumar, Y. Gaumont and R. L. Kisliuk, *J. Med. Chem.* **31** (1988) 181.
131. M. Przybylski, R. Renkel and P. Fonrobert, in *Chemistry and Biology of Petridines*, eds. B. A. Cooper and V. M. Whitehead, Walter de Gruyter, Berlin (1986) p. 65.
132. R. G. Moran, *Adv. Exp. Med. Biol.* **163** (1983) 327.
133. R. G. Moran and P. D. Colman, *Anal. Biochem.* **140** (1984) 326.
134. A. Rosowsky, H. Bader, C. A. Cucchi, R. G. Moran, W. Kohler and J. H. Freisheim, *J. Med. Chem.* **31** (1988) 1332.
135. D. H. Boschelli, S. Webber, J. M. Whiteley, A. L. Oronsky and S. S. Kerwar, *Arch. Biochem. Biophys.* **265** (1988) 43.

8 Inhibitors of steroid hormone biosynthesis and action

R. MCCAGUE

8.1 Introduction

This chapter covers the chemistry of those drugs that are aimed specifically at steroid hormone sensitive tumours. To date, the treatment of cancers of the breast and of the prostate have benefited from this approach, a consequence of the growth of a high proportion of such tumours being dependent on circulating oestrogens and androgens, respectively. Drugs in this class have the

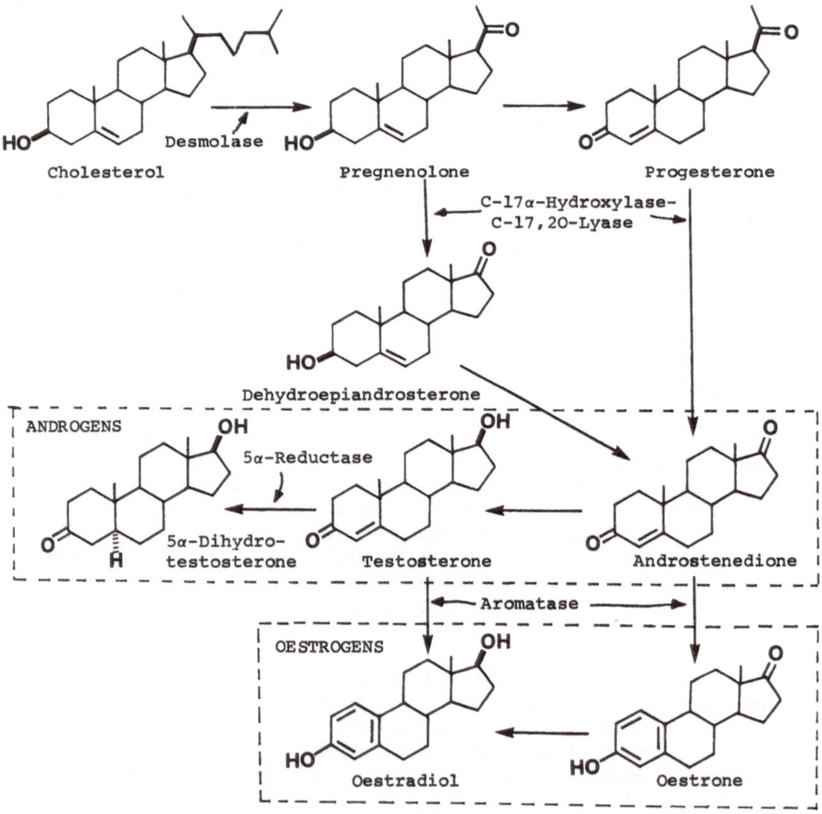

Figure 8.1 An outline of the biosynthesis of androgens and oestrogens from cholesterol.

advantage over conventional cytotoxic agents in that they are generally of low toxicity.

Figure 8.1 outlines the biosynthesis of androgens and oestrogens from cholesterol[1]. One basis for the treatment of hormone sensitive tumours is the inhibition of the biosynthesis of the hormones. Any of the steps can be a target but inhibition of the early steps has the disadvantage that essential corticosteroids are depleted and then replacement therapy is needed. Consequently, for the inhibition of oestrogen biosynthesis, aromatase is a particularly favourable target (see aminoglutethimide and 4-hydroxyandrostenedione), and in the case of prostatic cancer, agents that inhibit the 5α-reductase enzyme which reduces testosterone to a more potent androgen are becoming popular (see 4-azasteroids). Alternatively, prostatic cancer can be treated with luteinising hormone-releasing hormone (LHRH) agonists which disrupt the pituitary hormone stimulation of androgen biosynthesis in the testes. A second approach is the inhibition of the action of the steroids at the level of the cellular receptors to which they bind, as with anti-oestrogens (e.g. tamoxifen) and anti-androgens (e.g. flutamide, cyproterone acetate). Oestrogen action can also be diminished by administration of a progestin (e.g. medroxyprogesterone acetate).

The drugs are described in turn and are divided according to whether their principal use is for the treatment of breast cancer or for prostatic cancer.

8.2   Drugs for the treatment of breast cancer

8.2.1   *Aminoglutethimide*

Aminoglutethimide   **(1)**   (3-ethyl-3-(4-aminophenyl)piperidine-2,6-dione, [125-84-8]* Ormeten, Elipten, Ciba-Geigy Ltd) was originally developed as an anti-epileptic drug but was withdrawn from this use because it was found to cause inhibition of steroid hormone biosynthesis by the adrenals. It is its inhibition of the enzyme aromatase that has since led to its success in the treatment of hormone dependent breast cancer[2]. On the other hand, glutethimide (Doriden, Ciba-Geigy Ltd), which lacks the amino function, does not inhibit steroid hormone biosynthesis and is a clinically used sedative.

Aminoglutethimide   is   synthesised   via   glutethimide   as   illustrated   in

(1)

*Chemical Abstracts Service registry number.

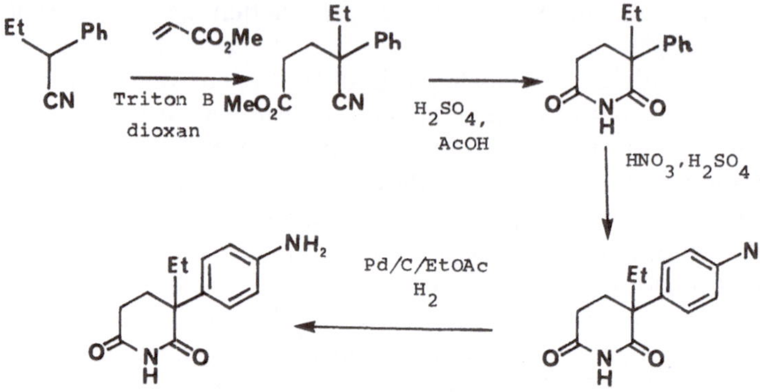

Scheme 8.1 Synthesis of aminogluthethimide.

Scheme 8.1[3,4] although the nitro group can optionally be introduced earlier in the sequence[5] and methyl acrylate can be replaced by acrylonitrile. A drawback of this synthesis is that the nitration of glutethimide gives a mixture of isomers from which the *para*-isomer must be isolated. The minor isomer was originally thought to be the *ortho*-nitro compound[5,6] but the nitration of glutethimide has since been shown to give essentially a 2:1 mixture of *para*- and *meta*-isomers, the structures of the products having been proved by permanganate oxidation to the known nitrobenzoic acids[7]. Consequently reduction of the minor nitroglutethimide product results in the *meta* and not the *ortho*-aminoglutethimide as originally stated[6]. Indeed a recent attempt to prepare *ortho*-aminoglutethimide gave a substituted 2-oxoindole via intramolecular ring opening of the glutarimide ring by the amino group[8].

Carbon-14 labelled aminoglutethimide has been prepared from labelled glutethimide in modifications of the route of Scheme 8.1; in both cases described the label originated from [^{14}C]cyanide[9,10].

Aminoglutethimide is chiral but the commercially used drug is a racemic mixture. The mixture has been resolved by repeated recrystallisation of the tartrate salt from methanol, and the absolute configurations of the resulting enantiomers were determined by comparison of their circular dichroism curves with that of *s*-2-phenyl-2-ethylsuccinimide which has been unambiguously prepared from natural amino acids[11]. The aromatase inhibitory activity was found to be mainly confined to the *R*-enantiomer (**1a**)[11,12].

A variety of compounds structurally related to aminoglutethimide have been shown to inhibit aromatase. An example is the simple compound 4-cyclohexylaniline (**2**). Therefore neither the functionality in the glutarimide ring nor the angular ethyl group is essential for activity. Like aminoglutethimide, acetylation of (**2**) abolished antitumour activity indicating the arylamine group to be essential[13]. Variation of the angular alkyl substituent has been accomplished using essentially the route in Scheme 8.1 following

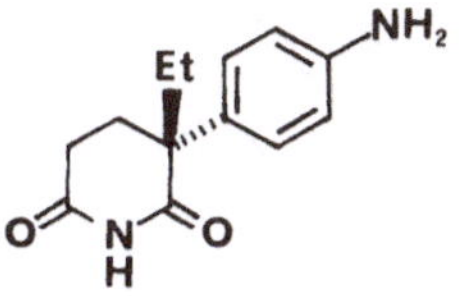

(1a) 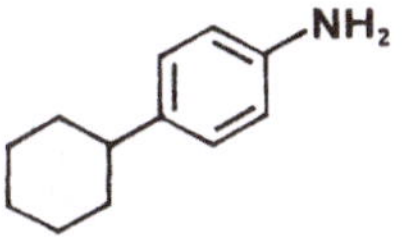   (2)

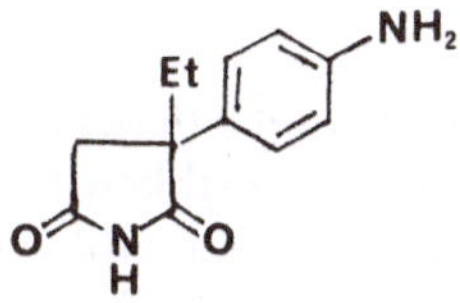

(3) 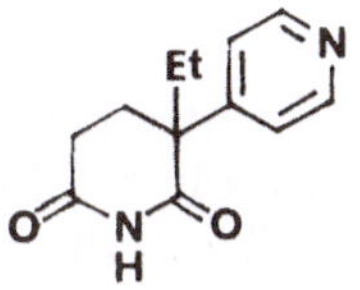   (4)

alkylation of phenylacetonitrile with the appropriate alkyl iodide. The 3-methylbutyl analogue has been shown to be a more potent inhibitor of aromatase than aminoglutethimide[14]. Replacement of the glutarimide ring by a 5-membered pyrrolidine-2,5-dione ring as in (3) gave a compound with similar inhibitory activity towards aromatase[15].

In addition to its action against aromatase, aminoglutethimide inhibits cholesterol sidechain cleavage by desmolase causing the requirement for corticosteroid replacement therapy. This is a disadvantage in breast cancer treatment but has led to the use of aminoglutethimide in prostate cancer treatment since the biosynthesis of androgens is inhibited[16,17]. A strong selective inhibition of desmolase is achieved by 1-amino-3-ethyl-3-phenylpiperidinedione which is formed from glutethimide by ring opening with hydrazine and then reclosure with loss of ammonia[18]. The pyrrolidinedione (3) is a more selective inhibitor of aromatase than is aminoglutethimide but better is 3-ethyl-3-(4-pyridyl)-piperidine-2,6-dione (4) which is a strong inhibitor of aromatase lacking desmolase activity[19]. This result is in contrast to previous indications that the amino group is essential for activity; rather it can be replaced by another function which is capable of coordinating to the cytochrome P-450 of the enzyme. The pyridyl analogue (4) was synthesised from 4-pyridylacetonitrile following ethylation and then essentially as for the preparation of aminoglutethimide in Figure 8.1. It is also unlike aminoglutethimide in that it does not suffer rapid metabolic deactivation by acetylation[20] and self-induced hydroxylamine formation[21].

8.2.2 4-Hydroxyandrost-4-ene-3,17-dione

4-Hydroxyandrostenedione (5) [566-48-3], simply a hydroxylated derivative of a natural steroid, is an irreversible inhibitor of the enzyme aromatase having much greater potency *in vitro* than aminoglutethimide and which has proven

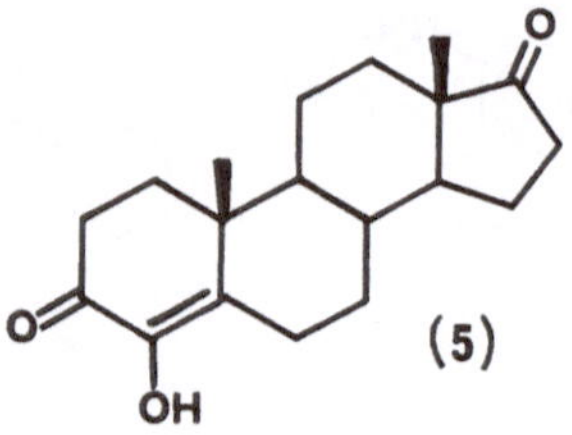

(5)

effectiveness in the treatment of metastatic hormone dependent breast cancer combined with low toxicity[22]. Formally an α-diketone, the preferred tautomeric structure of (5) is as the enol shown. It is readily synthesised from androstenedione by epoxidation under alkaline conditions and then acid catalysed rearrangement of the mixture of α and β epoxides to give the enol (Scheme 8.2)[23-25].

A drawback with 4-hydroxyandrostenedione is the need for parenteral administration. This can be solved by using the acetyl derivative formed from the enol by acetylation using acetic acid and pyridine[26], but this derivative has lower potency than the enol. A further problem of 4-hydroxyandrostenedione is the difficulty in maintaining a sufficiently high plasma concentration, a consequence of the reactivity of the hydroxyl group allowing metabolic conjugation to give a glucuronide ether[27].

Other related steroids possessing aromatase inhibitory activity are 4-hydroxyandrosta-4,6-diene-3,17-dione[28] and 1,4,6-androstatriene-3,17-dione[29]. Comparisons of 4-hydroxyandrostenedione with 4-hydroxy-4-estrene-3,17-dione, which lacks the 19-methyl group and which does not cause a time dependent decrease in aromatase activity, indicate that 4-hydroxy-androstenedione acts as a suicide inhibitor of the enzyme whereby the 19-methyl group is hydroxylated prior to irreversible binding to the enzyme[30].

Deuterium and tritium labelled 4-hydroxyandrostenedione, useful in metabolism and pharmacokinetic studies, have been prepared from 4-hydroxyandrosta-4,6-diene-3,17-dione by catalytic reduction using deuterium and tritium, respectively[31].

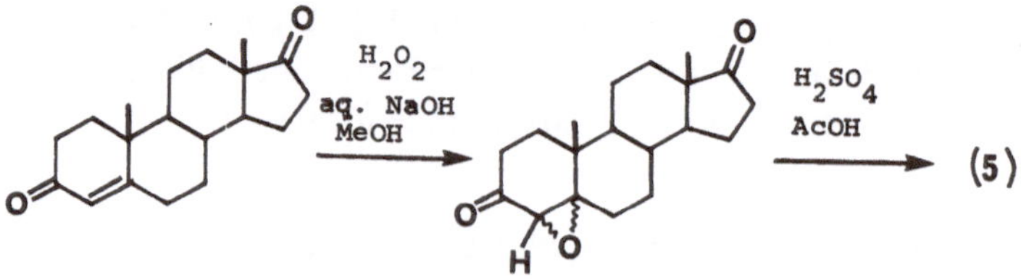

Scheme 8.2 Synthesis of 4-hydroxyandrost-4-ene-3,17-dione.

8.2.3 *Danazol*

Danazol (**6**) (17α-pregn-4-en-20-yno[2,3-*d*]isoxazol-17β-ol, [17230-88-5], Danol, Sterling Winthrop), an agent originally approved for the treatment of endometriosis and benign breast disease, has been shown to be an effective agent for the treatment of a proportion of patients with hormone dependent

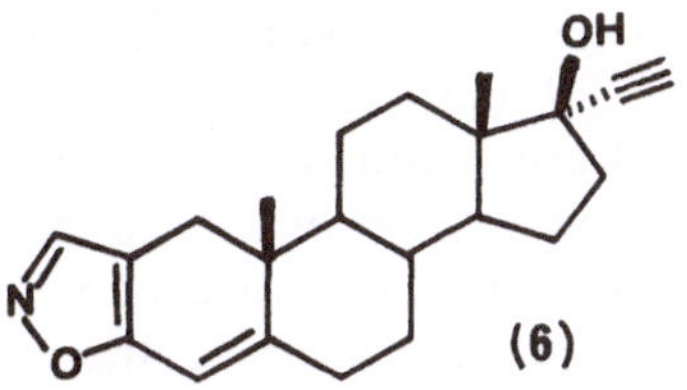

(6)

advanced breast cancer[32]. The principal action of danazol is possibly its inhibition of luteinising hormone (LH) and follicle-stimulating hormone (FSH) release by the pituitary (compare the use of LHRH agonists for the treatment of prostatic cancer), although it also inhibits various steps of steroid hormone biosynthesis and binds to androgen and progestin receptors.

The drug was developed by Sterling Winthrop who used the synthesis shown in Scheme 8.3[33]. The 2-hydroxymethylene precursor was prepared by treatment of ethisterone with base and methyl formate. It exists in the conjugated enolic form shown, stabilised by an intramolecular hydrogen bond. The reaction of the hydroxymethyl compound with hydroxylamine has been shown to proceed via initial formation of a mono-oxime and then cyclisation to a 5-hydroxyisoxazoline which finally undergoes dehydration to form danazol[34]. However, the reaction can give the isomeric [3,2-*c*]isoxazole in which the positions of the oxygen and nitrogen are interchanged, and careful

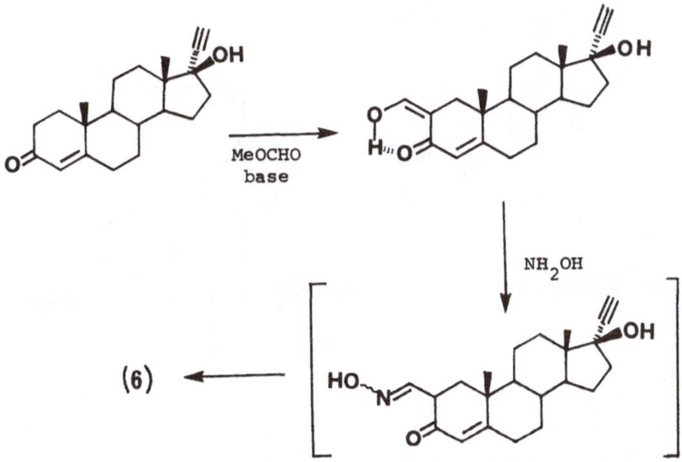

Scheme 8.3 Synthesis of danazol.

control of the pH of the reaction is required to achieve an optimum yield of the required [2,3-*d*]-isomer[33]. These isomers are readily distinguished since on treatment with strong base (sodium methoxide) the [2,3-*d*]-isoxazole undergoes fragmentation to give an isomeric 2-cyanosteroid. The [3,2-*c*]-isoxazole, which does not have a proton α to the nitrogen atom, is stable to these conditions[33].

The fragmentation has been employed in the synthesis of trilostane, a 2-cyanosteroid that inhibits 3β-hydroxysteroid dehydrogenase[35].

Large doses of danazol are required for treatment since it suffers rapid metabolism. All observed metabolites of danazol had undergone cleavage of the weak nitrogen–oxygen bond of the isoxazole[36,37], presumably following coordination of the basic nitrogen atom to the cytochrome *P*-450 of the metabolising enzymes. Thus, as with many drugs, the preferred site of metabolism coincides with the region of the molecule having highest chemical reactivity.

8.2.4 *Tamoxifen*

Tamoxifen (**7**) (*Z*-1-[4-[2-(dimethylamino)ethoxy]phenyl]-1,2-diphenyl-1-butene, [10540-29-1], Nolvadex, Imperial Chemical Industries) is an anti-oestrogen developed by Imperial Chemical Industries that has been highly successful in the treatment of hormone responsive advanced breast cancer[38]. Patients undergoing treatment with this drug can benefit from a period free from disease of up to about 18 months. It is thought to act principally by preventing oestradiol from binding to its protein receptor. The drug was developed[39] by modification of other triarylethylene derivatives such as clomiphene (**8**), an *E/Z* mixture of which is used as a fertility agent for anovulatory women[40].

(**7**)   (**8**)

Tamoxifen is the *Z*-geometrical isomer. The *E* isomer is oestrogenic and would oppose the action of tamoxifen[41]. Therefore chemical syntheses need to provide the pure *Z*-isomer. The classical synthesis developed by ICI[42] is illustrated in Scheme 8.4. The final dehydration of the tertiary alcohol (**9**) gives

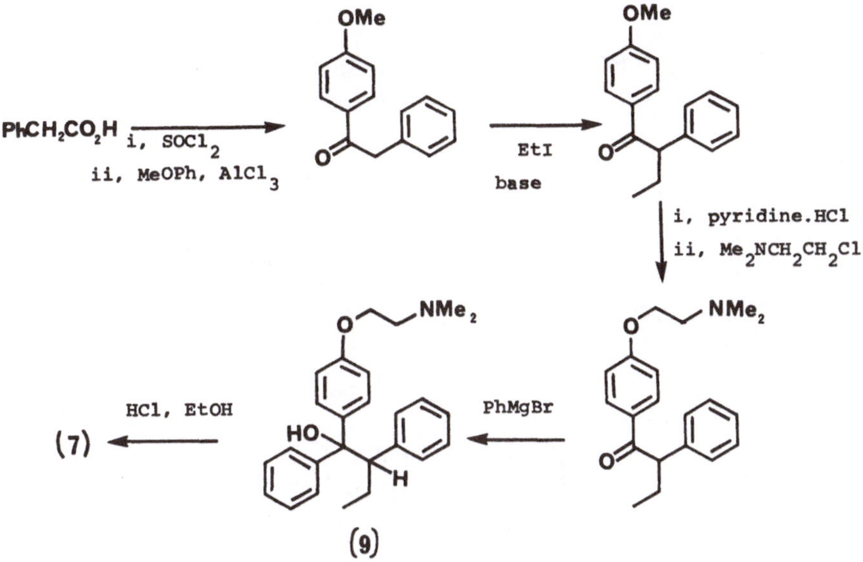

Scheme 8.4 The original synthesis of tamoxifen.

a mixture of Z and E isomers, the Z-isomer of which has been isolated by fractional recrystallisation of the citrate salt[43]. Other workers have claimed that the isomer separation is better accomplished using a dibenzoyltartrate salt[44]. On an analytical scale, separation of the isomers is accomplished by thin-layer silica chromatography using benzene-triethylamine as the eluant[45].

Although the acid catalysed dehydration of the tertiary alcohol (9) has been reported to give a 1:1 mixture of isomers[39], by selecting a suitably mild procedure, the reaction can be controlled to give a 2:1 mixture in favour of the Z-isomer[46]. Extended acid treatment equilibrates either isomer into a 1:1 mixture presumably via protonation of the central double bond. Studies with model compounds show that the isomer distribution is indepedent of the stereochemistry of the tertiary alcohol precursor[46], explainable by the dehydration proceeding via a common carbenium ion intermediate. This ion can be regenerated by fluorosulphonic acid and then quenched to give mainly Z-tamoxifen[46]. Alternatively, dehydration of the tertiary alcohol, which is formed from the ketone according to Cram's rule, can be dehydrated in a highly stereoselective manner with base and carbon disulphide to give tamoxifen precursors in up to 25:1 Z/E but in only moderate yield[46].

Although the separation of Z-tamoxifen from the Z/E mixture can prove difficult, separation of isomers of the chloroethoxy precursors (10) by recrystallisation from isopropanol is relatively straightforward. This compound is a convenient precursor for the preparation of tamoxifen as well as allowing variation of the basic sidechain by substitution of the chlorine with agents other than dimethylamine. This chloroethyl compound is readily

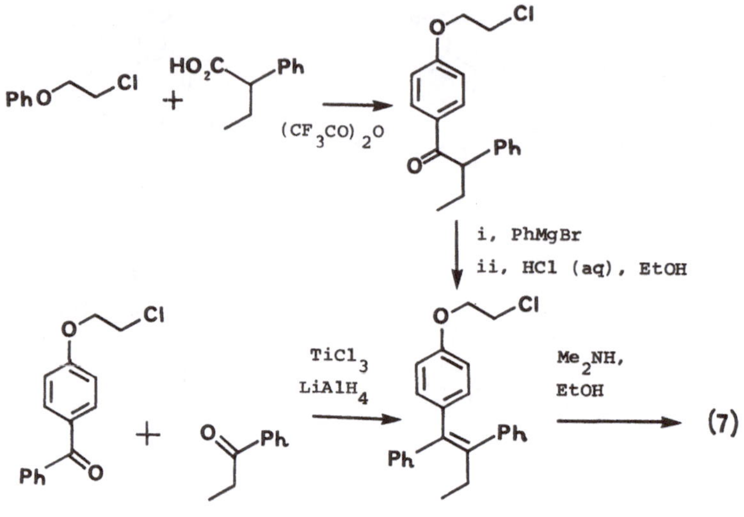

Scheme 8.5 New syntheses of tamoxifen.

prepared by one of two routes (Scheme 8.5); either by elaboration of the ICI route[46] or from a low valent titanium mediated coupling of the appropriate ketones[47]. This latter method is reported to give remarkably a 10:1 mixture of Z- and E-isomers if 4-hydroxybenzophenone is coupled with propiophenone[47] and has also been used for the synthesis of tamoxifen directly[48].

An alternative approach to the preparation of the pure Z-isomer that allows variation of the basic sidechain is via the mixed perfluorotolyl ethers (11) which are prepared from E/Z hydroxy precursors on reaction with octafluorotoluene. These isomers can be easily separated by chromatography, the perfluorotolyl group removed using sodium methoxide and the basic sidechain added without isomerisation[49]. This methodology has been extended to the synthesis of pure Z- and E-isomers of 4-hydroxytamoxifen (Scheme 8.6)[50]. The Z-isomer (12) is a metabolite of tamoxifen that has much higher *in vitro* potency than the parent drug[51] and probably contributes significantly to the

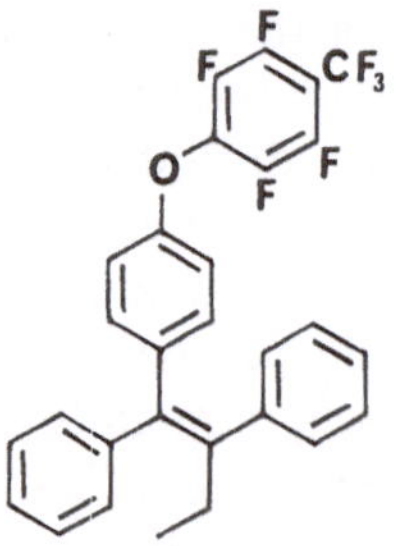

(11) 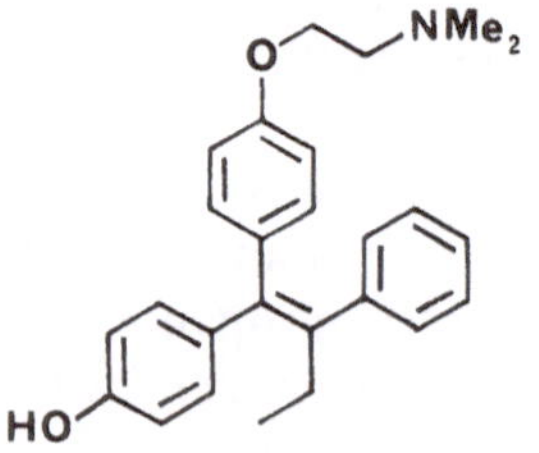   (12)

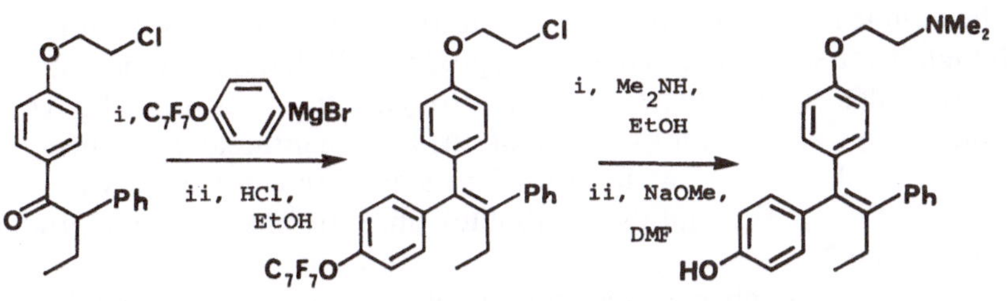

Scheme 8.6 A synthesis of 4-hydroxytamoxifen.

overall activity of tamoxifen. In binding to the oestrogen receptor, the 4-hydroxy group of (12) is thought to play the same role as the 3-hydroxyl group in oestradiol[52].

A novel stereospecific synthesis of tamoxifen (Scheme 8.7) has been reported by Miller and Al-Hassan based on a *cis*-selective carbometallation of an acetylene and the following stereospecific replacement of trimethylsilyl and bromo groups by different aryl groups[53].

The best method for routinely distinguishing the isomers of tamoxifen and related triarylethylenes is by proton NMR. The aryl rings in tamoxifen are not planar owing to non-bonded interactions between hydrogen atoms of adjacent rings; rather the rings are twisted out of the plane of the olefinic bond by more than 50° to give a propeller like conformation as revealed by X-ray crystallography[54]. Consequently in the NMR spectrum, phenyl rings shield adjacent rings and give an upfield shift of the resonances of attached proton groupings. Thus in the Z-isomer of tamoxifen, the AB quartet of the disubstituted ring protons is approx. 0.4 ppm upfield of the corresponding signals for the E-isomer[45,55], and the Z-isomer gives a triplet for the OCH_2 protons at δ 3.83 whereas that for the E-isomer is at δ 4.02[55,56].

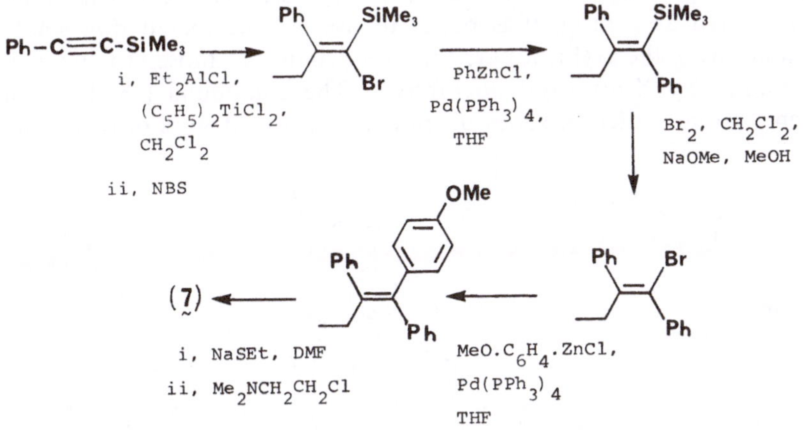

Scheme 8.7 A stereospecific route to tamoxifen.

The double bond in tamoxifen, being tetra-substituted, is fairly unreactive, although it can be hydrogenated[57] or epoxidised[58]. Irradiation of tamoxifen by ultraviolet light initiates a 6π-photocyclisation between the vicinal aromatic rings and, in the presence of air, a phenanthrene is formed. This photochemical reaction has been used to achieve greater sensitivity in the detection of tamoxifen and its metabolites since the phenanthrene products are fluorescent[59].

When one of the phenyl rings in tamoxifen bears a *para*-hydroxyl group, isomerisation of the individual isomers into an E/Z mixture becomes very facile taking place rapidly in the presence of acid or radical sources[45,60] (such as is present in solutions in chlorinated solvents). The radical catalysed isomerisation presumably takes place following abstraction of the hydroxyl hydrogen atom. Delocalisation of the unpaired electron reduces the bond order of the olefinic bond allowing its rotation. Acid catalysed isomerisation can take place by a mechanism analogous to that described for diethylstilboestrol[61]. In the case of 4-hydroxytamoxifen, isomerisation has been shown to take place under cell cuture conditions[62] thereby complicating the metabolism profile of tamoxifen and making it difficult to establish reliable structure-activity relationships, particularly for the less potent E-isomers. In 3-hydroxytamoxifen, isomerisation is not a problem[60] but potency is reduced relative to the 4-isomer.

The complication of isomerisation in 4-hydroxytamoxifen is largely overcome by placement of a 2-methyl group into the *para*-hydroxylated benzene ring. The resulting 2-methyl-4-hydroxytamoxifen (13) not only forms essentially the desired pure E-isomer on dehydration of the tertiary alcohol precursor (although the stereochemistry is E, the relative stereochemistry is the same as in Z-tamoxifen) but also this compound resists isomerisation[63,64]. Complete prevention of isomerisation can be accomplished by ring-fusion. Such is the situation in nafoxidine (14) but the flattening of the fused ring by the fusion[65] is possibly the cause of unwanted side effects if nafoxidine is administered as a drug. It is better to use a 7-membered ring since the orientations of the aryl rings are then very similar to those in tamoxifen, as determined by X-ray crystallography[66]. The compound (15) has similar potency to tamoxifen. A series of tricyclic fused analogues of type (16) had

(13)   (14)   (15)

(16)

X=O, S, CH$_2$CH$_2$, SCH$_2$; Y=H, OH   (17)   (18)

lower potency than corresponding tamoxifen derivatives[67], possibly since two of the rings are inclined towards one another rather than being in the propeller conformation of tamoxifen.

A further consequence of the propellor conformation of tamoxifen is that the molecule possesses helicity and hence there are enantiomers differing in the direction of wind of the helix. In the case of tamoxifen, the enantiomeric conformers (atropisomers) rapidly interconvert[52], even at $-75°C$[66]. However, in an *ortho*-methyl substituted 7-membered ring analogue, the conformers can be distinguished on the NMR timescale at room temperature[68]. If the isomers can be isolated individually they should be useful for a detailed structure-activity study of oestrogen receptor binding at a three-dimensional level.

Structure-activity studies of triarylethylenes reveal that considerable variation in structure maintains the anti-oestrogenicity. Many reports to this effect have appeared. Apart from variation of the basic sidechain which is tolerated so long as it can accept a hydrogen bond[69], the ethyl group can be replaced by such groups as chlorine (clomiphene)[40,70], nitro (nitromiphene)[71], trifluoro-methyl[72], or 2-chloroethyl (toremifene)[73]. Other structures based on triaryl-ethylenes that have anti-oestrogenicity include substituted 2,3-diphenyl-indoles[74], 2-benzoyl-3-phenylthiophene[75], and various acetoxy substituted triarylethylenes[76], 1,2-diphenylethanes[77] and 2-phenylindenes[78].

A variety of radiolabelled derivatives of tamoxifen have been reported. Tamoxifen labelled by tritium in the 2-phenyl ring has been prepared by tritiodehalogenation of the 2-(4-bromophenyl) derivative[45]. This method was also used to prepare tritium-labelled 4-hydroxytamoxifen[45]. Carbon-14 labelled tamoxifen has been prepared by the method of Scheme 8.4 but using [U-^{14}C]-bromobenzene as the source of the Grignard reagent[79]. Iodo[80] and fluorotamoxifens[81] (17) and (18) have been prepared from the corresponding amino compounds via the diazonium salt. Tamoxifen labelled by ^{125}I has been prepared from tamoxifen via ring substitution *ortho* to the dimethylaminoeth-oxy function with tri-*n*-butyltin[82]. When radiolabelled halogen is used, the compounds should be suitable for external imaging (^{121}I/^{123}I), site selective radiotherapy (^{125}I) or positron emission tomography (^{18}F).

In addition to its use for the treatment of hormone dependent breast cancer, tamoxifen has also been found to be of some benefit in the treatment of prostatic cancer[83], possibly because such tumours can have an oestrogen dependency in addition to their androgen dependency.

8.2.5 Medroxyprogesterone acetate

Medroxyprogesterone acetate (**19**) (17α-acetoxy-6α-methylprogesterone, [520-85-4]) is a potent progestin. It has been found to be useful in the treatment of certain advanced breast cancer patients[84], its mechanism of action possibly being a progestin induced inhibition of oestrogen receptor mediated processes.

(19)

Its synthesis from 17α-hydroxyprogesterone was published in 1958 and is outlined in Scheme 8.8[85]. The principal synthetic manipulation was introduction of the 6α-methyl substituent, which gives the steroid enhanced potency. Nucleophilic opening of the α-5,6-epoxide intermediate gave the expected diaxial substituted product, which after regeneration of the enone function, was epimerised under acidic catalysis to give the thermodynamically more stable equatorial 6α-methyl steroid.

8.3 Drugs for the treatment of prostatic cancer

8.3.1 Luteinising hormone-releasing hormone agonists[86–88]

Androgen biosynthesis in the testes is stimulated by luteinising hormone (LH). Production of this hormone by the pituitary gland is controlled by the decapeptide LHRH (Luliberin) which has the structure (**20**). LH is secreted only if LHRH levels are pulsatory and consequently either a LHRH agonist or antagonist should be capable of inhibiting the process. In practice, LHRH agonists have been chosen for prostate cancer treatment. In order to obtain LHRH analogues with greater potency than LHRH itself, approaches have aimed to increase the duration of action. This is achievable by replacement of the glycine residue at the 6-position by an unnatural (D stereochemistry) hydrophobic amino acid[89]. This inhibits metabolic degradation by hydrolysis of the 5–6 linkage and promotes deposition into fatty tissues by virtue of the hydrophobicity. Particularly effective amino acid residues are D-3-(2-naphthyl)alanine (D-Nal(2)) and 3-(2,4,6-trimethylphenyl)alanine (D-Tmp) which improve the potency by a factor of approx. 200. Heterocyclic amino

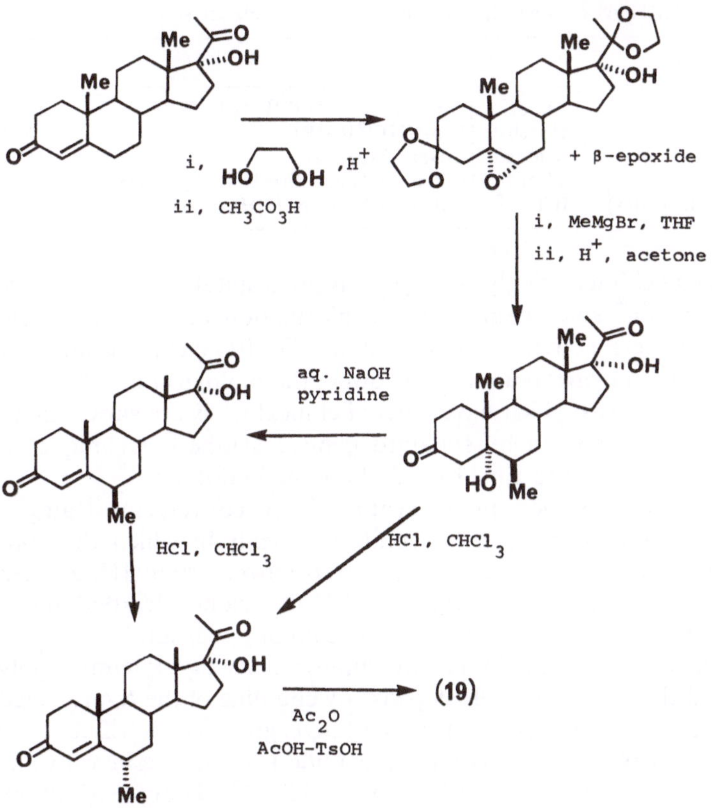

Scheme 8.8 Synthesis of medroxyprogesterone acetate.

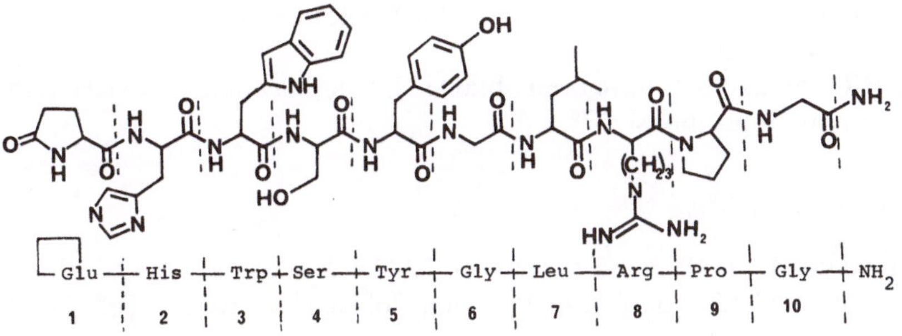

(20)

acids based on 3-(2-benzimidazolyl)alanine have also proved effective[90]. Modification at the terminal 10-glycine residue also affects activity and even greater potency is achieved if this residue is substituted by α-azaglycine. The improvement possibly arises through inhibition of metabolic cleavage of the

Table 8.1 LHRH agonists which have entered clinical trials are shown in Table 8.1.

Manufacturer			Registry no.
Leuprolide	Abbot	[D-Leu[6],Pro-NHEt[9]]LHRH	[53714-56-0]
Burserelin	Hoechst	[Pro-NHEt[9]]LHRH	[57982-77-1]
Nafarelin	Syntex	[D-Nal(2)[6]]LHRH	[80458-30-6]
Lutrelin	Wyeth	[D-Trp[6], MeLeu[7], Pro-NHEt[9]]LHRH	
Goserelin (Zoladex)	ICI	[D-Ser('Bu)[6], Pro-NHNHCONH$_2$[9]]LHRH	[65807-02-5]

9–10 linkage. Interestingly, azaglycine is not a suitable replacement for the 6-glycine residue since it alters the conformation of the peptide chain and thereby lowers activity[91]. Removal of the 10-glycine residue altogether increases the *in vitro* potency but lowers the biological half-life.

LHRH agonists which have entered clinical trials are shown in Table 8.1. They can be prepared by standard peptide synthesis techniques either in solution or by the Merrifield solid phase methodology.

In a typical solution phase synthesis[91], a convergent strategy is used employing the coupling of tripeptide fragments in which the appropriate amino groups are protected as the *N-α-tert*-butoxycarbonyl (BOC) derivatives and carboxyl groups as methyl esters. *N,N*-Dicyclohexylcarbodiimide (DCC)-hydroxybenzotriazole is chosen as the coupling reagent.

The solid phase method typically employs a (benzhydrylamino) polystyrene resin and the peptide is built stepwise by coupling of the *N*-protected (BOC) amino acids with DCC, deprotection of BOC groups with trifluoroacetic acid, and final cleavage of the complete peptide from the resin with anhydrous hydrofluoric acid prior to purification on HPLC[94]. The ease of automation of this process using a commercially available peptide synthesiser has made this the method of choice in recent years.

An improvement that is still sought for LHRH agonists is to achieve good oral bioavailability.

LHRH antagonists have been obtained by modification of LHRH at the 1, 2, 3, 6 and 10 positions[92–94].

8.3.2 *Ketoconazole*

Ketoconazole **(21)** (*cis*-1-acetyl-4-[4-[2-(2,4-dichlorophenyl)-2-(*1H*-imida-zol-1-ylmethyl-1,3-dioxolan-4-yl)methoxy]phenyl]piperazine, [65277-42-1],

(21)

Nizoral, Janssen Pharmaceuticals) was originally developed as an anti-fungal agent but has been found to decrease serum testosterone levels[95] leading to its use for the treatment of hormone dependent prostatic cancer[96,97]. However, it has now been withdrawn from this use owing to undesirable side effects. Since the drug caused accumulation of 17α,20α-dihydroxyprogesterone in a testicular cell model its likely site of action is the *P*-450 dependent C-17,20 lyase catalysed step[98].

The synthesis of ketoconazole[99,100] is outlined in Scheme 8.9. The product of ketalisation of 2,4-dichloroacetophenone with ethylene glycol possesses two chiral centres and is formed as a mixture of diastereoisomers. Following bromination of the methyl group, the required *cis*-isomer is isolated by fractional crystallisation of the benzoate ester from ethanol. After replacement of the bromine atom by imidazole and saponification of the benzoate ester, the acetylpiperazinophenyl moiety is introduced. Ketoconazole is a mixture of enantiomers. It would prove an interesting challenge to prepare these individually by a chiral synthesis.

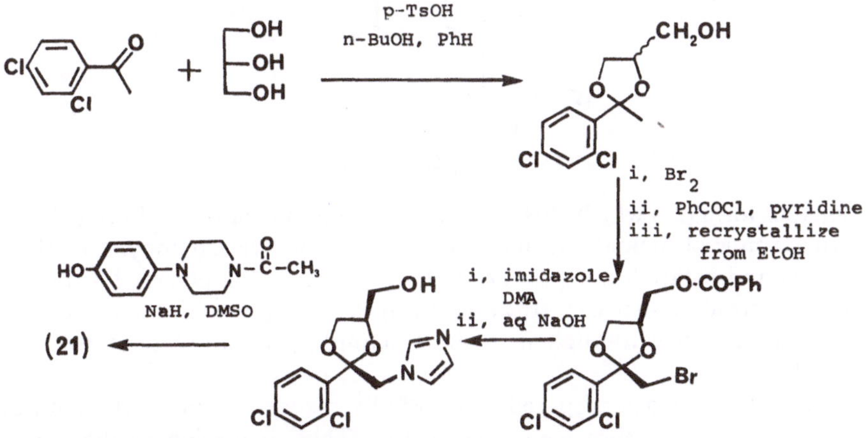

Scheme 8.9 Synthesis of ketoconazole.

8.3.3 *4-Azasteroids as 5α-reductase inhibitors*

5α-Dihydrotestosterone (DHT) is the hormone that is primarily responsible for promoting prostate growth, and inhibition of its biosynthesis represents a target for prostate cancer treatment as well as related benign conditions[101].

4-Azasteroids are rationally designed inhibitors of the 5α-reductase enzyme which converts testosterone into DHT. Scheme 8.10 illustrates the pathway for the conversion whereby the enzyme catalyses transfer of hydrogen from NADPH onto the 5α-position of testosterone. Ketone formation from the resulting enol (22) may be necessary before the steroid can dissociate from the enzyme binding site. The azasteroids inhibit the enzyme by virtue of the amide

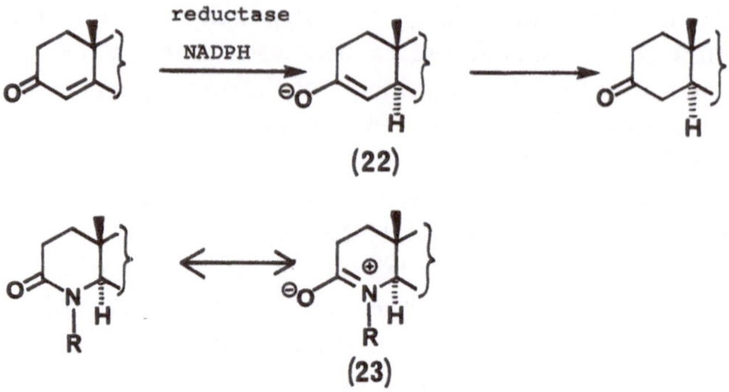

(22)

(23)

Scheme 8.10 Pathway for reduction of the enone function in testosterone showing how 4-azasteroids can mimic the enolate intermediate.

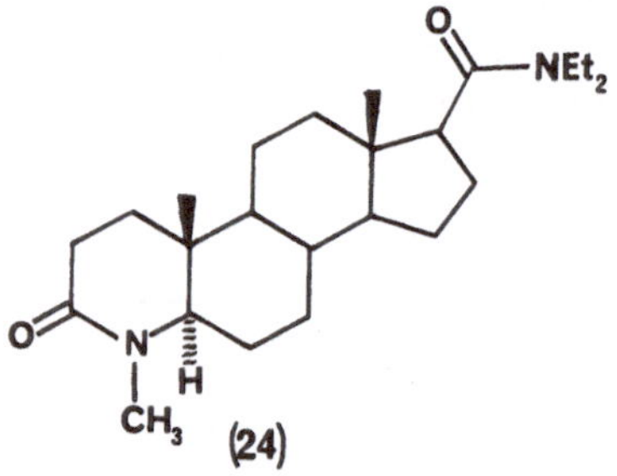

(24)

resonance form (23) which closely mimics the intermediate enol form (22)[102].

An example of such an azasteroid is 17β-N,N-diethylcarbamoyl-4-methyl-4-aza-5α-androstan-3-one [73671-86-0] (24) which has emerged from an extensive structure-activity study as being a very potent inhibitor of the reductase enzyme, although it is short acting[103], a feature of steroids in general.

The azasteroids are prepared efficiently by reconstruction of the natural steroid structure as shown in Scheme 8.11. Thus, oxidation of the enone containing ring opened this ring and removed the unwanted carbon atom at position-4. Condensation with an amine then gave recyclisation to an enelactam which was catalytically hydrogenated to give the required product[103]. Similar strategies, for instance involving Beckmann rearrangement to enable conversion of a cyclic ketone to a larger ring lactam, have allowed the formation of a variety of other azasteroids[104,105] and the azasteroid with a 7-membered A-ring also proved very potent[104]. However, interestingly, 2-oxa substitution which might have been expected to improve inhibition of the enzyme by virtue of having reduced bond order of the carbon-oxygen linkage (hence resembling the electronic structure of intermediate (22) more closely) had lower activity[105,106].

Considerable variation of the 17-substituent can retain activity but semi-polar substituents such as carboxylates, esters and amides are preferred[103,104].

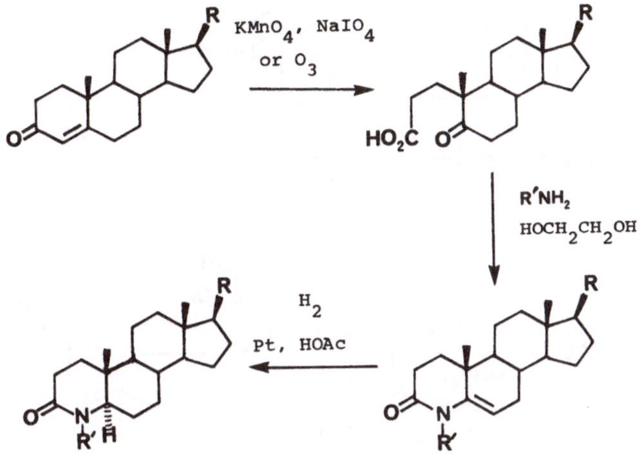

Scheme 8.11 Synthesis of 4-azasteroids.

The 4-*N*-methyl compounds were more potent than the 4-NH compounds and *N*-ethyl compounds were inactive. These observations indicate that there is a compromise between lipophilic preferences and steric requirements for enzyme binding.

Other structures that inhibit the 5α-reductase enzyme are 4-methylene-3-ketosteroids[107] and 4-cyano-$\Delta_{4,5}$-3-ketosteroids[105].

8.3.4   *Cyproterone acetate*

Cyproterone acetate **(25)** (17α-(acetyloxy)-6-chloro-1β,2β-dihydro-2'*H*-cyclopropa[1,2]pregna-1,4,6-triene-3,20-dione, [427-51-0], Schering A. G.) is an antiandrogen, acting by competitively blocking the binding of testosterone and 5α-dihydrotestosterone to the androgen receptor. It has proven effectiveness in the treatment of prostatic carcinoma[108]. Its development is the subject of an excellent review by Neumann and Wiechart[109]. Originally, the aim was to obtain an orally effective progestin and structural modifications were made to progesterone in order to inhibit rapid metabolism. Thus, introduction of the 17α-acetoxy substituent retarded metabolic reduction of the 20-keto function, and the presence of the 6,7-double bond and 6-chlorine substituent

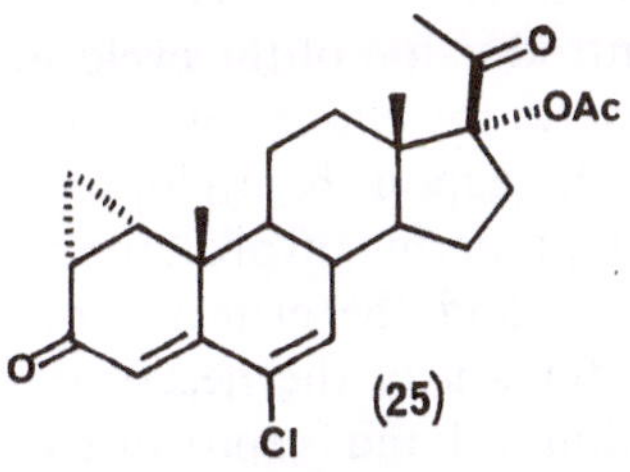

(25)

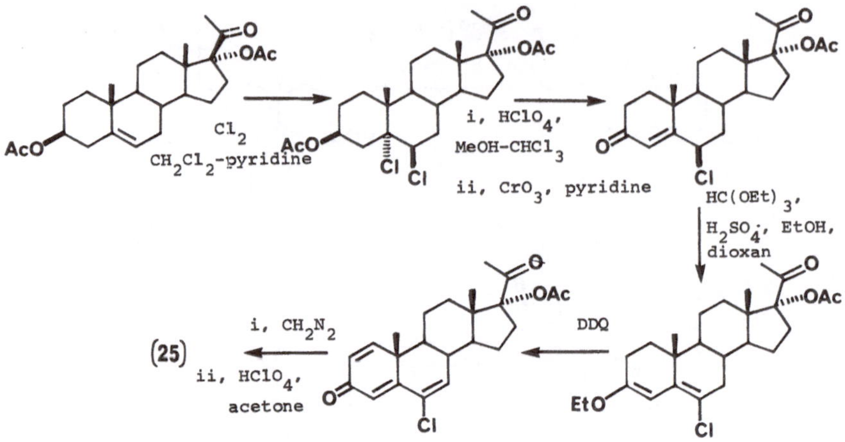

Scheme 8.12 A synthesis of cyproterone acetate.

was to block metabolism at the 6-position. Cyproterone acetate is indeed a very potent progestin but it was found that gestating rats gave birth to offspring bearing only female genitalia, a property revealing its anti-androgenic action and precluding its use for the prevention of miscarriage.

A synthesis of cyproterone acetate from the diacetylated derivative of the natural steroid 17α-hydroxypregnenolone is shown in Scheme 8.12[110]. The route illustrates some of the elegant functional group manipulation that can be made on the steroid framework. Chlorination of the 5,6-double bond gave specifically the 5α,6β-diaxial addition product and specific hydrolysis of the 3-acetoxy group was accomplished with perchloric acid in methanol. On oxidation of the resulting 3-hydroxy function to the ketone, elimination of the 5α-chlorine proceeded spontaneously and afforded the 6-chloro-Δ_4-3-ketosteroid. Formation of the ethyl enol ether with triethyl orthoformate and then oxidation with 2,3-dichloro-5,6-dicyano-4-benzoquinone (DDQ) introduced unsaturation both at C1–2 and C5–6. Fusion of the cyclopropyl group has been made by reaction of the product with diazomethane which undergoes a 1,3-dipolar cyclo-addition onto the less hindered α-face of 1,2-double bond resulting in a 1α-pyrrazoline that can be converted into the cyclopropyl compound either by treatment with strong acid[111], by thermolysis[112], or more mildly on Lewis acid treatment[109], in order to promote loss of nitrogen. An alternative method for introduction of the cyclopropyl ring, which can give improved yields[110], is to treat the enone with dimethylsulphoxenium methylide [$Me_2S^+(O^-)CH_2^-$] whereupon the cyclopropylketone forms directly.

An alternative way of preparing cyproterone acetate starts from 17α-hydroxyprogesterone[113,114] and the elements of this route are shown in Scheme 8.13. After introduction of the necessary unsaturation[115] and the cyclopropyl ring, the 5,6-double bond is epoxidised on the less hindered α-face

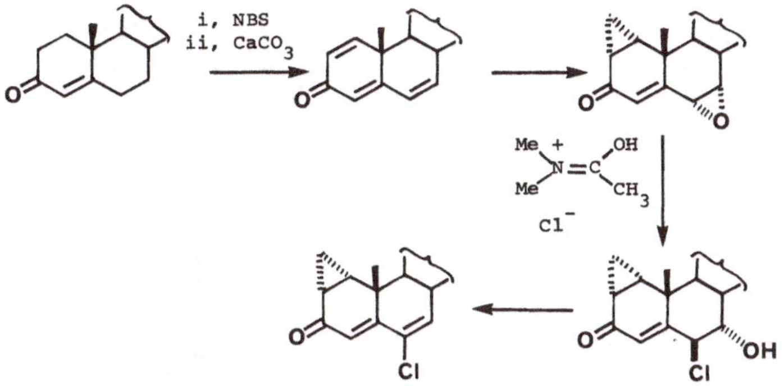

Scheme 8.13 An alternative synthesis of cyproterone acetate.

with peracid. Treatment of the α-epoxide with a carbinoliminium chloride, generated from dimethylacetamide and hydrogen chloride, opens the epoxide to give the diaxial chlorohydrin which undergoes dehydration with a second equivalent of reagent[115] to give the functionality of cyproterone acetate.

8.3.5 Bifluranol

Bifluranol (**26**) (erythro-3,3'-difluoro-4,4'-dihydroxy-α-methyl-α'-ethyl-bibenzyl, [34633-34-6], Biorex) was developed from the non-steroidal oestrogen diethylstilboestrol (**27**) but has one eighth of its oestrogenic potency. The antiprostatic action of bifluranol probably arises from its mimicing the negative feedback of testosterone on androgen biosynthesis so that luteinising hormone secretion is inhibited[116]. Only the erythro-isomer (shown) is active and it is interesting to compare hexoestrol for which only the comparable meso-diastereoisomer (**28**) is active. Conformation studies on hexoestrol reveal that the preferred conformation has the benzylic hydrogen atoms in an anti-periplanar orientation so that the meso-form has an overall structure resembling oestradiol[117].

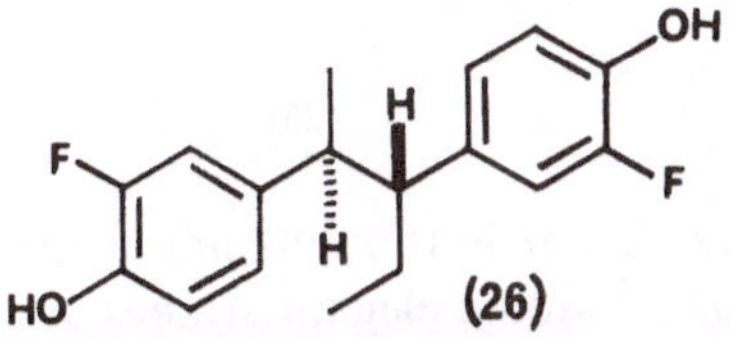

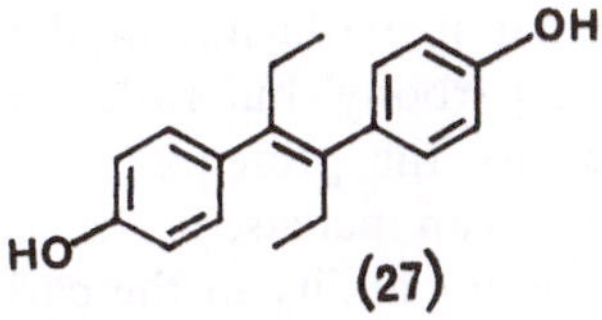

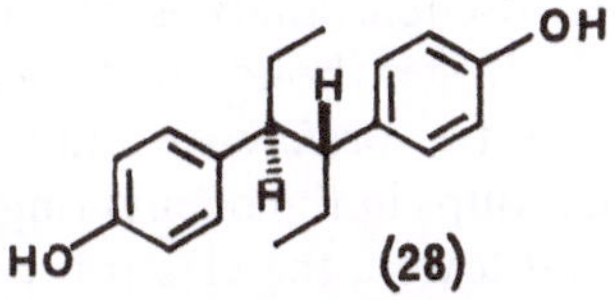

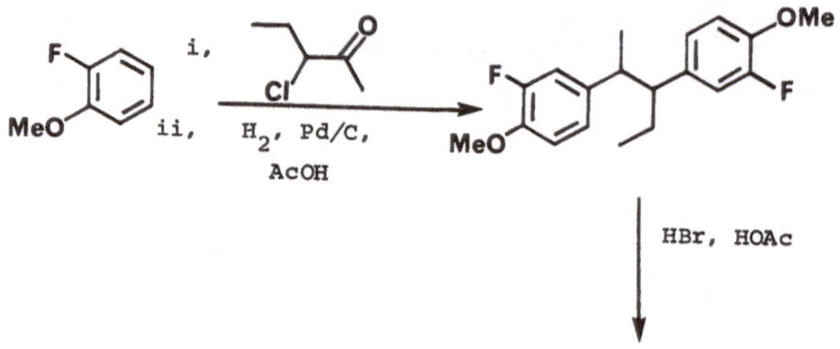

(26)

Scheme 8.14 Synthesis of bifluranol.

Bifluranol can be synthesised as shown in Scheme 8.14. Following condensation of *ortho*-fluoroanisole with 3-chloro-2-pentanone, the resulting olefinic product is hydrogenated. The *erythro*-isomer of the dimethyl ether of bifluranol is isolated by fractional crystallisation before demethylation with hydrobromic acid[118].

8.3.6   *Flutamide*

Flutamide (**29**) (4-nitro-3-trifluoromethylisobutyranilide, [1331-84-7]) was originally developed as a bacteriostatic agent[119] but was also found to be anti-androgenic[120] which has led to its use clinically for the treatment of prostatic cancer[121]. It is simply prepared by reaction of 4-nitro-3-trifluoromethylaniline with isobutyryl chloride[122], although optionally the nitro group can be introduced last by nitration[123]. Flutamide has also been prepared by Beckmann rearrangement of the appropriate oxime[124].

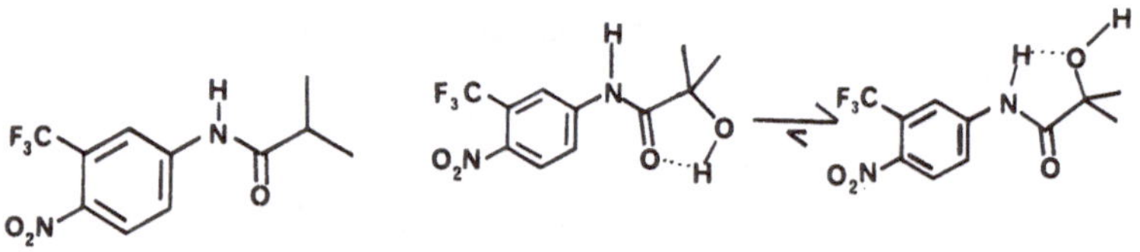

(29)   **(30)**   **(31)**

The biologically active agent is thought not to be flutamide itself but a hydroxylated metabolite[125]. Some elegant studies at ICI have identified the roles of the various functional groups present in hydroxyflutamide. Thus infrared spectroscopic analysis reveals that it exists not in the form (**30**) involving a hydrogen bond to the amide carbonyl but rather in the form (**31**)[126]. This is the preference only because the presence of the electron withdrawing groups in the phenyl ring causes an increase in the acidity of the amide NH and lowers the electron pair donor ability of the carbonyl. The

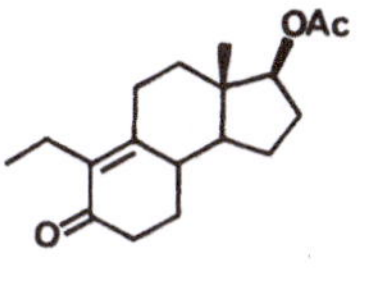

(32) 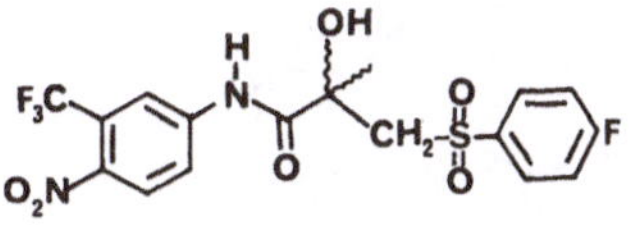   (33) 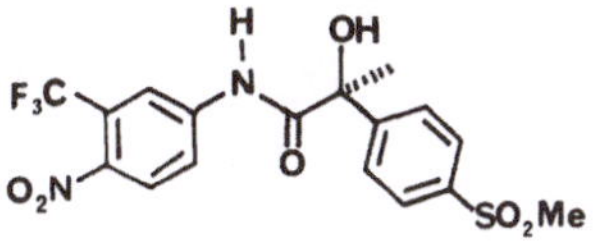   (34)

hydroxyl hydrogen atom is then free to form the important association with the androgen receptor, and there is presumably an analogy with the 17β-hydroxyl group in testosterone. It is worth comparing the similarity of the conformation (31) of hydroxyflutamide with that of the des-A-ring steroid (32) (RU38882)[127] which is also an anti-androgen, although compound (32) is intended for androgen dependent skin conditions.

One problem with flutamide is that in addition to binding to androgen receptors in prostate tissue, it blocks the negative feedback by testosterone on the hypothalamic-pituitary axis and consequently causes stimulation of testosterone biosynthesis. This problem appears to have been overcome in certain substituted derivatives of hydroxyflutamide, e.g. compound (33)[128], which is best prepared as shown in Scheme 8.15 and which in the rat does not block the feedback mechanism but retains anti-androgenic action on peripheral tissues. This compound has been selected for clinical development. Although it is racemic, in a related example only the S-enantiomer of compound (34) was active[126].

Another anti-androgen related to flutamide is 5,5-dimethyl-3-[4-nitro-3-(trifluoromethyl)phenyl]-2,4-imidazolidinedione ([63612-50-0], Anadron, Roussel Uclef) (35)[129]. Anandron is synthesised as shown in Scheme 8.16 from 4-nitro-3-trifluoromethylphenyl isocyanate by construction of the heterocyc-

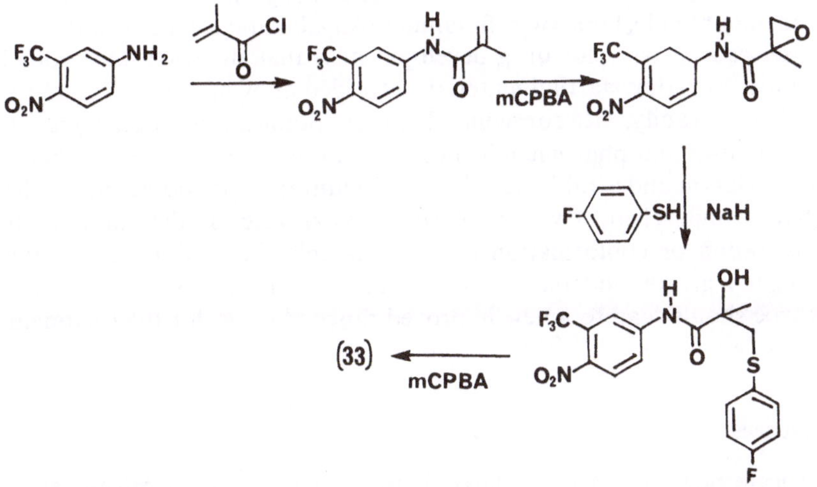

Scheme 8.15 The synthesis of a new peripherally selective analogue of flutamide.

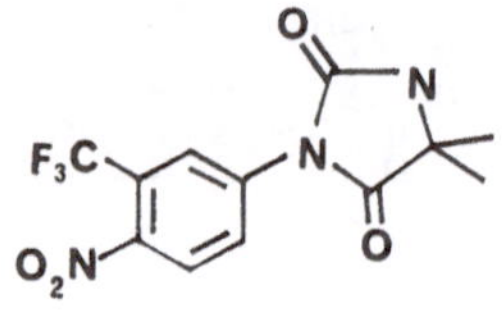

(35)

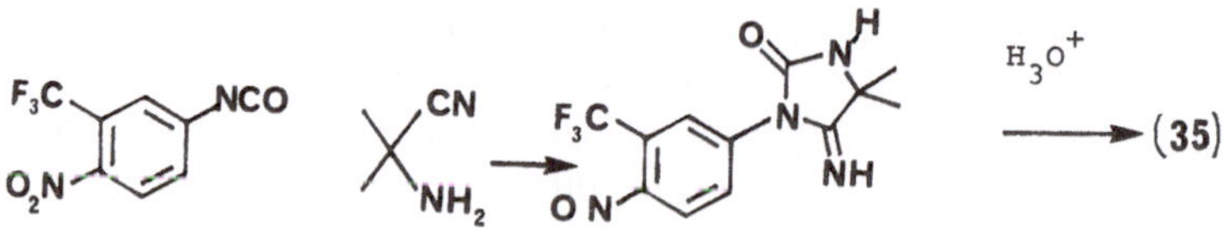

Scheme 8.16 Synthesis of Anandron.

lic ring[130]. Anandron can be considered a conformationally fixed equivalent of hydroxyflutamide (31) although it is now a hydrogen attached to nitrogen that forms a hydrogen bond to the receptor. In Anandron, the presence of the nitro and trifluoromethyl groups will enhance the acidity of the NH group by transmission of their inductive effect through the heterocyclic system and this will improve the strength of interaction with the receptor.

8.4 Concluding remarks

The search over the past 35 years for drugs that inhibit steroid hormone biosynthesis or action for the treatment of hormone dependent breast and prostatic cancers has led to compounds having diverse structural types. In addition to modified peptides (LHRH agonists) and modified steroids (e.g. cyproterone acetate), many of the present drugs are non-steroidal (e.g. aminoglutethimide, tamoxifen, flutamide). An advantage of not using steroids themselves as a basis for drug development is that the need for a complex naturally occurring starting material is avoided allowing the structure to be varied more readily, with consequently easier optimisation of potency, control of metabolism and pharmacokinetics, and improvement in the specificity of action to lessen undesirable side effects. In addition to providing the synthesis of new drugs, chemistry has played a vital role in determining drug configuration or conformation (see for example the section on flutamide) leading to a greater understanding of the interaction with its target. Studies of this type should lead to a new improved range of drugs for the treatment of cancers.

References

1. For a detailed treatment of the pathways and regulation of steroid hormone biosynthesis, see: D. Schulster, S. Burnstein and B. A. Cooke, *Molecular Endocrinology of the Steroid Hormones*, Wiley, London (1976).

INHIBITORS OF STEROID HORMONE BIOSYNTHESIS AND ACTION 257

2. I. E. Smith, B. M. Fitzharris, J. A. McKinna, D. R. Fahmy, A. G. Nash, A. M. Neville, J.-C. Gazet, H. T. Ford and T. J. Powles, *Lancet* ii (1978) 646; S. A. Wells Jr., R. J. Santen, A. Lipton, D. E. Haagensen, Jr., E. J. Ruby, H. Harvey and W. G. Dilley, *Ann. Surg.* 187 (1978) 475.
3. K. Hoffmann and E. Tagmann, U.S. Patent, 2673205 (1954).
4. K. Hoffmann and E. Urech, U.S. Patent, 2848455 (1958).
5. R. Paul, R. P. Williams and E. Cohen, *J. Med. Chem.* 17 (1974) 539.
6. H. Y. Aboul-Enain, C. W. Schauberger, A. R. Hansen and L. J. Fischer, *J. Med. Chem.* 18 (1975) 736.
7. G. Stajer, P. Nemeth, E. Vinkler, L. Lehotay and P. Sohar, *Arch. Pharm.* (*Weinheim*) 312 (1979) 1032.
8. M. Jarman, C.-S. Leung and D. Manson, *Synthesis* (1984) 1061.
9. V. Askam, N. Eweiss and P. J. Nicholls, *J. Labelled Compd. Radiopharm.* 20 (1983) 1331.
10. N. K. Chaudhuri, O. Servando, M. S. Sung and H. Baumann, *J. Labelled Compd. Radiopharm.* 21 (1983) 193.
11. N. Finch, R. Dziemian, J. Cohen and B. G. Steinetz, *Experientia* 31 (1975) 1002.
12. P. E. Graves and H. A. Salhanick, *Endocrinology* 105 (1979) 52.
13. J. T. Kellis Jr. and L. E. Vickery, *Endocrinology* 114 (1984) 2128.
14. R. W. Hartmann and C. Batzl, *J. Med. Chem.* 29 (1986) 1362.
15. M. J. Daly, G. W. Jones, P. J. Nicholls, H. J. Smith, M. G. Rowlands and M. A. Bunnett, *J. Med. Chem.* 29 (1986) 520.
16. E. J. Sanford, J. R. Drago, T. L. Rohner, Jr., R. Santen and A. Lipton, *J. Urol.* 115 (1976) 170.
17. B. A. J. Ponder, R. J. Shearer, R. D. Pocock, J. Miller, D. Easton, C. E. D. Chilvers, M. Dowsett and S. L. Jeffcoate, *Br. J. Cancer* 50 (1984) 757.
18. A. B. Foster, M. Jarman, C.-S. Leung, M. G. Rowlands and G. N. Taylor, *J. Med. Chem.* 26 (1983) 50.
19. A. B. Foster, M. Jarman, C.-S. Leung, M. G. Rowlands, G. N. Taylor, R. G. Plevey and P. Sampson, *J. Med. Chem.* 28 (1985) 200; A. B. Foster, M. Jarman and C.-S. Kwan, Eur. Pat. Appl. EP 169062 (1986) (*Chemical Abstracts* 105 (1986) 54601b).
20. J. S. Douglas and P. J. Nicholls, *J. Pharm. Pharmacol.* 24 (1972) 150.
21. M. Jarman, A. B. Foster, P. E. Goss, L. J. Griggs, I. Howe and R. C. Coombes, *Biomed. Mass Spectrum.* 10 (1983) 620.
22. R. C. Coombes, P. Goss, M. Dowsett, J.-C. Gazet and A. M. H. Brodie, *Lancet* ii (1984) 1237; A. M. H. Brodie, L.-Y. Wing, P. Goss, M. Dowsett and R. C. Coombes, *J. Steroid Biochem.* 24 (1986) 91.
23. A. M. H. Brodie, W. C. Schwarzel, A. A. Shaikh and H. J. Brodie, *Endocrinology* 100 (1977) 1684.
24. H. B. Henbest and W. R. Jackson, *J. Chem. Soc.* C (1967) 2459.
25. J. Mann and B. Pietrzak, *J. Chem. Soc., Perkin Trans.* 1 (1983) 2681.
26. A. M. H. Brodie, H. J. Brodie and D. A. Marsh U.S. Patent 4235893 (1980).
27. P. E. Goss, M. Jarman, J. R. Wilkinson and R. C. Coombes, *J. Steroid Biochem.* 24 (1986) 619.
28. D. A. Marsh, H. J. Brodie, W. Garratt, C.-H. Tasai-Morris and A. M. H. Brodie, *J. Med. Chem.* 28 (1985) 788.
29. W. C. Schwarzel, W. Kruggel and H. J. Brodie, *Endocrinology* 92 (1973) 866.
30. D. F. Covey and W. F. Hood, *Mol. Pharmacol.* 21 (1982) 173.
31. D. A. Marsh, L. Romanoff, K. I. H. Williams, H. J. Brodie and A. M. H. Brodie, *Biochem. Pharmacol.* 31 (1982) 701.
32. R. C. Coombes, D. Dearnaley, J. Humphreys, J. C. Gazet, H. T. Ford, A. G. Nash, K. Mashiter and T. J. Powles, *Cancer Treat. Rep.* 64 (1980) 1073; R. C. Coombes, D. Perez, J. C. Gazet, H. T. Ford and T. J. Powles, *Cancer Chemother. Pharmacol.* 10 (1983) 194.
33. A. J. Manson, F. W. Stonner, H. C. Neumann, R.G. Christiansen, R. L. Clarke, J. H. Ackerman, D. F. Page, J. W. Dean, D. K. Phillips, G. O. Potts, A. Arnold, A. L. Beyler and R. O. Clinton, *J. Med. Chem.* 6 (1963) 1.
34. A. Quilico, *Chem. Heterocycl. Compd.* 17 (1962) 1.
35. H. C. Neumann, G. O. Potts, W. T. Ryan and F. W. Stanner, *J. Med. Chem.* 13 (1970) 948.
36. D. Rosi, H. C. Neumann, R. G. Christiansen, H. P. Schane and G. O. Potts, *J. Med. Chem.* 20 (1977) 349.
37. C. Davidson, W. Banks and A. Fritz, *Arch. Int. Pharmacodyn. Ther.* 221 (1976) 294.

38. R. C. Heel, R. N. Brogden, T. M. Speight and G. S. Avery, *Drugs* **16** (1978) 1; V. C. Jordan, *Breast Cancer Res. Treat.* **3** (1983) 73.
39. (a) B. J. A. Furr and V. C. Jordan, *Pharmacol. Ther.* **25** (1984) 127; (b) V. C. Jordan, *Pharmacol. Rev.* **36** (1984) 245.
40. D. E. Holtkamp, S. C. Greslin, C. A. Root and L. J. Lerner, *Proc. Soc. Exp. Biol. Med.* **105** (1960) 197.
41. M. J. K. Harper and A. L. Walpole, *Nature* **212** (1966) 87; V. C. Jordan, B. Haldemann and K. E. Allen, *Endocrinology* **108** (1981) 1353.
42. M. J. K. Harper, D. N. Richardson and A. L. Walpole, British Patent 1013907 (1965).
43. M. J. K. Harper, D. N. Richardson and A. L. Walpole, British Patent 1064629 (1965).
44. L. Magdana, A. Hutak, E. Szatmari, I. Simoni, J. Halmos and F. Nemere, Belgium Patent 892662 (1982).
45. D. W. Robertson and J. A. Katzenellenbogen, *J. Org. Chem.* **47** (1982) 2387.
46. R. McCague, *J. Chem. Soc., Perkin Trans. 1* (1987) 1011.
47. P. L. Coe and C. E. Scriven, *J. Chem. Soc., Perkin Trans. 1* (1986) 475.
48. M. Jack (Bristol-Myers), Eur Patent, Al/0126 470/1984.
49. M. Jarman and R. McCague, *J. Chem. Res.* (1985) (S), 116; (M) 1342.
50. R. McCague, *J. Chem. Res.* (1986) (S) 58; (M) 771.
51. V. C. Jordan, M. M. Collins, L. Rowsby and G. Prestwich, *J. Endocrinol.* **75** (1977) 305; J.-L. Borgna and H. Rochefort, *Mol. Cell. Endocrinol.* **20** (1980) 71.
52. M. Pons, F. Michel, E. Bignon, A. Crastes de Paulet, J. Gilbert, J. F. Miquel, G. Precigoux, M. Hospital, T. Ojasco and J. P. Raynaud, *Prog. Cancer Res. Ther.* **31** (1984) 27.
53. R. B. Miller and M. I. Al-Hassan, *J. Org. Chem.* **50** (1985) 2121.
54. P. G. Precigoux, C. Courseille, S. Geoffre and M. Hospital, *Acta Crystallogr, Sect. B* **35** (1979) 3070; S. D. Cutbush, S. Neidle, A. B. Foster and F. Leclercq, *Acta Crystallogr. Sect. B.* **38** (1982) 1024.
55. G. R. Bedford and D. N. Richardson, *Nature* **212** (1966) 733.
56. D. J. Collins, J. J. Hobbs and C. W. Emmens, *J. Med. Chem.* **14** (1971) 952.
57. Imperial Chemical Industries, French Patent, 1 568 713 (1969).
58. R. McCague and A. Seago, *Biochem. Pharmacol.* **35** (1986) 827.
59. D. W. Menderhall, H. Kobayashi, F. M. L. Shih, L. A. Sternson, T. Higuchi and C. Fabin, *Clin. Chem.* **24** (1978) 1518; Y. Golander and L. A. Sternson, *J. Chromatogr.* **181** (1980) 41.
60. P. C. Ruenitz, J. R. Bagley and C. M. Mokler, *J. Med. Chem.* **25** (1982) 1056.
61. V. W. Winkler, M. A. Nyman and R. Egan, *Steroids* **17** (1971) 197.
62. B. S. Katzenellenbogen, M. J. Norman, R. L. Eckert, S. W. Peltz and W. F. Mangel, *Cancer Res.* **44** (1984) 112; J. A. Katzenellenbogen, K. E. Carlson and B. S. Katzenellenbogen, *J. Steroid Biochem.* **22** (1985) 589.
63. A. B. Foster, M. Jarman, O.-T. Leung, R. McCague, G. Leclercq and N. Devleeschouwer, *J. Med. Chem.* **28** (1985) 1491.
64. R. Kuroda, S. Cutbush, S. Neidle and O.-T. Leung, *J. Med. Chem.* **28** (1985) 1497.
65. N. Camerman, L. Y. Y. Chan and A. Camerman, *J. Med. Chem.* **23** (1980) 941.
66. R. McCague, R. Kuroda, G. Leclercq and S. Stoessel, *J. Med. Chem.* **29** (1986) 2053.
67. D. Acton, G. Hill and B. S. Tait, *J. Med. Chem.* **26** (1983) 1131.
68. R. McCague, *Tetrahedron Lett.* **28** (1987) 701.
69. D. W. Robertson, J. A. Katzenellenbogen, J. R. Hayes and B. S. Katzenellenbogen, *J. Med. Chem.* **28** (1982) 167.
70. P. C. Ruenitz, J. R. Bagley, C. K. W. Watts, R. E. Hall and R. L. Sutherland, *J. Med. Chem.* **29** (1986) 2511.
71. D. Lednicer, *Nature* **211** (1966) 539.
72. K. Tory, E. Csanyi, T. Horvath, G. Abraham, G. Cseh and J. Boryendeg, *Recent Adv. Chemother. Proc. Int. Congr. Chemother.* **14** (1985) 614.
73. L. Kangas, G. Blanco, A. Hajba, M. Heikkinen, L. Holsti, L. Kalliomaki, E. Nordman, S. Pyrhonen and P. Rissanen, *Recent Adv. Chemother. Proc. Int. Congr. Chemother.* Anticancer Sect. 1 **14** (1985) 612–613.
74. E. von Angerer, J. Prekajac and M. Berger, *Eur. J. Cancer Clin. Oncol.* **31** (1985) 531; E. von Angerer and J. Strohmeier, *J. Med. Chem.* **30** (1987) 131.
75. C. D. Jones, M. G. Jevnikar, A. J. Pike, M. K. Peters, L. J. Black, A. R. Thompson, J. L. Falcone and J. A. Clements, *J. Med. Chem.* **27** (1984) 1057.

76. M. R. Schneider, H. Ball and H. Schonenberger, *J. Med. Chem.* **25** (1982) 1070; *J. Med. Chem.* **28** (1985) 1880; M. R. Schneider, H. Ball and C.-D. Schiller, *J. Med. Chem.* **29** (1986) 1355.
77. R. W. Hartmann, H. Buchborn, G. Kranzfelder, H. Schonenberger and A. Bogden, *J. Med. Chem.* **24** (1981) 1192.
78. M. R. Schneider and H. Ball, *J. Med. Chem.* **29** (1986) 75.
79. J. Burns and D. Ruttler, *J. Labelled Compd. Radiopharm.* **19** (1981) 229.
80. D. H. Hunter, N. C. Payne, A. Rahien, J. F. Richardson and Y. Z. Pance, *Can. J. Chem.* **61** (1983) 42.
81. J. Shani, A. Gazit, T. Livshitz and S. Biran, *J. Med. Chem.* **28** (1985) 1504.
82. W. D. Bloomer, W. H. McLaughlin, R. R. Weichselbaum, G. L. Tonnesen, S. Hellman, D. E. Seitz, R. N. Hanson, S. J. Adelstein, A. L. Rosner, N. A. Burnstein, J. J. Nove and J. B. Little, *Int. J. Radiat. Biol.* **38** (1980) 197; D. E. Seitz, G. L. Tonnesen, S. Hellman, R. N. Hanson and S. Adelstein, *J. Organomet. Chem.* **186** (1980) C33.
83. A. Kocze and J. Szekely, *Lancet* i (1980) 539.
84. P. E. Goss, S. Ashley, T. J. Powles and R. C. Coombes, *Cancer Treat. Rep.* **70** (1986) 777.
85. J. C. Babcock, E. S. Gutsell, M. E. Herr, J. A. Hogg, J. C. Stucki, L. E. Barnes and W. E. Dulin, *J. Am. Chem. Soc.* **80** (1958) 2904.
86. A. S. Dutta and B. J. A. Furr, LHRH agonists, in *Annual Reports in Medicinal Chemistry*, Chap. 21, Academic Press, New York (1985).
87. J. J. Nestor, Jr., in *Proceedings of the Third SCI-RSC Medicinal Chemistry Symposium*, ed. R. W. Lambert, The Royal Society of Chemistry (1986) pp. 362–384.
88. U. K. Wenderoth and G. H. Jacobi, *World J. Urol.* **1** (1983) 40.
89. J. J. Nestor Jr, T. L. Ho, R. A. Simpson, B. L. Horner, G. H. Jones, G. I. McRae and B. H. Vickery, *J. Med. Chem.* **25** (1985) 795.
90. J. J. Nestor Jr, B. L. Horner, T. L. Ho, G. H. Jones, G. I. McRae and B. H. Vickery, *J. Med. Chem.* **27** (1984) 320.
91. A. S. Dutta, B. J. A. Furr, M. B. Giles and B. Valcaccia, *J. Med. Chem.* **21** (1978) 1018.
92. J. Humphries, Y.-P Wan, T. Wasiak, K. Folkers and C. Y. Bowers, *J. Med. Chem.* **22** (1979) 774.
93. G. W. Moersch, M. C. Rebstock, E. L. Wittle, F. J. Tinney, E. D. Nicolaides, M. P. Hutt, T. F. Mich, J. M. Vandenbelt, R. E. Edgren, J. R. Reel, W. C. Dermody and R. R. Humphrey, *J. Med. Chem.* **22** (1979) 935.
94. J. J. Nestor Jr, R. Tahilramani, T. L. Ho, G. I. McRae and B. H. Vickery, *J. Med. Chem.* **27** (1984) 1170.
95. R. J. Santen, H. Vanden Bossche, J. Symoens, J. Brugmons and R. De Coster, *J. Clin. Endocrinol. Metab.* **57** (1983) 732.
96. J. M. Allen, D. J. Kerle, H. Ware, A. Doble, G. Williams and S. R. Bloom, *Br. Med. J.* **287** (1983) 1766.
97. J. Trachtenberg and A. Pont, *Lancet* i (1984) 433.
98. W. F. J. Lauwers, L. Le Jeune, H. Vanden Bossche and G. Willernsens, *Biomed. Mass. Spectrom.* **12** (1985) 296.
99. J. Heeres, L. J. J. Backx, J. H. Mostmans and J. Van Cutsem, *J. Med. Chem.* **22** (1979) 1003.
100. For a detailed discussion on the synthesis of ketoconazole and related imidazole antifungal agents, see G. J. Ellames, Modern synthetic antifungal agents, in *Topics in Antibiotic Chemistry*, ed. P. G. Sammes, Vol. 6, Ellis-Horwood, Chichester (1982) pp. 9–98.
101. N. Kadohama, J.P. Karr, G. P. Murphy and A. A. Sandberg, *Cancer Res.* **44** (1984) 4947.
102. T. Liang, J. R. Brooks, A. Cheung, G. F. Reynolds and G. H. Rasmusson, *Prog. Cancer Res. Ther.* **31** (1984) 497.
103. T. Liang and C. E. Heiss, *J. Biol. Chem.* **256** (1981) 7998.
104. G. H. Rasmusson, G. F. Reynolds, T. Utne, R. B. Jobson, R. L. Primka, C. Berman and J. R. Brooks, *J. Med. Chem.* **27** (1984) 1690.
105. G. H. Rasmusson, G. F. Reynolds, N. G. Steinberg, E. Walton, G. F. Patel, T. Liang, M. A. Cascien, A. H. Cheung, J. R. Brooks and C. Berman, *J. Med. Chem.* **29** (1986) 2298.
106. P. M. Weintraub, T. R. Blohm and M. Laughlin, *J. Med. Chem.* **28** (1985) 831.
107. Y. Petrow, *J. Steroid Biochem.* **19** (1983) 1491.
108. G. H. Jacobi, U. Tunn and T. H. Senge, in *Prostate Cancer, International Perspectives in Urology*, eds. G. H. Jacobi and R. F. Hohenfellner, Williams and Wilkins, Baltimore (1982) pp. 305–319.

109. F. Neumann and R. Wiechart, *Chimicaoggi* **27** (1986) 10.
110. E. L. Shapiro, T. L. Popper, L. Weber, R. Neri and H. L. Herzog, *J. Med. Chem.* **12** (1969) 631.
111. G. W. Kakower and H. A. Van Dine, *J. Org. Chem.* **31** (1966) 3467.
112. R. Wiechert and E. Kaspar, *Chem. Ber.* **93** (1960) 1710.
113. H. Laurent, G. Schulz and R. Wiechert, *Chem. Ber.* **102** (1969) 2570.
114. H. E. Harris and C. J. Miskowicz (Schering A.-G), US Patent 3766225 (1973) (*Chemical Abstracts* **80** (1973) 27430d).
115. Z. Wang, Z. Yin and W. Zhow, *Youji Huaxue* (1984) 376 (*Chemical Abstracts* **102** (1984) 149597g).
116. J. B. Dekanski, *Br. J. Pharmacol.* **71** (1980) 11.
117. M. R. Kilbourn, A. J. Arduengo, J. T. Park and J. A. Katzenellenbogen, *Mol. Pharmacol.* **19** (1981) 388.
118. J. C. Turner and R. P.-K Chan, Ger. Offen, DE 2110428 (1971) (*Chemical Abstracts* **76** (1971) 14093p).
119. J. W. Baker, G. L. Bachman, I. Schumacher, D. P. Roman and A. L. Tharp, *J. Med. Chem.* **10** (1967) 93.
120. R. K. Neri, P. Florence, P. Koziol and S. VanCleave, *Endocrinology* **91** (1972) 427.
121. R. O. Neri and N. Kassem, *Prog. Cancer Res. Ther.* **31** (1984) 507.
122. R. O. Neri and J. G. Topliss (Schering) Ger. Offen DE 2130450 (1972) (*Chemical Abstracts* **78** (1972) 58091g).
123. L. Peer and J. Mayer, US Patent 4302599 (1981) (*Chemical Abstracts* **96** (1981) 103817).
124. R. O. Neri and J. G. Topliss, Canadian Patent 1012462 (1971) (*Chemical Abstracts* **88** (1971) 22155y).
125. B. Katchen and S. Buxbaum, *J. Clin. Endocrinol. Metab.* **41** (1975) 373.
126. A. T. Glen, L. R. Hughes, J. J. Morris and P. J. Taylor, in *Proceedings of the Third SCI-RSC Medicinal Chemistry Symposium*, ed. R. W. Lambert, The Royal Society of Chemistry, London (1986) pp. 345–361.
127. H. Morales-Alanis, M. J. Brienne, J. Jacques, M. M. Boulton, L. Nedelec., V. Torelli and C. Tournemine, *J. Med. Chem.* **28** (1985) 1796.
128. H. Tucker, J. W. Crook and G. J. Chesterson, *J. Med. Chem.* **31** (1988) 954.
129. M. Moguilewsky, J. Fiet, C. Tournemine and J. P. Raynaud, *J. Steroid Biochem.* **24** (1986) 139; J. P. Raynaud, C. Bonne, M. Moguilewsky, F. A. Lefebyre, A. Belanger and F. Labrie, *The Prostate* **5** (1984) 299.
130. J. Perronnet, P. Girault and B. C. Pierre, Ger. Offen. DE 2649925 (1977) (*Chemical Abstracts* **87** (1977) 85005z).

9 Anticancer pyrimidines, pyrimidine nucleosides and prodrugs

M. MacCOSS AND M. J. ROBINS

9.1 Introduction

Of the plethora of pyrimidines, pyrimidine nucleosides, analogues and prodrug derivatives that have been synthesised and evaluated as anticancer agents, only two are in general clinical use. The long-studied 5-fluorouracil (5-FUra (**1**), see Scheme 9.1), and some nucleoside and prodrug derivatives, are used regularly for the treatment of breast cancer, tumours of the gastrointestinal tract and other solid tumours. Acute leukaemias and lymphomas, especially acute myelogenous leukaemia, often respond success-fully to 1-(β-D-arabinofuranosyl)cytosine (araC (**2**)). The symmetrical triazine analogue of cytidine, 5-azacytidine (5-azaCyd (**3**)) and its 2′-deoxy, arabino and dihydro derivatives have been studied extensively, and 5-azaCyd is available as an investigational antitumour drug effective against leukaemias. This chapter presents outlines of basic synthetic methods which have been reported for the preparation of 5-FUra (**1**), araC (**2**), 5-azaCyd (**3**) and selected nucleoside and/or prodrug derivatives. Generally, only an initial type of synthetic approach is given. Subsequent work that employed the basic synthetic strategies is not usually quoted unless a significant conceptual contribution was made in addition to yield improvements and/or indus-trially useful modifications. Some discussion of molecular modes of anticancer action that involve defined chemical entities, certain metabolic conversions, and properties and interactions of prodrugs are presented. Three pyrimidine nucleoside analogues that are under current investigation as potential

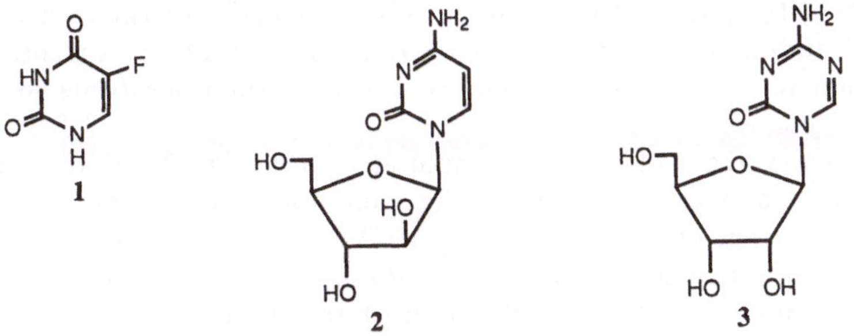

Scheme 9.1

anticancer drugs are considered briefly. No consideration of the use of pyrimidine-related compounds as radiation enhancers, or discussion of combination chemotherapy with deaminase inhibitors, biological response modifiers, etc. is given.

'Antimetabolite' action was recognised with 'prontosil', a dye that functioned as a prodrug form of the antibacterial 'sulphanilamide' (*p*-aminobenzenesulphonamide), which approximates the bacterial metabolite *p*-aminobenzoic acid required for biosynthesis of folate cofactors[1]. Since humans ingest completed folates, sulpha drugs selectively poison bacteria by antimetabolite inhibition of their *de novo* folate pathway. Empirical antimetabolite rationale has been used extensively in the design of potential anticancer agents, although the concept is intrinsically flawed for treatment of neoplasms. Since cancer usually represents a heterogeneous population of host-derived cells under varying degrees of aberrant genetic control, it is inherently unlikely that significant differences in the basic metabolic pathways or metabolite requirements will exist. Thus, although empirically discovered antimetabolite drugs provide significant palliation and/or cures for certain neoplastic diseases, the quantitative rather than qualitative differences in tumour versus normal host cell metabolism frequently result in serious drug toxicity for the patient.

Since cancer is a family of aberrant cellular proliferative diseases, chemo-therapeutic agents that interfere with replicative cellular processes are candidate drugs. Analogues of naturally occurring pyrimidines and pyr-imidine nucleosides are 'primary antimetabolite' candidates for potential incorporation into nucleic acids and/or interaction with crucial enzymes of nucleic acid metabolic pathways. Other useful anticancer agents exert their efficacy by 'fortuitous interaction' with nucleic acids and/or related enzymes, but were generally discovered in empirical screening programmes rather than by antimetabolite design.

The first synthesis of 5-FUra (**1**) and announcement of its antitumour activity were reported over 30 years ago[2]. Elucidation of its mode of action as a metabolically activated inhibitor of thymidylate synthase represents a classic in basic molecular biochemistry[3-5]. However, recent studies suggest that the primary mode of action may vary in different circumstances, and antimeta-bolite activity by fraudulent incorporation of 5-FUra (**1**) into ribonucleic acids (RNAs) and/or deoxyribonucleic acids (DNAs) may take precedence over inhibition of thymidylate synthase by 2'-deoxy-5-fluorouridylate in some cases[6].

The first synthesis of 1-(β-D-arabinofuranosyl)cytosine (**2**) was reported 30 years ago,[7] and early studies on this inhibitor of DNA synthesis were reviewed[8]. Inversion of configuration at C2' of araC (**2**) relative to cytidine gives an 'approximate antimetabolite' of 2'-deoxycytidine with the C2' hydrogen atom of araC on the α-face of the furanosyl ring. Successive phosphorylations of araC to the 5'-triphosphate level provides the proximate antimetabolite, araCTP. Interaction of DNA polymerases with araCTP including incorporation into DNA as well as enzyme inhibition occurs at differing levels, and ultimate biological consequences can depend on complementary sequences[8,9].

Studies on the cellular uptake, competing intracellular metabolic pathways, and ultimate molecular biochemical interactions involved with these two clinically effective anticancer agents (5-FUra and araC) provide exciting stories of chemical/biomedical drug development.

9.2   5-Fluorouracil and related 5-fluoropyrimidine compounds

Extensive studies on the biochemical mode of action, biological activities and clinical studies with 5-FUra have been reviewed[3-6]. In this section, methods for the chemical synthesis of 5-FUra and its nucleoside analogues are outlined. This is followed by a brief discussion of its molecular interaction with thymidylate synthase, a crucial rate-limiting enzyme for the synthesis of DNA, and its catabolism to α-fluoro-β-alanine and conjugates.

9.2.1   *Syntheses of 5-fluorouracil, nucleosides and derivatives*

The first reported synthesis of 5-FUra (1) utilised a 'chemical *de novo*' approach beginning with ethyl fluoroacetate (4)[2]. Mixed Claisen condensation of (4) with ethyl formate gave the potassium enolate of ethyl α-fluoroformylacetate (5) (see Scheme 9.2). Base-promoted condensation of (5) with the hydrobromide salt of *S*-ethylisothiourea gave 2-ethylthio-5-fluoropyrimidin-4-one (6). Hydrolysis of the 2-ethylthio group from (6) in aqueous hydrochloric acid gave 5-FUra (1). Chlorination of (6) gave

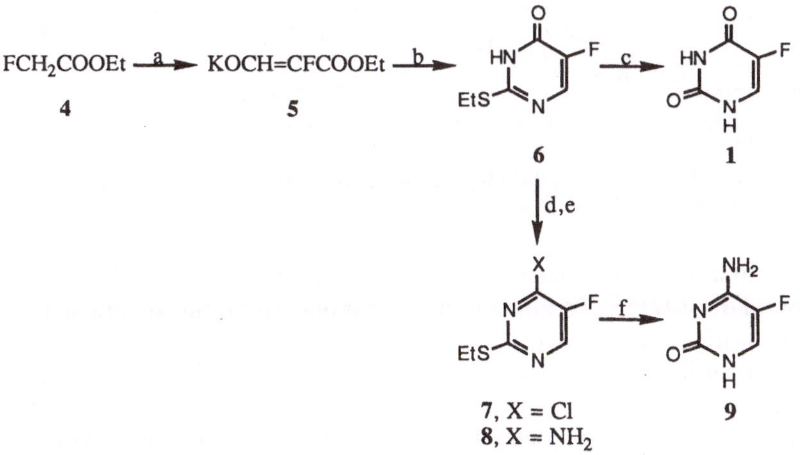

Scheme 9.2 (a) HCOOEt/KOEt/toluene; (b) EtSC(NH$_2$)$_2$Br/NaOMe/EtOH; (c) HCl/H$_2$O; (d) POCl$_3$; (e) NH$_3$ (liq.); (f) HBr/H$_2$O.

4-chloro-2-ethylthiopyrimidine (**7**) which was treated with liquid ammonia to give 4-amino-2-ethylthiopyrimidine (**8**). Hydrolysis of (**8**) in aqueous hydrobromic acid gave 5-fluorocytosine (**9**)[2].

A ring-transformation synthesis of 5-FUra (**1**) began with 1,3-dimethyl-5-azauracil (**10**) and utilised fluoroacetamide (**11**) as the C–C–N fragment for annulation of the pyrimidine ring with expulsion of N,N'-dimethylurea (see Scheme 9.3)[10]. The potent base, lithium diisopropylamide (LDA), was required.

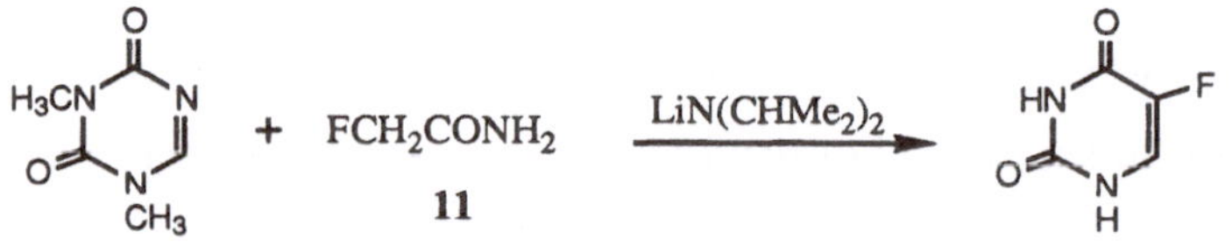

Scheme 9.3

A new total synthesis of 5-FUra that employed chloroperfluoroethylene chemistry was reported recently[11]. Trifluoroacrylic acid (**12**) was heated in ethanol containing a base such as triethylamine, sodium ethoxide, or lithium hydroxide (or alternatively the lithium salt of trifluoroacrylic acid prepared *in situ* from chlorotrifluoroethene (**13**), butyllithium, and carbon dioxide was heated in ethanol) to provide diethyl fluoromalonate (**14**) (see Scheme 9.4). Condensation of (**14**) with urea in ethanolic sodium methoxide gave 5-fluorobarbituric acid (**15**) which was treated with phosphoryl chloride and N, N-dimethylaniline to give 6-chloro-5-fluorouracil (**16**). Catalytic hydrogenolysis of (**16**) gave 5-FUra (**1**)[11].

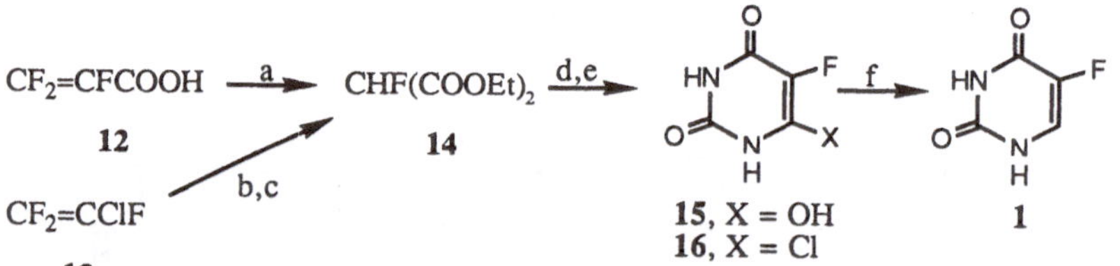

Scheme 9.4 (a) LiOH/EtOH/Δ; (b) (i) BuLi, (ii) CO$_2$; (c) EtOH/Δ; (d) CO(NH$_2$)$_2$/NaOMe/EtOH; (e) POCl$_3$/PhNMe$_2$/100°C; (f) H$_2$/Pd/C.

The first general procedure for direct fluorination of pyrimidines was reported in 1971[12]. Treatment of methanolic solutions of uracil (**17**), 1-methyluracil, or acetylated uracil nucleosides at $\leqslant -78°C$ with a similarly cooled solution of trifluoromethyl hypofluorite (fluoroxytrifluoromethane, CF$_3$OF) in trichlorofluoromethane resulted in rapid formation of *cis*-5-fluoro-5,6-dihydro-6-methoxyuracil adducts (**18**) (see Scheme 9.5). Noteworthy was the observation that the adducts had a highly predominant *cis* relationship of 5-fluoro- and 6-methoxy groups regardless of whether they were generated

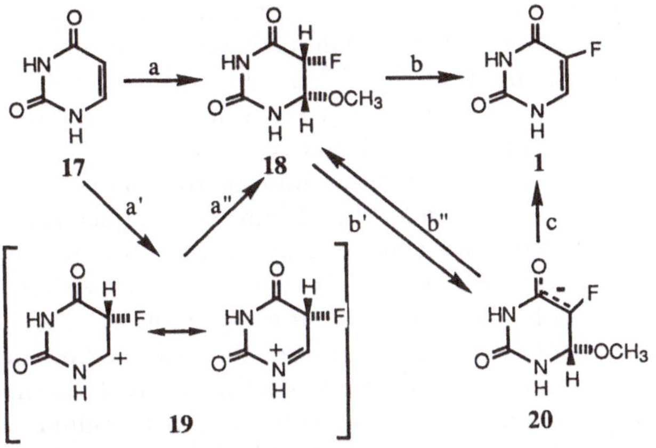

Scheme 9.5 (a) $CF_3OF/CCl_3F/CH_3OH$; (a') $F-OCF_3$; (a'') $HOCH_3-H^+$ (b) $Et_3N/CH_3OH/H_2O$; (b') base; (b'') H^+; (c) $-CH_3O^-$.

from cationic (**19**) or anionic (**20**) intermediates[13]. Thus, nucleophilic attack of C5 of the uracil system on an electrophilic fluorine species generated an enamide resonance stabilised cation (**19**) which underwent energetically preferential *syn* nucleophilic attack by solvent methanol at C6 followed by loss of a proton to give the *cis* enantiomeric adducts (**18**) at $\leqslant -78°C$.

Treatment of the *cis* adducts (**18**) with a solution of triethylamine in aqueous methanol (or other basic conditions) resulted in *anti* elimination of the elements of methanol from C5 and C6 to generate the 5-fluorouracil (**1**) products in high yields. Operation of an E_{1CB} type mechanism involving formation of the C5 enolate (**20**) was suggested by hydrogen/deuterium exchange at C5 of the re-isolated adducts after treatment with $Et_3N/MeOD/D_2O$ for a limited period. ^1H-NMR analysis of the 5-deutero adduct and its subjection to a second exchange (in $Et_3N/MeOH/H_2O$ solution) followed by ^1H-NMR analysis confirmed the highly predominant retention of stereochemistry even at ambient temperature in the re-isolated (*cis*) adducts. Therefore, capture of a proton/deuteron by the 5-fluoroenolate (**20**) occurred at the face opposite to the 6-methoxy group to produce *cis*-5-fluoro-5,6-dihydro-6-methoxyuracil (**18**). Repetition of reactions with other adducts reported in the literature[14] which could be interpreted as proceeding via enolate-like species also gave the *cis* dihydro products, with no peaks for *trans* isomers visible at the 100 MHz ^1H-NMR field available to us at that time[13]. It is known that ethane systems vicinally substituted with fluorine and/or oxygen preferentially populate *gauche* conformation ranges[13]. The dramatic *cis* stereoelectronic preference indicated by NMR spectra of these dihydro heterocycles was confirmed by single crystal X-ray analysis[15] of the 5-fluoro-5,6-dihydro-6-methoxy-1-methyluracil adduct. A *cis* 5-fluoro to 6-methoxy dihedral angle of 59.7° was found[15] in near perfect harmony with the energetically favoured 60° *gauche* stereochemistry[13].

Shortly after our initial reports[12,16], Barton and coworkers described fluorination of uracil and other pyrimidine bases with elemental fluorine or trifluoromethyl hypofluorite[17,18]. Their initially formed 5-fluoro-5,6-dihydropyrimidine adducts were converted to the unsaturated 5-fluoro products during sublimation. Cech and coworkers also developed direct fluorination procedures that employed a fluorine/pyridine complex[19], electrochemical halogenation techniques[20] and fluorine in aqueous/acid media[21]. Yurasova has described a low yield preparation of 5-fluorouracil by treatment of uracil with xenon difluoride (XeF_2)[22] and Yemul et al. have reported a high yield conversion with a xenon hexafluoride/graphite ($C_{19}XeF_6$) inclusion reagent[23]. A number of workers have utilised the basic methods developed by ourselves, Barton and Cech with different combinations of elemental flourine, fluoroxy compounds and various solvents to prepare 5-fluoropyrimidines, nucleosides and analogues for biological testing. Fluorine isotopes have been used to prepare diagnostic radiopharmaceuticals. Modifications have also utilised orotic acid (uracil 6-carboxylic acid)[24] and uracil 5-carboxylate[25] derivatives.

Stavber and Zupan noted that treatment of 1,3-dimethyluracil with $CsSO_4F$ in methanol at room temperature gave a mixture of 5-fluoro-5,6-dihydro-6-methoxy adducts with a cis/trans ratio of greater than 11:1[26]. It is noteworthy that this ratio was obtained under ambient conditions whereas the original studies with $CF_3OF/CCl_3F/CH_3OH$ were performed at $\leqslant -78°C$[12,13]. It might be informative to conduct studies on adduct formation between $-78°C$ and ambient temperature to see if cis/trans isomer ratios parallel stereoelectronic energy predictions. Stavber and Zupan treated uridine with $CsSO_4F/CH_3OH$ followed by $Et_3N/CH_3OH/H_2O$ and obtained 5-fluorouridine (79%)[26] in the same yield obtained with CF_3OF[13].

Visser et al. have recently reported studies on fluorination of uracil and cytosine with ^{18}F-labelled elemental fluorine and acetyl hypofluorite[27]. They have discussed mechanistic and stereochemical considerations and work by others with fluorine radioisotopes. A speculative oxidative cation radical mechanism was proposed on the basis of limited small scale experiments conducted at unspecified temperatures on two compounds. ^1H-NMR evidence was presented for the formation of 5-fluoro-5,6-dihydro-6-oxy adducts with cis/trans ratios of approx. 10:1. If these fluorinations were performed at room temperature, the cis/trans ratios are in harmony with those of Stavber and Zupan with $CsSO_4F/MeOH$ at ambient temperature[26]. This is not substantially different[27] from the results at $\leqslant -78°C$ (with $CF_3OF/CCl_3F/MeOH$)[13] since energetically favoured cis/trans ratios would be expected to be greater at the lower temperatures and observation of the minor trans isomer at ratios of $\geqslant 95:5$ might not have been detected at 100 MHz.

Earlier preparations of 5-fluoropyrimidine nucleosides that utilised coupling of pyrimidine metal salts and protected glycosyl halides have been

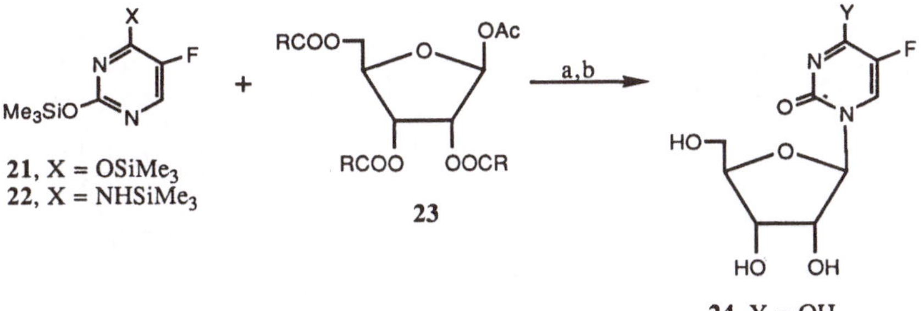

21, X = OSiMe$_3$
22, X = NHSiMe$_3$

23

24, Y = OH
25, Y = NH$_2$

Scheme 9.6 (a) CF$_3$SO$_3$SiMe$_3$/solvent/Δ; (b) base/MeOH.

superseded by the Vorbruggen glycosylation procedure[28]. Thus, treatment of solutions of pertrimethylsilylated-5-fluorouracil (**21**) or -5-fluorocytosine (**22**) and 1-*O*-acetyl-2,3,5-tri-*O*-acyl-β-D-ribofuranoses (**23**) with an appropriate Lewis acid catalyst (e.g. tin(IV) chloride (SnCl$_4$), trimethylsilyl trifluoromethanesulfonate (CF$_3$SO$_3$SiMe$_3$), etc.) gives acylated 5-fluoropyrimidine ribonucleosides that can be deprotected to give good yields of 5-fluorouridine (**24**) or 5-fluorocytidine (**25**)[29,30] (see Scheme 9.6).

Hoffer *et al.* first reported coupling of mercury salts of 5-FUra and 5-fluorocytosine with 3,5-di-*O*-*p*-chloro(or *p*-methyl)benzoyl-2-deoxy-α-D-*erythro*-pentofuranosyl chloride to give anomeric 2′-deoxynucleosides[31]. Other coupling syntheses have been described, but the best stereoselectivity for pyrimidine 2′-deoxynucleoside β-anomers was claimed recently by Aoyama[32]. According to this report, a solution of 5-fluoro-2,4-bis(trimethylsilyloxy)pyrimidine (**21**), 3,5-di-*O*-*p*-chlorobenzoyl-2-deoxy-α-D-*erythro*-pentofuranosyl chloride (**26**) and *p*-nitrophenol was stirred in dry chloroform to yield a mixture of 92%/4.5% of the protected β/α anomers of 2′-deoxy-5-fluorouridine (**27**)[32] (see Scheme 9.7).

Holy and Cech first described a procedure for selective 2′-deoxygenation of 5-fluorouridine (**24**)[33]. Treatment of (**24**) with diphenyl carbonate according to Hampton and Nichol[34] gave the 2,2′-anhydroarabinosyl derivative (**28**) which

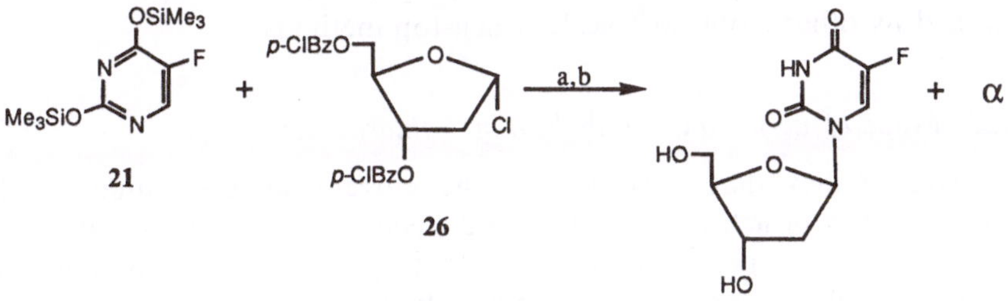

21

26

27

Scheme 9.7 (a) *p*-O$_2$NC$_6$H$_4$OH/CHCl$_3$; (b) base/MeOH.

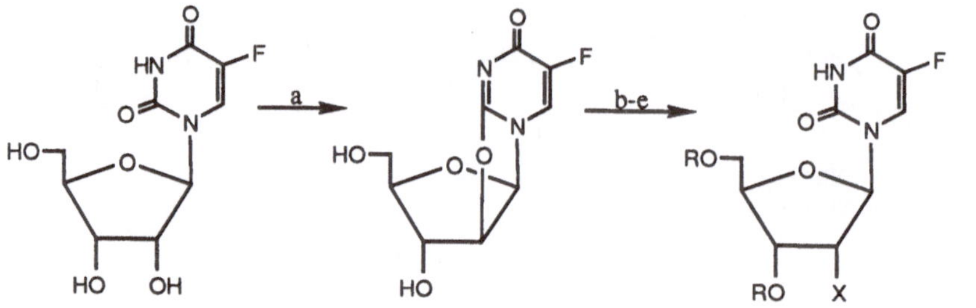

Scheme 9.8 (a) $(PhO)_2CO/NaHCO_3/HMPA/\Delta$; (b) BzCN; (c) HCl/DMF; (d) $Bu_3SnH/AIBN/PhH/\Delta$; (e) base/MeOH.

was benzoylated and then the anhydro ring was opened by treatment with HCl/DMF to give the protected 2'-chloro-2'-deoxyuridine (29) (see Scheme 9.8). Radical mediated hydrogenolysis of the 2'-chloro function with tributyltin hydride followed by deprotection gave 2'-deoxy-5-fluorouridine (27)[33]. Modifications of this strategy have been noted[35,36].

Our general procedure for direct fluorination with CF_3OF provides high yield syntheses of 5-fluoropyrimidine nucleosides and analogues (in addition to the 5-fluorouracil and 5-fluorocytosine bases) conveniently on a laboratory scale[13]. Reasonable yields of the corresponding nucleotides also can be obtained by this method[37]. As noted above, Barton et al. and Cech et al. pioneered the use of elemental fluorine in various solvents and others have used modified conditions with elemental fluorine or fluoroxy reagents to accomplish analogous direct syntheses of fluorinated nucleosides and analogues. Undoubtedly the basic de novo heterocyclic condensation methods, or fluorination of uracil with elemental fluorine, followed by functional group transformations and Vorbruggen-type couplings with carbohydrate derivatives are the most practical industrial routes to provide antitumour drugs in large quantities. Preferences vary for different methods according to reagent availability, laboratory equipment and experience of investigators. However, the high yields and versatility of base, nucleoside and nucleotide substrate structures successfully fluorinated by our procedure[13,37] have not been exceeded by other laboratory scale bench-top methods.

9.2.2 Mode of action and catabolism of 5-fluorouracil

Extensive studies that have led to the current understanding of the primary modes of action of 5-FUra and nucleoside derivatives have been reviewed[3-6, 38]. It is presently considered that a general mode (and until quite recently thought by most investigators to be the only significant mode) of tumour inhibitory action by 5-FUra involves its metabolic activation by cells to the level of 2'-deoxy-5-fluorouridine 5'-monophosphate (5-FdUMP). This

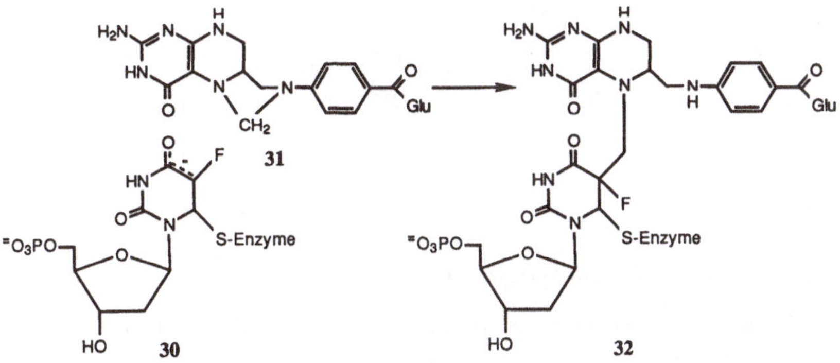

Scheme 9.9

5-fluoro analogue of 2′-deoxyuridine 5′-monophosphate (dUMP) functions as an initially reversible-binding alternative substrate of thymidylate synthase (a rate-limiting enzyme for DNA synthesis, and thereby cell division, which affects production of the obligatory DNA pyrimidine building block, thymidylate (dTMP)). Reversible binding of 5-FdUMP to the enzyme is followed by Michael-type nucleophilic attack of an enzyme cysteine sulphydryl group to give a 6-(enzyme)thio-substituted 5-fluoroenolate intermediate (30) (see Scheme 9.9). Nucleophilic attack of this 5-fluoroenolate at the methylene carbon of the bound cofactor 5,10-methylenetetrahydrofolate (31) gives a covalently-linked ternary enzyme-inhibitor-cofactor complex (32) analogous to the normal biosynthetic route to thymidylate[4-6, 38]. However, the 5-fluoro substituent on the dihydrouracil moiety of (32) cannot be removed by the enzyme (removal of 'F[+]' would be required by analogy with the normal removal of H[+]) resulting in essentially[4,5,38] irreversible stoichiometric inhibition of one crucial macromolecular enzyme and one essential cofactor per molecule of inhibitor.

As noted in section 9.1, it is now considered that incorporation of 5-FUra or 5-fluorouridine (as the RNA polymerase alternative substrate 5-FUTP) into specific RNAs may sometimes be the most important event[6]. The primary inhibitory mode of action in that case would correspond to antimetabolite activity at the macromolecular (RNA) level. Analogous incorporation of 5-FUra or 2′-deoxy-5-fluorouridine (as the DNA polymerase alternative substrate 5-FdUTP) into DNA[6] could lead to antimetabolite action at the primary genetic level.

Nucleotides and nucleosides of 5-FUra (1) are catabolised to the base level by various phosphoesterase, phosphorylase and hydrolase activities. Reduction of (1) to 5-fluoro-5,6-dihydrouracil (33) is followed by hydrolysis to α-fluoro-β-ureidopropionate (34) (see Scheme 9.10). Further hydrolysis of (34) gives α-fluoro-β-alanine (35) which is excreted directly as well as conjugated with bile acids to give other catabolites[39-41].

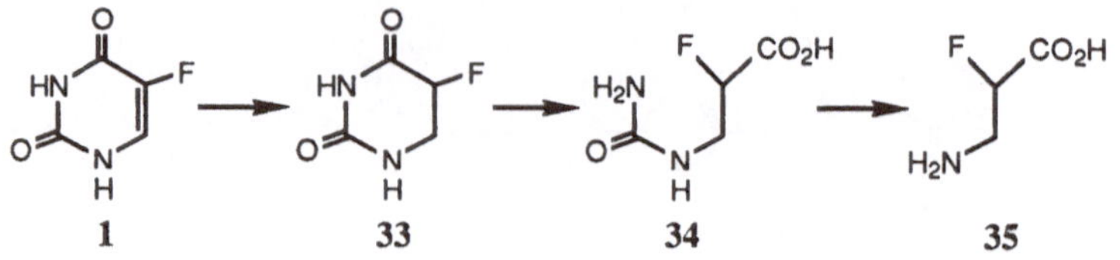

Scheme 9.10

9.3   1-(β-D-Arabinofuranosyl)cytosine

It has been known for over 20 years that 1-(β-D-arabinofuranosyl)cytosine (araC, cytosine arabinoside, cytarabine, cytosar (2)) is a potent inhibitor of DNA synthesis[8,42]. AraC has antiviral activity (DNA viruses) and inhibits the growth of various mammalian cell lines and rodent tumours[8]. It presently is used as a drug for acute leukaemias and lymphomas, particularly acute myelogenous leukaemia in adults[43]. The O2,C2'-cyclonucleoside, 2,2'-anhydro-1-(β-D-arabinofuranosyl)cytosine (42) (see Scheme 9.12) undergoes spontaneous chemical hydrolysis to araC (2) in aqueous solution with pH $\geqslant 8$[44]. Therefore, synthetic routes to (42) also constitute syntheses of (2). Syntheses of (42), which could also be included in the present discussion, are presented in sections 9.6.1.4 and 9.6.1.5 on prodrugs of araC.

9.3.1   Syntheses of 1-(β-D-arabinofuranosyl)cytosine

The first synthesis of araC (2) proceeded via formation of a 2,2'-anhydro-araC intermediate (37) (see Scheme 9.11). Dekker et al.[7] heated cytidine (36) in

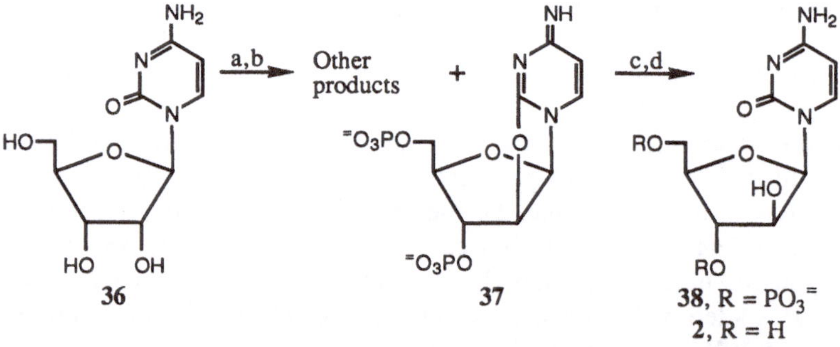

Scheme 9.11 (a) Polyphosphoric acid/Δ; (b) H_2O/Δ; (c) LiOH; (d) phosphatase.

polyphosphoric acid followed by addition of water and further heating to give a mixture of products including the 3',5'-diphosphate (37) of 2,2'-anhydro-araC. Anhydro ring opening of (37) followed by phosphatase treatment gave araC (2) in low overall yield. Later modifications of this procedure were noted[45]. Many subsequent syntheses of araC have utilised analogous inversions of configuration at C2' of cytidine or uridine with intermediate formation of 2,2′-anhydro derivatives. A number of reagents that produce better leaving groups at C2′ (in place of the presumed 2′-pyrophosphates of the Dekker synthesis[7]) and react under milder conditions to give araC in higher yields have been devised. 'Partially hydrolysed' phosphoryl chloride $(POCl_3)^{46}$, the Vilsmeier–Haack reagent ($POCl_3$ or $SOCl_2$ plus $HCONMe_2$)[47], the diphenyl carbonate method (($PhO)_2CO/NaHCO_3/$ DMF)[48,49] of Hampton and Nichol[34], the α-acyloxyisobutyryl halide reaction of nucleosides $(Me_2C(OCOR)COX)^{50}$ discovered with diols by Mattocks[51], and acetylbromide $(MeCOBr)^{52}$, among others, have been used to effect *in situ* transformations of cytidine to 2,2′-anhydro-araC (**42**) and/or araC (**2**) products.

Derivatives of 1-(β-D-arabinofuranosyl)uracil have been converted into araC and related compounds by 2,4-dithiation and amination[53], 4-chlorination ($POCl_3/Et_2NPh/EtOAc$) and amination[54], 4-thiation and amination[55], and 4-phosphorylation/triazolation (POX_3/triazole or 3-nitro-triazole) and amination[56].

Mercury-catalysed coupling of a protected cytosine with a 2-O-acetyl-3-O-tosylxylofuranose derivative followed by treatment of the resulting 1-(2-O-acetyl-3-O-tosyl-β-D-xylofuranosyl)cytosine derivative with base gave araC, presumably via formation of a 2′,3′-anhydro (*ribo*-epoxide) intermediate,

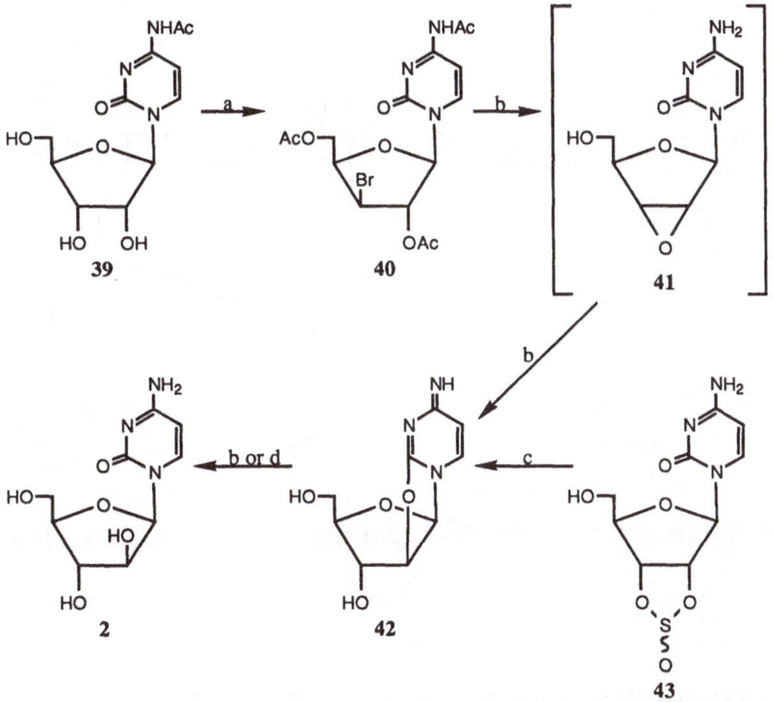

Scheme 9.12 (a) CH_3COBr/Δ; (b) KOH/EtOH; (c) H_2O/Δ; (d) H_2O (pH 10).

spontaneous epoxide ring opening by the 2-oxo function, and hydrolysis of the resulting 2,2'-anhydro-araC to araC[57]. Analogous sequential double ring closure/ring opening was presumed[52] to occur (via (41) and (42)) in the conversion of 4-N-acetyl-1-(2,5-di-O-acetyl-3-bromo-3-deoxy-β-D-xylofuranosyl)cytosine (40) to araC (2) upon treatment with potassium hydroxide in ethanol (see Scheme 9.12). Acetylation of the 4-amino group of cytidine apparently reduces the nucleophilicity of the 2-oxo function sufficiently to permit external bromide to nucleophilically attack C3' of an intermediate acetoxonium ion formed by heating 4-N-acetylcytidine (39) with acetyl bromide[52]. Treatment of uridine or cytidine (36) with thionyl chloride gave stable 2',3'-O-sulfinyl derivatives[58]. The cytidine 2',3'-cyclic sulfite (43) was heated in water to give 2,2'-anhydro-araC (42) via intramolecular displacement of alkyl sulfite at C2' by the 2-oxo function. Mild hydrolysis of (42) gave araC (2) in good overall yield[58].

A novel approach to araC involved construction of the pyrimidine ring onto a carbohydrate. Treatment of D-arabinose (44) with cyanamide in aqueous solution gave the bridged aminooxazoline (45)[59] (see Scheme 9.13). Conjugate addition of (45) to propiolonitrile gave a cyanovinyl intermediate[59] (46)[60] which underwent cyclisation[59] to a salt of 2,2'-anhydro-araC (42) upon treatment with aqueous acetic acid, and araC (2) if treated with aqueous ammonia. The Shannahoff and Sanchez synthesis was modified for industrial

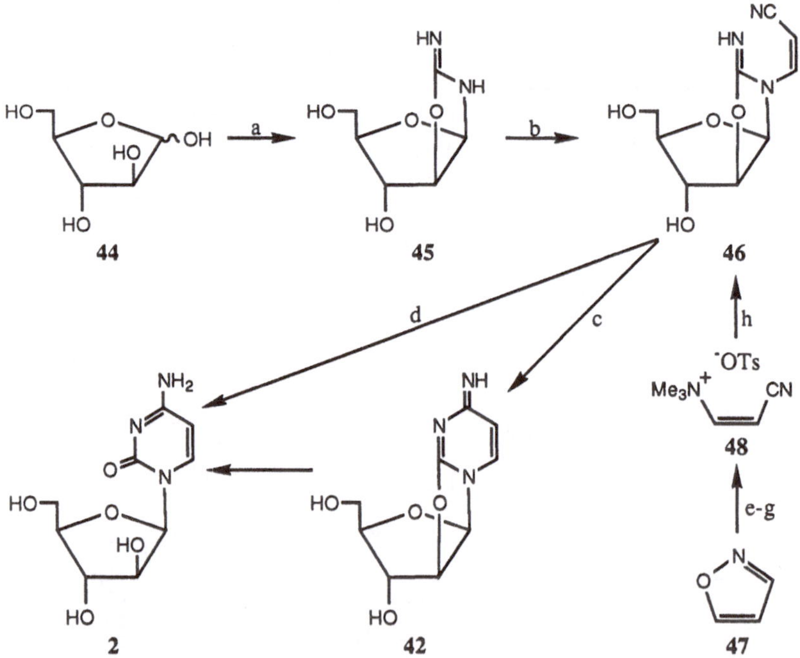

Scheme 9.13 (a) H_2NCN/H_2O; (b) HCCCN; (c) $AcOH/H_2O$; (d) NH_3/H_2O; (e) tBuOK/THF/$-40°C$; (f) TsCl; (g) NMe_3; (h) (45).

production of araC by substitution of *cis*-β-trimethylammoniumacrylonitrile tosylate (**48**)[60] for propiolonitrile[59]. Treatment of isoxazole (**47**) with potassium *tert*-butoxide in tetrahydrofuran at low temperature resulted in deprotonation/ring opening to give the *cis* acrylonitrile-β-oxide (enolate of cyanoacetaldehyde) which was tosylated *in situ* and treated with trimethylamine to give (**48**) by addition/elimination with retention of configuration about the double bond[60]. Condensation of (**45**) with (**48**) occurred to give a very efficient and economical route to 2,2'-anhydro-araC (**42**) and araC (**2**).

As noted in section 9.1, araC undergoes activation by successive phosphorylations to araCTP. Inhibition of DNA polymerase by araCTP as well as incorporation into DNA and sequence specific effects are involved in its biological activity[8,9,61]. Enzymatic deamination of araC to the inactive araU occurs readily as discussed in section 9.6.1, and the majority of administered araC is excreted as araU[61].

9.4 5-Azacytidine and related nucleosides

The first synthesis of 4-amino-1-(β-D-ribofuranosyl)-1,3,5-triazin-2-one (5-azacytidine, 5-azaCyd (**3**)) was reported 25 years ago[62]. Clinical activity of 5-azaCyd against leukaemias has been demonstrated[61] and 5-azacytosine nucleotides are incorporated into RNA and DNA. However, (**3**) exhibited serious toxicity which necessitated protracted infusion therapies, and its triazine ring system is unstable in aqueous (especially basic) solutions. Therefore, 5-azaCyd is available as an investigative anticancer drug, but does not enjoy widespread use. Precise mode(s) of anticancer action, uniform pathways of chemical and/or enzymatic degradation/metabolism, etc. have not been established for 5-azaCyd (**3**).

9.4.1 *Syntheses of 5-azacytidine and analogues*

The first synthesis of 5-azacytidine (**3**) involved construction of the triazine ring onto a sugar derivative[62] (see Scheme 9.14). Treatment of 2,3,5-tri-*O*-acetyl-D-ribofuranosyl chloride (**49**) with silver cyanate gave the glycosylisocyanate (**50**). Condensation of (**50**) with *O*-methylisourea gave the glycosyl *O*-methylisobiuret derivative (**51**) which was heated with triethyl orthoformate to give the 4-methoxy-1,3,5-triazin-2-one nucleoside (**52**). Treatment of (**52**) with methanolic ammonia under carefully controlled conditions (to minimise triazine ring opening) effected ammonolysis of the acetyl protecting groups and replacement of the 4-methoxy function to give 5-azaCyd (**3**)[62].

A straightforward synthesis of 5-azacytosine (**55**) involved partial hydrolysis of dicyandiamide (*N*-cyanoguanidine (**53**)) with methanolic hydrochloric acid to give 'guanylurea' (**54**) which was treated with 'dimethylformamide dimethylacetal' to give 5-azacytosine (**55**) in good yield[63]. Coupling of the bis(trimethylsilyl) derivative (**56**) of 5-azacytosine with 2,3,5-tri-*O*-acetyl-D-

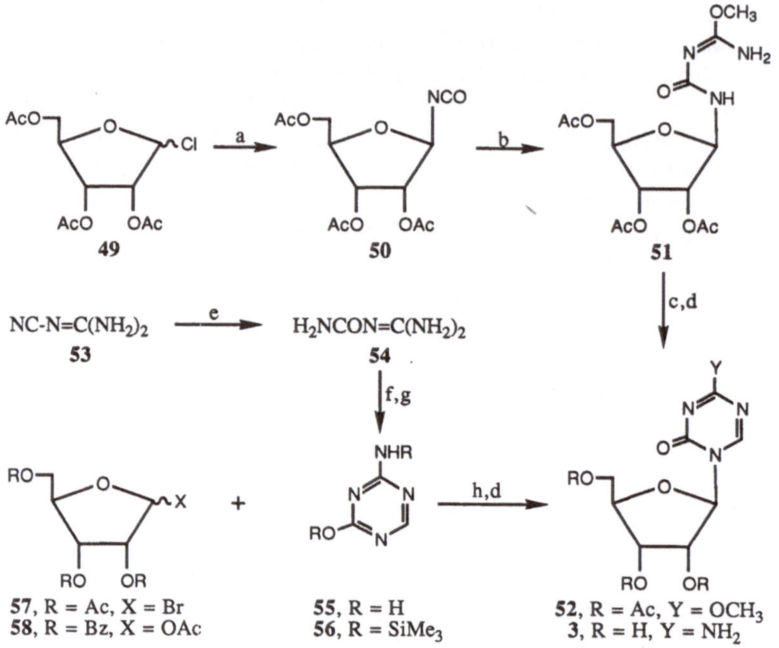

Scheme 9.14 (a) AgNCO; (b) $CH_3OC(NH)NH_2$; (c) $HC(OC_2H_5)_3$; (d) $NH_3/MeOH$; (e) $HCl/H_2O/MeOH$; (f) $HC(OMe)_2NMe_2$; (g) $(Me_3Si)_2NH$; (h) MeCN (X = Br); $SnCl_4$ (X = OAc).

ribofuranosyl bromide (**57**) followed by deprotection with methanolic ammonia gave 5-azaCyd (**3**) in low yield[64]. Analogous coupling with protected D-arabinofuranosyl and 2-deoxy-D-*erythro*-pentofuranosyl halides gave ara-5-azaCyd and 5-aza-2′-deoxyCyd derivatives. However, the labile glycosyl-4-amino-1,3,5-triazin-2-one ring underwent heterocyclic modifications during attempted deprotection. Vorbruggen coupling of (**56**) and 1-O-acetyl-2,3,4-tri-O-benzoyl-D-ribofuranose (**58**) gave 2′,3′,5′-tri-O-benzoyl-5-azaCyd in high yield[65]. Deprotection according to the original Piskala and Sorm conditions[62] was required to give 5-azaCyd (**3**).

Reduction of the 5,6 double bond of 5-azaCyd (**3**) to give dihydro-5-azaCyd has been effected with sodium borohydride or catalytic hydrogenation[66,67]. The 4-amino-1-glycosyl-5,6-dihydro-1,3,5-triazin-2-one compounds are more stable than the parent system. Driscoll *et al.*[68] employed this enhanced stability to effect reductive deprotection of the Winkley and Robins[64] arabinosyl intermediate. Novel air oxidation of silylated dihydro compounds to the triazin-2-one nucleosides was effected with bis(trimethylsilyl)trifluoroacetamide[67,68]. The greater stability of dihydro-5-azaCyd (as a presumed prodrug of 5-azaCyd) is countered by its lower antileukaemia activity. Coupling of (**56**) with 3,5-bis-O-(9-fluorenylmethoxycarbonyl)-D-*erythro*-pentofuranosyl chloride followed by careful eliminative deprotection with triethylamine gave an improved synthesis of 5-aza-2′-deoxyCyd and its α-anomer[69]. It remains to be seen if anticancer efficacies of arabinosyl, 2′-deoxy, 5,6-dihydro, etc. modifications of 5-azaCyd (**3**) will lead to an agent with sustained clinical use.

9.5 Selected experimental agents

A number of pyrimidines, nucleosides and analogues have been prepared and subjected to biological studies including preclinical pharmacology and initial clinical trials. However, most compounds that were seriously toxic and/or exhibited no significant clinical or combination chemotherapeutic benefits are no longer under investigation as anticancer drugs. Three currently studied pyrimidine nucleoside analogues have been chosen arbitrarily for inclusion in this section.

The synthesis of 4-hydroxy-1-(β-D-ribofuranosyl)pyridin-2-one (3-deaza-uridine (**59**)) (by conversion of the known 4-hydroxy-2-pyridinone, prepared by straightforward ring closure chemistry, to 2,4-bis(trimethylsilyloxy)-pyridine, condensation with a protected ribofuranosyl halide, and deprotection) was reported over 20 years ago[70]. Other '3-deazapyrimidine, base and nucleoside analogues were prepared and tested, but only (**59**) and '3-deazacytidine' exhibited significant anticancer activity[71]. No incorporation of 3-deazaUrd (**59**) into DNA and RNA was detected at the level of the radioactively labelled sample available[72], but potent inhibition of cytidine triphosphate synthetase (the unique enzyme for *de novo* biosynthesis of cytosine metabolites) by 3-deazaUrd (**59**) triphosphate was observed[73]. This is presumed to be the significant mode of action, although other enzymes are inhibited by (**59**) and other metabolites[61]. Clinical trials have shown limited response to 3-deazaUrd as a single agent, but its toxicity is manageable and it remains an available investigational agent for combination chemotherapy as a biological response modifier[74]. It is noteworthy that 3-deazaUrd (**59**) apparently represents an 'antimetabolite inhibitor' of crucial enzymes involved in the biosynthesis of nucleic acids without being incorporated into DNA or RNA[72,73]. Such a compound, if clinically effective, could offer significant

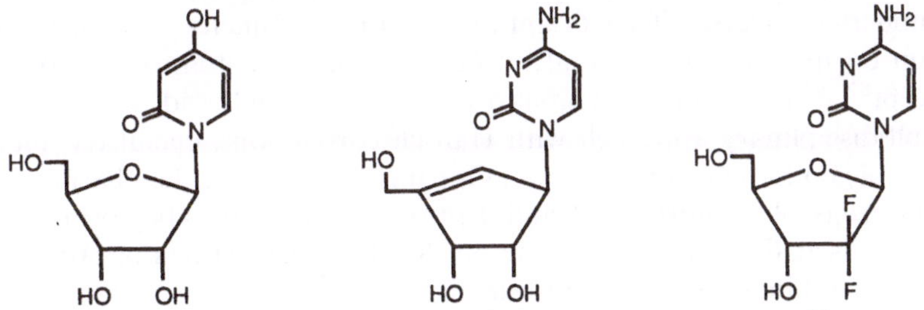

Scheme 9.15

advantages in terms of long-term toxicity effects.

Marquez *et al.* have reported efficient multistep total syntheses of 'carbocyclic nucleoside' analogues of the antibiotic neplanocin A (including the adenine antibiotic)[75]. Their cytosine analogue (60) has significant antineoplastic activity in cell culture and animal models. The activity level of (60) was greater than, but parallel with, that of 3-deazaUrd (59) in several comparative screens[75].

Hertel *et al.*[76] have devised a total synthesis of 2-deoxy-2,2-difluoro-D-*erythro*-pentose from readily available precursors. Coupling of protected derivatives with per(trimethylsilyl)bases provided 2'-deoxy-2',2'-difluoronucleoside analogues of which 2'-deoxy-2',2'-difluorocytidine (61) exhibits potent antitumour activity[77]. This new class of nucleoside analogues, although in preliminary stages of clinical trial at this time, is noteworthy owing to its reported effectiveness against solid tumours. The *gem*-difluoromethylene unit at C2' confers pronounced stability against glycosyl cleavage.

9.6　Prodrugs of pyrimidines and pyrimidine nucleosides

Albert is generally credited with coining the term 'prodrug'[78] which he defined as 'a substance which is converted after administration to the actual substance which combines with receptors'[79]. In the strictest terms the prodrug should have no biological activity. Conversion to the parent drug can be by enzymatic or non-enzymatic methods. Harper introduced the term 'drug latentiation' for enzymatic release mechanisms[80] and this was expanded by Kupchan *et al.* to include non-enzymatic processes as well[81]. The need for a prodrug arises in instances when the parent drug has undesirable physicochemical or pharmacological properties that limit its usefulness *in vivo*. Properties that can be modified using the prodrug approach include (i) altered solubility, (ii) enhanced bioavailability, (iii) modified transport, tissue distribution (including selective targeting), metabolism and excretion, (iv) increased patient acceptance (e.g. reduced pain on injection, modified taste, gastrointestinal irritation and toxic effects in general), (v) enhanced stability and (vi) improved drug formulation properties. Thus, at the cellular level, prodrugs can be utilised to increase efficacy of the chemotherapeutic agent and to protect it from biodegradation and excretion, and at the clinical level they can be utilised to moderate adverse effects of antitumour drug regimens which arise from action on normal cells. Several reviews have been written on the prodrug concept[82-91] including discussions of aspects of rational prodrug design using a multidisciplinary approach with Hansch correlations, chemistry, biology and clinical pharmacokinetics[85,91]. One has appeared on the pharmacology of prodrugs of 5-fluorouracil and 1-(β-D-arabinofuranosyl)cytosine[92]. The present discussion will concentrate on the chemistry aspects of prodrugs of cytosine arabinoside and 5-fluorouracil.

The efficacy of these agents depends on several issues including intact delivery to the site of the tumour cell, rate of degradation (catabolism) of the active moiety, rate of enzymatic activation of the parent drug (the actual cytotoxic agent is often a higher metabolite), toxicity of the agent, size of dose and dosage regimen, and acquired drug resistance. In certain instances, prodrugs of nucleoside analogues have alleviated some of these problems.

9.6.1 Prodrugs of 1-(β-D-arabinofuranosyl)cytosine

One shortcoming with the clinical use of araC (2) is its good substrate activity for the catabolic enzyme deoxycytidine deaminase resulting in its rapid deamination to the biologically inactive 1-(β-D-arabinofuranosyl)uracil (araU). This enzyme is found in many tissues[93] and high levels are present in human liver and kidney. Thus, araC has a plasma half-life of approximately 12 min in man[94], resulting in rapid appearance of araU in human urine after injection of araC. Infusion therapies were developed to keep blood levels of araC sufficiently high for antineoplastic activity[95] and prodrugs of araC were designed that are not substrates for deoxycytidine deaminase and which slowly release araC to maintain adequate plasma levels of the drug.

9.6.1.1 Oligonucleotides containing 1-(β-D-arabinofuranosyl)cytosine. The first prodrugs of araC were described by Wechter and co-workers[96,97] who prepared dinucleoside monophosphates containing araC to evaluate alternative mechanisms of action and/or transport systems. These dinucleotides showed minor increases in potency relative to araC, but no apparent improvement in specificity. The synthetic approach used standard blocking groups and the phosphodiester method of Khorana et al.[98]. A more recent approach to deoxynucleotidyl-(3',5')-arabinofuranosylcytosines has been described[99] with NMR analyses of their solution conformations[96,100].

9.6.1.2 Lipophilic carboxylic acid esters and amides of 1-(β-D-arabinofuranosyl)cytosine. The first dramatic improvement in efficacy of a prodrug of araC was demonstrated with 5'-O-adamantoyl-1-(β-D-arabinofuranosyl)cytosine ((65) R = adamantan-1-yl)[101]. Thus, (65) was shown to possess superior therapeutic properties for treatment of L 1210 leukaemic mice[101] and greater immunosuppressive activity in rats[102]. Unlike araC, the 5'-adamantoate (65) had long-lasting immunosuppressive and antileukaemic activity with single injections[101,102]. These observations prompted preparation and evaluation of 2'-, 3'- or 5'-acylated and di- and tri-acylated araC derivatives[103-107].

The earliest preparation of (65)[101] involved synthesis of the 4-N-5'-O-bis-adamantoyl derivative of araC followed by N-deacylation. A subsequent route that was used for a series of 5'-O-acylates is shown in Scheme 9.16[103]. Conversion of 5'-O-triphenylmethyl-araC (62) into its 4-N-(2,2,2-trichloroethoxycarbonyl) derivative (63) was followed by detritylation of the 5'-hydroxyl group. Acylation with acid chlorides or anhydrides gave (64) which was converted to the desired 5'-esters (65) by reductive cleavage of the trichloroethoxycarbonyl group. Overall yields (from araC) were low (3–5%),

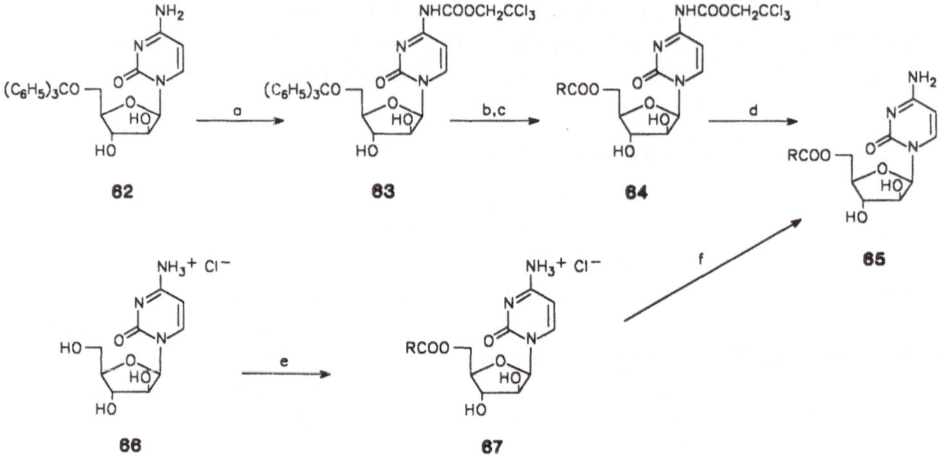

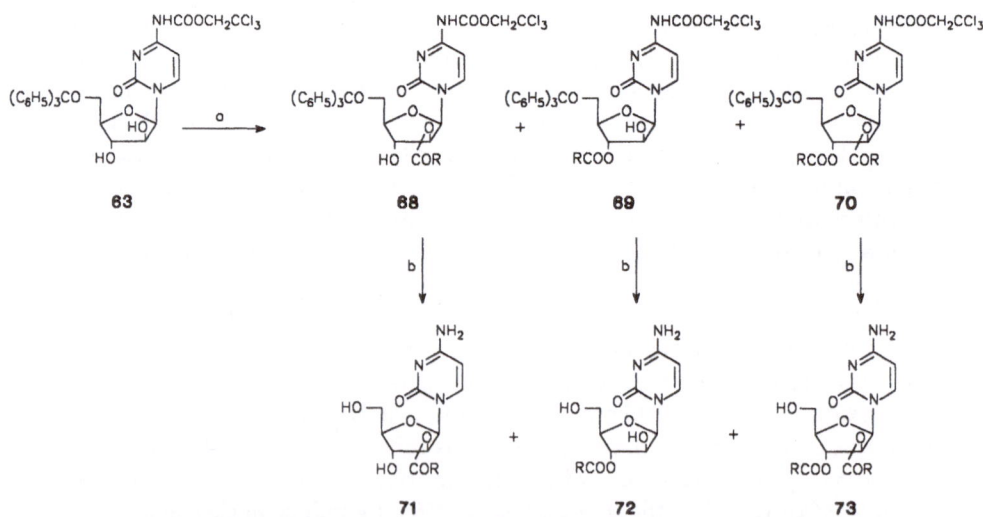

Scheme 9.16 (a) Cl₃CCH₂OCOCl/pyridine; (b) 80% AcOH; (c) RCOCl/pyridine; (d) Zn dust/MeOH; (e) RCOCl/DMF; (f) NaHCO₃.

and an improved single-step sequence is shown in Scheme 9.16 in which the protonated araC hydrochloride (**66**) provided effective protection against *N*-acylation[103] (overall yields 40–80%).

The first preparation of 2'- (**71**) and 3'-monoesters (**72**) and 2',3'-diesters (**73**) of araC is shown in Scheme 9.17[105]. This utilised the 4-*N*-5'-*O*-bis blocked derivative (**63**) and involved simple acylation followed by separation of the esters (**68–70**) and deblocking. Final products were obtained in poor overall yields (2–5%) and a more efficient direct route to the active 3'-esters (**72**) from

Scheme 9.17 RCOCl or (RCO)₂O in pyridine; (b) 80% AcOH.

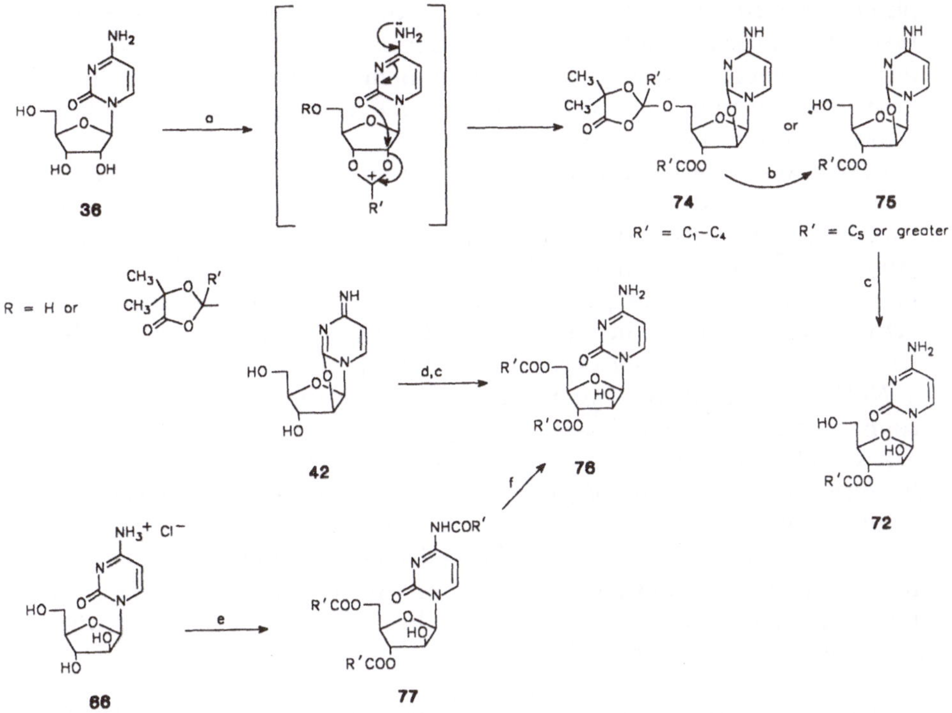

Scheme 9.18 (a) $(CH_3)_2C(OCOR')COCl/CH_3CN$; (b) dil. HCl; (c) MeOH/pyridine; (d) R'COCl/DMA; (e) R'COCl/pyridine; (f) picric acid/MeOH.

cytidine (36) was developed by Moffatt and co-workers[50, 108] (see Scheme 9.18). This approach involved acyloxonium ion formation[50, 51] with 2-acyloxyisobutyryl chlorides on a vicinal diol system to prepare 2,2'-anhydro-1-(3'-O-acyl-β-D-arabinofuranosyl)cytosine derivatives (75)[106,108]. Hydrolytic cleavage of the 2,2'-anhydro linkage of (75) to give (72), without saponification of the esters, was accomplished with pyridine/methanol (1:1) at 75–80°C[106]. The same group described preparation of 3',5'-diesters (76) by direct acylation of 2,2'-anhydro araC (42) followed by selective cleavage of the 2,2'-anhydro bond[106, 109]. An alternative sequence to 3',5'-di-O-acyl-araC' derivatives (76) was described by initial preparation of 1-(3,5-di-O-acyl-β-D-arabinofuranosyl)-4-N-acylcytosines (77)[107] followed by selective N-deacylation. Similar methods were utilised for preparation of 2',3',5'-tri-O-acyl derivatives from 1-(2,3,5-tri-O-acyl-β-D-arabinofuranosyl)-4-N-acylcytosines[107].

Wechter et al.[110] analysed the structure-activity relationships of these O-acylated derivatives. Aqueous solubilities (Hansch partition values) and rates of release of the parent drug (enzymatic as well as alkaline hydrolysis) were considered. From this study, two derivatives were chosen for clinical trials, and 5'-O-palmitoyl-araC was evaluated extensively[111]. When administered to patients either intramuscularly or orally, 5'-O-palmitoyl-araC was incompletely excreted. One patient retained 92% of the unrecovered material in tissue at the injection site[111]. This poor absorption correlated with the lack of improved efficacy (relative to araC) in humans. Thus, although these derivatives function well as depot forms of araC in mice, they have not been advantageous for human therapy.

Derivatives di- or tri-acylated on the sugar moiety with shorter chain acyl groups (e.g. acetyl and butyryl) showed some activity against L 1210 leukaemia in mice. Activity was lost upon additional acylation of the 4-amino group[106,107] and 4-N-acetyl-araC[112] and 4-N-adamantoyl-araC[107] were no more efficacious than the parent compound. However, good antineoplastic activity consistent with a depot effect was observed with araC derivatives that had longer chain acyl groups such as stearoyl (C_{18}) or behenoyl (C_{22}) attached at $N4$[113]. These derivatives were more water soluble than their sugar acylated counterparts and were resistant to deoxycytidine deaminase.

A straightforward synthesis of these derivatives ($> 90\%$ yields) was effected by selective N-acylation of araC with the appropriate carboxylic acid anhydride in aqueous dioxane at room temperature[113].

Recently, a new 4-N amide derivative, 4-N-[N-(cholesteryl-oxycarbonyl)glycyl]-araC, has been shown to be well incorporated into liposomes and to be more potent than other 4-N amide derivatives when administered in this manner[114].

9.6.1.3 *Phosphate and pyrophosphate diesters of 1-(β-D-arabinofuranosyl)-cytosine.* In addition to rapid catabolism of araC by deoxycytidine deaminase, an additional clinical problem is tumour resistance because of decreased metabolism of araC to it's 5'-triphosphate (araCTP)[115,116]. Treatment with araC 5'-nucleotides is not feasible due to (i) lack of a transport pathway for pyrimidine nucleotides into cells, (ii) inability of a doubly negatively charged and non-lipophilic nucleotide to penetrate cellular membranes by passive diffusion and (iii) rapid cleavage of 5'-nucleotides to parent nucleosides by serum phosphatases. Therefore, attention has been focused on prodrugs of araCMP that incorporate 5'-phosphate and pyrophosphate ester derivatives with increased lipophilicity which might allow cellular penetration followed by cleavage to give intracellular release of araCMP[117]. Bennett et al.[118] recently noted that such an approach, for the treatment of resistant tumours with diminished kinase activity, is unlikely to be effective unless there is selective uptake of the prodrug into the tumour or selective release of the parent drug in the tumour relative to normal (sensitive) cells. Normal cells will be susceptible to intracellularly released nucleotide drugs, and additionally will be susceptible to nucleoside drugs released by extra-cellular enzymatic cleavage. Since resistant tumour cells will not be susceptible to the extracellularly released nucleoside drugs, selective tumour cell kill would not be expected to occur[118].

Lipophilic 5'-(alkyl phosphate) esters (**78**) of araC have been prepared as

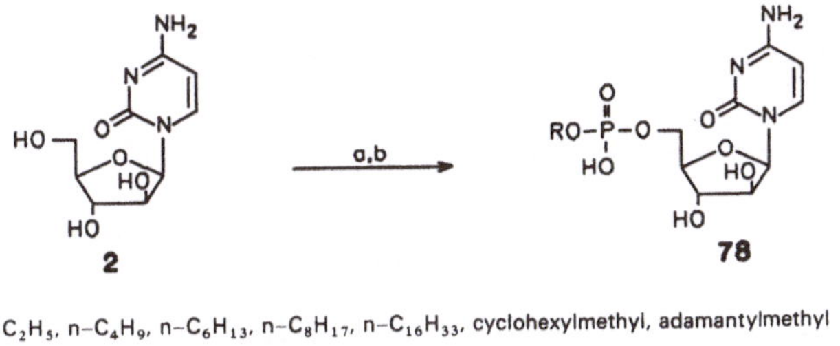

R=C_2H_5, n-C_4H_9, n-C_6H_{13}, n-C_8H_{17}, n-$C_{16}H_{33}$, cyclohexylmethyl, adamantylmethyl

Scheme 9.19 (a) POCl$_3$/(CH$_3$O)$_3$PO; (b) ROH.

shown in Scheme 9.19[117,119]. AraC was phosphorylated (POCl$_3$/(RO)$_3$PO)[120] and the resulting phosphorodichloridate was quenched *in situ* with an anhydrous alcohol. The phosphodiesters (**78**) were obtained in yields of 20–56%. Some products showed antileukaemic activity in mice comparable to or greater than araC[117]. However, the desired activity against a kinase-deficient resistant cell line *in vivo* was not observed.

Another approach for phosphate esters involves steroid conjugates of araC. The first derivatives in this area were described by Hong and co-workers[121,122] who utilised conjugates of cortisol (**80**), cortisone (**81**), corticosterone (**82**), prednisolone (**83**) and prednisone (**84**) linked from position 21 of the steroid to the 5'-position of araC via a phosphodiester bond. The latter two derivatives (**83, 84**) incorporated steroids which had been used in combination with other agents for treatment of human lymphoid leukaemias and lymphomas[123] and had exerted synergistic effects in combination chemotherapy[124]. Thus, these are prodrugs in which both portions could be expected to show antineoplastic activity upon enzymatic cleavage of the conjugate. These derivatives showed excellent activities against L 1210 leukaemia both *in vitro* and in mice, and led to the preparation of additional corticosteroid conjugates (**85–89**)[125]. The efficacy observed was dependent on the nature of the steroid since the fludrocortisone (**87**) and dexamethasone (**88**) conjugates were less active than araC. The more potent derivatives showed better antineoplastic activity than either component (steroid or araC) or an equimolar mixture. These conjugates were resistant to cytidine deaminase from human liver[126] and alkaline phosphatase[122,125] but were sensitive to phosphodiesterase I, 5'-nucleotidase and acid phosphatase. Incubation with baboon[122], mouse or human[125] plasma for 24 h left 50–85% of the conjugates intact, implying that they should function as sustained release forms of araC. Phosphodiesterase action gave araCMP and steroid (no steroid-phosphate was detected), but the araCMP formed was rapidly converted to araC by serum phosphatases. Like the simple alkyl phosphate esters (**78**), the conjugates (**80**) and (**82**) showed only marginal activity against a kinase deficient L 1210 leukaemia resistant to araC[126]. Of particular note for these prodrugs was the observation that several conjugates showed excellent

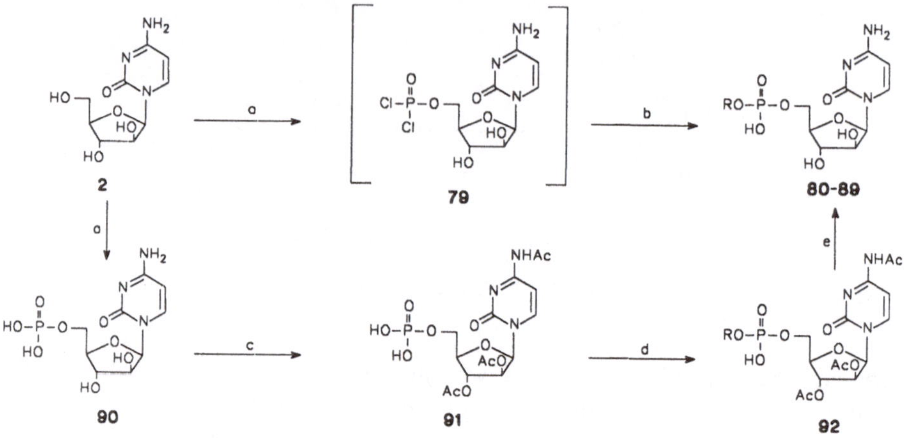

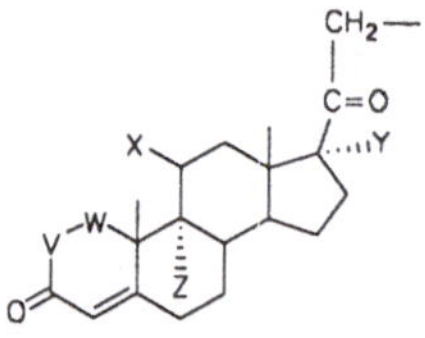

80 V = CH$_2$, W = CH$_2$, X = OH, Y = OH, Z = H
81 V = CH$_2$, W = CH$_2$, X = =O, Y = OH, Z = H
82 V = CH$_2$, W = CH$_2$, X = OH, Y = H, Z = H
83 V–W = CH=CH, X = OH, Y = OH, Z = H
84 V–W = CH=CH, X = =O, Y = OH, Z = H
85 V = CH$_2$, W = CH$_2$, X = H, Y = H, Z = H
86 V = CH$_2$, W = CH$_2$, X = H, Y = OH, Z = H
87 V = CH$_2$, W = CH$_2$, X = OH, Y = OH, Z = F
88 V = CH$_2$, W = CH$_2$, X = H, Y = H, Z = CH$_3$
89 V = CH$_2$, W = CH$_2$, X = F, Y = CH$_3$, Z = H

Scheme 9.20 (a) POCl$_3$/(CH$_3$O)$_3$PO; (b) ROH; (c) Ac$_2$O/pyridine; (d) ROH/DCC/pyridine; (e) NH$_3$/MeOH.

activities when administered intraperitoneally against L 1210 leukaemia implanted intracerebrally[126]. Such efficacy was not observed with araC and implies transport of the conjugate into the brain.

Synthetic routes[125] to these steroid conjugates (**80–89**) are shown in Scheme 9.20 and involve condensation of the steroid 21-hydroxyl group and 4-N-2′, 3′-di-O-triacetyl-araCMP (**91**) with dicyclohexylcarbodiimide (DCC), followed by deacetylation of (**92**), or by a procedure similar to that described (Scheme 9.19)[117] in which araC is phosphorylated with POCl$_3$ and the intermediate reacted *in situ* with the 21-hydroxyl group of the steroid.

A prodrug type (**94**) closely related to the monophosphate conjugates has the steroid and araC joined by a pyrophosphate linkage[119,127]. These derivatives also showed good activity in mice but were generally less effective than the monophosphate derivatives. Treatment of these conjugates with phosphodiesterase I, 5′-nucleotidase and acid phosphatase gave araCMP and steroid 21-phosphate which were further degraded to araC and steroid[127]. Incubation with human plasma gave cleavage over 24 h (28–62% depending on the steroid) with the final mixture containing conjugate, araCMP, steroid and araU[119,127]. Comparison of the antitumour activites of the lipophilic 5′-(alkyl phosphate) esters, mono and diphosphate linked steroid esters has been made[119]. Diphosphate linked conjugates were prepared[119] as shown in

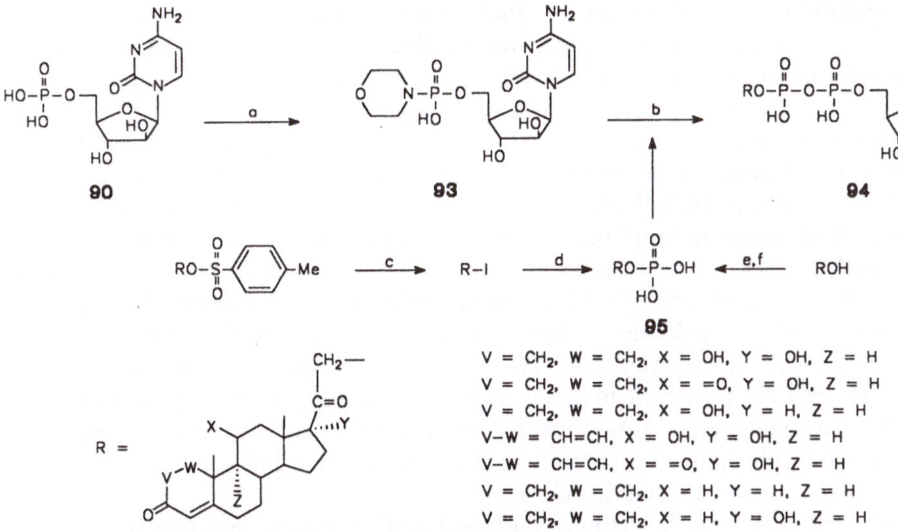

Scheme 9.21 (a) Morpholine/DCC; (b) ROPO(OH)$_2$; (c) NaI/acetone; (d) H$_3$PO$_4$/Et$_3$N/CH$_3$CN; (e) (HO)$_2$POOCH$_2$CH$_2$CN/DCC/pyridine; (f) base.

Scheme 9.21 by condensation of the steroid 21-phosphate (95) and the monophosphoromorpholidate of araCMP (93)[128].

Another series of araC conjugates linked via mono or diphosphate moieties have araC linked to a phospholipid. The monophosphate linked derivative, araCMP-L-dipalmitin (98), was initially prepared[129] (see Scheme 9.22) by condensation of 4-N-2',3'-di-O-trilevulinyl-araC (96) with L-α-phosphatidic acid using DCC. The levulinyl blocking group was chosen for deprotection

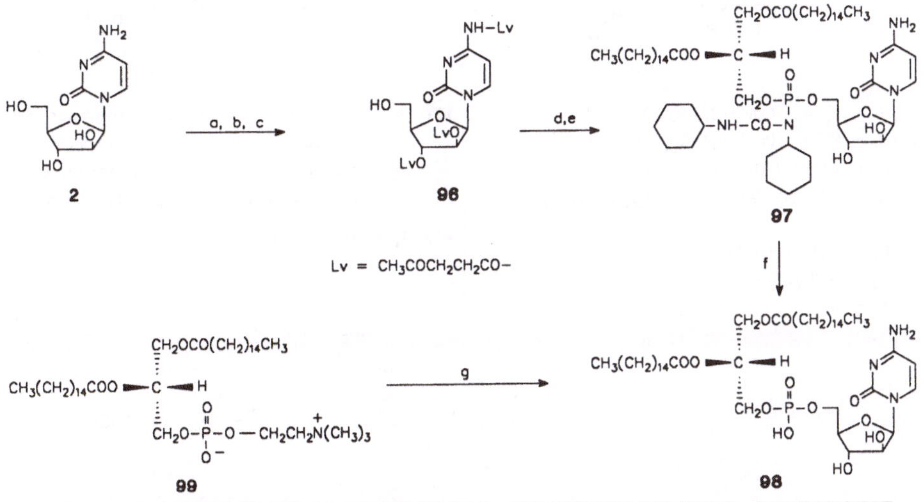

Scheme 9.22 (a) TBDMS-Cl/imidazole/DMF; (b) Lv-OH/DCC/DMAP/EtOAc; (c) TBAF/THF; (d) L-phosphatidic acid/DCC/pyridine; (e) NH$_2$NH$_2$/pyridine/AcOH; (f) CHCl$_3$/MeOH/H$_2$O; (g) (2)/phospholipase D/aq. buffer/CHCl$_3$.

compatibility with the final product which could tolerate hydrazine in pyridine/acetic acid. Interestingly, the product isolated was shown to be a diastereomeric mixture of the N-phosphoryldicyclohexylurea adducts (97). These adducts are generally unstable in pyridine/H$_2$O, and the unusual stability in this instance was attributed to formation of aggregates that precluded penetration by H$_2$O[129]. Hydrolysis was achieved in a denaturing solvent mixture (CHCl$_3$/MeOH/H$_2$O, 2:3:1). A recent preparation of (98) and other nucleoside monophosphate-linked diacylglycerols (including 5-fluorouridine, 2′-deoxy-5-fluorouridine and 5-fluorocytosine arabinoside) using an enzymatic transphosphatidylation reaction has been described[130]. The enzyme and nucleoside were mixed in a buffer and L-α-dipalmitoylphosphatidyl choline (99) in CHCl$_3$ was added. This two phase system gives the conjugate in excellent yields in one step without preparation of blocked precursors. Monophosphate linked derivative (98) gave significant increases in life span of mice bearing P388 leukaemia and was much more effective in this assay than araC[130].

More attention has been directed toward araC diphosphate derivatives that are 2′-epimeric analogues of ribo-CDP-L-diacylglycerols, the precursors of certain cell membrane constituents. Initial studies involved araCDP-DL-dipalmitin[131] (100), araCDP-L-diacylglycerol[132] ((101), in which the diacyl species was derived from egg lecithin), araCDP-L-dipalmitin[128] (102), ether linked diastereomeric derivatives araCDP-DL-bis(hexadecyloxy)propyl

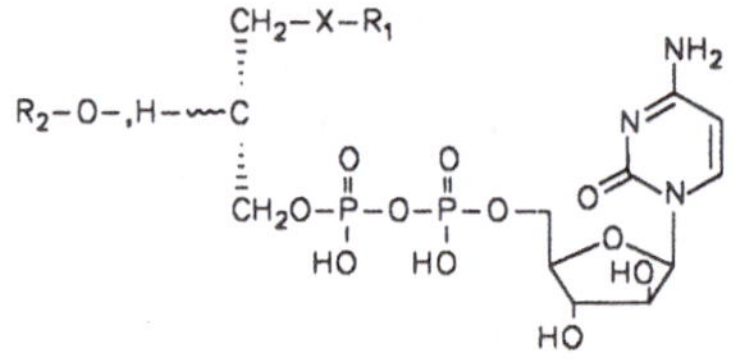

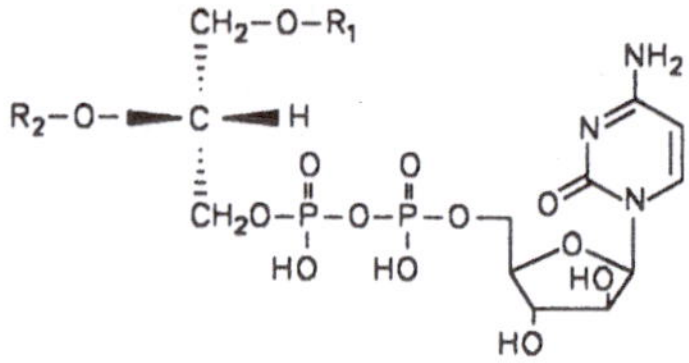

100 R$_1$ = R$_2$ = —CO(CH$_2$)$_{14}$CH$_3$, X = O
103 R$_1$ = R$_2$ = —CH$_2$(CH$_2$)$_{14}$CH$_3$, X = O
104 R$_1$ = —CH$_2$(CH$_2$)$_{16}$CH$_3$,
    R$_2$ = —CH$_2$(CH$_2$)$_7$(CH=CH—CH$_2$)$_2$(CH$_2$)$_3$CH$_3$
    X = O
105 R$_1$ = —CH$_2$(CH$_2$)$_{16}$CH$_3$
    R$_2$ = —CO(CH$_2$)$_{14}$CH$_3$
    X = S

101 R$_1$ = predominantly multispecies saturated and monosaturated long fatty acyl chains
    R$_2$ = predominantly multispecies mono-, di- and higher unsaturated long fatty acyl chains

102 R$_1$ = R$_2$ = —CO(CH$_2$)$_{14}$CH$_3$

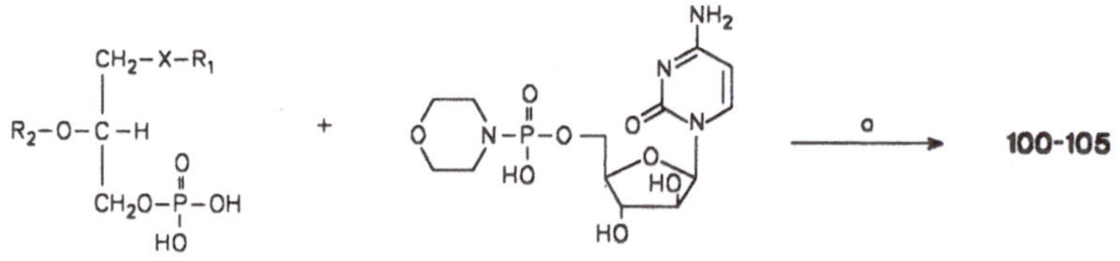

106   93   100-105

Scheme 9.23 (a) Pyridine.

ether[132,133] (103) and araCDP-DL-O^1-(octadecyloxy)-O^2-(octadeca-dienyloxy)propyl ether[132,133] (104) (see Scheme 9.23). These derivatives showed excellent activity against neoplastic *in vivo* models[131-134]. The observed efficacy depended on the length and nature of the acyl group sidechains[129] and the chirality of the glycerol moiety[134,135]. AraCDP-L-dipalmitin (102) was shown to be resistant to cytidine deaminase and to act as a sustained release drug in mice[136]. This conjugate (102) was shown to be effective in mice when administered i.p. against i.c. implanted L 1210 leukaemia[134], implying transport of the conjugate across the blood-brain barrier since araC was ineffective.

Early work indicated the importance of solubility characteristics of the conjugates, and satisfactory dissolution of araCDP-L-dipalmitin (102) in aqueous systems could not be obtained unless the suspended material was sonicated prior to testing[128]. Examination of the aggregation characteristics of sonicated solutions of a series of araCDP-L-diacylglycerols showed that the morphological nature of the aggregate formed (i.e. whether it was present as a large bilayer sheet, a stable three-dimensional array of interlocking bilayer stacks, small disc-shaped micelles or multilamellar liposome type structures) was dependent upon the temperature at which the sonication was carried out, the ionic strength of the solution, and the chain length and amount of unsaturation in the acyl sidechain[137]. Ready uptake of these conjugates by human and canine lipoproteins under physiological conditions was demonstrated, implying that they might be transported in this fashion[138] in serum.

Recent developments have involved phospholipids bearing ether or thioether functionalities in place of one or both fatty acyl groups[139-142]. Of particular note is work with araCDP-β-palmitoyl-D,L-thiobatyl alcohol (araCDP-D,L-PTBA, Cytoros (105))[140-142], a conjugate in which *both* the phospholipid and nucleoside moieties have independent antineoplastic activities. This prodrug generates two cytotoxic principles with different targets inside a neoplastic cell[142]. Cytoros possesses excellent activity against murine L 1210 and P 388 leukaemias[140] and Lewis lung carcinoma[141,142], including inhibition of metastasis of Lewis lung carcinoma when (105) was administered as adjuvant therapy shortly after surgical removal of the primary tumour. Synthesis of this type of pyrophosphate linked conjugate proceeds by condensation of the appropriate phospholipid (106) and the monophosphomorpholidate of araCMP[128] (93) as shown in Scheme 9.23.

9.6.1.4 *2,2'-Anhydro-1-(β-D-arabinofuranosyl)cytosine.* In contrast with lipophilic prodrugs of araC, 2,2'-anhydro-1-(β-D-arabinofuranosyl)cytosine (cyclocytidine (42)) is a hydrophilic prodrug which has been investigated extensively. This compound is resistant to enzymatic deamination[143] and is hydrolysed to give araC by a non-enzymatic process[44]. Ho showed that cyclocytidine is not a substrate for nucleoside kinases, but is hydrolysed to araC, which is then phosphorylated[143]. Cyclocytidine had a higher therapeutic index than araC and was active against a variety of animal tumour models[144,145]. In general, cyclocytidine was less toxic in animals[146] than araC and was retained by tissues and organs in a more localised fashion[147]. Unfortunately, the promising animal model profile of cyclocytidine was not duplicated in humans. Phases I and II clinical trials of cyclocytidine showed greater toxicity than araC[148,149].

The initial preparation of (42) (see Scheme 9.11) involved treatment of cytidine (36) with polyphosphoric acid to give a mixture of phosphorylated products, one of which was identified as the 3′, 5′-diphosphate of cyclocytidine (37). Enzymatic dephosphorylation of (37) gave cyclocytidine in low overall yield. Other preparations of (42) have been described in section 9.3.1.

Russell et al.[50] utilised the reaction of 2-acetoxyisobutyryl chloride with cytidine ((36) see Scheme 9.18) to give 2,2′-anhydro-1-(3-O-acetyl-β-D-arabinofuranosyl)cytosine ((75) R′ = CH₃) hydrochloride via the 5′-O-(2, 5, 5-trimethyl-1, 3-dioxolan-4-on-2-yl) derivative ((74) R′ = CH₃). The anhydro ring is produced by intramolecular nucleophilic attack of the cytosine oxygen at C2′ of the intermediate 2′,3′-acyloxonium ion. The 3′-acetate ((75) R′ = CH₃) could be converted into cyclocytidine or araC to provide efficient and economical syntheses of both compounds. A similar acyloxonium ion approach was described[108] in which treatment of cytidine with O-acetylsalicyloyl chloride gave the 3′-acetate which was immediately deblocked to give cyclocytidine.

9.6.1.5 *Esters of 2, 2′-anhydro-1-(β-D-arabinofuranosyl)cytosine.* In addition to the use of cyclocytidine as a prodrug of araC, several groups have investigated 3′-esters and 3′,5′-diesters of cyclocytidine as 'second generation' prodrugs[108,109,150–152]. These were designed to further alter metabolic and pharmacological profiles of araC. Both series of esters exhibited good activity against L 1210 leukaemia in mice. Esters having chain lengths of C_{11} to C_{22} gave good increases in life span and high percentages of 30-day survivors after single injections[108]. Unsaturation in the sidechain gave better results, possibly due to solubility characteristics[109]. The 3′-esters were prepared (see Scheme 9.18) from cytidine and 2-acyloxyisobutyryl halides[108]. The homogeneous 3′, 5′-diesters were prepared by acylation of cyclocytidine,

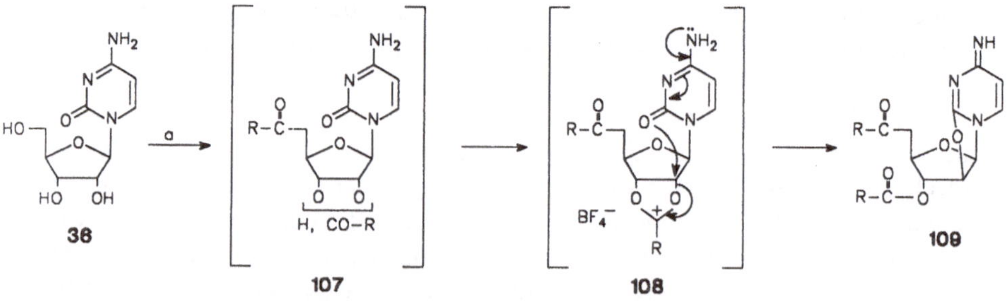

Scheme 9.24 (a) (RCO)₂O or RCOCl/BF₃·Et₂O/CH₃CN.

and acylation of a 3'-ester gave different substitution at the 3'- and 5'-positions[109].

Kondo and Inoue described acylation of cytidine in acetonitrile with catalysis by boron trifluoride etherate[151,152] (see Scheme 9.24). Presumably, this gave 2'(3'),5'-di-O-acylated nucleosides (107) which, in the presence of BF$_3$·Et$_2$O, formed the 2',3'-acyloxonium ion intermediates (108) that were opened by intramolecular nucleophilic attack to give the cyclocytidine diesters (109).

Other types of 'multiply latentiated' prodrugs in this area include 5'-phosphorylated 3'-O-acyl-2,2'-anhydro-araC derivatives, which were prepared to increase water solubility and ease of formulation[151], and the lipophilic 5'-(alkyl phosphate) esters of 3'-O-acyl-2,2'-anhydro-araC[117]. The preparation of the latter series was carried out in a similar fashion to that described (see Scheme 9.24) except a CMP alkylphosphate derivative was used as the starting material instead of cytidine.

9.6.1.6 1-(β-D-Arabinofuranosyl)-2-amino-1,4(2H)-4-iminopyrimidine.

The title compound ((110) araAIPy) also represents a 'multiply latentiated' prodrug that is readily converted into another prodrug form (cyclocytidine (42)) of araC (2). It has been proposed that araAIPy also can be hydrolysed directly to araC, thus providing dual pathways to the parent drug[153]. AraAIPy is resistant to deamination and has improved activity over cyclocytidine against L 1210 leukaemia in mice.

Preparation of araAIPy (110) hydrochloride by nucleophilic attack of ammonia at C2 of the anhydro ring of cyclocytidine (42) hydrochloride[154] is shown in Scheme 9.25.

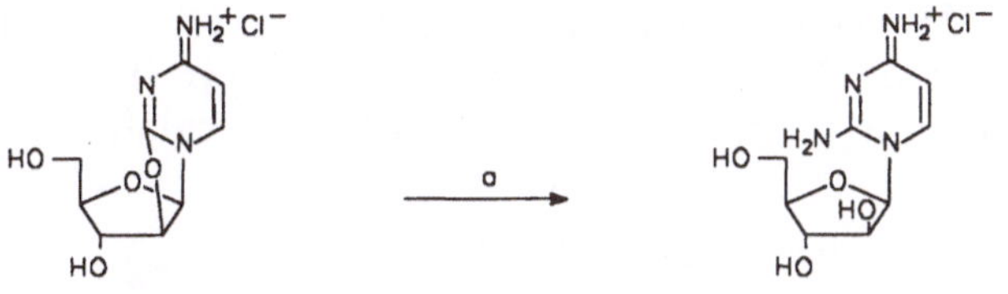

Scheme 9.25 (a) Liquid NH$_3$.

9.6.2 Prodrugs of 5-fluorouracil and 2'-deoxy-5-fluorouridine

The chief disadvantages of the clinical use of 5-FUra are its toxicity to rapidly dividing normal cells and onset of resistance. Early attempts to circumvent resistance centred around use of 2'-deoxy-5-fluorouridine (27). While (27) shows superior activity in some systems, it appears to have no clinical advantages over 5-FUra, presumably due to its rapid enzymatic cleavage to 5-FUra[155]. Utilisation of phosphate derivatives of (27) is thwarted by the problems of plasma stability and transport across cell membranes discussed earlier for araC nucleotides (see section 9.6.1.3). Thus, efforts in the prodrug area with 5-FUra and (27) have centred around selective release of the parent drug in tumour cells (as a means to alleviate toxicity) and the development of masked 5′-phosphate derivatives (5′-FdUMP analogues) as a means to overcome resistance.

9.6.2.1 *1-(2-Tetrahydrofuranyl)-5-fluorouracil (ftorafur) and 1,3-bis(2-tetrahydrofuranyl)-5-fluorouracil.* 1-(2-Tetrahydrofuranyl)-5-fluorouracil (ftorafur (113)) is a prodrug of 5-FUra that has been examined extensively in humans. It has a spectrum of antitumour activity similar to 5-FUra against solid tumours of the breast, stomach, colon, brain and rectum[156–158] but is less myelosuppressive and causes less mucositis. Ftorafur is well absorbed from the gastrointestinal tract and high sustained plasma levels of drug have been observed as well as accumulation of drug in tumour tissue. However, at high doses both gastrointestinal and central nervous system toxicity is observed[158,159]. Ftorafur as a single agent exhibits clinical responses similar to 5-FUra, but in combination protocols in place of 5-FUra it produces inferior results[160]. Metabolism of ftorafur occurs primarily in the liver[159,161] by a cytochrome *P*-450 mediated system[161] and several hydroxylated metabolites have been identified. Two metabolic pathways in which oxidation at C-5′ is effected by liver microsomal enzymes, and oxidation at C-2′ and/or C-5′ is accomplished by cytosolic enzymes have been proposed[162]. Both hydroxylated intermediates are expected to produce 5-FUra.

Ftorafur (113) is a racemate and both enantiomers show similar antitumour activities[163,164]. The first synthesis of (113) by Hiller *et al.*[165] involved condensation of chloromercuri-5-FUra or 5-fluoro-2,4-bis(trimethylsilyloxy) pyrimidine (21) with 2-chlorotetrahydrofuran (111) at low temperature. Condensations of (21) with 2-acetoxytetrahydrofuran (112) or 2-methoxy-tetrahydrofuran in the presence of a Friedel-Crafts catalyst also have been

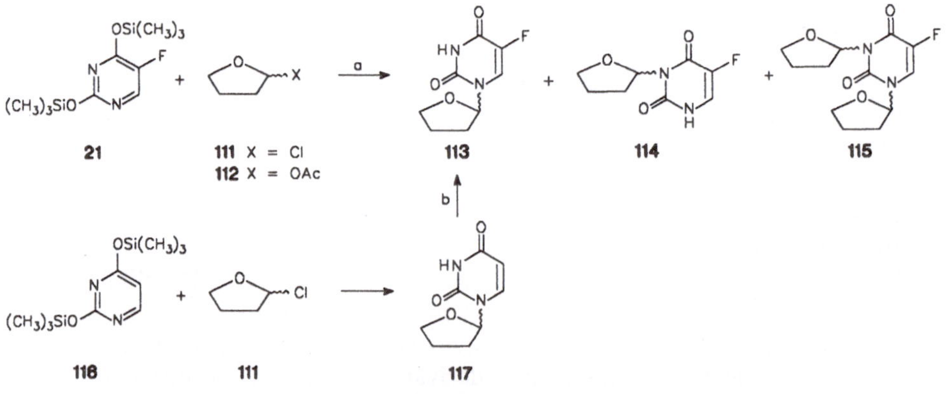

Scheme 9.26 (a) $BF_3 \cdot Et_2O$ or $SnCl_4$ or NaI in CH_2Cl_2 or CH_3CN; (b) CF_3OF.

used to prepare (113)[166] (see Scheme 9.26). Ftorafur (113), 3-(2-tetrahydro-furanyl)-5-fluorouracil (114) and 1,3-bis(2-tetrahydrofuranyl)-5-fluorouracil (115) were obtained depending on reaction conditions. Ftorafur has been prepared[167] by condensation of 2,4-bis(trimethylsilyloxy)pyrimidine (116) and (111), followed by fluorination of 1-(2-tetrahydrofuranyl)uracil (117) with trifluoromethyl hypofluorite[12].

The second generation prodrug 1,3-bis(2-tetrahydrofuranyl)-5-fluorouracil (115) can undergo metabolism to 5-FUra by two pathways. Enzymatic cleavage of the tetrahydrofuranyl group from the 1-position occurs followed by chemical collapse of the unstable 3-(2-tetrahydrofuranyl)-5-fluorouracil (114) to give 5-FUra, or initial cleavage of the tetrahydrofuranyl moiety from the 3-position to give ftorafur (113) which then proceeds to 5-FUra enzymatically[168]. Accordingly, (115) exhibits complex pharmacodynamics and appears to give higher levels of 5-FUra in plasma and tumour tissue than ftorafur. In a comparison with ftorafur, (115) was deposited less in bone marrow but higher levels were attained in kidney and much higher levels of 5-FUra were observed in the brains of rats treated with (115)[168].

9.6.2.2 *1-Alkylcarbamoyl-5-fluorouracil derivatives.* Another series of pro-drugs which have been examined clinically are 1-alkylcarbamoyl derivatives of 5-FUra[169]. These have been prepared[170] with chain lengths ranging from C_1 to C_8 and most were active against transplantable tumours in mice[171]. The 1-hexylcarbamoyl-5-fluorouracil (HCFU) had a better therapeutic effect than 5-FUra and ftorafur in mice and was selected for clinical trials. HCFU showed toxic effects not seen with 5-FUra and ftorafur, but some toxic aspects of therapy with 5-FUra and ftorafur occurred less frequently[169]. Metabolism of HCFU occurs primarily in liver and major metabolites appear to be derivatives with alkyl side chain oxidation[172], giving ultimate release of 5-FUra. This slow release of drug gave levels of 5-FUra in mouse plasma of $2\,\mu g/ml$ for up to 6 h, whereas similar treatment with 5-FUra showed no drug present after 2 h[173]. Synthesis of these 1-alkylcarbamoyl-5-fluorouracils involves treatment of 5-FUra with the appropriate isocyanate in pyridine in the presence of an organic base[170].

9.6.2.3 *5'-Deoxy-5-fluorouridine and related analogues.* 5'-Deoxy-5-fluorouridine ((120) 5'-dFUrd) is a prodrug of 5-FUra which enters cells via a transport system and then releases the heterocycle upon cleavage by uridine phosphorylase. Uridine phosphorylase is distributed unequally among vari-ous tissues and organs[174] and this activity is considerably higher in certain tumours than in surrounding tissue[174,175]. Tumour selective generation of 5-FUra is not seen with the previously discussed prodrugs and explains the enhanced therapeutic index of 5'-dFUrd relative to 5-FUra, 5-FdUrd and ftorafur[176].

The preparation of 5'-dFUrd from 5-fluorouridine (24) is shown in Scheme 9.27[177,178]. Isopropylidination of the 2'- and 3'-hydroxyls of (24) followed by

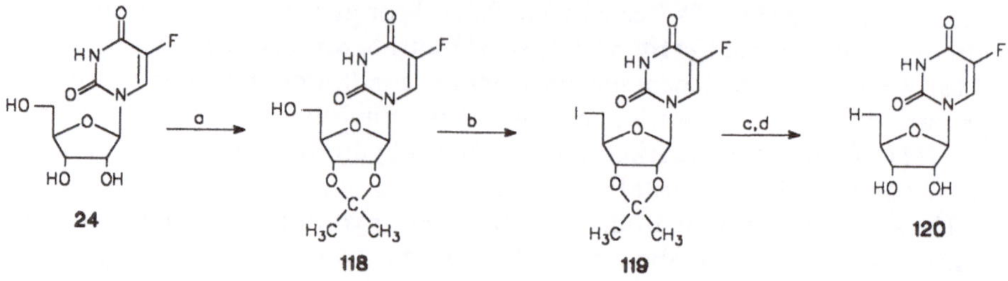

Scheme 9.27 (a) Acetone/H$_2$SO$_4$; (b) CH$_3$(C$_6$H$_5$O)$_3$P$^+$I$^-$/DMF; (c) H$_2$/5% Pd on C; (d) CF$_3$COOH.

halogenation of the 5′-position of (**118**) gives (**119**) which is reduced to the 5′-deoxy derivative and deblocked to give 5′-dFUrd (**120**).

Because 5′-dFUrd is a relatively poor substrate for uridine phosphorylase, several efforts have been made to prepare other 5′-substituted-5′-deoxy-5-fluorouridine derivatives that might give enhanced rates of release of 5-FUra in tumour tissue[179]. Although some compounds had lower K_m values than 5′-dFUrd, none had appreciably different V_{max} values for glycosyl bond cleavage[180] or exhibited enhanced antitumour activities[179]. Ajmera et al.[180] prepared 5′-deoxy-4′,5-difluorouridine (**124**) (see Scheme 9.28) which has a V_{max}/K_m ratio 50 times greater than 5′-dFUrd with uridine phosphorylase and is 10 times more effective in inhibiting the growth of L 1210 cells. Benzylation of 5-fluoro-5′-O-trityluridine (**121**) at O-2′ and O-3′ was followed by detrityl-ation and mesylation to give (**122**). Elimination of mesylate with KOtBu gave the 4′,5′-unsaturated derivative which was treated with pyridinium poly(hydrogen fluoride) at $-50°$C to give (**123**) as the sole stereoisomer. Fluorination at 0°C gave a 1:1 mixture of β-D-ribo and α-L-lyxo isomers. Reductive cleavage of the benzyl groups from (**123**) gave (**124**).

9.6.2.4   *2′,3′-Dideoxy-2′,3′-didehydro-5-fluorouridine.*   2′,3′-Dideoxy-2′,3′-didehydro-5-fluorouridine (1-(2,3-dideoxy-β-D-*glycero*-pent-2-enofuranosyl)-5-fluorouracil) was prepared in an attempt to improve the chemothera-peutic efficacy of 5-FdUrd[181]. This compound showed good activity against mouse leukaemias including some resistant to 5-FdUrd. However, it was not

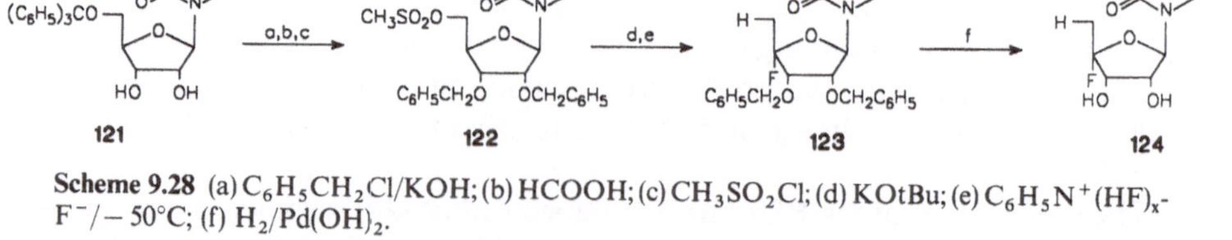

Scheme 9.28 (a) C$_6$H$_5$CH$_2$Cl/KOH; (b) HCOOH; (c) CH$_3$SO$_2$Cl; (d) KOtBu; (e) C$_6$H$_5$N$^+$(HF)$_x$-F$^-$/$-50°$C; (f) H$_2$/Pd(OH)$_2$.

phosphorylated by thymidine or uridine kinases and was not a substrate for nucleoside phosphorylase. Subsequent studies showed that it was a prodrug of 5-FUra with release by a non-enzymatic process[182].

9.6.2.5 *Lipophilic esters of 2′-deoxy-5-fluorouridine.* Twenty-seven esters of 5-FdUrd were prepared[183] to evaluate slow release forms. Cleavage to give 5-FdUrd was necessary before phosphorylase action could release 5-FUra[155]. Efficacy and toxicity were related to chain length of the ester, and 3′,5′-diesters bearing C_4 to C_{16} chains were more efficacious than equivalent doses of 5-FdUrd.

A recent application of this type of prodrug[184] showed that 2′-deoxy-3′,5′-di-*O*-octanoyl-5-fluorouridine administered by intrahepatic arterial injection in the tumoritropic lymphographic agent, Lipiodol, has excellent selective localisation in a rabbit hepatoma. The active metabolites, 5-FdUrd and 5-FdUMP, were also localised and little or no drug was observed in the stomach, duodenum, kidneys or bone marrow.

Syntheses of these esters utilised direct acylation of 5-FdUrd with acyl anhydrides or chlorides[183]. Equimolar amounts of nucleoside and acyl anhydride were used for selective 5′-esterification. Acylation of 2′-deoxy-5-fluoro-5′-*O*-trityluridine by standard methods followed by deprotection with acid gave the 3′-esters.

9.6.2.6 *Masked phosphates of 2′-deoxy-5-fluorouridine.* As discussed in sections 9.6.1.3 and 9.6.2, utilisation of nucleoside phosphates as drugs to overcome resistance are impractical due to the poor stability of nucleotides in plasma and inability to transport these negatively charged species across cell membranes. Several attempts have been made to utilise phosphate esters and masked neutral prodrugs of 5-FdUrd.

Heidelberger *et al.*[185] and Montgomery *et al.*[186] prepared phosphate diesters and triesters of 5-FdUMP with varying lipophilicities and evaluated them in cell culture. Included in this work were 3′,5′-dinucleoside monophosphates containing 5-FdUrd and also some mixed anhydrides of 5-FdUMP and L-alanine. No facilitation of transport of 5-FdUMP into cells was demonstrated. Subsequent attempts involved preparation of the symmetrical dinucleoside monophosphate 5-fluorodeoxyuridylyl-(5′,5′)-5-fluorodeoxyuridine[187,188]. However, this derivative was less effective than 5-FdUMP and did not inhibit cells resistant to 5-FdUrd. More recent studies have examined the neutral triester bis(pivaloyloxymethyl)-3′-*O*-acetyl-2′-deoxy-5-fluorouridine-5′-monophosphate. Release of the parent nucleotide would require cleavage of the acyloxy groups followed by elimination of formaldehyde[189]. In cell culture, this derivative appeared to be as active as 5-FdUrd.

Another approach involved synthesis of the neutral cyclic phosphoramidate derivative (**125**)[190,191] which has the same phosphorus containing ring

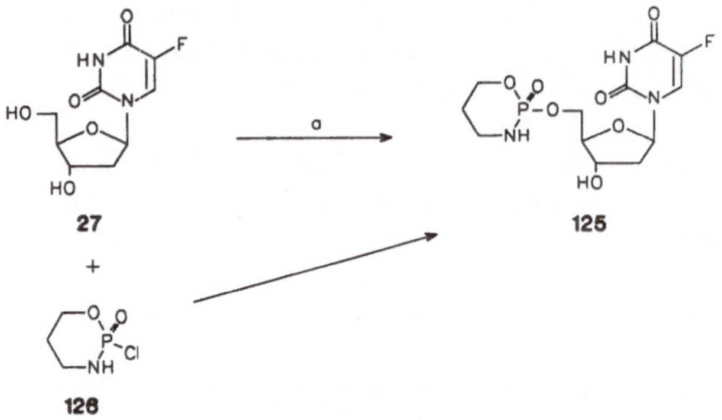

Scheme 9.29 (a) $POCl_3/(EtO)_3PO/HOCH_2CH_2CH_2NH_2$.

as the clinical anticancer drug cyclophosphamide. Oxidative metabolism of cyclophosphamide by hepatic cytochrome *P*-450 mixed-function oxidase gives 4-hydroxycyclophosphamide[192] which subsequently degrades to a phosphoramidate species[193]. Similar metabolism of (**125**) would give the 5'-phosphoramidate of 5-FdUrd which would be expected to hydrolyse readily to 5-FdUMP[190,191]. Unfortunately, (**125**) had poor activity *in vivo* and *in vitro*. The synthesis of (**125**) (Scheme 9.29) involved a one-pot reaction of 5-FdUrd (**27**) with $POCl_3$, $(EtO)_3PO$ and 3-amino-1-propanol[191], or treatment of 5-FdUrd (**27**) with 2-chloro-1,3,2-oxazaphosphacyclohexane 2-oxide (**126**)[190].

Since steroid and phospholipid conjugates of araC showed improved efficacy in animal models (see section 9.6.1.3), similar conjugates of 5-FdUrd and 5-Urd have been prepared and evaluated[130,185,194]. The 5-FUMP-L-1,2-dipalmitin was superior to 5-FUrd in the treatment of P388 leukaemia in mice, but 5-FdUMP-L-1,2-dipalmitin possessed only moderate activity[130]. Preparations of these derivatives were carried out essentially as described for the araC analogues (see section 9.6.1.3).

9.6.3   *2,2'-Anhydro-1-(β-D-arabinofuranosyl)-5-fluorocytosine*

2,2'-Anhydro-1-(β-D-arabinofuranosyl)-5-fluorocytosine (AAFC) is a deaminase resistant prodrug form of 1-(β-D-arabinofuranosyl)-5-fluorocytosine (AFC), which is formed from AAFC by non-enzymatic hydrolytic opening of the 2,2'-anhydro ring similar to formation of araC from cyclocytidine[195] (see section 9.6.1.4). AFC is susceptible to deamination[196] which gives another active compound 1-(β-D-arabinofuranosyl)-5-fluorouracil (AFU)[197] and thus AAFC can potentially exert biological activity through two different metabolites[198]. Activities of AAFC more closely resemble those of cyclocytidine than 5-FUra[199], but its tissue distribution is quite different[195]. AAFC causes different toxicities than AFC, but it produced clinical results similar to araC without the postural hypotension and jaw pain exhibited by cyclocytidine.

The synthesis of AAFC[198] followed the procedures described by Kiku-gawa and Ichino[47] for conversion of cytidine to cyclocytidine, except 5-fluorocytidine was used as starting material.

9.7 Conclusions

Early studies in the laboratories of Heidelberger, Fox and others resulted in synthesis and biomedical evaluation of 5-fluorouracil (5-FUra), 5-fluorocyto-sine and a variety of nucleosides and derivatives. Of these 5-fluoropyrimidine compounds, only 5-FUra is in general clinical use. The use of 2'-deoxy-5-fluorouridine in place of 5-FUra has been promoted, but clinical effects are not dramatically different with the two agents. Arabinosyl analogues and other nucleoside derivatives have been studied, but have not provided markedly improved results. The tetrahydrofuran-2-yl derivative of 5-FUra (ftorafur) and 5'-deoxy-5-fluorouridine have been prepared and studied as prodrugs of 5-FUra. However, it appears unlikely that their initially reported benefits will translate into general clinical acceptance. Other prodrugs of 5-FUra either have not provided adequate advantages for clinical acceptance or are in preliminary trials.

The use of 1-(β-D-arabinofuranosyl)cytosine (araC) in the treatment of acute leukaemias and lymphomas is well established. The 5-fluoro and 5-aza analogues of araC show pronounced activity in model systems, but do not show sufficient advantages relative to their individual disadvantages in synthesis, toxicity and instability to justify major clinical usage at present. A large number of prodrugs of araC including directly protected araC deriva-tives, 2, 2'-anhydro-araC, protected derivatives of the latter and pyrimidine ring-modified compounds that can be transformed into araC and/or its 2, 2'-anhydro derivative have been prepared and tested. Although animal model studies were promising in several instances, the human clinical results were less favourable, owing primarily to side effects that resulted from different tissue distributions and metabolic profiles of the prodrugs relative to the parent agent. Hopefully, as multidisciplinary efforts expand the body of knowledge in this field, the basis for translation of successful animal model results into human clinical efficacy will be provided.

The use of 5-azacytidine for treatment of leukaemias and lymphomas has been successful in some cases, and 5-azaCyd remains an investigational drug. However, problems with its intrinsic chemical instability and toxic side effects on patients will likely continue to restrict its acceptance as a clinical drug.

Several new pyrimidine nucleoside analogues have been reported, and 2'-deoxy-2', 2'-difluorocytidine is noteworthy for its reported activity against solid tumours. It is our opinion that advancements in basic biochemical knowledge coupled with the intuitive molecular insight of synthetic organic chemists, possibly aided by computer-assisted modelling, will result in the discovery of new pyrimidine-related agents that will be more effective anticancer drugs than those presently available.

Acknowledgements

The authors thank B. Hajner, D. Faibes and S. Bach of the Merck Sharp and Dohme Information Research and Analysis Department for their efficient literature searching for this chapter. M. J. R. thanks the American Cancer Society (Grant no. CH-405) for generous research support.

References

1. A. Burger, in *Burger's Medicinal Chemistry*, ed. M. E. Wolff, 4th edn., Part 1, Wiley-Interscience, New York (1980) p. 1.
2. R. Duschinsky, E. Pleven and C. Heidelberger, *J. Am. Chem. Soc.* **79** (1957) 4559.
3. C. Heidelberger, *Prog. Nucleic Acid Res. Mol. Biol.* **4** (1965) 1.
4. C. Heidelberger, P. V. Danenberg and R. G. Moran, *Adv. Enzymol.* **54** (1983) 57.
5. D. V. Santi, *J. Med. Chem.* **23** (1980) 103.
6. A. D. Broom, *J. Med. Chem.* **32** (1989) 2.
7. E. R. Walwick, W. K. Roberts and C. A. Dekker, *Proc. Chem. Soc.* (1959) 84.
8. S. S. Cohen, *Prog. Nucleic Acid Res. Mol. Biol.* **5** (1966) 2.
9. Y. Ohno, D. Spriggs, A. Matsukage, T. Ohno and D. Kufe, *Cancer Res.* **48** (1988) 1494.
10. W. K. Chung, J. H. Chung and K. A. Watanabe, *J. Heterocycl. Chem.* **20** (1983) 457.
11. T. Fuchikami, A. Yamanouchi and Y. Suzuki, *Chem. Lett.* (1984) 1573.
12. M. J. Robins and S. R. Naik, *J. Am. Chem. Soc.* **93** (1971) 5277.
13. M. J. Robins, M. MacCoss, S. R. Naik and G. Ramani, *J. Am. Chem. Soc.* **98** (1976) 7381.
14. R. Duschinsky, T. Gabriel, W. Tautz, A. Nussbaum, M. Hoffer. E. Grunberg, J. H. Burchenal and J. J. Fox, *J. Med. Chem.* **10** (1967) 47.
15. M. N. G. James and M. Matsushima, *Acta Crystallogr., Sect. B* **32** (1976) 957.
16. M. J. Robins and S. R. Naik, *J. Chem. Soc., Chem. Commun.* (1972) 18.
17. D. H. R. Barton, R. H. Hesse, H. T. Toh and M. M. Pechet, *J. Org. Chem.* **37** (1972) 329.
18. D. H. R. Barton, W. A. Bubb, R. H. Hesse and M. M. Pechet, *J. Chem. Soc., Perkin Trans. 1* (1974) 2095.
19. H. Meinert and D. Cech, *Z. Chem.* **12** (1972) 292.
20. H. Meinert and D. Cech, *Z. Chem.* **12** (1972) 335.
21. D. Cech, H. Meinert, G. Etzold and P. Langen, *J. Prakt. Chem.* **315** (1973) 149.
22. T. I. Yurasova, *Zh. Obshch. Khim.* **44** (1974) 956.
23. S. S. Yemul, H. B. Kagan and R. Setton, *Tetrahedron Lett.* **21** (1980) 277.
24. S. Morikawa, K. Uchida, S. Yonemori, Y. Oda, A. Yamazaki and H. Morisawa, *Nippon Kagaku Kaishi* (1985) 2185 (*Chemical Abstracts* **105** (1986) 172147f).
25. O. Miyashita, K. Matsumura, H. Shimadzu and N. Hashimoto, *Chem. Pharm. Bull.* **29** (1981) 3181.
26. S. Stavber and M. Zupan, *J. Chem. Soc., Chem. Commun.* (1983) 563.
27. G. W. M. Visser, S. Boele, B. W. v. Halteren, G. H. J. N. Knops, J. D. M. Herscheid, G. A. Brinkman and A. Hoekstra, *J. Org. Chem.* **51** (1986) 1466.
28. H. Vorbruggen, in *Nucleoside Analogues: Chemistry Biology, and Medical Applications*, eds. R. T. Walker, E. De Clercq and F. Eckstein, NATO Advanced Study Institute Series, Vol. 26A, Plenum Press, New York (1979) p. 35.
29. M. Saneyoshi, M. Inomata and F. Fukuoka, *Chem. Pharm. Bull.* **26** (1978) 2990.
30. M. Sharma and J. L. Alderfer, *Nucleosides Nucleotides* **2** (1983) 189.
31. M. Hoffer, R. Duschinsky, J. J. Fox and N. Yung, *J. Am. Chem. Soc.* **81** (1959) 4112.
32. H. Aoyama, *Bull. Chem. Soc. Jpn.* **60** (1987) 2073.
33. A. Holy and D. Cech, *Collect. Czech. Chem. Commun.* **39** (1974) 3157.
34. A. Hampton and A. W. Nichol, *Biochemistry* **5** (1966) 2076.
35. S. Ozaki, T. Katakami and M. Saneyoshi, *Bull. Chem. Soc. Jpn.* **50** (1977) 2197.
36. J. Brokes, H. Hrebabecky and J. Beranek, *Collect. Czech. Chem. Commun.* **44** (1979) 439.
37. M. J. Robins, G. Ramani and M. MacCoss, *Can. J. Chem.* **53** (1975) 1302.
38. R. J. Cisneros, L. A. Silks and R. B. Dunlap, *Drugs Future* **13** (1988) 859.
39. G. D. Heggie, J.-P. Sommadossi, D. S. Cross, W. J. Huster and R. B. Diasio, *Cancer Res.* **47** (1987) 2203.
40. W. E. Hull, R. E. Port, R. Herrmann, B. Britsch and W. Kunz, *Cancer Res.* **48** (1988) 1680.

41. D. J. Sweeny, S. Barnes and R. B. Diasio, *Cancer Res.* **48** (1988) 2010.
42. M. Y. Chu and G. A. Fisher, *Biochem. Pharmacol.* **11** (1962) 423.
43. B. D. Clarkson, *Cancer* **30** (1972) 1572.
44. M. C. Wang, R. A. Sharma and A. Bloch, *Cancer Res.* **33** (1973) 1265.
45. W. K. Roberts and C. A. Dekker, *J. Org. Chem.* **32** (1967) 816.
46. T. Kanai, T. Kojima, O. Maruyama and M. Ichino, *Chem. Pharm. Bull.* **18** (1970) 2569.
47. K. Kikugawa and M. Ichino, *J. Org. Chem.* **37** (1972) 284.
48. K. K. Ogilvie, *Carbohydr. Res.* **24** (1972) 210.
49. J. Beranek, T. J. Delia and P. Drasar, *Collect. Czech. Chem. Commun.* **42** (1977) 1588.
50. A. F. Russell, M. Prystasz, E. K. Hamamura, J. P. H. Verheyden and J. G. Moffatt, *J. Org. Chem.* **39** (1974) 2182.
51. A. R. Mattocks, *J. Chem. Soc.* (1964) 1918.
52. R. Marumoto and M. Honjo, *Chem. Pharm. Bull.* **22** (1974) 128.
53. W. V. Ruyle and T. Y. Shen, *J. Med. Chem.* **10** (1967) 331.
54. M. Kaneko and B. Shimizu, *Chem. Pharm. Bull.* **20** (1972) 1050.
55. J. Brokes and J. Beranek, *Collect. Czech. Chem. Commun.* **39** (1974) 3100.
56. K. J. Divakar and C. B. Reese, *J. Chem. Soc., Perkin Trans. 1* (1982) 1171.
57. B. Shimizu and F. Shimizu, *Chem. Pharm. Bull.* **18** (1970) 1060.
58. T. Sowa and K. Tsunoda, *Bull. Chem. Soc. Jpn.* **48** (1975) 505.
59. D. H. Shannahoff and R. A. Sanchez, *J. Org. Chem.* **38** (1973) 593.
60. E. J. Hessler, *J. Org. Chem.* **41** (1976) 1828.
61. B. A. Chabner, in *Cancer and Chemotherapy*, Vol. III, eds. S. T. Crooke and A. W. Prestayko, Academic Press, New York (1981) p. 3.
62. A. Piskala and F. Sorm, *Collect. Czech. Chem. Commun.* **29** (1964) 2060.
63. A. Piskala, *Collect. Czech. Chem. Commun.* **32** (1967) 3966.
64. M. W. Winkley and R. K. Robins, *J. Org. Chem.* **35** (1970) 491.
65. U. Niedballa and H. Vorbruggen, *J. Org. Chem.* **39** (1974) 3672.
66. J. A. Beisler, M. M. Abbasi, J. A. Kelley and J. S. Driscoll, *J. Med. Chem.* **20** (1977) 806.
67. J. A. Beisler, M. M. Abbasi, J. A. Kelley and J. S. Driscoll, *J. Carbohydr., Nucleosides, Nucleotides* **4** (1977) 281.
68. J. A. Beisler, M. M. Abbasi and J. S. Driscoll, *J. Med. Chem.* **22** (1979) 1230.
69. J. Ben-Hattar and J. Jiricny, *J. Org. Chem.* **51** (1986) 3211.
70. M. J. Robins and B. L. Currie, *Chem. Commun.* (1968) 1547.
71. A. Bloch, G. Dutschman, B. L. Currie, R. K. Robins and M. J. Robins, *J. Med. Chem.* **16** (1973) 294.
72. M. C. Wang and A. Bloch, *Biochem. Pharmacol.* **21** (1972) 1063.
73. R. P. McPartland, M. C. Wang, A. Bloch and H. Weinfeld, *Cancer Res.* **34** (1974) 3107.
74. W. J. Moriconi, M. Slavik and S. Taylor, *Invest. New Drugs* **4** (1986) 67.
75. V. E. Marquez, M.-I. Lim, S. P. Treanor, J. Plowman, M. A. Priest, A. Markovac, M. S. Khan, B. Kaskar and J. S. Driscoll, *J. Med. Chem.* **31** (1988) 1687.
76. L. W. Hertel, J. S. Kroin, J. W. Misner and J. M. Tustin, *J. Org. Chem.* **53** (1988) 2406.
77. V. Heinemann, L. W. Hertel, G. B. Grindey and W. Plunkett, *Cancer Res.* **48** (1988) 4024.
78. A. Albert, *Nature (London)* **182** (1958) 421.
79. A. Albert, *Selective Toxicity*, 2nd edn., Wiley, New York (1960) p. 30.
80. N. J. Harper, *J. Med. Pharm. Chem.* **1** (1959) 467.
81. S. M. Kupchan, A. F. Casy and J. V. Swintosky, *J. Pharm. Sci.* **54** (1965) 514.
82. R. E. Notari, *J. Pharm. Sci.* **62** (1973) 865.
83. R. E. Notari, *Pharmacol. Ther.* **14** (1981) 25.
84. A. A. Sinkula, *Annu. Rep. Med. Chem.* **10** (1975) 306.
85. A. A. Sinkula and S. H. Yalkowsky, *J. Pharm. Sci.* **64** (1975) 181.
86. T. Higuchi and V. Stella, eds., *Prodrugs as Novel Drug Delivery Systems*, ACS Symp. Ser. Vol. 14, American Chemical Society, Washington, DC (1975).
87. V. J. Stella and K. L. Himmelstein, *J. Med. Chem.* **23** (1980) 1275.
88. V. J. Stella, T. J. Mikkelson and J. D. Pipkin, in *Drug Delivery Systems*, ed. R. L. Juliano, Oxford University Press, London and New York (1980) pp. 112–176.
89. N. J. Harper, *Prog. Drug Res.* **4** (1962) 221.
90. E. J. Ariens, in *Drug Design*, Academic Press, New York (1971) 1.
91. A. A. Sinkula, in *Prodrugs as Novel Drug Delivery Systems*, ACS Symp. Ser. Vol. 14, eds. T. Higuchi and V. Stella, American Chemical Society, Washington DC (1975).

92. A. F. Hadfield and A. C. Sartorelli, *Adv. Pharmacol. Chemother.* **20** (1984) 21.
93. D. H. W. Ho, *Cancer Res.* **33** (1973) 2816.
94. D. H. W. Ho and E. Frei, *Clin. Pharm. Ther.* **12** (1971) 944.
95. D. H. W. Ho, *Cancer Treat. Rep.* **61** (1977) 717.
96. W. J. Wechter, *J. Med. Chem.* **10** (1967) 762.
97. C. G. Smith, H. H. Buskirk and W. L. Lumis, *J. Med. Chem.* **10** (1967) 774.
98. H. G. Khorana, *Some Recent Developments in the Chemistry of Phosphate Esters of Biological Interest*, Wiley, New York and London (1961).
99. S. H. Gray, C. F. Ainsworth, C. L. Bell, S. S. Danyluk and M. MacCoss, *Nucleosides Nucleotides* **2** (1983) 435.
100. S. H. Gray, C. F. Ainsworth, C. L. Bell, S. S. Danyluk and M. MacCoss, *Nucleosides Nucleotides* **5** (1986) 481.
101. G. L. Neil, P. F. Wiley, R. C. Manak and T. E. Moxley, *Cancer Res.* **30** (1970) 1047.
102. G. D. Gray and M. M. Mickelson, *Immunology* **19** (1970) 419.
103. D. T. Gish, R. C. Kelly, G. W. Camiener and W. J. Wechter, *J. Med. Chem.* **14** (1971) 1159.
104. G. D. Gray, F. R. Nicol, M. M. Mickelson, G. W. Camiener, D. T. Gish, R. C. Kelly, W. J. Wechter, T. E. Moxley and G. L. Neil, *Biochem. Pharmacol.* **21** (1972) 465.
105. D. T. Warner, G. L. Neil, A. J. Taylor and W. J. Wechter, *J. Med. Chem.* **15** (1972) 790.
106. E. K. Hamamura, M. Prystasz, J. P. H. Verheyden, J. G. Moffatt, K. Yamaguchi, N. Uchida, K. Sato, O. Shiratori, S. Takase and K. Katagiri, *J. Med. Chem.* **19** (1976) 667.
107. J. A. Montgomery and H. J. Thomas, *J. Med. Chem.* **15** (1972) 116.
108. E. K. Hamamura, M. Prystasz, J. P. H. Verheyden, J. G. Moffatt, K. Yamaguchi, N. Uchida, K. Sato, A. Nomura, O. Shiratori, S. Takase and K. Katagiri, *J. Med. Chem.* **19** (1976) 654.
109. E. K. Hamamura, M. Prystasz, J. P. H. Verheyden, J. G. Moffatt, K. Yamaguchi, N. Uchida, K. Sato, A. Nomura, O. Shiratori, S. Takase and K. Katagiri, *J. Med. Chem.* **19** (1976) 663.
110. W. J. Wechter, M. A. Johnson, C. M. Hall, D. T. Warner, A. E. Berger, A. H. Wenzel, D. T. Gish and G. L. Neil, *J. Med. Chem.* **18** (1975) 339.
111. D. H. W. Ho and G. L. Neil, *Cancer Res.* **37** (1977) 1640.
112. M. R. Dollinger, J. H. Burchenal, W. Kreis and J. J. Fox, *Biochem. Pharmacol.* **16** (1967) 689.
113. M. Akiyama, J. Oh-ishi, T. Shirai, K. Akashi, K. Yoshida, J. Nishikido, H. Hayashi, Y. Usubuchi, D. Nishimura, H. Itoh, C. Shibuya and T. Ishida, *Chem. Pharm. Bull.* **26** (1978) 981.
114. Y. Tokunaga, T. Iwasa, J. Fugisaki, S. Sawai and A. Kagayama, *Chem. Pharm. Bull.* **36** (1988) 3574.
115. T.-C. Chou, Z. Arlin, B. D. Clarkson and F. Philips, *Cancer Res.* **37** (1977) 3561.
116. Y. Rustum and H. D. Preisler, *Cancer Res.* **39** (1979) 42.
117. A. Rosowsky, S.-H. Kim, J. Ross and M. M. Wick, *J. Med. Chem.* **25** (1982) 171.
118. L. L. Bennett Jr., R. W. Brockman and J. A. Montgomery, *Nucleosides Nucleotides* **5** (1986) 117.
119. C. I. Hong, A. J. Kirisits, A. Nechaev, D. J. Buchheit and C. R. West, *J. Med. Chem.* **28** (1985) 171.
120. M. Yoshikawa, T. Kato and T. Takenishi, *Bull. Chem. Soc. Jpn.* **42** (1969) 3505.
121. C. I. Hong, A. Nechaev and C. R. West, *Biochem. Biophys. Res. Commun.* **88** (1979) 1223.
122. C. I. Hong, A. Nechaev and C. R. West, *J. Med. Chem.* **22** (1979) 1428.
123. E. S. Henderson and R. J. Samaha, *Cancer Res.* **29** (1969) 2272.
124. F. Rosner, Y. Hirshaut, H. W. Grunwald and M. Dietrich, *Cancer Res.* **35** (1975) 700.
125. C. I. Hong, A. Nechaev, A. J. Kirisits, D. J. Buchheit and C. R. West, *J. Med. Chem.* **23** (1980) 1343.
126. C. I. Hong, A. Nechaev, A. J. Kirisits, D. J. Buchheit and C. R. West, *Eur. J. Cancer Clin. Oncol.* **19** (1983) 1105.
127. C. I. Hong, A. Nechaev and C. R. West, *Biochem. Biophys. Res. Commun.* **94** (1980) 1169.
128. M. MacCoss, E. K. Ryu and T. Matsushita, *Biochem. Biophys. Res. Commun.* **85** (1978) 714.
129. E. K. Ryu, R. J. Ross, T. Matsushita, M. MacCoss, C. I. Hong and C. R. West, *J. Med. Chem.* **25** (1982) 1322.
130. S. Shuto, H. Itoh, S. Ueda, S. Imamura, K. Fukukawa, M. Tsujino, A. Matsuda and T. Ueda, *Chem. Pharm. Bull.* **36** (1988) 209.
131. C. R. H. Raetz, M. Y. Chu, S. P. Srivastava and J. G. Turcotte, *Science* **196** (1977) 303.
132. J. G. Turcotte, S. P. Srivastava, W. A. Meresak, B. A. Riskalla, F. Louzon and T. P. Wunz, *Biochim. Biophys. Acta* **619** (1980) 604.

133. J. G. Turcotte, S. P. Srivastava, J. M. Steim, P. Calabresi, L. M. Tibbetts and M. Y. Chu, *Biochim. Biophys. Acta* **619** (1980) 619.
134. C. I. Hong, S.-H. An. D. J. Buchheit, A. Nechaev, A. J. Kirisits, C. R. West, E. K. Ryu and M. MacCoss, *Cancer Drug Delivery* **1** (1984) 181.
135. C. I. Hong, S.-H. An, L. Schliselfeld, D. J. Buchheit, A. Nechaev, A. J. Kirisits and C. R. West, *J. Med. Chem.* **31** (1988) 1793.
136. T. Matsushita, E. K. Ryu, C. I. Hong and M. MacCoss, *Cancer Res.* **41** (1981) 2707.
137. M. MacCoss, J. J. Edwards, T. M. Seed and S. P. Spragg, *Biochim. Biophys. Acta* **719** (1982) 544.
138. M. MacCoss, J. J. Edwards, P. Lagocki and Y.-E. Rahman, *Biochem. Biophys. Res. Commun.* **116** (1983) 368.
139. C. I. Hong, S.-H. An, D. J. Buchheit, A. Nechaev, A. J. Kirisits, C. R. West and W. E. Berdel, *J. Med. Chem.* **29** (1986) 2038.
140. C. I. Hong, A. J. Kirisits, D. J. Buchheit, A. Nechaev and C. R. West, *Cancer Drug Delivery* **3** (1986) 101.
141. W. E. Berdel, S. Danhauser, H. D. Schick, C. I. Hong, C. R. West, M. Fromm, U. Fink, A. Reichert and J. Rastetter, *Lipids* **22** (1987) 943.
142. W. E. Berdel, S. Danhauser, C. I. Hong, H. D. Schick, A. Reichert, R. Busch, J. Rastetter and W. R. Vogler, *Cancer Res.* **48** (1988) 826.
143. D. H. W. Ho, *Drug Metabol. Dispos.* **1** (1973) 752.
144. J. M. Venditti, M. C. Baratta, N. H. Greenberg, B. J. Abbott and I. Kline, *Cancer Chemother. Rep.* **56** (1972) 483.
145. R. Tokuzen, K. Kuretani and W. Nakahara, *Gann* **65** (1974) 89.
146. H. Hirayama, K. Sugihara, K. Wakigawa, J. Hikita, T. Sugihara, T. Kubota, H. Ohkuma and K. Oguro, *Oyo Yakuri* **6** (1972) 1259.
147. W. H. Schrier, R. H. Hayashikawa and J. Nagyvary, *Biochem. Pharmacol.* **26** (1977) 2375.
148. R. A. Minow, J. A. Gottlieb, R. M. O'Bryan and B. Hoogstraten, *Cancer Treat. Rep.* **59** (1976) 219.
149. T. F. Burks, T. L. Loo and M. N. Grubb, *Proc. Soc. Exp. Biol. Med.* **159** (1978) 374.
150. J. G. Moffatt, E. K. Hamamura, A. F. Russell, M. Prystasz, J. P. H. Verheyden, K. Yamaguchi, N. Uchida, K. Sato, S. Nishiyama, O. Shiratori and K. Katagiri, *Cancer Chemother. Rep.* **58** (1974) 451.
151. K. Kondo, T. Nagura, Y. Arai and I. Inoue, *J. Med. Chem.* **22** (1979) 639.
152. K. Kondo and I. Inoue, *J. Org. Chem.* **42** (1977) 2809.
153. T. A. Khwaja, L. J. Kigwana and A. M. Mian, *Cancer Res.* **39** (1979) 3129.
154. I. L. Doerr and J. J. Fox, *J. Org. Chem.* **32** (1967) 1462.
155. G. D. Birnie, H. Kroeger and C. Heidelberger, *Biochemistry* **2** (1963) 566.
156. N. G. Blokhina, E. K. Vozny and A. M. Garin, *Cancer* **30** (1972) 390.
157. S. W. Hall, M. Valdivieso and R. S. Benjamin, *Cancer Treat. Rep.* **61** (1977) 1495.
158. T. Buroker, F. Padilla, C. Groppe, G. Guy, J. Quagliana, J. McCracken, V. K. Vaitkevicius, B. Hoogstraten and L. Heilbrun, *Cancer* **44** (1979) 48.
159. J. A. Benvenuto, K. Lu, S. W. Hall, R. S. Benjamin and T. L. Loo, *Cancer Res.* **38** (1978) 3867.
160. G. N. Hortobagyi, G. R. Blumenschein, C. K. Tashima, A. U. Buzdar, M. A. Burgess, R. B. Livingston, M. Valdivieso, J. U. Gutterman, E. M. Hersh and G. P. Bodey, *Cancer* **44** (1979) 398.
161. H. Joide, H. Akiyoshi, Y. Minato, H. Okuda and S. Fujii, *Gann* **68** (1977) 553.
162. J. L. Au and W. Sadee, *Cancer Res.* **40** (1980) 2814.
163. J. P. Horwitz, J. J. McCormick, K. D. Philips, V. M. Maher, J. R. Otto, D. Kessel and J. Zemlicka, *Cancer Res.* **35** (1975) 1301.
164. M. Yasumoto, A. Moriyama, N. Unemi, S. Hashimoto and T. Suzue, *J. Med. Chem.* **20** (1977) 1592.
165. S. A. Hiller, R. A. Zhuk and M. Y. Lidak, *Dokl. Acad. Nauk. USSR* **176** (1967) 332.
166. M. Yasumoto, I. Yamawaki, T. Marunaka and S. Hashimoto, *J. Med. Chem.* **21** (1978) 738.
167. R. A. Earl and L. B. Townsend, *J. Heterocycl. Chem.* **9** (1972) 1141.
168. S. Fujii, Y. Nakamura, S. Takeda, K. Morita, T. Sato, T. Marunaka, Y. Kawaguchi and N. Unemi, *Gann* **71** (1980) 30.
169. Y. Koyama, *Cancer Treat. Rev.* **8** (1981) 147.
170. S. Ozaki, Y. Ike, H. Mizuno, K. Ishikawa and H. Mori, *Bull Chem. Soc. Jpn.* **50** (1977) 2406.

171. M. Iigo, A. Hoshi, A. Nakamura and K. Kuretani, *Cancer Chemother. Pharmacol.* **1** (1978) 203.
172. T. Kobari, Y. Iguro, A. Ujiie and H. Namekawa, *Xenobiotica* **11** (1981) 57.
173. M. Iigo, K. Kuretani and A. Hoshi, *J. Natl. Cancer Inst.* **66** (1981) 345.
174. H. Ishitsuka, M. Miwa, K. Takemoto, K. Fukuoka, A. Itoga and H. B. Maruyama, *Gann* **71** (1980) 112.
175. R. D. Armstrong and R. B. Diasio, *Cancer Res.* **40** (1980) 3333.
176. S. Suzuki, Y. Hongu, H. Fukazawa, S. Ichihara and H. Shimizu, *Gann* **71** (1980) 238.
177. A. F. Cook, M. J. Holman, M. J. Kramer and P. W. Trown, *J. Med. Chem.* **22** (1979) 1330.
178. H. Hrebabecky and J. Beranek, *Collect. Czech. Chem. Commun.* **43** (1978) 3268.
179. S. Ajmera and P. V. Danenberg, *J. Med. Chem.* **25** (1982) 999.
180. S. Ajmera, A. R. Bapat, E. Stephanian and P. V. Danenberg, *J. Med. Chem.* **31** (1988) 1094.
181. T. A. Khwaja and C. Heidelberger, *J. Med. Chem.* **10** (1967) 1066.
182. R. J. Kent and C. Heidelberger, *Biochem. Pharmacol.* **19** (1970) 1095.
183. S. W. Anderson and C. Heidelberger, *Biochem. Pharmacol.* **14** (1965) 1605.
184. T. Kawaguchi, S. Fukushima, Y. Hayashi, M. Kaneko and M. Nakano, *Cancer Res.* **48** (1988) 4179.
185. D. C. Remy, A. V. Sunthankar and C. Heidelberger, *J. Org. Chem.* **27** (1962) 2491.
186. J. A. Montgomery, H. J. Thomas and H. J. Schaeffer, *J. Org. Chem.* **26** (1961) 1929.
187. D. G. Parsons and C. Heidelberger, *J. Med. Chem.* **9** (1966) 159.
188. J. A. Montgomery and H. J. Thomas, *J. Med. Chem.* **10** (1967) 1163.
189. D. Farquhar, D. N. Srivastava, N. J. Kuttesch and P. P. Saunders, *J. Pharm. Sci.* **72** (1983) 324.
190. R. N. Hunston, M. Jehangir, A. S. Jones and R. T. Walker, *Tetrahedron* **36** (1980) 2337.
191. D. Farquhar, N. J. Kuttesch, M. G. Wilkerson and T. Winkler, *J. Med. Chem.* **26** (1983) 1153.
192. D. L. Hill, W. R. Laster and R. F. Struck, *Cancer Res.* **32** (1972) 658.
193. M. Brock, *Cancer Treat. Res.* **60** (1976) 301.
194. S.-H. An, C. R. West and C. I. Hong, *Steroids* **47** (1986) 413.
195. T. Chou, P. Vidal and F. S. Philips, *Cancer Treat. Rep.* **61** (1977) 617.
196. M. R. Dollinger, J. H. Burchenal, W. Kreis and J. J. Fox, *Biochem. Pharmacol.* **16** (1967) 689.
197. J. J. Fox, N. Miller and I. Wempen, *J. Med. Chem.* **9** (1966) 101.
198. J. J. Fox, E. A. Falco, I. Wempen, D. Pomeroy, M. D. Dowling and J. H. Burchenal, *Cancer Res.* **32** (1972) 2269.
199. M. Yoshida, A. Hoshi, K. Kuretani, T. Kanai and M. Ichino, *Gann* **66** (1975) 561.

10 Purines and purine nucleoside analogues as antitumour agents

R. K. ROBINS AND G. D. KINI

10.1 Introduction

Purines, purine nucleosides and their analogues have been extremely useful as anticancer agents. Robins[1], in 1964, reviewed the antitumour activity of purines and purine nucleosides from a structure-activity standpoint. Subsequently, several reviews on this subject have been published[2-7]. The chemotherapeutic uses of purines and purine analogues have prompted tremendous synthetic efforts towards them, both in academia and in the pharmaceutical industry. Concurrently, studies on mechanism of action, metabolites, etc., of purines and purine nucleosides have also been developed significantly. While several fine reviews on the antitumour activity of purines and purine analogues have appeared earlier, this chapter is an attempt to focus on the synthesis and chemistry of purines, purine nucleosides and their analogues that have shown clinical promise as well as those currently in use clinically as antitumour agents.

10.2 Purine analogues

6-Mercaptopurine (2) was the first purine to be used in cancer chemotherapy in 1954. Marketed by Burroughs Wellcome Co. under the trade name Purinethol, the synthesis of 6-mercaptopurine was first reported by Elion and Hitchings[8]. This synthesis was accomplished by treatment of hypoxanthine (1) with phosphorus pentasulphide in tetralin at elevated temperature.

Another purine analogue, 6-thioguanine (4), is frequently used in the treatment of childhood acute myeloid leukaemia (AML)[9]. The synthesis of 6-thioguanine was also reported by Elion and Hitchings[10] under conditions similar to those for the synthesis of 6-mercaptopurine. Treatment of guanine (3) with phosphorus pentasulphide in refluxing pyridine afforded 6-thioguanine.

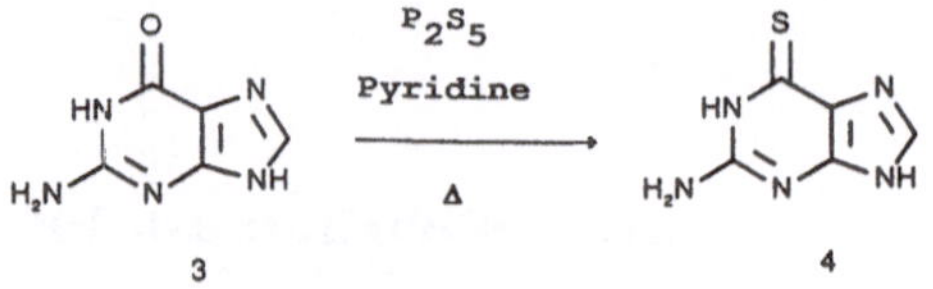

Both 6-mercaptopurine and 6-thioguanine have been studied from a mechanism of activity standpoint[11].

Leopold et al.[12] have recently reported the completion of preclinical development of 3-deazaguanine (8), a purine analogue, as an antitumour agent. Khwaja and co-workers[13,14] have reported in vivo studies on the action of 3-deazaguanine against a variety of animal breast tumour models. 3-Deazaguanine is currently in clinical trials[15]. Synthetic studies of this analogue were reported by Robins and his colleagues[16,17]. Prior to these reports, syntheses in the imidazo[4,5-c]pyridine ring system[18-26] were generally accomplished by ring closure of appropriately substituted 3,4-diaminopyrimidines. This general method had been successfully used in the synthesis of such purine analogues as 3-deazaadenine[19], 3-deazahypoxanthine[19] and 3-deaza-6-mercaptopurine[20b]. Attempts to synthesise 3-deazaguanine by the method used for the synthesis of 3-deazaadenine and 3-deazahypoxanthine were unsuccessful[27]. Robins et al.[18] reported an alternative approach to the imidazo[4,5-c]pyridines via base catalysed cyclisation of methyl 4(5)-carbamoylmethylimidazole-5(4)-carboxylates.

A slightly modified procedure was used in the successful synthesis of 3-deazaguanine[16,17]. Methyl 5(4)-carbamoylmethylimidazole-4(5)-carboxylate (5) was dehydrated with phosphorus oxychloride to methyl 5(4)-cyanomethylimidazole-4(5)-carboxylate (6) which on treatment with liquid ammonia cyclised to 3-deazaguanine (8) (Scheme 10.1).

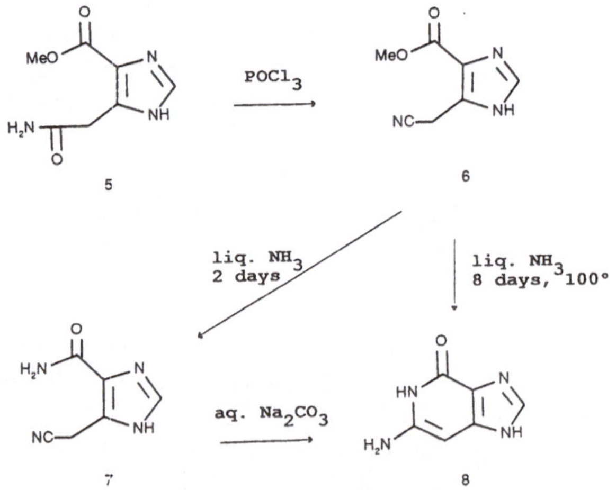

Scheme 10.1

Alternatively, interruption of the reaction with liquid ammonia gave the intermediate carboxamide (7), which subsequently cyclised to 3-deazaguanine upon treatment with aqueous Na_2CO_3. This method provided the compound in good yield. A modification of this procedure also led to the synthesis of the corresponding nucleoside, 3-deazaguanosine.

Srivastava and Robins[28] reported an improved synthesis of 3-deazaguanine subsequently using an intermediate already described in the previous synthesis, but with a novel ring closure. Treatment of methyl 5(4)-cyanomethylimidazole-4(5)-carboxylate (6) with hydroxylamine in refluxing ethanol led to the corresponding amidoxime (9), which on hydrogenation with Raney nickel yielded 3-deazaguanine via the amidine intermediate (10) as shown in the scheme. The advantages of this procedure are better yields and avoidance of the use of a high pressure vessel (Scheme 10.2).

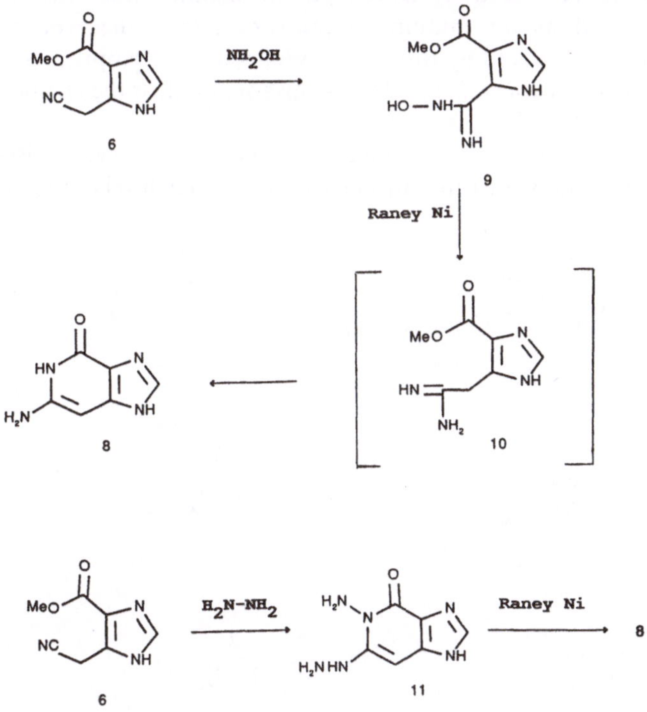

Scheme 10.2

Revankar et al.[29] reported yet another route to 3-deazaguanine. Reaction of methyl 5(4)-cyanomethylimidazole-4(5)-carboxylate (6) with hydrazine hydrate in ethanol at reflux temperature produced 5-amino-6-hydrazino-1,5-dihydroimidazo[4,5-c]pyridin-4-one (11) in yields in excess of 90%. Hydrogenation of (11) with Raney nickel catalyst afforded 3-deazaguanine (8) in 80% yield. This procedure, in addition to not using a high pressure vessel, provides the target compound in yields better than the previously discussed methods.

A formulation of 3-deazaguanine used in the treatment of cancer is its methanesulphonic acid salt, deazaguanine mesylate[30].

Two purine antimetabolites, azathioprine[31] and allopurinol[32], although not antitumour agents by themselves, are in clinical use as adjuvants for cancer chemotherapy.

12   13

Azathioprine (12), sold by Burroughs Wellcome under the trade name Imuran, is used as an immunosuppressant after cancer surgery, while allopurinol (13), also sold by Burroughs Wellcome as Zyloprim, is used in the prevention of hyperuricaemia and uricosuria that follow the chemotherapy of leukaemias and lymphomas.

A number of other purine analogues have been synthesised in our laboratory and show antitumour activity at various levels of testing.

14   15

16   17

18   19

8-[(2-Hydroxyethyl)thioazo]hypoxanthine (14), synthesised in 1961[33], 2-aminopurine-6,8-dithione (15)[1], 2-amino-6-chloropurine (16)[34], purine-6-N-methylsulphonamide (17)[35], purine-6-sulphonamide (18)[35] and 9-[bis(β-chloroethyl)aminoethyl]adenine (19)[36] are some other purine analogues of interest as potentially useful antitumour agents. Their antitumour activity in animals has recently been reviewed in detail by Robins and Revankar[37].

10.3   Purine nucleosides

6-Methylmercaptopurine riboside (**21**) is a purine nucleoside analogue that was found to be useful in systems resistant to 6-mercaptopurine[38,39]. Hampton *et al.*[40] reported the first synthesis of 6-methylmercaptopurine riboside by condensing 6-methylmercaptopurine (**20**)[8] with 2,3,5-tri-*O*-acetyl-D-ribo-furanosyl chloride using mercuric chloride as catalyst, followed by deacetylation (Scheme 10.3).

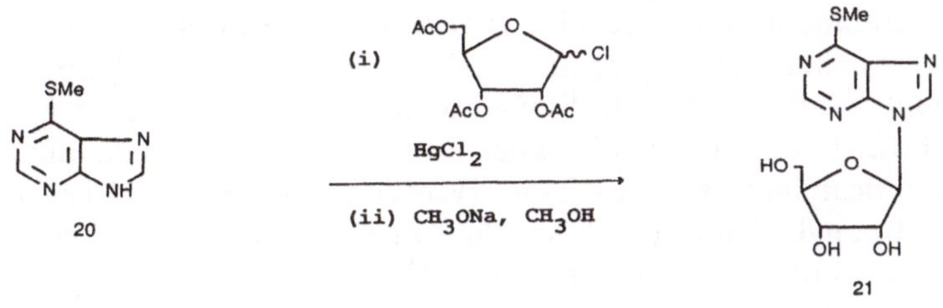

Scheme 10.3

Fox *et al.*[41] subsequently reported the synthesis of 6-methylmercaptopurine riboside by direct methylation of 6-thioinosine. Inosine (**22**) was converted to its tri-*O*-benzoyl derivative (**23**) by treatment with benzoyl chloride in pyridine. Treatment with phosphorus pentasulphide in pyridine resulted in 2′,3′,5′-*O*-benzoyl-6-thioinosine (**24**), which was deblocked with sodium methoxide and subsequently methylated under basic conditions to yield 6-methylmercaptopurine riboside (**21**) (Scheme 10.4).

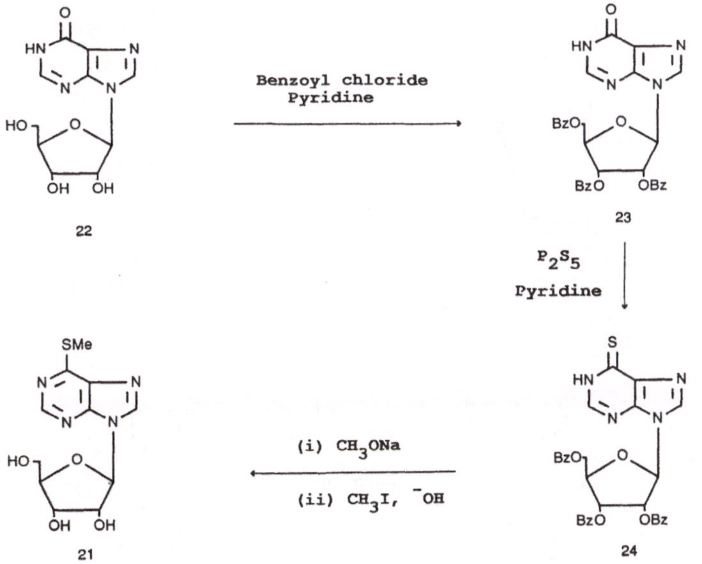

Scheme 10.4

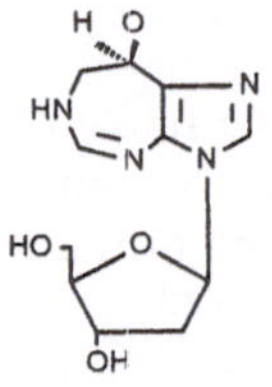

25

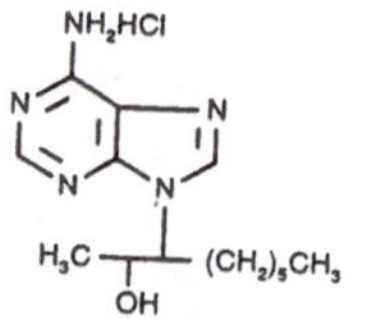

26

9-β-D-Arabinofuranosyl adenine (ara-A) (31) is a purine nucleoside that exhibits antiviral activity against DNA viruses of the herpes group. Suhadolnik[42] has reviewed the biological activity of ara-A. Ara-A is curative against L 1210 leukaemia in mice when used in combination with inhibitors of adenosine deaminase such as 2'-deoxycoformycin (25)[43] and *erythro*-9-(2-hydroxy-3-nonyl)adenine (EHNA) (26). Ara-A has been studied clinically in combination with 2'-deoxycoformycin[44,45].

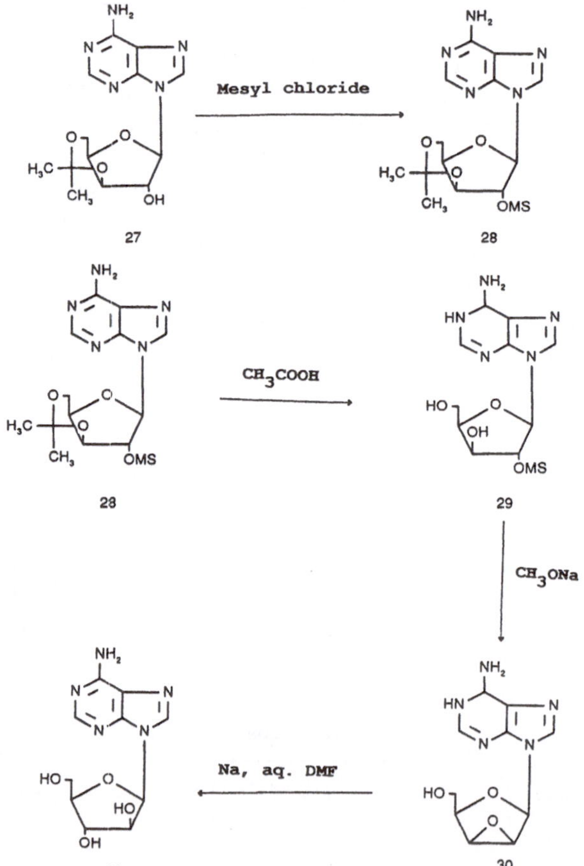

Scheme 10.5

Synthetic studies on ara-A were first reported by Reist *et al.*[46] Their synthesis of ara-A was accomplished by modfication of the nucleoside, 9-(β-D-xylofuranosyl)adenine[47], as shown in Scheme 10.5.

The 3',5'-di-*O*-isopropylidene derivative of 9-(β-D-xylofuranosyl)adenine (**27**) was converted to the corresponding 2'-*O*-mesyl derivative (**28**) by treatment with methanesulphonyl chloride. Acidolytic cleavage with acetic acid resulted in 9-(2'-*O*-mesyl-β-D-xylofuranosyl)adenine (**29**), which on treatment with sodium methoxide in methanol afforded the 2',3'-epoxide (**30**). Ring opening of the epoxide (**30**) with sodium in aqueous dimethylformamide resulted in ara-A (**31**), which was purified by crystallisation.

Glaudemans and Fletcher[48] reported the synthesis of ara-A by condens-ation of 2,3,5-tri-*O*-benzyl-D-arabinofuranosyl chloride (**32**) with N^6-benzoyladenine (**33**). Treatment with barium methoxide to remove the *N*-benzoyl group and subsequent hydrogenolysis to effect the removal of the benzyl protecting groups resulted in ara-A (Scheme 10.6). More recently,

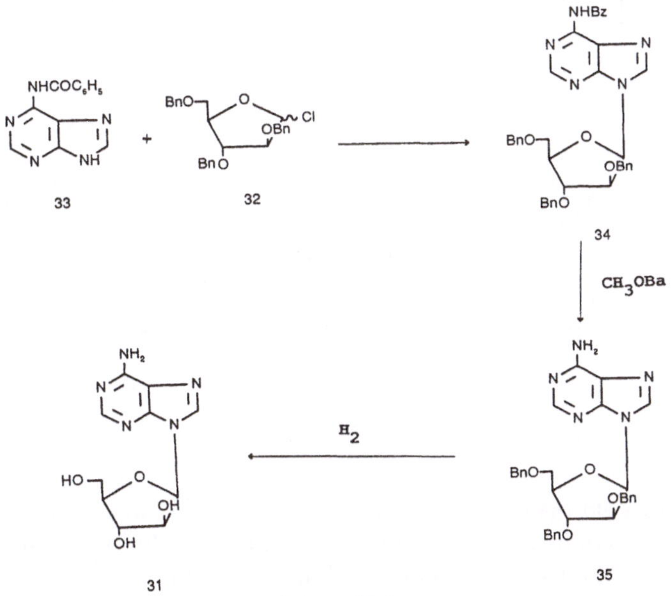

Scheme 10.6

Ranganathan[49] reported a novel synthesis of ara-A via a thionoxazolidine derivative of arabinofuranose (**36**) (Scheme 10.7).

Treatment of arabinofuranothionoxazolidine (**36**)[50] with sodium hydride in dimethylformamide solution followed by the addition of mercuric bromide gave the bromo-mercury intermediate (**37**), which was subsequently reacted with 4-amino-6-chloro-5-nitropyrimidine (**38**), leading to the intermediate (**39**). Reduction of the 5-nitro group of (**39**) to the 5-amino group was accomplished by catalytic hydrogenation over Pd/C or with aluminium

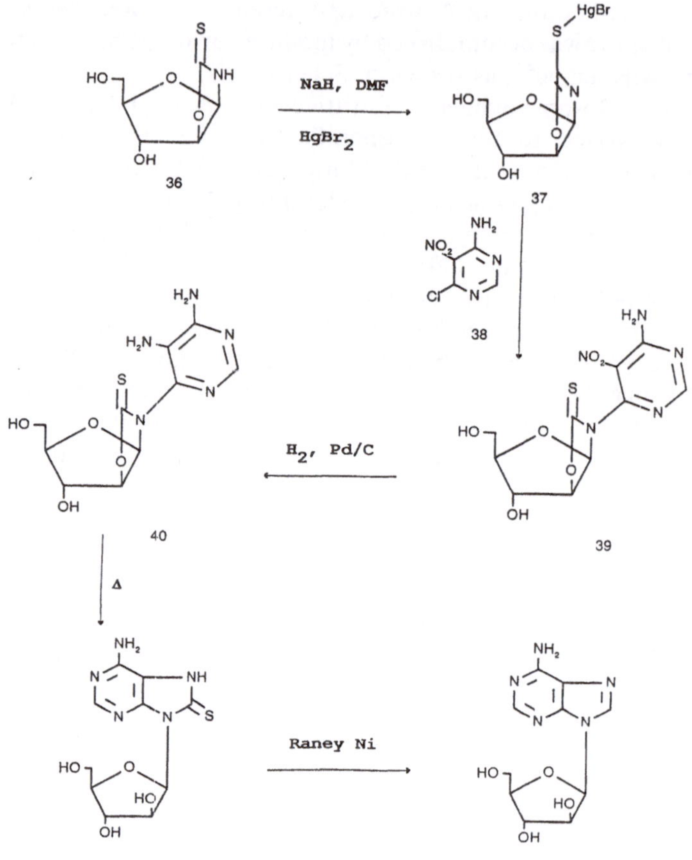

Scheme 10.7

amalgam to yield the diaminopyrimidine (**40**). Cyclisation to 8-mercapto-ara-A (**41**) was accomplished by heating an aqueous solution of the diamino intermediate (**40**). Raney nickel desulphurisation of (**41**) resulted in ara-A.

Robins *et al.*[51] reported the synthesis of ara-A monophosphate and ara-A 3′,5′-cyclic phosphate.

9-β-D-Arabinofuranosyl-2-fluoroadenine (2-fluoro-ara-A) is an adenosine analogue resistant to deamination by adenosine deaminase[52]. Consequently, it is active by itself against several ara-A resistant tumours. Animal studies with 2-fluoro-ara-A have been reported[53,54]. 2-Fluoro-ara-A monophosphate, named fludarabine phosphate, is under clinical investigation for non-Hodgkin's lymphoma[55].

Montgomery and Hewson[56] reported synthetic approaches to 2-fluoro-ara-A via a five-step sequence starting with 2,6-dichloropurine. In 1979,

Montgomery et al.[57] reported an improved synthesis of 2-fluoro-ara-A from 2,6-diaminopurine.

2,6-Diaminopurine (42)[58] was acetylated with acetic anhydride in refluxing pyridine to 2,6-diacetamidopurine (43). Condensation of (43) with 2,3,5-tri-O-benzyl-α-D-arabinofuranosyl chloride (44) resulted in the blocked nucleoside (45) as the β-anomer. Deacetylation of (45) was accomplished by treatment with methylamine in ethanol to yield the diamino nucleoside (46). Treatment of (46), in a mixture of fluoboric acid and tetrahydrofuran, with aqueous sodium nitrite led to 2′,3′,5′-tri-O-benzyl-β-D-arabinofuranosyl-2-fluoroadenine (47) which was subsequently deprotected with boron trichloride to 2-fluoro-ara-A (48) (Scheme 10.8).

Montgomery et al.[59] have more recently reported an improvement of this procedure, with essentially the same chemistry, for the synthesis of 2-fluoro-ara-A.

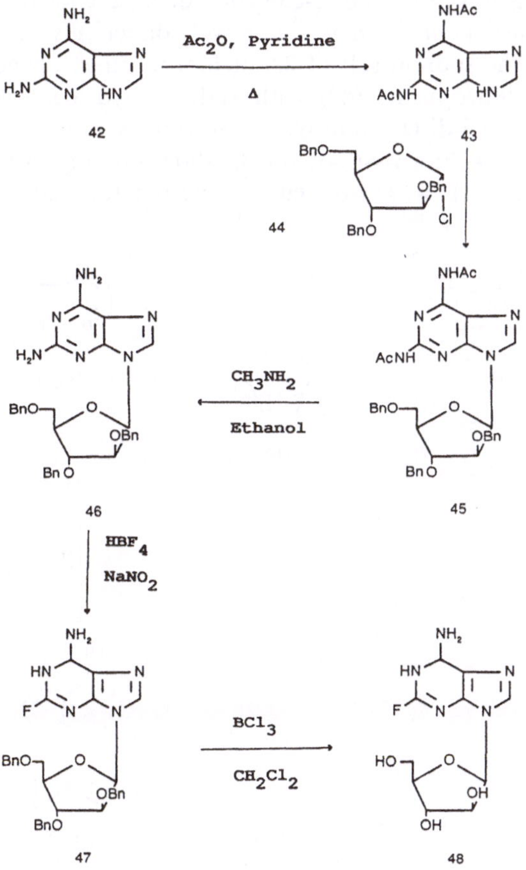

Scheme 10.8

2-Chloro-2'-deoxyadenosine is a nucleoside currently in clinical trials against human leukaemia. Two reports of its antileukaemic activity have been published by Carson and colleagues[60,61]. It is reported that 2-chloro-2'-deoxyadenosine is resistant to adenosine deaminase[62] and has been shown to be more active than the corresponding 2-fluoro and 2-bromo derivatives against L 1210 leukaemia in mice[63].

Several reports of synthetic studies leading to 2-chloro-2'-deoxyadenosine have appeared in the literature. Christensen *et al.*[64] reported that fusion of appropriately protected 2'-deoxyribose with 2,6-dichloropurine gave a mixture of α- and β-anomers, which was subsequently aminated to a mixture of α- and β-anomers of 2-chloro-2-deoxyadenosine. Montgomery and Hewson[56] also obtained a mixture of α- and β-anomers upon fusion of 2,6-dichloropurine with 1,3,5-tri-*O*-acetyl-2-deoxy-D-*erythro*-pentofuranose. Using a similar procedure, Robins and Robins[65] synthesised the α anomer of 2-chloro-2-deoxyadenosine.

The most elegant synthesis of 2-chloro-2'-deoxyadenosine was reported in 1984 by Robins *et al.*[66] using a novel direct glycosylation method (Scheme 10.9). The sodium salt of 2,6-dichloropurine[67], generated by treatment of 2,6-dichloropurine (**49**) with sodium hydride, was reacted with 1-chloro-2-deoxy-3,5-di-*O*-*p*-toluoyl-α-D-*erythro*-pentofuranose (**50**)[68] in acetonitrile. 2,6-Dichloro-9-(2-deoxy-3,5-di-*O*-*p*-toluoyl-β-D-*erythro*-pentofuranosyl)purine (**51**) was isolated by chromatography in 59% yield.

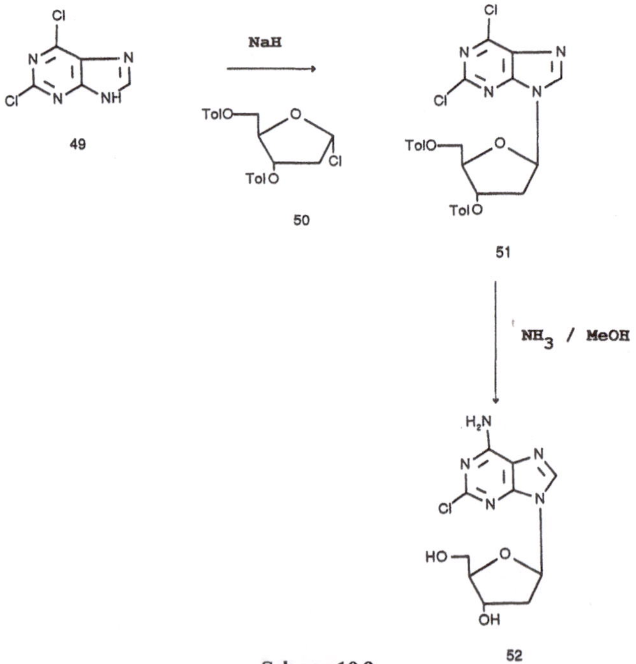

Scheme 10.9

Amination with methanolic ammonia resulted in 2-chloro-2'-deoxyadenosine (52).

10.4 Purine cyclic nucleotides

In 1982, Revankar and Robins[69] published an extensive review on the chemistry and synthesis of cyclic nucleotides and cyclic nucleotide analogues. Meyer and Miller[70] reviewed general synthetic approaches towards analogues of cyclic AMP and cyclic GMP.

8-Chloro-cAMP is of considerable interest as an antitumour agent, and several recent reports on the antitumour activity of 8-chloro-cAMP have appeared[71-74] in the literature. Avery and colleagues[75-77] have recently reported *in vivo* antitumour studies with 8-chloro-cAMP. 8-Chloro-cAMP is under preclinical evaluation at the National Cancer Institute. Muneyama and coworkers[78,79] reported the synthesis of 8-chloro-cAMP.

Bromination of cAMP (53) with bromine in the presence of aqueous sodium hydroxide resulted in 8-bromo-cAMP (54). Treatment of (54) with thiourea at elevated temperature resulted in 8-thio-cAMP (55). Chlorination of (55) with chlorine gas in a solution of methanolic HCl produced 8-chloro-cAMP (56) (Scheme 10.10).

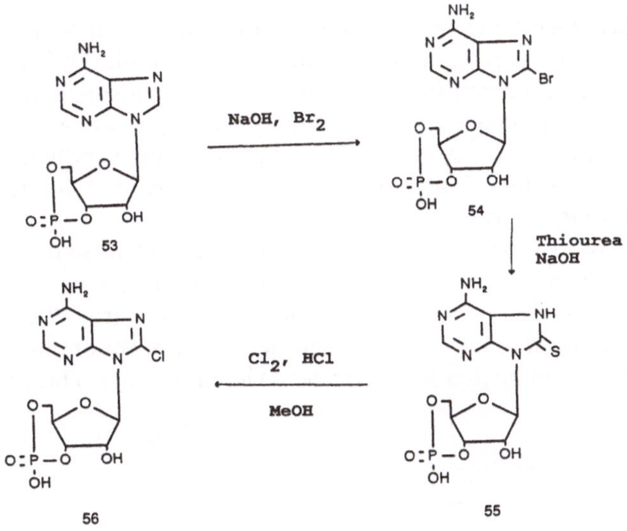

Scheme 10.10

10.5 Certain other nucleosides biochemically related to purines

10.5.1 *Tiazofurin*

Tiazofurin (2-β-D-ribofuranosyl-thiazole-4-carboxamide) is a nucleoside of considerable current clinical interest as an antitumour agent[80]. Robins *et al.*[81]

reported the chemical synthesis in 1977. This synthesis was accomplished as shown in Scheme 10.11.

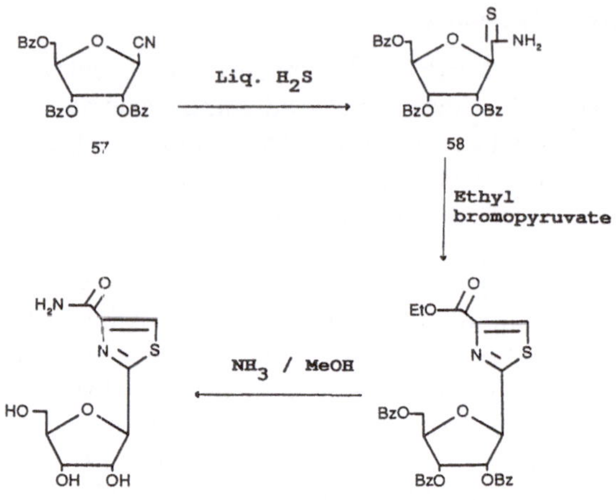

Scheme 10.11

2,5-Anhydro-3,4,6-tri-O-benzoyl-D-allonothioamide (**58**) was synthesised by treatment of the corresponding nitrile (**57**) with liquid H_2S, a procedure superior to those reported in the literature[82]. Ring closure[83] by treatment with ethylbromopyruvate led to the ethyl 2-(2,3,5-tri-O-benzoyl-β-D-ribofuranosyl)thiazole-4-carboxylate (**59**) as the major product, with small amounts of the α isomer. The major product was separated and treated with methanolic ammonia to give tiazofurin (**60**).

Subsequently, large scale synthesis of tiazofurin, using essentially the same chemistry, has been reported[84]. Robins et al.[85] have also recently reported a new synthesis of tiazofurin. Mechanistic studies towards the activity of tiazofurin by Kuttan et al.[86] have shown that it is phosphorylated to the 5'-monophosphate and further forms the NAD analogue (**61**). This dinucleotide

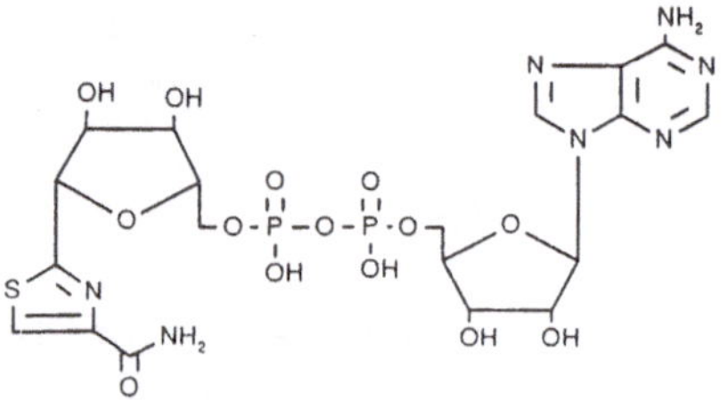

61

(**61**) (acronymed TAD, tiazofurin adenosine dinucleotide), is a potent inhibitor of IMP dehydrogenase. Weber[87] has discussed the significance of IMP dehydrogenase to cancer chemotherapy. Robins and Revankar[37] have discussed tiazofurin's antitumour activity in detail.

10.5.2 Selenazofurin

The seleno analogue of tiazofurin, selenazofurin (64), is also of considerable interest as an antitumour agent. It was synthesised by Srivastava and Robins[88] under conditions strikingly similar to those used in the synthesis of tiazofurin (Scheme 10.12). Treatment of 2,5-anhydro-3,4,6-tri-O-benzoyl-D-allononitrile (57) with liquid H_2Se led to the corresponding selenamide (62). Ring closure with ethylbromopyruvate led to the ethyl 2-(2,3,5-tri-O-benzoyl-β-D-ribofuranosyl)selenazole-4-carboxylate (63). Methanolic ammonia treatment of (63) yielded selenazofurin (64).

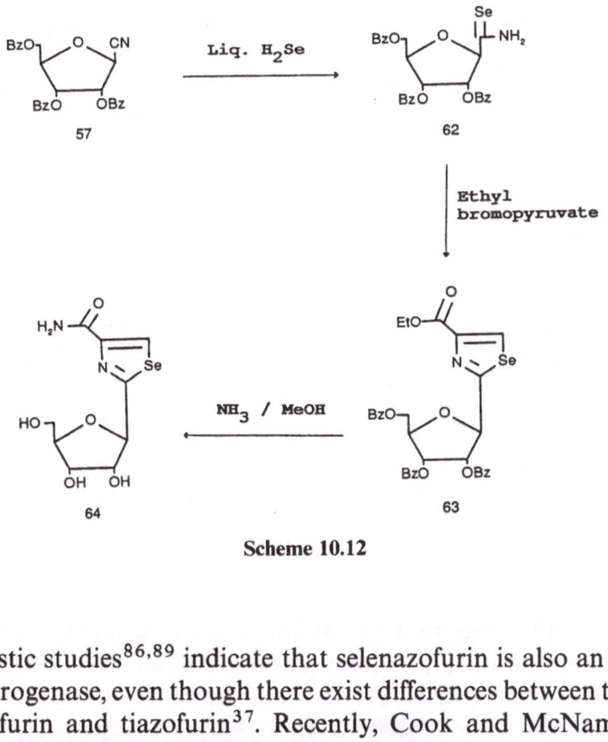

Scheme 10.12

Mechanistic studies[86,89] indicate that selenazofurin is also an inhibitor of IMP dehydrogenase, even though there exist differences between the activities of selenazofurin and tiazofurin[37]. Recently, Cook and McNamara[90] have published a large scale synthesis of selenazofurin.

X-ray crystallographic studies have been performed on both tiazofurin[91] and selenazofurin[92]. These studies show that the sulphur atom in the thiazole ring and the selenium atom in the selenazole ring are both in close proximity to the oxygen atom of the ribofuranose ring. This interaction would probably favour the stability of the anticonformation.

10.5.3 Pyrazofurin

Pyrazofurin (3-(1-β-D-ribofuranosyl)-4-hydroxyprazole-5-carboxamide) is a naturally occurring carbon linked nucleoside that is of considerable interest because of its antitumour and antiviral properties. Formerly known as pyrazomycin, it was isolated from *Streptomyces candidus* in 1969 by Gerzon *et al.*[93] and by Williams and Hoehn[94] in 1974. Reviews on pyrazofurin have appeared in the literature[95-97], and it has been clinically evaluated as an antitumour agent[98,99].

Farkas *et al.*[100] reported the synthesis of pyrazofurin in 1972.

Treatment of the α-keto ester (**65**)[101] with the hydrazine derivative (**66**) resulted in the diazo intermediate (**67**). Heating (**67**) with acetic anhydride and sodium acetate led to the cyclised hydroxypyrazole nucleoside (**68**). Amination of (**68**) with methanolic ammonia followed by debenzylation gave pyrazofurin (**69**) (Scheme 10.13).

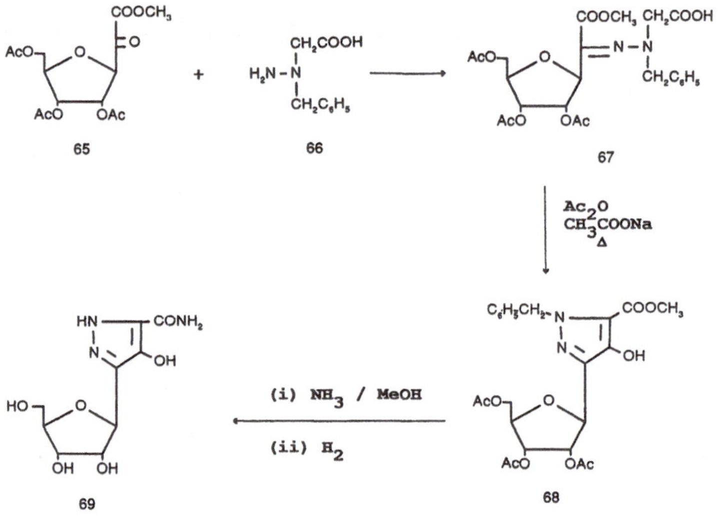

Scheme 10.13

Gutowski *et al.*[102] reported the isolation of the α-epimer of pyrazofurin from the same strain of *Streptomyces candidus* that afforded pyrazofurin. The α-form, also known as pyrazomycin B, was characterised by proton NMR, X-ray crystallography and ^{13}C-NMR[103].

De Bernardo and Weigele[104] reported a synthetic route to both the α- and β-epimers of pyrazofurin (Scheme 10.14). 2,3-Di-*O*-isopropylidene-D-ribofuranose (**70**), as a mixture of α and β anomers, was treated with *p*-nitrobenzoyl chloride to give a mixture of the corresponding *p*-nitrobenzoates (**71**). Treatment of this mixture with hydrogen bromide in dichloromethane resulted in the bromoriboside (**72**) as the β-anomer. Reaction

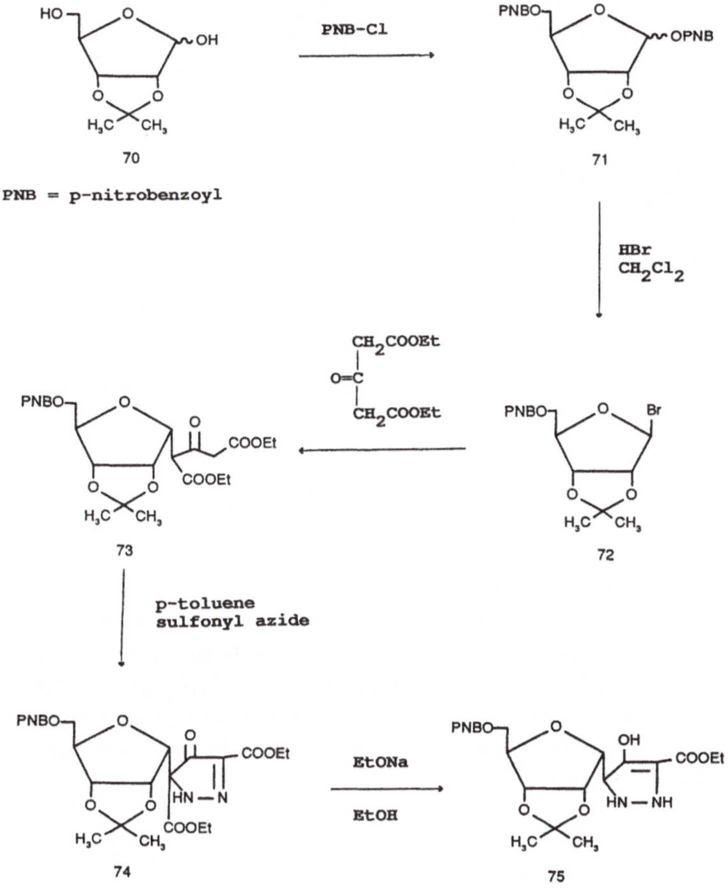

Scheme 10.14

of this bromide with diethyl acetonedicarboxylate as its potassium salt resulted in (73) as the major product. Treatment of (73) with p-toluene sulphonyl azide resulted in the cyclic intermediate (74) directly. Reaction of (74) with sodium ethoxide in ethanol produced the hydroxy ester (75) which on amination gave the α- and β-anomers of 2',3'-di-O-isopropylidene pyrazofurin (76 and 77) (Scheme 10.15). Interestingly, treatment of (75) with methanolic ammonia for 3 h at 100 °C resulted in the α-pyrazofurin as major product while increasing the reaction time to 12 h gave the β- as the sole product. Removal of the isopropylidene groups by treatment with trifluoroacetic acid resulted in α- and β-pyrazofurin.

Buchanan et al.[105] published an account of the biosynthesis of pyrazofurin. Buchanan et al.[106] also reported yet another synthesis of pyrazofurin. 4-Nitro-5-(2',3',5'-tri-O-acetyl-β-D-ribofuranosyl)pyrazole-3-carbonitrile (78)[107]

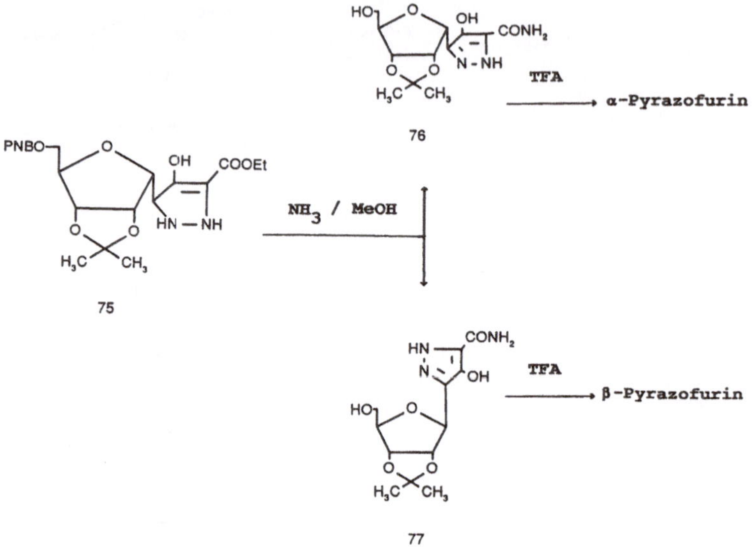

Scheme 10.15

was reduced to the corresponding 4-amino derivative (**79**) using the dithionite reduction procedure. Diazotisation with sodium nitrite followed by neutralisation resulted in the hydroxynitrile (**80**). Hydrolysis of the nitrile in glacial acetic acid with nickel acetate tetrahydrate gave the tri-*O*-acetyl pyrazofurin (**81**), deblocking of which with methanolic ammonia yielded pyrazofurin (Scheme 10.16).

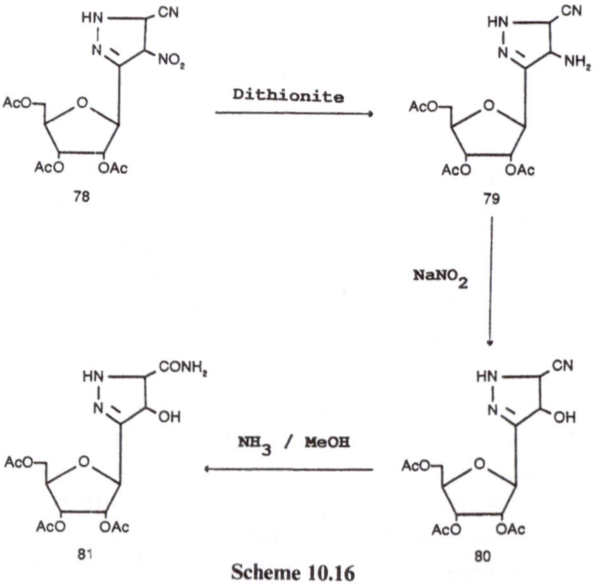

Scheme 10.16

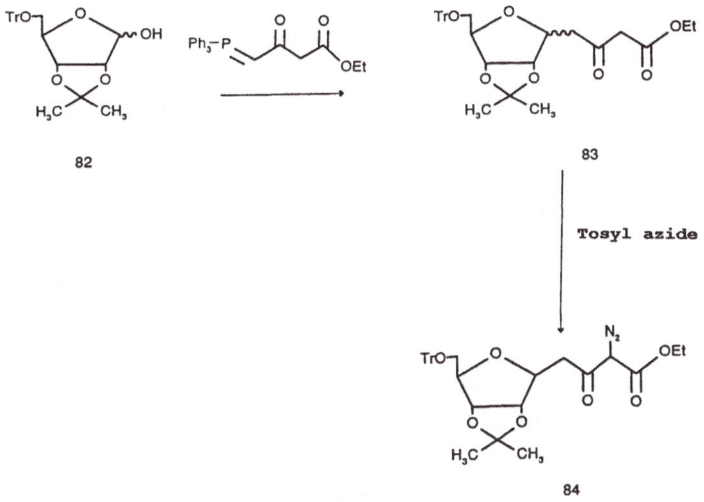

Scheme 10.17

Katagiri et al.[108] reported a novel synthesis of pyrazofurin from ribosyl-β-keto acid derivatives. The β-keto acid derivatives were synthesised by reaction of appropriately protected ribose with phosphoranes. Treatment of 2,3-di-O-isopropylidene-5-O-trityl-D-ribose (82)[109] with 3-ethoxycarbonyl-2-oxopropylidene-triphenylphosphorane[110] resulted in ethyl 4-(2′,3′-di-O-isopropylidene-5′-O-trityl-D-ribofuranosyl)-3-oxobutanoate (83). Reaction of (83) with tosyl azide and triethylamine resulted in the diazo intermediate (84) (Scheme 10.17). Cyclisation of (84) with sodium hydride in dimethoxyethane resulted in ethyl 4-hydroxy-3-(2′,3′-di-O-isopropylidene-5′-O-trityl-D-ribofuranosyl)pyrazole-5-carboxylate (85) as a mixture of α- and β-anomers with a β:α ratio of 2:1. The anomers were separated by chromatography and treated with methanolic ammonia to yield the amide (86). Deprotection of (86) with trifluoroacetic acid resulted in pyrazofurin (69) and the α-isomer (Scheme 10.18).

Herrera and Baelo[111] have recently reported a synthesis similar to the one described above along with an extensive proton NMR study of pyrazofurin.

10.5.4 2′-Deoxycoformycin (pentostatin)

Pentostatin has shown significant antitumour activity in combination with ara-A in vitro[112] and in vivo[113]. A recent report on clinical trials of pentostatin by itself against hairy cell leukaemia (HCL) has been published by Golomb et al.[114].

The total synthesis of pentostatin, (8R)-3-(2-deoxy-β-D-erythro-pentofuranosyl)-3,6,7,8-tetrahydroimidazo[4,5-d][1,3]-diazepin-8-ol (25), was reported by Chan et al.[115]. The synthesis was accomplished by glycosyl-

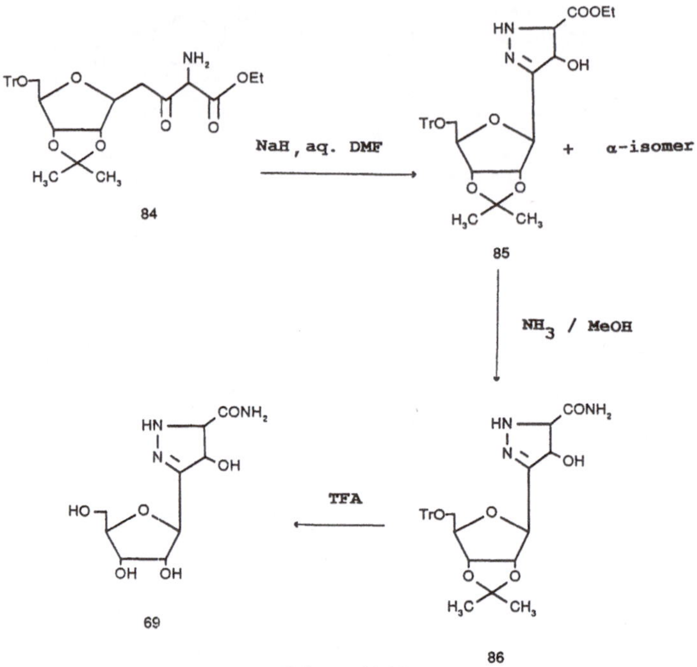

Scheme 10.18

ation of the aglycon, 6,7-dihydroimidazo[4,5-d][1,3]diazepin-8(3H)-one (**93**), with 2-deoxy-3,5-di-O-p-toluoyl-D-*erythro*-pentofuranosyl chloride (**50**). The synthesis of the aglycon (**93**) of pentostatin started with 4-methyl-5-nitro-1H-imidazole (**87**)[116].

5-Nitro-4-styrylimidazole (**88**) was synthesised from (**87**) by a reported

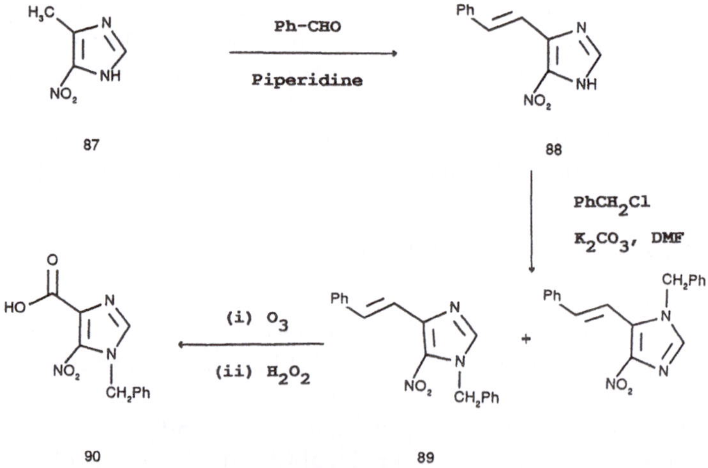

Scheme 10.19

procedure[116], with modifications. Treatment of (88) with benzyl chloride resulted in the *N*-benzylimidazole (89) and its other isomer in a 3:1 ratio. This mixture, upon ozonolysis followed by peroxide oxidation, resulted in 1-benzyl-5-nitro-1*H*-imidazole-4-carboxylic acid (90) and its other isomer

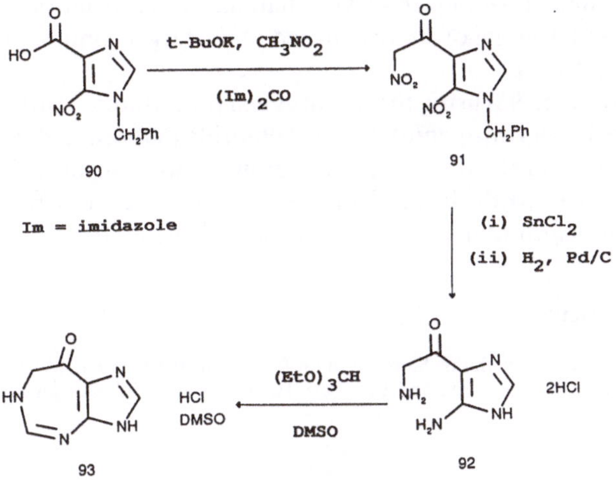

Scheme 10.20

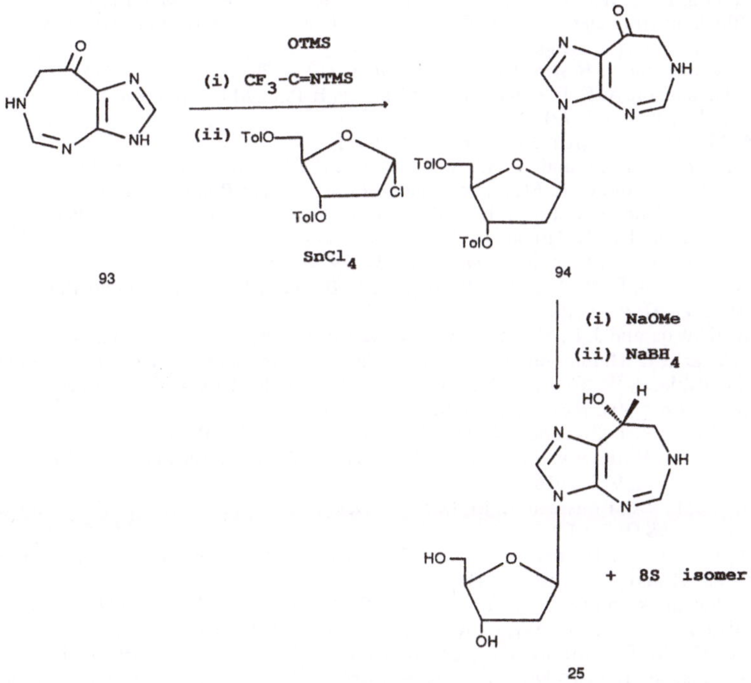

Scheme 10.21

(Scheme 10.19). The isomers were separated, and **(90)** was converted to 2-nitro-1-(1-benzyl-5-nitro-1H-imidazole-4-yl)ethanone **(91)** by a reported procedure[117].

Treatment of **(91)** with stannic chloride led to reduction of the nitro groups to amino groups; subsequent hydrogenolysis of the benzyl group resulted in 2-amino-1-(4-amino-1H-imidazol-4-yl)-ethanone dihydrochloride **(92)**. Treatment of **(92)** with triethyl orthoformate led to the target compound **(93)** in 81% yield (Scheme 10.20).

Glycosylation of **(93)** using the Vorbrüggen procedure[118] with 2-deoxy-3,5-di-O-p-toluoyl-D-erythropentofuranosyl chloride **(50)** resulted in the blocked nucleoside **(94)** and its α-anomer. The anomers were separated and treated with sodium methoxide in methanol, followed by sodium borohydride to produce pentostatin **(25)** and the (8S) isomer (Scheme 10.21).

Acknowledgements

The authors would like to thank Sandy Young for typing the chapter, Michael L. Black for drawing the structures in the text and Elizabeth Henry for editorial assistance.

References

1. R. K. Robins, *J. Med. Chem.* **7** (1964) 186.
2. A Goldin, H. B. Wood, Jr. and R. R. Engle, *Cancer Chemother. Rep.* **1** (1968) 1.
3. A. Bloch, in *Drug Design*, ed. E. S. Ariens, Academic Press, New York (1973) p. 285.
4. A. Bloch, *Ann. N.Y. Acad. Sci.* **255** (1975) 576.
5. J. A. Montgomery, *Handb. Exp. Pharmacol.* **38** (1974) 76.
6. J. F. Henderson, A. R. P. Paterson, I.C. Caldwell, B. Paul, M.C. Chan and K. F. Lau, *Cancer Chemother. Rep.* **3** (1972) 71.
7. J. A. Montgomery, *Med. Res. Rev.* **2** (1982) 271.
8. G. B. Elion, E. Burgi and G. H. Hitchings, *J. Am. Chem. Soc.* **74** (1952) 411.
9. B. A. Chabner and C. E. Meyer, in *Cancer: Principles and Practice of Oncology*, eds. V. T. DeVita, S. Hellman and S. A. Rosenberg, J. B. Lippincott, Philadelphia (1985) p. 302.
10. G. B. Elion and G. H. Hitchings, *J. Am. Chem. Soc.* **77** (1955) 1676.
11. H. J. Breter and R. K. Zahn, *Cancer Res.* **39** (1979) 3744.
12. W. R. Leopold, D. W. Fry, T. J. Boritzki, J. A. Besserer, J. C. Pattison and R. C. Jackson, *Invest. New Drugs* **3** (1985) 223.
13. T. A. Khwaja and J. C. Varven, *Proc. Am. Assoc. Cancer Res.* **17** (1976) 200.
14. P. Schwartz, D. Hammond and T. A. Khwaja, *Proc. Am. Assoc. Cancer Res.* **18** (1977) 153.
15. L. Pendyala, J. W. Cowens, J. E. Plager, L. R. Whitfield, A. J. Grillo-Lopez and P. J. Creaven. *Proc. Am. Assoc. Cancer Res.* **28** (1987) 192.
16. P. D. Cook, R. J. Rousseau, A. M. Mian, R. B. Meyer, Jr., P. Dea, G. Ivanovics, D. G. Streeter, J. T. Witkowski, M. G. Stout, L. N. Simon, R. W. Sidwell and R. K. Robins, *J. Am. Chem. Soc.* **97** (1975) 2916.
17. P. D. Cook, R. J. Rousseau, A. M. Mian, P. Dea, R. B. Meyer. Jr. and R. K. Robins, *J. Am. Chem. Soc.* **98** (1976) 1492.
18. R. K. Robins, J. H. Horner, C. V. Greco, C. W. Noell and C. G. Beames, Jr., *J. Org. Chem.* **28** (1963) 3041 and Refs. cited therein.
19. C. A. Salemink and G. M. Van Der Want, *Recl. Trav. Chim. Pays-Bas* **68** (1949) 1013.
20. (a) R. J. Rousseau, J. A. May, Jr., R. K. Robins and L. B. Townsend, *J. Heterocycl. Chem.* **11** (1974) 233; (b) J. A. Montgomery and K. Hewson, *J. Med. Chem.* **8** (1965) 708.
21. R. J. Rousseau, L. B. Townsend and R. K. Robins, *Biochemistry* **5** (1966) 756; Y. Mizuno, S. Tazawa and K. Kaguera, *Chem. Pharm. Bull.* **16** (1968) 2011.

22. D. G. Markees and G. W. Kidder, *J. Am. Chem. Soc.* **78** (1956) 4130; B. S. Gorton and W. Shive, *J. Am. Chem. Soc.* **79** (1957) 670.
23. G. S. Gorton, J. M. Ravel and W. Shive, *J. Biol. Chem.* **233** (1957) 331.
24. G. W. Kidder and V. C. Dewey, *Arch. Biochem. Biophys.* **66** (1957) 486.
25. K. B. de Roos and C. A. Salemink, *Recl. Trav. Chim. Pays-Bas* **90** (1971) 1166.
26. (a) J. A. May, Jr. and L. B. Townsend, Abstract, 167th National Meeting of the American Chemical Society, Los Angeles, CA, March (1974) Abstract No. 55; (b) J. A. Montgomery and K. Hewson, *J. Med. Chem.* **9** (1966) 105; (c) P. C. Jain, S. K. Chatterjee and N. Anand, *Indian J. Chem.* **1** (1965) 30.
27. R. de Bode and C. A. Salemink, *Recl. Trav. Chim. Pays-Bas.* **93** (1974) 3.
28. P. C. Srivastava and R. K. Robins, *J. Heterocycl. Chem.* **16** (1979) 1063.
29. G. R. Revankar, P. K. Gupta, A. D. Adams, N. K. Dalley, P. A. McKernan, P. D. Cook, P. G. Canonico and R. K. Robins, *J. Med. Chem.* **27** (1984) 1389.
30. D. A. Berry, R. B. Gilbertsen and P. D. Cook, *J. Med. Chem.* **29** (1986) 2034.
31. G. Hitchings, *Cancer Res.* **29** (1969) 1895.
32. R. K. Robins, *J. Am. Chem. Soc.* **78** (1956) 784.
33. G. A. Usbeck, J. W. Jones and R. K. Robins, *J. Am. Chem. Soc.* **83** (1961) 1113.
34. G. D. Daves, Jr., C. W. Noell, R. K. Robins, H. C. Koppel and A. G. Beaman, *J. Am. Chem. Soc.* **82** (1960) 2635.
35. L. R. Lewis, C. W. Noell, A. G. Beaman and R. K. Robins, *J. Med. Pharm. Chem.* **5** (1962) 607.
36. D. E. O'Brien, R. K. Robins, J. D. Westover and C. C. Cheng, *J. Med. Chem.* **12** (1969) 540.
37. R. K. Robins and G. R. Revankar, *Med. Res. Rev.* **5** (1985) 273.
38. G. P. Bodey, H. S. Brodovsky, A. A. Isassi, M. L. Samuels and E. L. Freireich, *Cancer Chemother. Rep.* **52** (1968) 315.
39. J. K. Luce, E. P. Frenkel, T. J. Vietti, A. A. Isassi, K. W. Hernandez and J. P. Howard, *Cancer Chemother. Rep.* **51** (1967) 535.
40. A. Hampton, J. J. Biesele, A. E. Moor and G. B. Brown, *J. Am. Chem. Soc.* **78** (1956) 5695.
41. J. J. Fox, I. Wempen, A. Hampton and I. L. Doerr, *J. Am. Chem. Soc.* **80** (1958) 1669.
42. R. J. Suhadolnik, in *Nucleosides as Biological Probes*, Wiley-Interscience, New York (1979) p. 217.
43. G. A.LePage, L. S. Worth and A. P. Kimball, *Cancer Res.* **36** (1976) 1481.
44. P. P. Major, R. P. Agarwal and D. Kufe, *Cancer Chemother. Pharmacol.* **10** (1983) 125.
45. D. P. Gray, M. R. Grever, M. F. E. Siaw, M. S. Coleman and S. P. Balcerzak, *Cancer Treat. Rep.* **66** (1982) 253.
46. E. L. Reist, A. Benitez, L. Goodman, B. R. Baker and W. W. Lee, *J. Org. Chem.* **27** (1980) 3274.
47. A. Benitez, O. P. Crews, Jr., L. Goodman and B. R. Baker, *J. Org. Chem.* **25** (1960) 1946.
48. C. P. J. Glaudemans and H. G. Fletcher, Jr., *J. Org. Chem.* **28** (1963) 3004.
49. R. Ranganathan, *Tetrahedron Lett.* **13** (1975) 1185.
50. W. H. Bromund and R. M. Herbst, *J. Org. Chem.* **10** (1945) 267.
51. G. R. Revankar, J. H. Huffman, L. B. Allen, R. W. Sidwell, R. K. Robins and R. L. Tolman, *J. Med. Chem.* **18** (1975) 721.
52. R. W. Brockman, F. M. Schabel and J. A. Montgomery, *Biochem. Pharmacol.* **26** (1977) 2193.
53. J. R. Barrueco, D. M. Jacobsen, C. H. Chang, R. W. Brockman and F. M. Sirotnak, *Proc. Am. Assoc. Cancer Res.* **27** (1986) 300.
54. V. I. Avramis and W. Plunkett, *Cancer Res.* **42** (1982) 2587.
55. J. M. Leiby, K. M. Snider, E. H. Kraut, E. N. Metz, L. Malspeis and M. R. Grever, *Cancer Res.* **47** (1987) 2719.
56. J. A. Montgomery and K. Hewson, *J. Med. Chem.* **12** (1969) 498.
57. J. A. Montgomery, S. D. Clayton and A. T. Shortnacy, *J. Heterocycl. Chem.* **16** (1979) 157.
58. R. K. Robins, K. J. Dille, C. H. Willits and B. E. Christensen, *J. Am. Chem. Soc.* **75** (1953) 264.
59. J. A. Montgomery, S. D. Clayton and A. T. Shortnacy, in *Nucleic Acid Chemistry*, eds. L. B. Townsend and R. S. Tipson, Wiley-Interscience, New York (1986) Part III, p. 156.
60. D. A. Carson, D. B. Wasson and E. Beuther, *Proc. Natl. Acad. Sci. (USA)* **81** (1984) 2232.
61. L. D. Piro, C. J. Carrera, E. Beutler and D. A. Carson, *Blood* **72** (1988) 1069.
62. D. A. Carson, D. B. Wasson, J. Kaye, B. Ullman, D. W. Martin, Jr., R. K. Robins and J. A. Montgomery, *Proc. Natl. Acad. Sci. (USA)* **77** (1980) 6865.

63. J. A. Montgomery, *Cancer Res.* **42** (1982) 3911.
64. L. F. Christensen, A. D. Broom, M. J. Robins and A. Bloch, *J. Med. Chem.* **15** (1972) 735.
65. M. J. Robins and R. K. Robins, *J. Am. Chem. Soc.* **87** (1965) 4934.
66. Z. Kazimierczuk, H. B. Cottam, G. R. Revankar and R. K. Robins, *J. Am. Chem. Soc.* **106** (1984) 6379.
67. A. G. Beaman and R. K. Robins, *J. Appl. Chem.* **12** (1962) 432.
68. M. Hoffer, *Chem. Ber.* **93** (1960) 2777.
69. G. R. Revankar and R. K. Robins, in *Handbook of Experimental Pharmacology*, eds. J. A. Nathanson and J. W. Kebabian, Springer-Verlag, Heidelberg (1982) Vol. 58/I, p. 17.
70. R. B. Meyer, Jr. and J. P. Miller, *Life Sci.* **14** (1974) 1019.
71. D. Katsaros, S. Ally and Y. S. Cho-Chung, *Int. J. Cancer* **41** (1988) 863.
72. S. Ally, G. Tortora, T. Clair, D. Grieco, G. Merlo, D. Katsaros, D. Ogreid, S. O. Doskeland, T. Jahnsen and Y. S. Cho-Chung, *Proc. Natl. Acad. Sci.* **85** (1988) 6319.
73. P. Tagliaferri, D. Katsaros, T. Clair, S. Ally, G. Tortora, L. Neckers, B. Rubalcava, Z. Parandoosh, Y. Chang, G. R. Revankar, G. W. Crabtree, R. K. Robins and Y. S. Cho-Chung, *Cancer Res.* **48** (1988) 1642.
74. D. Katsaros, G. Tortora, P. Tagliaferri, T. Clair, S. Ally, L. Necker, R. K. Robins and Y. S. Cho-Chung, *FEBS Lett.* **223** (1987) 97.
75. S. Ally, T. Clair, D. Katsaros, G. Tortora, R. A. Finch, T. L. Avery and Y. S. Cho-Chung, *Proc. Am. Assoc. Cancer Res.* **29** (1988) 56.
76. T. L. Avery, R. A. Finch, R. K. Robins, T. Arefieg, M. Mukutmoni, K. M. Vasquez, Y. A. Chang and G. R. Revankar, *Proc. Am. Assoc. Cancer Res.* **29** (1988) 354.
77. Z. Parandoosh, T. L. Avery, R. A. Finch, G. W. Crabtree, R. K. Robins and B. Rubalcava, *Proc. Am. Assoc. Cancer Res.* **29** (1988) 354.
78. K. Muneyama, R. J. Bauer, D. A. Shuman, R. K. Robins and L. N. Simon, *Biochemistry* **10** (1971) 2390.
79. K. Muneyama, D. A. Shuman, K. H. Boswell, R. K. Robins, L. N. Simon and J. P. Miller, *J. Carbohydr. Nucleosides Nucleotides* **1** (1974) 55.
80. G. J. Tricot, H. N. Jayaram, C. R. Nichols, K. Pennington, E. Lapis, G. Weber and R. Hoffman, *Cancer Res.* **47** (1987) 4988.
81. P. C. Srivastava, M. V. Pickering, L. B. Allen, D. G. Streeter, M. T. Campbell, J. T. Witkowski, R. W. Sidwell and R. K. Robins, *J. Med. Chem.* **20** (1977) 256.
82. M. Feurtes, M. T. Garcia-Lopez, G. Garcia Munoz and R. Madronero, *J. Carbohydr. Nucleosides Nucleotides* **2** (1975) 277.
83. J. M. Sprague and A. H. Land, *J. Heterocycl. Compd.* **5** (1957) 624.
84. J. L. Parsons, D. Vizine, M. Summer, S. Marathe and D. Henryk, European Patent 0072977 A1 (1983).
85. W. J. Hennen, B. C. Hinshaw, T. A. Riley, S. A. Wood and R. K. Robins, *J. Org. Chem.* **50** (1985) 1741.
86. R. Kuttan, R. K. Robins and P. P. Saunders, *Biochem. Biophys. Res. Commun.* **107** (1982) 862.
87. G. Weber, *Cancer Res.* **43** (1983) 3466.
88. P. C. Srivastava and R. K. Robins, *J. Med. Chem.* **26** (1983) 445.
89. H. N. Jayaram, R. L. Dion, R. I. Glazer, D. G. Johns, R. K. Robins, P. C. Srivastava and D. A. Cooney, *Biochem. Pharmacol.* **31** (1982) 2371.
90. P. D. Cook and D. J. McNamara, *J. Heterocyl. Chem.* **23** (1986) 155.
91. B. M. Goldstein, F. Takusagawa, H. M. Berman, P. C. Srivastava and R. K. Robins, *J. Am. Chem. Soc.* **105** (1983) 7416.
92. B. M. Goldstein, F. Takusagawa, H. M. Berman, P. C. Srivastava and R. K. Robins, *J. Am. Chem. Soc.* **107** (1985) 1394.
93. K. Gerzon, R. H. Williams, M. Hoehn, M. Gorman and D. C. Delong, Abstracts, 2nd International Congress of Heterocyclic Chemistry, Montpellier, France (1969) Abstract C-30.
94. R. H. Williams and M. M. Hoehn, U.S. Patent, 3,802,999 (1974).
95. R. J. Suhadolnik, in *Nucleoside Antibiotics*, Wiley-Interscience, New York (1970) p. 390.
96. L. B. Townsend, in *Handbook of Biochemistry and Molecular Biology, Nucleic Acids*, ed. G. D. Fasman, CRC Press, Cleveland (1975) Vol. I, p. 271.
97. A. Bloch, in *Encyclopedia of Chemical Technology*, Wiley-Interscience, New York (1978) Vol. 2, p. 962.

98. G. E. Gutowsky, M. J. Sweeney, D. C. Delong, R. L. Hamill, K. Gerzon and R. W. Dyke, *Ann. N.Y. Acad. Sci.* **225** (1975) 544.
99. T. Ohnuma and J. F. Holland, *Cancer Treat. Rep.* **61** (1977) 389.
100. J. Farkas, L. Flegelova and F. Sorm, *Tetrahedron Lett.* **22** (1972) 2279.
101. L. Kalvoda, J. Farkas and F. Sorm, *Tetrahedron Lett.* (1970) 2297.
102. G. E. Gutowski, M. O. Chaney, N. D. Jones, R. L. Hamill, F. A. Davis and R. D. Miller, *Biochem. Biophys. Res. Commun.* **51** (1973) 312.
103. E. Wenkert, E. W. Hagaman and G. E. Gutowski, *Biochem. Biophys. Res. Commun.* **51** (1973) 318.
104. S. De Bernardo and M. Weigele, *J. Org. Chem.* **41** (1976) 87.
105. J. G. Buchanan, M. R. Hamblin, G. R. Sood and R. H. Wightman, *J. Chem. Soc. Chem. Commun.* (1980) 917.
106. J. G. Buchanan, A. Stobie and R. H. Wightman, *J. Chem. Soc., Perkin Trans.* (1981) 1191.
107. J. G. Buchanan, A. Stobie and R. H. Wightman, *Can. J. Chem.* **58** (1980) 2624.
108. N. Katagiri, K. Takashima, T. Haneda and T. Kato, *J. Chem. Soc. Perkin Trans. 1* (1984) 553.
109. H. Ohrui and J. J. Fox, *Tetrahedron Lett.* (1973) 1951.
110. F. Serratosa and E. Sole, *An. R. Soc. Esp. Fis. Quim. Ser B.* **62** (1966) 431.
111. F. J. L. Herrera and C. U. Baelo, *Carbohydr. Res.* **143** (1985) 161.
112. C. E. Cass and T. H. Au-Yeung, *Cancer Res.* **36** (1976) 1486.
113. F. J. Cummings, G. W. Crabtree, W. Spremulli, T. Rogler-Brown, R. E. Parks, Jr. and P. Calabresi, *Proc. Am. Soc. Clin. Oncol.* **21** (1980) 332.
114. H. M. Golomb, M. J. Ratain and J. Moormeier, *J. Clin. Oncol.* **7** (1989) 156.
115. E. Chan, S. R. Putt, H. D. H. Showalter and D. C. Baker, *J. Org. Chem.* **47** (1982) 3457.
116. W. E. Allsebrook, J. M. Gulland and L. F. Story, *J. Chem. Soc.* (1942) 232.
117. D. C. Baker and S. R. Putt, *Synthesis* (1978) 478.
118. U. Niedballa and H. Vorbrüggen, *J. Org. Chem.* **39** (1974) 3654.

11 Methylmelamines

D. E. V. WILMAN

11.1 Introduction

The 1,3,5-triazine, *sym-* or *s*-triazine nucleus, a 6-membered heterocyclic ring with alternating carbon and nitrogen atoms, has been incorporated into a number of compounds which have evoked interest as antitumour agents. The compound substituted at the 2,4 and 6 positions by amino groups is melamine (2,4,6-triamino-1,3,5-triazine (**1**), this structure also shows the conventional numbering of the *s*-triazine ring), which has formed the basis of a number of clinically active agents. The first of these was the alkylating agent triethylenemelamine (TEM (**2**)), which is discussed in detail in Chapter 3. An extensive series of active-site-directed irreversible inhibitors of dihydrofolate reductase incorporating the *s*-triazine ring was developed by Baker *et al.*[1]. However, of most significance from the current clinical point of view and the subject of this chapter are the methylated melamines and their derivatives.

1   2

The parent compound in this series is hexamethylmelamine (HMM, 2,4,6-*tris*-(dimethylamino)-1,3,5-triazine, **3**). The original biological use of HMM was as an inactive analogue of TEM[2]. Whilst it does not have the direct alkylating ability of TEM, HMM was found to have antitumour activity against the Walker 256 carcinosarcoma in the rat. This discovery led to many investigations of the mechanism of the antitumour action of the methylmelamines, however, it has so far evaded elucidation. The work which has been undertaken has led to a greater understanding of the metabolism of these compounds and to the design of more clinically useful agents.

This chapter describes the three drugs in this class which have been investigated clinically, hexamethylmelamine, pentamethylmelamine (PMM (**5**)) and N^2,N^4,N^6-*tris*-(hydroxymethyl)-N^2,N^4,N^6-trimethylmelamine (trimelamol (**8**)). The objective is to indicate the importance of an interdisciplinary approach to anticancer drug design.

11.2 Hexamethylmelamine

The antitumour activity of hexamethylmelamine (**3**), the initial clinical agent in this series, was discovered independently by Buckley *et al.*[2] and by Hendry *et al.*[3]. Despite the fact that its activity against the standard rodent tumours is at best marginal, it was introduced into the clinic[4] and has proved to have a wide spectrum of moderate activity against solid tumours[5]. Activity in early clinical trial was most marked against ovarian cancer, lymphomas and carcinoma of the cervix, and to a lesser extent bronchogenic and breast carcinomas. A more recent review is critical of the clinical trials of this drug[7], in that none have sought to define the use of hexamethylmelamine relative to other treatments for the diseases against which it shows activity, or to define its contribution to combination regimens.

Initial theories concerning the mechanism of action of hexamethylmelamine noted its apparent structural similarity to triethylenemelamine (**2**) and suggested that its role was that of an alkylating agent. However, the fact that it fails to react with 4-(*p*-nitrobenzyl)pyridine confirms that this is not the case. Moreover, its metabolism differs markedly from TEM[6]. Indeed it was originally synthesised as a non-alkylating analogue of TEM[3]. A number of early reports on the metabolism of hexamethylmelamine, in relation to its antitumour activity[6,8,9], indicated that enzymic oxidative demethylation led to the production of a series of lower homologues. Much recent work in relation to the metabolism and mechanism of action of the methylmelamines has been undertaken by Rutty and her colleagues. These studies have been intimately involved with the search for second and third generation analogues of hexamethylmelamine.

Metabolic studies (Scheme 11.1) have shown that hexamethylmelamine (**3**) undergoes oxidative metabolism to *N*-hydroxymethylpentamethylmelamine (**4**), which by chemical loss of formaldehyde forms pentamethylmelamine (**5**). Further metabolism results in progressive symmetrical loss of the remaining methyl groups, in a similar fashion.

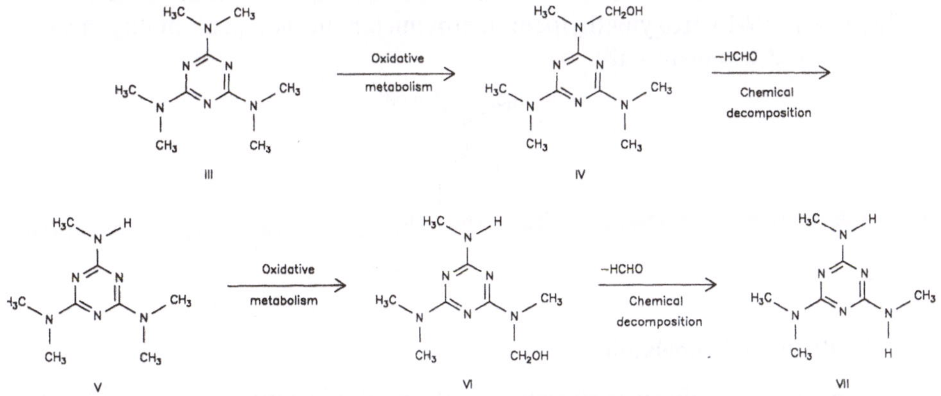

Scheme 11.1 Metabolism of hexamethylmelamine.

Hexamethylmelamine has two major disadvantages from the clinical point of view. It is extremely insoluble in aqueous media and is therefore not suitable for intravenous administration, and has of necessity always been administered orally. Early attempts to produce a parenteral formulation of HMM were unsuccessful. Gentisic acid complexes with a pH range from 3.5 to 4.5 were prepared[10], but unfortunately these were unstable at physiological pH resulting in precipitation of the free drug at the injection site. More recently, Ames and Kovach[11] have demonstrated the potential of a formulation obtained by dissolving the hydrochloride salt of hexamethylmelamine in a fat emulsion.

The second clinical problem is the severe nausea and vomiting encountered in all patients. The necessity for oral administration meant that it was not possible to determine whether this toxicity was a local or centrally mediated effect, as the dose limiting toxicity is neurological together with some myelosuppression.

The search for a suitable second generation methylmelamine, therefore, centred on compounds with at least the same antitumour activity as hexamethylmelamine, but also increased aqueous solubility to facilitate parenteral administration. The design of such a compound was hindered by the lack of a sufficiently sensitive test system. Only one rodent tumour, the mouse ADJ/PC6A plasmacytoma, shows any significant response, but even in this system the activity of hexamethylmelamine is in no way comparable to that of other agents to which the tumour is sensitive; difunctional alkylating agents such as melphalan, for example.

The problem was fortuitously overcome when it was found that certain human tumour xenografts were particularly sensitive to hexamethylmelamine[12]. One of these, a human lung adenocarcinoma (P246) growing originally in immune-deprived mice and latterly in athymic nude mice, was chosen for a structure-activity study initiated by Ross[13–15]. All the analogues in this study were designed to have greater water solubility than hexamethylmelamine and, in addition, three had equal or greater antitumour activity. These were N-hydroxymethylpentamethylmelamine (4), pentamethylmelamine (5) and trimelamol (8).

11.3 Pentamethylmelamine

Pentamethylmelamine was chosen from the work of Cumber and Ross[13] as a successor to HMM as it has equal experimental antitumour activity but PMM is 23 times more soluble than HMM at pH 7 and 25°C. The poor stability of the hydroxymethyl analogues was regarded as a major obstacle to formulation at this time. The solubility of PMM (2.16 mg/ml) was adequate for intravenous administration[15], and use of this route would determine whether the severe gastrointestinal toxicity seen with oral HMM was locally or centrally mediated.

The results of the various Phase I clinical trials, both in Europe and North America, were disappointing. Not only was PMM without significant antitumour effect in man, it was as emetic as HMM and at elevated doses caused dose related sedation and even coma in some patients[16-20], demonstrating that the gastrointestinal toxicity is central nervous system mediated. The clinical results obtained with pentamethylmelamine have been reviewed by Foster et al.[21].

All Phase I trials undertaken by the Institute of Cancer Research in conjunction with the Royal Marsden Hospital are closely monitored by pharmacokinetic studies. The main purpose of this is to establish whether or not the drugs are metabolised in man in the same way as they are in the rodent species used for the preclinical investigations. It was therefore quickly established[22] that the clinical failure of PMM was due to inadequate metabolism of the drug to cytotoxic N-hydroxymethyl species, such metabolism being readily performed by mice but less so by rats, paralleling the observed antitumour activity in these species, and even more poorly in man. Rutty had previously shown that the N-hydroxymethylmelamine metabolites are a prerequisite for antitumour activity[23].

11.4   Trimelamol

Following the clinical failure of pentamethylmelamine and the reasons for this failure having been determined[22], it became apparent that if derivatives of hexamethylmelamine were to have clinical utility they would need to carry ready formed N-hydroxymethylamino groups or be capable of producing them without the need for oxidative metabolism. The large number of N-hydroxymethyl groups present in the molecule and substantial solubility (9 mg/ml) therefore made trimelamol (**8**) an ideal candidate for clinical investigation. An additional point in its favour is that with a ten-fold lower octanol-water partition coefficient than PMM[24], trimelamol enters the central nervous system very poorly. The $\log P$ value for trimelamol is 0.4 which falls outside the criteria established by Levin[25] (molecular weight below 400 and high $\log P$) for drugs to cross the 'blood-brain' barrier.

Trimelamol does, however, have one major drawback from the clinical point of view. This is its inherent chemical instability. However, a suitable formulation was achieved by Rutty and her colleagues[26,27] and trimelamol has undergone Phase I[28] and II[29] clinical trials, where it showed particular activity against recurrent stage III/IV ovarian carcinoma. Pharmacokinetic studies[30] have shown that plasma levels of trimelamol may be achieved clinically, which compare favourably with those observed in mice at dose levels which caused substantial tumour regression.

The toxicity observed with trimelamol was much less than with the other methylmelamines, both in terms of dose-related vomiting and sedation. This was, of course, to be expected if the drug's penetration into the human brain paralleled that observed in mice.

11.5   Chemistry

11.5.1   Synthesis of alkylmelamines

Alkylmelamines are best synthesised by the stepwise replacement of the chlorine atoms of cyanuric chloride (2,4,6-trichloro-1,3,5-triazine (9)) with appropriate primary and secondary amines (Scheme 11.2). This reaction

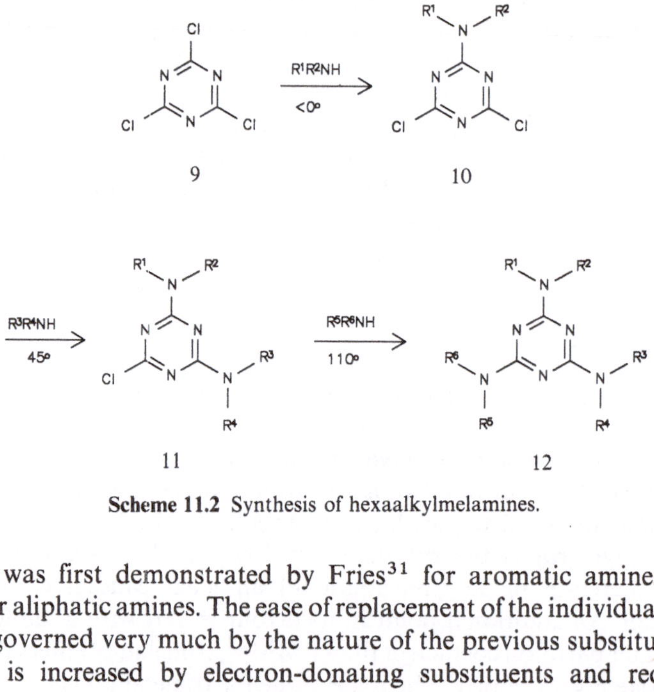

Scheme 11.2 Synthesis of hexaalkylmelamines.

sequence was first demonstrated by Fries[31] for aromatic amines and by Diels[32] for aliphatic amines. The ease of replacement of the individual chlorine atoms is governed very much by the nature of the previous substituents. The reactivity is increased by electron-donating substituents and reduced by electron-withdrawing ones. Thus all three chlorine atoms may be replaced by secondary amines such as morpholine and piperidine at 25°C[33]. The mono- and diaminated triazines can be obtained at suitably low temperatures. The generality of this approach was demonstrated by Thurston and his colleagues[34,35] and further developed by Bořkovec and DeMilo[36] and by Wörle and Lindinger[37], who demonstrated the utility of this method as a one-pot synthesis using toluene as solvent.

The route from cyanuric chloride has been used exclusively for the alkylmelamines synthesised at the Institute of Cancer Research[13,38]. Thus hexamethylmelamine is prepared from cyanuric chloride in aqueous media with intermediate isolation of 2-chloro-4,6-bis-(dimethylamino)-1,3,5-triazine ((11) $R^1 = R^2 = R^3 = R^4 = CH_3$). The conversion of (11) into HMM requires the gradual addition of an equal molar quantity of aqueous sodium hydroxide solution to a refluxing mixture of (11) and aqueous dimethylamine for the reaction to go to completion. If the addition of alkali is too rapid and the solution becomes severely alkaline, hydrolysis may occur at either the chloro or amino groups to form hydroxy derivatives such as (13), which exists as the keto tautomer (14). This degradation may continue to cyanuric acid.

13   14

PMM is prepared in a similar fashion via either the same intermediate or 2,4-dichloro-6-methylamino-1,3,5-triazine[36].

The alkylmelamines are stable crystalline solids, but are hydrolysed by heating in alkali to hydroxy derivatives and eventually cyanuric acid. In general, the amino groups in melamines behave very much as acid amides rather than amines due to the powerful electron withdrawing properties of the heterocyclic ring.

11.5.2   Synthesis of N-hydroxymethylmelamines

There is a great fund of chemical literature relating to the reaction of melamine with formaldehyde arising from the production of plastics and resins. In particular, numerous publications relate the synthesis of melamines with various degrees of hydroxymethylation. However, in almost all cases the only method used to identify the products has been by determination of the total available formaldehyde. HPLC analysis of these 'purified' reaction products shows them to be a mixture of at least three components[39]. Coupled with the general instability of the hydroxymethyl species in this type of compound it has so far proved impossible to prepare individual pure hydroxymethyl-melamines which retain an amino proton (N–H) in the molecule. Thus the oxidative metabolite of PMM, N^2-hydroxymethyl-N^2,N^4,N^6-tetramethylmelamine (6) is not known synthetically.

In view of the instability of the hydroxymethylamino group, the synthesis of trimelamol in purified form is fortuitous. It is dependent on the solubility difference between N^2,N^4,N^6-trimethylmelamine (15) and trimelamol. An

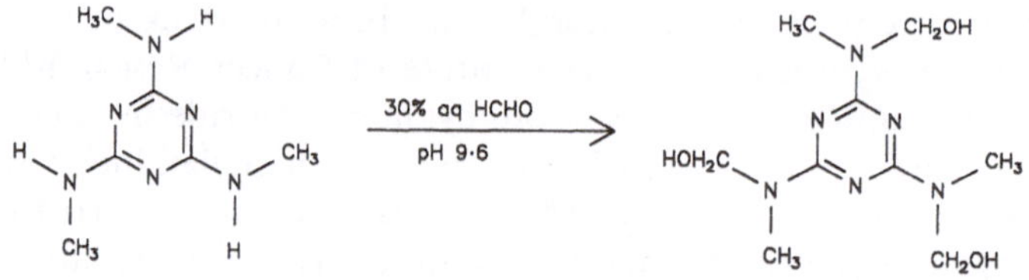

Scheme 11.3 Synthesis of trimelamol.

aqueous solution of (**15**) is mixed with formalin and the pH adjusted to 9.6. Stirring at room temperature for 16 h then produces a copious white precipitate which, if well washed with ice-cold water to remove formaldehyde, produces when dry a solid which is better than 96% pure trimelamol by HPLC analysis[38]. Although Cumber and Ross[13] report recrystallisation of trime-lamol this was found not to be practical on a large scale.

11.5.3 Radiolabelled syntheses

Ring ^{14}C-labelled hexamethylmelamine was prepared by Nam et al.[40] (Scheme 11.4). [^{14}C]Urea was cyclised in o-dichlorobenzene to form [2,4,6-^{14}C]cyanuric acid (**16**) which, following purification, was aminated directly with dimethylamine in the presence of hexamethylphosphoric triamide (HMPT). The [ring-^{14}C]HMM (**17**) was obtained in 18% overall yield from [^{14}C]urea with a specific activity of 32 mCi/mmol.

The same group subsequently prepared two radiolabelled forms of pen-tamethylmelamine. [ring-4,6-^{14}C]Pentamethylmelamine (**20**) was synthesised from barium [^{14}C]carbonate, which was initially converted to dimethyl-

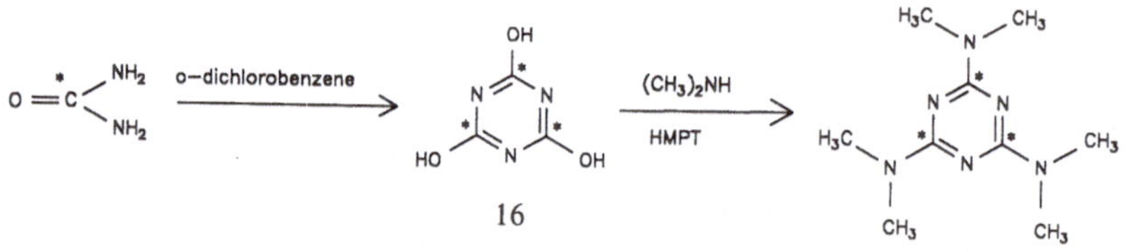

Scheme 11.4 Synthesis of ring ^{14}C-labelled hexamethylmelamine (**17**).

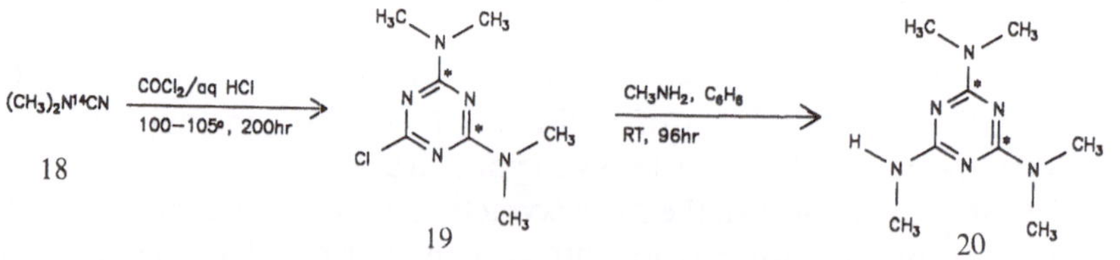

Scheme 11.5 Synthesis of [ring-4,6-^{14}C]pentamethylmelamine (**20**).

[^{14}C]-cyanamide (**18**). Cyclocondensation of this intermediate with phosgene in hydrochloric acid at 100–105°C for 200 h gave 2-chloro-4,6-*bis*-(dimethylamino)-[*ring*-4,6-^{14}C]-*s*-triazine (**19**) (Scheme 11.5), which was reacted further with methylamine to give the [*ring*-4,6-^{14}C]penta-methylmelamine (**20**) with a specific activity of 38 mCi/mmol[41].

Pentamethylmelamine with ^{14}C-labelled *bis*-(dimethylamino) groups was prepared by the condensation of (^{14}CH$_3$)$_2$NH with (**21**) (Scheme 11.6)[42]. The (^{14}CH$_3$)$_2$NH was prepared by methylation of *p*-toluenesulphonamide with ^{14}CH$_3$I, followed by HBr and NaOH cleavage of the amide. The overall yield of N^2,N^2,N^4,N^4-tetra-[^{14}C]methyl-N^6-methylmelamine (**22**) from ^{14}CH$_3$I was 8.5% with a radiochemical purity of 99% and specific activity of 78.5 mCi/mmol.

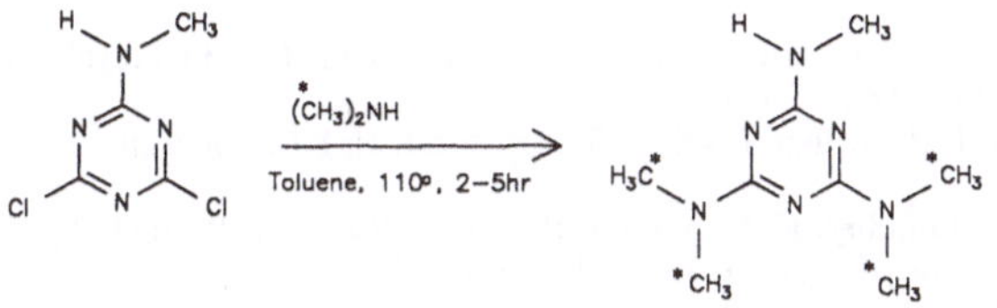

Scheme 11.6 Synthesis of N^2,N^2,N^4,N^4-tetra-[^{14}C]methyl-N^6-methylmelamine (**22**).

11.5.4   *Crystal structure*

Hexamethylmelamine is the only one of the compounds considered here for which the crystal structure has been determined. It forms hexagonal crystals and the whole of the molecule is planar apart from the hydrogen atoms[43]. This planarity coupled with the shortness of the ring carbon to exocyclic nitrogen bonds shows that the π-system extends into the exocyclic bonds accounting for the lack of true amine character of the amino groups.

11.6   Concluding remarks

The clinical development of the methylmelamine series of drugs, from hexamethylmelamine through pentamethylmelamine to trimelamol, has been based on a day to day collaboration between scientists and clinicians from many disciplines. It is also an example of the necessity of establishing the metabolism and pharmacokinetic profile of a compound before it undergoes clinical evaluation. It is then possible to observe at an early stage adverse or quantitative species differences in metabolism leading to lack of clinical activity. Once these are explained, a more appropriate drug may be designed.

References

1. B. R. Baker, *Design of Active-Site-Directed Irreversible Enzyme Inhibitors*, Wiley, New York (1967) p. 192.

2. S. M. Buckley, C. C. Stock, M. L. Crossley and C. P. Rhoads, *Cancer Res.* **10** (1950) 207.
3. J. A. Hendry, R. F. Homer, F. L. Rose and A. L. Walpole, *Br. J. Pharmacol.* **6** (1951) 357.
4. W. L. Wilson and J. G. de la Garza, *Cancer Chemother. Rep.* **48** (1965) 49.
5. S. S. Legha, M. Slavik and S. K. Carter, *Cancer* **38** (1976) 27.
6. J. F. Worzalla, B. M. Johnson, G. Ramirez and G. T. Bryan, *Cancer Res.* **33** (1973) 2810.
7. B. J. Foster, B. J. Harding, B. Leyland-Jones and D. Hoth, *Cancer Treat. Rev.* **13** (1986) 197.
8. J. F. Worzalla, B. D. Kaiman, B. M. Johnson, G. Ramirez and G. T. Bryan, *Cancer Res.* **33** (1974) 2669.
9. L. M. Lake, E. E. Grunden and B. M. Johnson, *Cancer Res.* **35** (1975) 2858.
10. B. Kreilgard, T. Higuchi and A. J. Repta, *J. Pharm. Sci.* **64** (1975) 1850.
11. M. M. Ames and J. S. Kovach, *Cancer Treat. Rep.* **66** (1982) 1579.
12. B. C. V. Mitchley, S. A. Clarke, T. A. Connors and A. M. Neville, *Cancer Res.* **35** (1975) 1099.
13. A. J. Cumber and W. C. J. Ross, *Chem.-Biol. Interact.* **17** (1977) 349.
14. B. C. V. Mitchley, S. A. Clarke, T. A. Connors, S. M. Carter and A. M. Neville, *Cancer Treat. Rep.* **61** (1977) 451.
15. T. A. Connors, A. J. Cumber, W. C. J. Ross, S. A. Clarke and B. C. V. Mitchley, *Cancer Treat. Rep.* **61** (1977) 927.
16. D. A. van Echo, D. F. Chuitden, M. Whitacre, J. Aisner, J. L. Lichteyeld and P. H. Wiernik, *Cancer Treat. Rep.* **64** (1980) 1325.
17. R. S. Goldberg, J. P. Griffen, J. W. McSherry and I. H. Krakoff, *Cancer Treat. Rep.* **64** (1980) 1319.
18. P. C. Ihde, J. S. Dutcher, K. C. Young, R. S. Corales, A. L. Barlock, S. M. Hubbard, R. B. Jones and M. R. Boyd, *Cancer Treat. Rep.* **65** (1981) 755.
19. E. S. Casper, R. J. Gratta, R. L. Garrett, B. R. Jones, T. M. Woodcock, C. Gordon, D. P. Kelson and C. W. Young, *Cancer Res.* **41** (1981) 1402.
20. J. R. F. Muindi, D. R. Newell, I. E. Smith and K. R. Harrap, *Br. J. Cancer* **47** (1983) 27.
21. B. J. Foster, K. Clagett-Carr, D. Hoth and B. Leyland-Jones, *Cancer Treat. Rep.* **70** (1986) 383.
22. C. J. Rutty, D. R. Newell, J. R. F. Muindi and K. R. Harrap, *Cancer Chemother. Pharmacol.* **8** (1982) 105.
23. C. J. Rutty and T. A. Connors, *Biochem. Pharmacol.* **26** (1977) 2385.
24. I. R. Judson, C. J. Rutty, G. Abel and M. A. Graham, *Br. J. Cancer* **53** (1986) 601.
25. V. A. Levin, *J. Med. Chem.* **23** (1980) 682.
26. C. J. Rutty, I. R. Judson, G. Abel, P. M. Goddard, D. R. Newell and K. R. Harrap, *Cancer Chemother. Pharmacol.* **17** (1986) 251.
27. A. H. Calvert, C. J. Rutty and I. R. Judson, Brit. Pat., 2170710, 1986.
28. I. R. Judson, A. H. Calvert, C. J. Rutty, G. Abel, L. Gumbrell, M. A. Graham, B. D. Evans, D. E. V. Wilman, S. E. Ashley and F. Cairnduff, *Cancer Res.* **49** (1989) 5475.
29. I. R. Judson, M. E. Gore, L. A. Gumbrell, K. Balmanno, D. I. Jodrell, T. J. Perren, P. Blaked and A. H. Calvert, *Br. J. Cancer* **58** (1988) 273.
30. I. R. Judson, C. J. Rutty, L. Gumbrell, G. Abel, K. R. Harrap and A. H. Calvert, *Br. J. Cancer* **54** (1986) 162.
31. H. H. Fries, Ber. **19** (1886) 2055.
32. O. Diels, Ber. **32** (1899) 691.
33. W. M. Pearlman and C. K. Banks, *J. Am. Chem. Soc.* **70** (1948) 3726.
34. J. T. Thurston, J. R. Dudley, D. W. Kaiser, I. Hechenbleikner, F. C. Schaefer and D. Holm-Hansen, *J. Am. Chem. Soc.* **73** (1951) 2981.
35. D. W. Kaiser, J. T. Thurston, J. R. Dudley, F. C. Schaefer, I. Hechenbleikner and D. Holm-Hansen, *J. Am. Chem. Soc.* **73** (1951) 2984.
36. A. B. Bořkovec and A. B. DeMilo, *J. Med. Chem.* **10** (1967) 457.
37. R. Wörle and H. Lindinger, *Ann. Chem.* (1973) 1430.
38. D. E. V. Wilman, Unpublished data.
39. D. E. V. Wilman and B. E. Heales, Unpublished data.
40. N.-H.-Nam, H. Hoellinger and L. Pichat, *J. Labelled Compd. Radiopharm.* **12** (1976) 517.
41. D.-C.-Thang, N.-H.-Nam, H. Hoellinger and L. Pichat, *J. Labelled Compd. Radiopharm.* **18** (1981) 1009.
42. D.-C.-Thang, N.-H.-Nam, H. Hoellinger and L. Pichat, *J. Labelled Compd. Radiopharm.* **20** (1983) 779.
43. G. J. Bullen, D. J. Corney and F. S. Stephens, *J. Chem. Soc., Perkin Trans.* 2 (1972) 642.

12 The chemistry of platinum antitumour agents

M. J. ABRAMS

12.1 Introduction

The vast majority of clinically used antitumour drugs are either synthetic or natural product based organic compounds. The platinum antitumour agents are unique in that they are coordination complexes. The parent compound of this class, *cis*-diamminedichloroplatinum(II) (cisplatin), contributes to the curative treatment of testicular teratoma and has significant activity against ovarian, head and neck, bladder, cervical and lung cancers. The compound is also highly toxic with the list of side effects including renal damage, severe nausea and vomiting, myelosuppression, ototoxicity and neurotoxicity. Since the antitumour properties of cisplatin were reported by Rosenberg and coworkers in 1969, well over 1000 platinum analogues have been tested for antitumour activity[1,2]. The goal of this work has been to find analogues with reduced toxicity and/or wider spectrum of activity. Active cisplatin analogues are generally neutral platinum complexes containing two ammine (NH_3) or amine ligands in a *cis* configuration in concert with two (for Pt(II) analogues) or four (for Pt(IV) compounds) relatively labile halide or oxygen ligands[3]. Complexes with amine groups in a *trans* configuration are invariably inactive[4]. Structure-activity relationships (in animal tumour models) and the clinical aspects of platinum drugs have been reviewed[2-6].

This paper concentrates on the synthesis, characterisation and chemistry of the parent compound (cisplatin) and two representative analogues, carboplatin (a Pt(II) malonate complex) and iproplatin (a Pt(IV) species) (Figure 12.1). Complexes of the 1,2-diaminocyclohexane(DACH) ligand are also discussed.

12.2 Cisplatin

12.2.1 *Synthesis and characterisation*

Cisplatin (or Peyrones chloride) and its *trans* isomer (or Reisets second chloride) are important complexes in the history of coordination chemistry. Studying this isomeric pair led Werner to the assignment of a square planar structure for Pt(II) complexes[7].

Cisplatin is routinely prepared using one of three procedures, all of which use some salt of the tetrachloroplatinate(II) $[PtCl_4]^{2-}$ ion as a starting material. In the procedure described by Kauffman and Cowan, $K_2[PtCl_4]$ is reacted with slightly over two equivalents of aqueous NH_3 in NH_4Cl

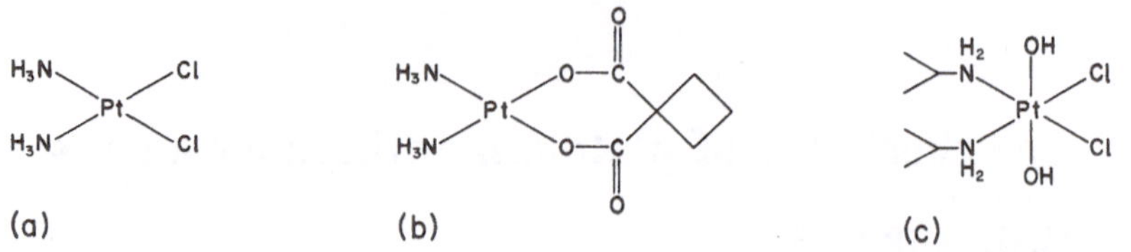

Figure 12.1 The structures of (a) cisplatin, (b) carboplatin, and (c) iproplatin.

solution[8]. Chilling this reaction mixture for 24–48 h yields cisplatin in 60% yield. The product is contaminated with small amounts of green insoluble $[Pt(NH_3)_4][PtCl_4]$ or Magnus's green salt, which is readily eliminated by recrystallising the product. In the procedure employed by Lebedinskii and Golovnya, an aqueous solution of $K_2[PtCl_4]$, $CH_3CO_2NH_4$ and KCl is heated at reflux temperature for 1.5 h yielding cisplatin upon cooling[9,10]. Perhaps the most versatile method for preparing cisplatin is the Dhara route, in which cis-$[Pt(NH_3)_2I_2]$ is prepared by reacting $K_2[PtI_4]$ (formed in situ by adding I^- to a solution of $[PtCl_4]^{2-}$) with two equivalents of aqueous ammonia (Figure 12.2)[11]. The diiodide complex is treated with $AgNO_3$ to yield a solution of cis-$[Pt(NH_3)(OH_2)_2]^{2+}$ and a precipitate of AgI. Treating the diaquo species with HCl yields cisplatin in high yield. The Dhara route is generally useful for preparing analogues of the type cis-$[Pt(RNH_2)_2Cl_2]$.

High purity cisplatin can be obtained by recrystallising the crude product from dimethyformamide/HCl mixtures[12].

Cisplatin has been characterised by a variety of physical methods including X-ray crystallography and infrared, electronic and ^{195}Pt-NMR spectroscopy[13-16].

Cisplatin can be readily distinguished from the trans isomer by the Kurnokov test, i.e. the reaction of cisplatin with excess thiourea to yield yellow $[Pt(thiourea)_4]^{2+}$ while the trans isomer gives colourless trans-$[Pt(NH_3)_2(thiourea)_2]^{2+}$ under the same conditions[17]. A paper chromatographic method for separating the two isomers has been reported[18].

12.2.2 Reaction chemistry and mechanism of action

With respect to its biological activity, the most important aspect of the reactivity of cisplatin is its hydrolysis chemistry, which has been extensively

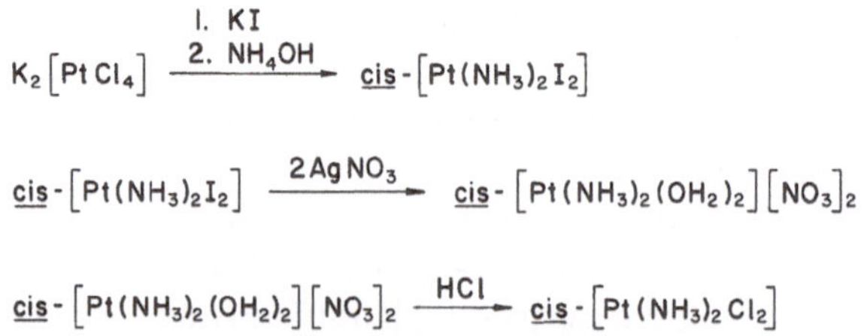

$$K_2[PtCl_4] \xrightarrow[\text{2. NH}_4\text{OH}]{\text{1. KI}} cis\text{-}[Pt(NH_3)_2I_2]$$

$$cis\text{-}[Pt(NH_3)_2I_2] \xrightarrow{\text{2 AgNO}_3} cis\text{-}[Pt(NH_3)_2(OH_2)_2][NO_3]_2$$

$$cis\text{-}[Pt(NH_3)_2(OH_2)_2][NO_3]_2 \xrightarrow{\text{HCl}} cis\text{-}[Pt(NH_3)_2Cl_2]$$

Figure 12.2 The Dhara synthesis of cisplatin.

studied[10,19]. Cisplatin dissolves in water to yield a solution of initially low conductivity, consistent with its formulation as a non-electrolyte. As the labile chloro ligands on the molecule are replaced by water, the conductivity of the solution rises to reach an equilibrium value in approximately 15 h at $25°C$[20]. At equilibrium the solution contains a mixture of cisplatin and cationic mono and diaquo species which retain the *cis*-diammine geometry[21]. The hydrolysis of cisplatin (in the dark) can be almost completely suppressed by the chloride concentration found in normal saline solution[20]. At physiological pH the equilibria associated with proton loss from the coordinated water molecules become important resulting in the formation of Pt aquo/hydroxo species[22]. These species are believed to be the actual active form of cisplatin in that they can bind to biological targets. Also, under certain conditions they can go on to form a variety of oxo bridged dimeric, trimeric and oligomeric species[2]. This process has been studied by ^{195}Pt-NMR and the structural aspects of cisplatin hydrolysis products have been reviewed[23,24].

The biological target involved with the antitumour effects of cisplatin is believed to be DNA. Cisplatin interactions with DNA and DNA constituents have been intensively studied and are the subject of several reviews[24-26]. Briefly, the intrastrand cross-link formed by the diammineplatinum unit and the N-7 positions on adjacent guanines is the most frequently found Pt-DNA binding mode (both *in vitro* and *in vivo*) and may be the important cytotoxic interaction[26].

Pt(II) is a soft Lewis acid and readily forms complexes with sulphur containing ligands[27]. Cisplatin reacts extensively with the endogenous sulphur ligands in biological systems and it is these interactions that may be responsible for the toxic effects of the drug[28]. Also, high levels of sulphur containing compounds such as glutathione may contribute to cellular resistance to platinum drugs[29]. However, compared to the extensive work done on the chemistry of Pt-purine and pyrimidine model systems, relatively little has been done to chemically characterise cisplatin-sulphur interactions.

The reactions of cisplatin with a number of sulphur containing amino acids and peptides have been studied. Cisplatin reacts with a stoichiometric amount of cysteine to form a poorly defined polymeric material[30]. Cisplatin and two equivalents of glutathione react to yield a material that gives elemental analysis results consistent with a formulation of bis(glutathionato)Pt(II)[30]. The loss of the NH_3 groups bound to the platinum in this reaction is typical of the reactions of cisplatin with sulphur containing ligands, which have a strong *trans*-labilising effect. Pt-methionine complexes have been proposed as possible metabolites of cisplatin[31].

Although the inactivation of important proteins caused by Pt binding to sulphur residues has been proposed as a mechanism of cisplatin toxicity, the details of cisplatin-protein interactions have been studied in only a few cases[32,33]. Cisplatin binds albumin at Cys-34 and this adduct retains the ability to interact with other proteins[34]. Both cisplatin and its *trans* isomer inactivate the plasma proteinase inhibitor α_1 PI by modifying the active site methionine, Met-358[35]. Platinum can also bind to proteins by coordinating to both sulphur atoms of a disulphide bond[36]. Although cisplatin does not induce metallothionein synthesis in rats it readily binds to the sulphur rich protein[37].

The high affinity of Pt(II) for sulphur ligands has led to the use of sulphur compounds such as sodium diethyldithiocarbamate(DDTC) as 'rescue' agents to mitigate the toxic effects of cisplatin[38,39]. DDTC has been shown to restore cisplatin-inhibited enzyme activity but it is incapable of reacting with bis-complexes of platinum and guanine bases. This is consistent with the observation that DDTC reduces cisplatin toxicity without reducing its antitumour properties. DDTC reacts with cisplatin to form the neutral, stable chelate complex [Pt(DDTC)$_2$][40]. Complexes of the type [Pt(diamine)(DDTC)]NO$_3$ have also been prepared[41].

Dimethylsulphoxide (DMSO) has often been employed as a solvent for Pt complexes in biological studies. Unfortunately, DMSO is also a good ligand for platinum and cisplatin dissolved in DMSO rapidly reacts to form cis-[Pt(NH$_3$)$_2$(DMSO)Cl]$^+$. This solvolysis reaction was recently studied by Sundquist et al.[42]. This solvolysis chemistry raises questions about the validity of biological results obtained using DMSO solutions of platinum complexes.

The chemistry of cisplatin with the amino acids glycine and histidine has recently been studied[43].

Cisplatin is a starting material for mixed amine/ammine Pt(II) species. The reactions of cisplatin with aqueous HCl in the presence of a Pt metal catalyst, or with [(C$_2$H$_5$)$_4$N]Cl in hot dimethylacetamide yields the [Pt(NH$_3$)Cl$_3$]$^-$ ion from which cis-[Pt(NH$_3$)ACl$_2$] (A = amine) complexes can be made[44-46].

12.3 Carboplatin

12.3.1 Synthesis and characterisation

Carboplatin (diammine(1,1-cyclobutanedicarboxylato)-platinum(II)) is a second generation platinum drug with roughly the same spectrum of activity as cisplatin but reduced toxicity[2]. The complex was selected for clinical development from a number of cisplatin analogues due to its activity in the

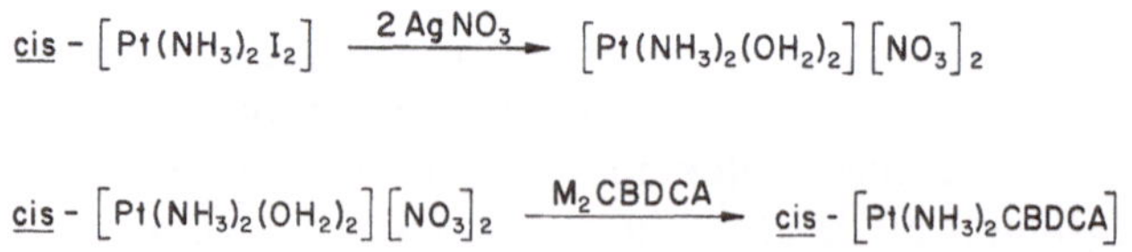

$$cis - \left[Pt(NH_3)_2 I_2 \right] \xrightarrow{\text{2 Ag NO}_3} \left[Pt(NH_3)_2(OH_2)_2 \right] \left[NO_3 \right]_2$$

$$cis - \left[Pt(NH_3)_2(OH_2)_2 \right] \left[NO_3 \right]_2 \xrightarrow{\text{M}_2\text{CBDCA}} cis - \left[Pt(NH_3)_2 CBDCA \right]$$

Figure 12.3 The synthesis of carboplatin starting from cis-[Pt(NH$_3$)$_2$I$_2$].

P246 mouse xenograft model and its relatively low toxicity[47]. Carboplatin is now commercially available in a number of countries including Canada and the UK.

Carboplatin is readily prepared via the reaction of the diaquo species *cis*-$[(NH_3)_2Pt(OH_2)_2]^{2+}$ with an alkali metal salt of 1,1-cyclobutanedicarboxylic acid (CBDCA) in aqueous solution (Figure 12.3)[48]. Analogous diammine (malonato)platinum(II) species can be prepared by this route. Alternatively, interaction of the sulphate complex *cis*-$[(NH_3)_2Pt(SO_4)]$ with the Ba^{2+} salt of the malonate ligand yields the Pt-malonate complex and insoluble $BaSO_4$[49]. Pasini and Caldirola have recently reported a novel synthesis of carboplatin that involves the direct reaction of cisplatin with K_2CBDCA in dimethylformamide[50].

Carboplatin has been characterised by X-ray crystallography and ^1H-, ^{13}C-, ^{15}N- and ^{195}Pt-NMR spectroscopy[51]. An interesting aspect of the crystal structure is that the cyclobutane ring is puckered such that the ring is over an axial site of the platinum atom. This feature may contribute to the relatively low reactivity of carboplatin.

The stability of carboplatin (as well as that of cisplatin, iproplatin and tetraplatin) in commonly used intravenous solutions has recently been reported[52].

12.3.2 *Reaction chemistry and mechanism of action*

In carboplatin, as in cisplatin, the unit that binds to DNA is the *cis*-diammineplatinum moiety. However, because of its stable bidentate malonate ring carboplatin forms reactive aquated platinum species much more slowly than cisplatin. The kinetics of the displacement of the CBDCA ligand from carboplatin have recently been described[53]. Roberts *et al.*[54] found the half-life of carboplatin in chloride-free phosphate buffer (pH 7, 37°C) to be 268 h as opposed to 24 h for cisplatin under the same conditions. This lower reactivity is manifest in the lower potency and toxicity of carboplatin with respect to cisplatin. Another consequence of this reduced reactivity is that carboplatin is less rapidly protein bound in serum than cisplatin and is excreted largely unchanged in the urine[55].

Mong *et al.*[56] found that carboplatin is more than 100 times less reactive towards PM-2 DNA than cisplatin but only four times less potent *in vivo*. Observations such as these suggest that some specific mechanism is involved in the removal of the CBDCA ligand in biological systems. Cleare and Hoeschele proposed that carboplatin might undergo enzymatic activation *in vivo*[57]. In a study using ^{14}C-labelled carboplatin Roberts *et al.*[54] found evidence for a platinum species containing both the CBDCA ligand and DNA that slowly reacts to lose CBDCA. A structurally characterised model complex containing a monodentate CBDCA ligand and a purine base on the *cis*-$[Pt(NH_3)_2]^{2+}$ core would be of interest.

Recently a 1:1 adduct of carboplatin with alpha-cyclodextrin has been prepared[58,59]. This material has improved solubility and stability with respect to the free drug and has been characterised by microcalorimetry, ^{1}H-NMR spectroscopy and X-ray crystallography.

12.4   Iproplatin

12.4.1   *Synthesis and characterisation*

The antitumour activity of Pt(IV) complexes prepared from cis-PtA$_2$Cl$_2$ (A = amine or ammine) species was discovered relatively early in the search for cisplatin analogues[46]. These complexes are generally prepared by oxidative addition of Cl$_2$ or H$_2$O$_2$ to the Pt(II) starting compound[8,46]. One such complex that has been extensively evaluated clinically is iproplatin ($cis,trans,cis$-bis(isopropylamine)dihydroxo-dichloroplatinum(IV)[2]. Iproplatin has a spectrum of activity similar to that of carboplatin, but may be more toxic[6]. There is preliminary evidence that iproplatin has some activity against breast cancer[60].

Iproplatin is synthesised by the reaction of cis-[Pt(i-PrNH$_2$)$_2$Cl$_2$] (prepared via the Dhara synthesis described earlier) with H$_2$O$_2$ in water, as shown in Figure 12.4[61]. The material obtained by this procedure is a perhydrate adduct, $cis,trans,cis$-[Pt(i-PrNH$_2$)$_2$(OH)$_2$Cl$_2$]·0.5H$_2$O$_2$. The presence of H$_2$O$_2$ in iproplatin was not recognised until 1983 and this oversight led to some incorrect conclusions about its biological properties[62]. Recrystallisation from dimethylacetamide yields the dimethylacetamide adduct from which unsolvated iproplatin can be prepared by heating in vacuum. The perhydrate, hydrate and dimethylacetamide adducts of iproplatin have been characterised by X-ray crystallography[61]. Iproplatin has also been characterised by ^{195}Pt-NMR spectroscopy[16]. The reaction of iproplatin with several perfluoro acid anhydrides yields volatile derivatives that have been studied by mass spectroscopy.[63]

12.4.2   *Reaction chemistry and mechanism of action*

There is strong evidence that Pt(IV) antitumour agents are reduced *in vivo* to their Pt(II) analogues which are the active species. Work done by Mong *et al.*[56] showed that iproplatin reacts directly with PM-2 DNA to cause strand breakage. This result was interesting because it suggested that Pt(IV) species (which, due to their low-spin d^6 electronic configuration, are very inert kinetically) could react directly with DNA. Later work by Vollano *et al.*[64] showed that this effect was caused by the H$_2$O$_2$ impurity and not iproplatin.

Iproplatin is readily reduced by ascorbic acid or ferrous ion to yield the Pt(II) species, cis-[Pt(i-PrNH$_2$)$_2$Cl$_2$][65]. This compound has also been identi-

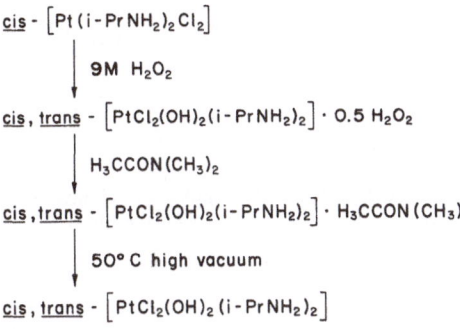

Figure 12.4 The synthesis of unsolvated iproplatin.

fied as a metabolite in the urine and plasma of patients receiving iproplatin[66]. The reactions of iproplatin (and several other Pt(IV) complexes) with 5'-guanosinemonophosphate yield only Pt(II)-nucleotide adducts[67]. The reactions of iproplatin with sulphur containing ligands have not been extensively studied although these interactions may have biological importance. The reaction of iproplatin with cysteine yields an insoluble material that was not extensively characterised[65].

12.5 Compounds containing the 1,2-diaminocyclohexane ligand

Platinum complexes of the 1,2-diaminocyclohexane (DACH) ligand have been extensively studied over the last 10 years. The major cause of this interest is that some of these compounds are active in murine leukaemia models that are resistant to cisplatin[68,69]. Examples of this class include [Pt(DACH)(4-carboxyphthalato)], [Pt(DACH)(oxalato)] and [Pt(DACH)Cl$_4$] (or tetraplatin).

The DACH ligand has two chiral centres and upon complexation of the various isomeric forms (*RR* or *SS* DACH are referred to as the *trans* form of the ligand, whereas the meso compound is referred to as the *cis* form) to platinum, specific conformers are formed[70]. A number of workers have found that platinum complexes containing the various isomeric forms of DACH have different antitumour activities but this effect is tumour line dependent[70,71]. The separation of *cis* and *trans* DACH can be achieved by preparation of their Ni(II) complexes[72].

The three DACH complexes discussed below are shown in Figure 12.5. [Pt(DACH)(4-carboxyphthalato)] (JM 82) is prepared by reacting a solution of the nitrato complex [Pt(DACH)(NO$_3$)$_2$] with the potassium salt of the ligand in water[73]. The material obtained by this procedure is soluble in aqueous base. The structure of JM 82 has not been unambiguously

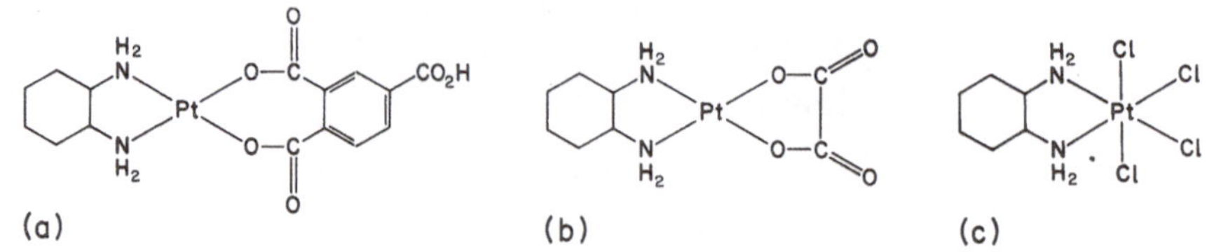

Figure 12.5 The structures of some Pt(DACH) complexes of clinical interest: (a) [Pt(DACH)(4-carboxypthalato)]; (b) [Pt(DACH)(oxalato)]; (c) [Pt(DACH)(tetrachloro)].

established. This is important since the complex as formulated contains a 7-membered phthalato chelate ring, an unusual structural feature. JM 82 has been evaluated in a Phase I clinical trial, but poor solution stability halted further clinical development[74,75].

The [Pt(DACH)(oxalato)] complex is prepared similarly by reacting the DACH/diaquo species (generated from [Pt(DACH)Cl$_2$] and two equivalents of AgNO$_3$) with potassium oxalate in water[71]. The structure of this material has been established by X-ray crystallography[76]. The analogous [Pt(DACH)(malonato)] species can be readily prepared via the reaction of the carbonato complex [Pt(DACH)CO$_3$] with malonic acid in water[77]. [Pt(DACH)(oxalato)] is reasonably soluble in water (7.9 mg/ml) and is currently being evaluated in Phase I and II clinical trials[78].

Tetraplatin can be prepared either by chlorine oxidation of [Pt(DACH)Cl$_2$] or by the direct reaction of DACH with K$_2$PtCl$_6$[79,80]. Infrared spectra and ^{1}H-NMR spectra have been reported for tetraplatin[81]. As with iproplatin, tetraplatin is almost certainly reduced from Pt(IV) to an active Pt(II) species *in vivo*. Eastman has published a study showing that tetraplatin is activated by glutathione to form a species that binds DNA[82].

Tetraplatin is currently being developed for clinical trial[79].

Lastly, a number of other Pt(DACH) complexes have been described recently. Some notable examples include a crystallographically characterised ascorbato complex with a Pt-carbon bond and a lipophilic bis(cyclopentenecarboxylato) species that can be encapsulated in multi-lamellar vesicles[83,84].

12.6 Concluding remarks

All of the platinum antitumour agents discussed above share some key characteristics. Most are simple Pt(II) or Pt(IV) coordination complexes containing two *cis* ammine (NH$_3$) or amine ligands with the rest of the coordination sphere filled by relatively labile, halide or oxygen donors. The introduction of carboplatin, which retains the spectrum of antitumour activity of cisplatin with reduced toxicity, represents an important development in the evolution of platinum agents. The next major accomplishment will be to find a complex with clinical activity against cisplatin resistant tumours. Given that the utility of murine leukaemia models for predicting activity against solid tumours in humans has been called into question, the progress of the various DACH complexes now in clinical trials will be watched closely[85].

Acknowledgements

The author thanks Susan O'Hara and Charlotte Anne Abrams for their help in preparing the manuscript.

References

1. B. Rosenberg, L. Van Camp. J. E. Trosko and V. H. Mansour, *Nature* **222** (1969) 385.
2. J. C. Dabrowiak and W. T. Bradner, *Prog. Med. Chem.* **24** (1987) 129.
3. A. Pasini and F. Zunino, *Angew. Chem. Int. Ed. Engl.* **26** (1987) 615.
4. B. Rosenberg, *Cancer* **55** (1985) 2303.
5. W. C. Rose, J. E. Shurig, J. B. Huftalen and W. T. Bradner, *Cancer Treat. Rep.* **66** (1982) 135.
6. C. F. J. Barnard, M. J. Cleare and P. C. Hydes, *Chem. Brit.* **22** (1986) 1001.
7. A. Werner, *Z. Anorg. Chem.* **3** (1893) 267.
8. G. B. Kauffman and D. O. Cowan, *Inorg. Synth.* **7** (1963) 239.
9. V. V. Lebedinskii and V. A. Golovnya, *Izv. Sektora Platiny Drug Blagorod. Metal., Inst. Obshch. Neorg. Khim. Akad. Nauk. S.S.S.R* **20** (1947) 95.
10. J. W. Reishus and D. S. Martin Jr., *J. Am. Chem. Soc.* **83** (1961) 2457.
11. S. C. Dhara, *Indian J. Chem.* **8** (1970) 193.
12. G. Raudaschl, B. Lippert, J. D. Hoeschele, H. E. Howard-Lock, C. J. L. Lock and P. Pilon, *Inorg. Chim. Acta* **106** (1985) 141.
13. G. H. W. Milburn and M. R. Truter, *J. Chem. Soc. (A)* (1966) 1609.
14. K. Nakamoto, P. J. McCarthy, J. Fujita, R. A. Condrate and G. T. Behnke, *Inorg. Chem.* **4** (1965) 36.
15. J. R. Perumareddi and A. W. Adamson, *J. Phys. Chem.* **72** (1968) 414.
16. F. Macdonald and P. J. Sadler, in *Biochemical Mechanisms of Platinum Antitumour Drugs*, eds. D. C. H. McBrien and T. F. Slater, IRL Press, Oxford (1986) p. 199.
17. N. S. Kurnakov, *J. Russ. Phys. Chem. Soc.* **25** (1893) 565.
18. F. Basolo, M. Lederer, L. Ossicini and K. H. Stephen, *J. Chromatogr.* **10** (1963) 262.
19. F. Aprile and D. S. Martin Jr., *Inorg. Chem.* **1** (1962) 551.
20. C. Barnard, Unpublished results.
21. K. W. Lee and D. S. Martin Jr., *Inorg. Chim. Acta* **17** (1976) 105.
22. M. C. Lim and R. B. Martin, *J. Inorg. Nucl. Chem.* **38** (1976) 1911.
23. T. G. Appleton, R. D. Berry, C. A. Davis, J. R. Hall and H. A. Kimlin, *Inorg. Chem.* **23** (1984) 3514.
24. C. J. L. Lock, in *Inorganic Chemistry in Biology and Medicine*, ed. A. E. Martell, ACS Symp. Ser. Vol. 140 (1980) p. 209.
25. B. de Castro, T. J. Kistenmacher and L. G. Marzilli, *Agents Actions* Suppl. 8 (1981) 435.
26. S. E. Sherman and S. J. Lippard, *Chem. Rev.* **87** (1987) 1153.
27. C. A. McAulliffe and S. H. Murray, *Inorg. Chim. Acta Rev.* **6** (1972) 103.
28. R. F. Borch and M. E. Pleasants, *Proc. Natl. Acad. Sci. USA* **76** (1979) 6611.
29. V. M. Richon. N. Schulte and A. Eastman, *Cancer Res.* **47** (1987) 2056.
30. B. Odenheimer and W. Wolf, *Inorg. Chim. Acta* **66** (1982) L41.
31. P. T. Daley-Yates and D. C. H. McBrien, *Biochem. Pharmacol.* **33** (1984) 3063.
32. J. L. Aull, R. L. Allen, A. R. Bapat, H. H. Daron, M. E. Friedman and J. F. Wilson, *Biochem. Biophys. Acta* **571** (1979) 352.
33. P. T. Daley-Yates and D. C. H. McBrien, *Chem. Biol. Interact.* **40** (1982) 325.
34. S. L. Gonias and S. V. Pizzo, *J. Biol. Chem.* **258** (1983) 5764.
35. S. V. Pizzo, M. W. Swaim, P. A. Roche and S. L. Gonias, *J. Inorg. Biochem.* **33** (1988) 67.

36. M. E. Howe-Grant and S. J. Lippard, in *Metal Ions in Biological Systems*, ed. H. Sigel, Marcel Dekker, New York (1980) p. 63.
37. A. J. Zelazowski, J. S. Garvery and J. D. Hoeschele, *Arch. Biochem. Biophys.* **229** (1984) 246.
38. P. C. Dedon, R. Qazi and R. F. Borch, in *Biochemical Mechanisms of Platinum Antitumour Drugs*, eds. D. C. H. McBrien and T. F. Slater, IRL Press, Oxford (1986) p. 199.
39. D. L. Bodenner, P. C. Dedon, P. C. Keng and R. C. Borch, *Cancer Res.* **46** (1986) 2745.
40. H. Iwabuchi, H. Nagashima and K. Nakamura, *Bunseki Kagaku* **37** (1988) 88.
41. N. Jain, T. S. Srivastava, K. Satymoorthy and M. P. Chitnis, *J. Inorg. Biochem.* **33** (1988) 1.
42. W. I. Sundquist, K. J. Ahmed, L. S. Hollis and S. J. Lippard, *Inorg. Chem.* **26** (1987) 1524.
43. V. Savdek, H. Pivcova, D. Noskova and J. Drobnik, *J. Inorg. Biochem.* **24** (1985) 13.
44. T. S. Elleman, J. W. Reishus and D. S. Martin Jr., *J. Am. Chem. Soc.* **80** (1958) 536.
45. M. J. Abrams, C. M. Giandomenico, J. F. Vollano and D. A. Schwartz, *Inorg. Chim. Acta* **131** (1987) 3.
46. P. D. Braddock, T. A. Connors, M. Jones, A. R. Khokhar, D. H. Melzack and M. L. Tobe, *Chem.-Biol. Interact.* **11** (1975) 145.
47. K. R. Harrap, M. Jones, C. R. Wilkinson, H. McD. Clink, S. Sparrow, B. C. V. Mitchley, S. Clarke and A. Veasey, in *Cisplatin: Current Status and New Developments*, eds. A. W. Prestayko, S. T. Crooke and S. K. Carter, Academic Press, New York (1980) p. 193.
48. M. J. Cleare, J. D. Hoeschele, B. Rosenberg and L. Van Camp, U. S. Patent. 4,140,707, 1979.
49. R. C. Harrison, C. A. McAuliffe and A. M. Zaki, *Inorg. Chim. Acta* **46** (1980) L15.
50. A. Pasini and C. Caldirola, *Inorg. Chim Acta* **151** (1988) 19.
51. S. Neidle. I. M. Ismail and P. J. Sadler, *J. Inorg. Biochem.* **13** (1980) 205.
52. Y. W. Cheung. J. C. Cradock. B. R. Vishnuvajjala and K. P. Flora, *Am. J. Hosp. Pharm.* **44** (1987) 124.
53. L. Canovese, L. Cattalini, G. Chessa and M. L. Tobe. *J. Chem. Soc., Dalton Trans.* (1988) 2135.
54. J. J. Roberts, R. J. Knox, F. Friedlos and D. A. Lydall, in *Biochemical Mechanisms of Platinum Antitumor Drugs*, eds. D. C. H. McBrien and T. F. Slater, IRL Press, Oxford (1986) p. 29.
55. S. J. Harland, D. R. Newell, Z. H. Siddik, R. Chadwick, A. H. Calvert and K. R. Harrap, *Cancer Res.* **44** (1984) 1693.
56. S. Mong, C. H. Huang, A. W. Prestayko and S. T. Crooke, *Cancer Res.* **40** (1980) 3318.
57. M. J. Cleare and J. D. Hoeschele, *Bioinorg. Chem.* **2** (1973) 187.
58. D. R. Alston, T. H. Lilley and J. F. Fraser, *J. Chem. Soc. Chem. Commun.* **22** (1985) 1600.
59. D. R. Alston, A. M. Z. Slawin, J. F. Fraser and D. J. Williams, *J. Chem. Soc. Chem. Commun.* **22** (1985) 1602.
60. G. Hortobagyi, F. Holmes, D. Frye, V. Hug and G. Fraschini, *Proc. Am. Assoc. Cancer Res.* **28** (1987) 198.
61. C. F. J. Barnard, P. C. Hydes, W. P. Griffiths and O. S. Mills, *J. Chem. Res. (S)* (1983) 302.
62. E. Smith and A. H. W. Nias, *Br. J. Radiol.* **59** (1986) 189.
63. J. W. Cowens, F. A. Stevie, J. L. Alderfer, G. E. Hansen, L. Pendyala and P. J. Creaven, *Int. J. Mass Spectrom. Ion Phys.* **48** (1983) 177.
64. J. F. Vollano, E. E. Blatter and J. C. Dabrowiak, *J. Am. Chem. Soc.* **106** (1984) 2732.
65. E. E. Blatter, J. F. Vollano, B. S. Krishnan and J. C. Dabrowiak, *Biochemistry* **23** (1984) 4817.
66. L. Pendyala, L. Cowens and S. Madajewicz, in *Platinum Coordination Complexes in Chemotherapy*, eds. M. P. Hacker. E. B. Douple and I. R. Krakoff, Martinus Nijhoff, Boston, MA (1984) p. 114.
67. J. L. Van der Veer, A. R. Peters and J. Reedijk, *J. Inorg. Biochem.* **26** (1986) 137.
68. J. H. Burchenal, K. Kalaher, K. Dew, L. Lokys and G. Gale, *Biochimie* **60** (1978) 961.
69. A. Eastman and S. Illenye, *Cancer Treat. Rep.* **68** (1984) 1189.
70. J. F. Vollano, S. Al-Baker, J. C. Dabrowiak and J. E. Schurig, *J. Med. Chem.* **30**, (1987) 716.
71. Y. Kidani, K. Inagaki, M. Iigo, A. Hoshi and K. Kuretani, *J. Med. Chem.* **21** (1978) 1315.
72. R. Saito and Y. Kidani, *Chem. Lett.* (1976) 123.
73. P. Schwartz, S. J. Meischen, G. R. Gale, L. M. Atkins, A. B. Smith and E. M. Walker Jr., *Cancer Treat. Rep.* **61** (1977) 1519.
74. D. P. Kelsen, H. Scher, N. Alcock, B. Leyland-Jones, A. Donner, L. Williams, G. Greene, J. H. Burchenal, C. Tan, F. S. Philips and C. W. Young, *Cancer Res.* **42** (1982) 4831.
75. P. J. Andrulis Jr., J. Biswas, A. Troy and P. Andrulis III, in *Platinum and Other Metal Coordination Compounds in Cancer Chemotherapy*, ed. M. Nicolini, Martinus Nijhoff, Boston, MA (1988) p. 450.

76. M. A. Bruck, R. Bav, M. Noji, K. Inagaki and Y. Kidani, *Inorg. Chim. Acta* **92** (1984) 279.
77. A. R. Khokhar, G. Lumetta and S. L. Doran, *Inorg. Chim. Acta* **151** (1988) 87.
78. Y. Kidani, in *Platinum and Other Metal Coordination Compounds in Cancer Chemotherapy*, ed. M. Nicolini, Martinus Nijhoff, Boston, MA (1988) p. 555.
79. W. K. Anderson, D. A. Quagliato, R. D. Haugwitz, V. L. Narayanan and M. K. Wolpert-DeFilippes, *Cancer Treat. Rep.* **70** (1986) 997.
80. B. R. Vishnuvajjala, U. S. Patent Applic. 780,932, 1985.
81. W. K. Anderson and D. A. Quagliato, U. S. Patent. 4,550,187, 1985.
82. A. Eastman, *Biochem. Pharmacol.* **36** (1987) 4177.
83. L. S. Hollis, A. R. Amundsen and E. W. Stern, *J. Am. Chem. Soc.* **107** (1985) 274.
84. R. Perez-Soler, A. R. Khokhar, M. P. Hacker and G. Lopez-Berestein, *Cancer Res.* **46** (1986) 6269.
85. T. H. Corbett, F. A. Valeriote and L. H. Baker, *Invest. New Drugs* **5** (1987) 3.

13 Hypoxia-selective agents: radiosensitisers and cytotoxins

T. C. JENKINS

13.1 Introduction

There is now considerable evidence from radiobiological and clinical studies that the majority of solid human tumours contain a population of clonogenic cells which is poorly oxygenated as a result of inefficient blood supply[1-3]. The presence of such oxygen-deprived hypoxic (or anoxic) cells probably represents a limiting factor in the successful local control of these tumours by classical therapeutic treatment[4,5]. Although recent studies suggest that partial reoxygenation of such cells may occur between the radiation fractions used in current clinical radiotherapy, it is clear that hypoxia continues to pose a significant problem with certain tumours[2-4].

Hypoxic cells develop essentially as a result of rapid tumour growth outpacing the available vascular supply, thus leading to a tumour fraction of oxygen-deprived cells. Protracted oxygen starvation, particularly within cells developing more than 150–200 μm from the nearest blood vessel, ultimately leads to cell death and the focal necrosis observed in most solid tumours[1,3]. Since both oxygen and nutrient gradients are necessarily formed, it is clear that such tumours will comprise a heterogeneous population of cells with differing metabolic status.

Viable hypoxic cells are considerably more resistant to the lethal effects of ionising radiation than normal aerobic (or oxic) cells and it is this factor that largely influences local tumour control by radiation. Hypoxia is also recognised as an important factor in the response of solid tumours to cytotoxic chemotherapy since hypoxic cells are relatively inaccessible to cytotoxic drugs and also have prolonged cell cycle times[1,5,6]. Further, the oxygen status of these cells renders them resistant to anticancer agents such as bleomycin and streptonigrin where cytotoxic efficacy requires the formation of reactive O_2-derived species[5]. Chronically hypoxic tumour cells also have altered chemoresistance and proliferative capacity as a consequence of nutrient depletion. Anaerobic glycolysis also leads to a lower intracellular pH and the necessary involvement of reductive biochemical processes[7]. More disturbingly, cells exposed to hypoxia have been shown to develop chemotherapeutic or multidrug resistance as a result of gene amplification following exposure to anticancer agents such as methotrexate or adriamycin[8,9].

There is thus a clear requirement to develop new agents with selective activity towards hypoxic cells and hence to overcome the dose-limiting problems associated with tumour sterilisation by radiotherapy and/or chemotherapy. The last two decades have seen considerable progress and success in the development of compounds targetted towards hypoxic cells; Denny and Wilson[6] have termed such drugs or adjuncts as hypoxia-selective agents (HSA).

Primary effort has largely focused on chemical agents which specifically enhance the radiosensitivity of hypoxic cells to radiation, in a manner analogous to oxygen, but which have no effect upon oxic cells. Such radiosensitisers, typified by the nitroxyl free radicals and nitroarenes, thus have oxygenmimetic properties. There is substantial evidence that radiosensitivity can also be increased, under both oxic and hypoxic conditions, by suppressing the intracellular level of reduced glutathione (GSH) with thiol-reactive agents such as diethyl maleate[10] and diamide[11], or by inhibition of GSH biosynthesis using buthionine sulphoximine[12], a known inhibitor of γ-glutamylcysteine synthetase. Biaglow et al.[13] have shown that GSH-depleted hypoxic cells are less radioresistant in the presence of radiosensitisers. The role of endogenous and exogenous thiols as modulators of cellular radio- and chemoresponse, by altering the intrinsic capacity for repair, has been discussed[13,14].

Recent effort has largely been directed towards the development of hypoxia-mediated agents which are differentially toxic towards hypoxic cells as a result of metabolic bioreductive activation[3,6,15]. This chemotherapeutic strategy exploits the capacity of hypoxic cells for reductive biochemistry which is either inhibited or prevented in the presence of oxygen. Such cytotoxic agents are typified by bioreducible nitroarenes and quinones, particularly where secondary chemical functionality is present. This selective behaviour has recently been used to potentiate the activity of classical alkylating or carbamoylating chemotherapeutic agents (chemosensitisation) towards hypoxic cells by inducing a level of accumulated sub-lethal chemical damage prior to treatment[3].

13.2 Radiosensitisers

The development of chemical agents which selectively enhance the radiation sensitivity of hypoxic cells has largely stemmed from investigations of the effects of molecular oxygen itself[1]. Radiobiological and radiation chemical studies, particularly using the technique of pulse radiolysis, provide conclusive evidence that the cellular oxygen effect involves fast free radical rather than slow biochemical processes and that oxygen serves to 'fix' radiation-induced chemical damage, particularly at the DNA level. Although the precise nature of the 'oxygen effect' is still not completely resolved, it is apparent that at least two distinct components are involved in its mechanism(s) of action.

Dioxygen is a bifunctional species in that it has both free radical properties due to its unpaired spins and an electron affinity which leads to an involvement in redox-type chemistry. These two characteristics have been extensively exploited in the development of oxygen-mimetic chemical radio-sensitisers; such agents may thus be classified as free radical or electron-affinic, depending on their essential chemical basis. There is now considerable evidence that the different properties are implicated in independent events of radiobiological significance.

Studies of free radical oxygen-mimetic radiosensitisers have primarily focused upon nitroxyls due to the inherent stability (persistence) of this class of organic compound and their ready suitability for chemical modification. However, most active development in the evolution of hypoxic cell radiosen-sitisers has involved the electroreducible or electron-affinic class of agent, since such compounds have more predictable *in vivo* stability.

1 : R = t-Bu
2 : R = 2,6-(MeO)$_2$C$_6$H$_3$

3 : X = O
4 : X = OH

13.2.1 *Nitroxyl free radicals*

Nitroxyl (or nitroxide) radicals are long-lived paramagnetic compounds which contain the N÷O molecular fragment and may thus be viewed as *N,N*-disubstituted derivatives of nitric oxide[16]. The nitroxyl moiety contains a two-centred three-electron bond in which the unpaired spin is effectively de-localised $(R_2\dot{N}-O \leftrightarrow R_2N^{\cdot+}-O^-)$ between the two heteroatoms. Radical persistence[17] is further attributed to steric inhibition of dimerisation, particularly when the nitrogen centre is flanked by bulky substituents[18,19]. The chemistry and applications of nitroxyls has been extensively reviewed[16,20]. Nitroxyls have been used to develop spin-labelled analogues of anticancer agents, including thio-TEPA and cyclophosphamide, as reporter probes for pharmacokinetic and metabolic studies by ESR spectrometry[21].

Radiosensitisation of hypoxic bacteria to ionising radiation by a nitroxyl was first reported by Emmerson and Howard-Flanders[22]. Di-t-butyl nitroxyl (1) and the aryl nitroxyl (2), like nitric oxide itself, were shown to selectively sensitise anoxic bacteria to the lethal effects of X-rays whereas treated aerobic bacteria remained unaffected. *In vivo* experiments, however, revealed that (1) and (2) were too toxic for application in mammalian systems and prompted a quest for active nitroxyls of lower inherent toxicity.

Later studies revealed that the piperidin-1-oxyl (3) (TAN), prepared by H$_2$O$_2$–Na$_2$WO$_4$ oxidation[16,23] of the corresponding piperidone, is an efficient radiosensitiser for hypoxic bacteria[24] and mammalian cells[25] *in vitro*.

Further, (3) proved to have low toxicity when administered i.p. to mice[24]. The bicyclic nitroxyl (7) (9-azabicyclo[3.3.1]nonan-3-one-9-oxyl; NPPN) proved to be a yet more efficient agent[26] for hypoxic *E. coli* bacteria than either (3) or the structurally similar alcohol (4) (TMPN).

6a : R=H
6b : R=CH$_2$Ph
6c : R=Me
6d : R=CO$_2$Ph
7 : R=O·

The synthesis of (7) involves Schöpf–Robinson condensation of 3-pentanone-1,5-dioic acid with 1,5-pentandial, formed by acid hydrolysis[27] of the dihydro-2*H*-pyran (5), in the presence of ammonia[28]. Catalytic oxidation of the formed amine (6a) with H$_2$O$_2$-phosphotungstic acid gives (7) in 74% yield[29]. Alternative synthesis of (6a) via hydrogenolysis of the N-protected benzyl derivative (6b) was suggested to provide a more reliable route[29]. However, cleavage of tertiary amine (6c) with ClCO$_2$Ph and subsequent hydrolysis of the formed carbamate (6d) provides a facile route to amine (6a) which avoids the hydrogenation step[30].

The presence of one (or more) hydrogen(s) at the α-carbon position(s) in a *sec*-alkyl nitroxyl normally results in facile bimolecular dismutation[16,20]. Nitroxyl (7) is stable, however, as dismutation is not possible since nitrone formation would involve the generation of a bridgehead carbon atom, in violation of Bredt's rule[29,31]; (7) is thus persistent and long-lived.

There is clear evidence that the indirect chemical damage induced during radiolysis of DNA in aqueous media resides chiefly on the nucleotide bases[32] and involves ˙OH-adduct formation via the primary water radiolysis product. Pulse radiolytic studies reveal that nitroxyls react readily with the bioradical adducts, B(OH)˙, formed by interaction of deoxyribonucleotides or DNA with hydroxyl radicals[26,33,34]:

$$R_2N\text{-}O + B(OH)\cdot \xrightarrow{k_2} \text{products}$$

where B represents a base residue on the DNA biotarget molecule. The radiation-modifying effect of nitroxyls is thus ascribed to their bimolecular reactivity towards radiation-induced bioradicals formed at biologically sensitive sites such as DNA and to effect 'fixation' of irreversible chemical damage[22]. Two modes of radical-radical coupling via the nitroxyl are feasible, leading to diamagnetic products. Thus coupling via the nitrogen centre will generate B(OH)N(O)R$_2$ amine-*N*-oxides, whilst the alternative mode of coupling involving the oxygen centre will afford B(OH)ONR$_2$ hydroxylamine derivatives. In the presence of molecular oxygen, as in aerobic cells, such processes will necessarily compete with the formation of peroxyl B(OH)O$_2$˙ adducts. The fates and possible radiobiological significance of the various products have been discussed[34].

Fast kinetic studies of the reactivity of (3), (4) and (7) towards such ˙OH-radical adducts using the pulse radiolysis method reveal[32] a steric kinetic factor in their reaction with bioradicals (e.g. $10^{-8} k_2[R_2N\dot{-}O + DNA(OH)\dot{\,}]$ = 2.0, 0.82 and $0.59\,dm^3\,mol^{-1}\,s^{-1}$ for (7), (3) and (4), respectively). Thus nitroxyl (7), where the radical centre is less sterically hindered, is typically 2–4 times more reactive towards such bioradicals than either (3) or (4). A similar dependence upon steric bulk has been implicated in the mechanism of radiosensitisation by nitroxyl radicals to account for the observed (7) > (3) > (4) ranking order of efficiency[26,33].

Nitroxyl (4) was found to marginally radiosensitise the hypoxic gastrointestinal tract of mice *in vivo*, provided that irradiation was performed soon after administration[35]. In contrast, other workers failed to achieve *in vivo* radiosensitisation with (3), (4) or (7)[36–38]. Quantitative electron spin resonance (ESR) studies reveal that the nitroxyl moieties of (4) and (7) are rapidly degraded in whole blood, with the more hindered radical (4) having the longer half-life[37,38]. *In vitro* studies of cells pretreated with reduced glutathione (GSH) or cysteine[39] indicate that the radiosensitising efficiency is removed by *in situ* reduction of the nitroxyl to an inactive hydroxylamine via a redox-coupled process:

$$R_2N\dot{-}O \quad\quad RSH \quad (e.g.\ GSH)$$
$$R_2N-OH \quad\quad \tfrac{1}{2}RS-SR$$

The one-electron redox properties of nitroxyl radicals and their derived hydroxylamines have been examined[40]. Clearly, the metabolic reactivity of nitroxyl free radicals suggests only a limited prospect of therapeutic utility, particularly with regard to short-lived, sterically-unhindered radicals (e.g. (7)).

$$4 \xrightarrow[25-60\%]{(ClCO)_2X - pyridine}$$

8 : X = $(CH_2)_2$
9 : X = cis-(C_3H_4)

Nitroxyls, in common with oxygen and most radiosensitisers, behave as radiation dose-modifying adjuncts in their effect upon hypoxic cells, i.e. sensitisation is shown only by an increase in the slope of the exponential part of the radiation survival curve. Cooke *et al.*[41], however, showed that the succinate ester biradical (8), derived from alcohol (4), is not only a better hypoxic dose-modifying agent on a molar basis than any mono-nitroxyl, but also suppresses the survival of hypoxic cells at low doses of radiation (shoulder removal) without undue toxicity. Indeed, (8) is a more efficient radiosensitiser than oxygen at radiation doses < 10 Gy. No effects were found for oxic cells. The shoulder modification reported for thiol-reactive agents such as diamide[42] suggested that depletion of the intracellular non-protein thiols essential for biochemical repair may be important[13,14].

Studies of the interaction of (8) with reducing species showed a usefully slow rate of cellular reduction and that sensitisation of hypoxic cells is maintained in conditions where the nitroxyl monoradicals are no longer effective. Interestingly, the valuable shoulder-suppressing property is lost following partial reduction to the mononitroxyl-hydroxylamine[39]. The similarity between biradical (8) and the triplet ground-state electronic configuration of molecular dioxygen ($^3O_2$), a stable biradical ($^.O-O^.$), was noted. These observations thus stimulated renewed impetus for the development of active nitroxyls, particularly nitroxyl biradicals, with useful application as hypoxia-selective radiosensitisers.

Millar et al.[43] showed that a shoulder-modifying response is also given by the cis-diester (9) and the succinamide (11) formed, respectively, by acylation of (4) and succinoylation of the amino-nitroxyl (10)[44]; both are effective radiosensitisers in vitro. The radiation dose-modifying effects are dependent upon both the cell-drug contact time and temperature, suggesting that nitroxyl biradicals affect cellular biochemistry in such a way as to alter the repair capacity of the cells and, further, that drug permeation may be difficult[43].

The homologous series of bifunctional nitroxyls (14), prepared by superior bis-acylation of 4-aminopiperidine (12) and subsequent oxidation of the formed carboxamides (13), was examined to establish the optimal spatial separation of the radical centres. The data obtained for (14), where $m = 0-7$, support the hypothesis that a lack of interaction between the two nitroxyl centres, as revealed by ESR, is important for events leading to shoulder modification. Further studies with the piperazine biradical (15) and the derived diquaternary salt (16) revealed that molecular charge is also an important factor which limits radiosensitising efficiency[45,46].

The glucocorticoid hormone dexamethasone inhibits shoulder modification by uncharged nitroxyl biradicals[46] but does not influence their cellular accumulation, as determined by quantitative ESR spectrometry. Examination of the data obtained with several nitroxyl mono- and biradicals provides

HNCOCH₂N⌒NCH₂CONH

12 $\xrightarrow[\text{31\%}]{\text{4 steps}}$

15: free base
16: $(\text{MeI})_2$ salt

compelling evidence that the cellular effects leading to dose and shoulder modification are mediated by different events. Further, it is suggested[46] that the accumulation of nitroxyls at specific intracellular sites, particularly cell membranes, and not thiol depletion plays a major role in determining shoulder modification.

Supporting evidence for this hypothesis was obtained with the two conformationally restricted isomeric biradicals (17a) and (17b), prepared by reductive amination of (3) with cis- and trans-1,2-diaminocyclopropane[47]. Quantitative ESR revealed that >70% of the intracellular concentration of (17b) is localised in the plasma membrane and that cellular integrity is essential for drug uptake. The dependence upon drug-cell contact time noted for nitroxyl biradicals (Table 13.1) is entirely in accord with their inherent lipophilic nature and, hence, rates of membrane permeation[47]. Nitroxyls are known to bind strongly to model bilayers in a manner which is dependent upon the lipid composition. Indeed, nitroxyls are often used as spin-labels to provide spectroscopic probes for the ultrastructure and integrity of cell membranes[48].

The current model[47] for radiosensitisation by nitroxyl biradicals predicts that these lipophilic agents penetrate cells more slowly than mono-nitroxyls. The plasma membrane acts as a barrier to drug uptake and accumulates a large proportion of the incorporated compound. Radiosensitisation at low

Table 13.1 Partition properties and uptake of nitroxyls (4), (17) and (18) and (27) (misonidazole) into Chinese hamster cells (from Refs. 46, 47); extracellular concentration 1 mmol dm^{-3}

Compound	Lipophilicity P (1-octanol/H₂O)	Contact time (h)	Intracellular concentration (μmol dm^{-3})
(4)	4.0 ± 0.1	1.0	10.3 ± 1.2
(17)	1.2 ± 0.1	0.25	1.6 ± 1.0
		1.0	3.6 ± 0.5
		3.0	96.5 ± 1.5
(18)	4.3 ± 0.2	0.25	<0.3
		1.0	4.0 ± 0.5
		3.0	630
(27)	0.43	1.0	680

extracellular nitroxyl concentrations and short cell-drug contact times is thus a membrane-related phenomenon and results in a dose-modifying response. Higher extracellular concentrations or longer contact times enable permeation to less accessible sites, which results in shoulder-modifying behaviour[47]. The influence of a nitroxyl (or dinitroxyl) hypoxic cell sensitiser upon radiation response will thus be largely dependent upon its individual physicochemical characteristics (i.e. lipophilicity, basicity, charge and aqueous solubility).

17a : cis
17b : trans

18a : R=O·
18b : R=OC(CN)Me₂

The effect of using nitroxyls in combination with 'electron-affinic' hypoxic cell radiosensitisers has been examined in many systems, and it is clear that the two classes of agent operate by different and independent mechanisms[49-51]. Indeed, there is evidence from studies with the mixed-functional agent (18a) and the derived O-alkylhydroxylamine (18b), formed by homolytic alkylation during the thermolysis of 2,2'-azobis(2-methylpropionitrile), that there is active competition between the two sensitiser moieties for radiation-induced lesions[51]. Nitroxyl (18a) is prepared by condensation of (10) with the 2-nitroimidazole oxirane (32).

The current generation of available nitroxyls is unlikely to be of real clinical advantage in fractionated radiotherapy due to their finite lifetimes *in vivo* as a result of metabolic removal. Attempts to minimise such reductive degradation by, for example, alterations in the route and vehicle of drug administration have thus far failed to reveal substantial *in vivo* activity in animal tumour systems[30].

It remains a fact, however, that, on a molar basis, nitroxyls are considerably more effective in mediating changes in the radiation response of hypoxic cells than electron-affinic agents which reach much higher intracellular concentrations[46,47]. Indeed the nitroxyls currently rank as the most efficient chemical modifiers of hypoxic cell radiation response *in vitro*, and the nitroxyls as a class appear to be essentially non-toxic *in vivo*.

The development of prodrug forms (e.g. amine-N-oxides, O-alkyl- or O-acylhydroxylamines) designed to improve the general bioavailability, where the biologically active nitroxyl species are generated as a result of metabolic conversion at or near the sensitive biotarget, clearly merits consideration.

On a lesser level, persistent nitroxyl free radicals continue to provide elegant mechanistic tools for probing the many chemical processes involved in radiobiological events[52,53].

13.2.2 *Electron-affinic radiosensitisers*

Many compounds other than nitroxyl free radicals are now known to be capable of differentially sensitising hypoxic cells to the lethal effects of ionising radiation. Such active compounds, typically quinones or nitro-arenes or -heteroarenes, must be present in the cells at the time of irradiation; addition of these agents within only milliseconds after radiation exposure greatly reduces or eliminates the sensitising efficiency[54]. The observation that the more powerful oxidizing agents are the most efficient radiosensitisers prompted the suggestion that the ease of one-electron reduction or electron affinity of such compounds is the basis for radiosensitisation[55]. This thesis has provided the basis for the development of electron-affinic agents which are (i) essentially non-toxic, (ii) able to diffuse into hypoxic cells, and (iii) not rapidly metabolised *in vivo*.

19 : R = Me
20 : R = (CH$_2$)$_2$NMe$_2$

21

22

Two independent groups reported efficient radiosensitisation of hypoxic mammalian cells *in vitro* by the nitrobenzene (**19**) without any radiobiological effect upon normal aerobic cells[56,57]. The water-soluble Mannich condensation product (**20**) was subsequently found to be superior, on a molar basis, by an order of magnitude[58,59]. Raleigh *et al.*[60] examined a series of ring-substituted nitrobenzenes *in vitro* and were able to establish an unambiguous structure-activity relationship for radiation dose modification, using Hammett σ-substituent constants.

In vitro radiosensitisation of hypoxic cells has since been found with a wide spectrum of quinones[55], including isoindole-4,7-diones[61], and nitroarenes and nitroheteroarenes, including simple nitro-benzene, -furan, -pyridine, -pyrrole, -thiophene, -pyrazole, -thiazole and -imidazole derivatives[62,63]. Many such compounds have clinical application as antimicrobial or antiprotozoal agents[63-65]. Subsequent assessment of selected compounds as *in vivo* radiosensitisers, however, revealed strictly limited efficacy due to metabolic inactivation, unfavourable pharmacokinetics or dose-limiting aerobic (or oxic) cytotoxicity[1,62,63]. Most attention has consequently focused upon 1-alkylnitroimidazole derivatives where biological effectiveness is less compromised by untoward cytotoxic effects.

The 5-nitroimidazole (**21**) (metronidazole) is an antiprotozoal agent which radiosensitises both *in vitro*[66] and *in vivo*[67]; clinical radiotherapeutic evidence exists for the radiosensitising activity of (**21**) towards human glioblastomas at elevated dose levels[68]. The maximum dose of (21) which could be tolerated following oral administration was, however, found to limit the maximum drug concentration achievable within a human tumour and hence the radiosensitising response[1]. The large dose of (21) required to effect appreciable enhancement of radiosensitivity *in vivo* thus promoted a search for more active nitroimidazoles[1,2,63,69].

Nitroarene agents alter the radiation response of mammalian cells by increasing the slope of the exponential portion of the survival curve following irradiation under hypoxic conditions. No sensitising effects are seen with aerobic cells. Only very few nitro-compounds influence the initial shoulder region, where the radiation response is inherently poor due to cellular repair processes. A measure of the radiosensitising efficiency or dose modification factor (enhancement ratio or ER) is given by the ratio of the ultimate slopes obtained in hypoxia in the presence and absence of the added sensitiser. The ER increases with drug concentration and may often approach or exceed the value of the oxygen enhancement ratio (OER) found in aerobic cells. Since the ER value is concentration-dependent, it is usual to compare the molar concentrations required to give an ER of 1.6 ($C_{1.6}$). Thus, the lower the $C_{1.6}$ value the greater the sensitising efficiency.

Adams *et al.*[69] examined the radiosensitising properties of several 1-alkyl-2- and -5-nitroimidazoles *in vitro* and showed that the $C_{1.6}$ values obtained correlate well with the electron affinities of the compounds, determined as their one-electron reduction potentials, E_7^1. This correlation extended beyond nitroimidazoles to include nitro-benzenes and -furans, and dioxygen itself. The lipophilicity, P, of the compounds has only negligible influence upon the sensitising efficiency *in vitro*.

The position of the one-electron transfer equilibrium:

$$S^{\cdot -} + Q \underset{}{\overset{K}{\rightleftharpoons}} S + Q^{\cdot -}$$

where Q is a redox indicator such as a viologen or quinone and $S^{\cdot -}$ is the radical anion of a sensitiser. S can be determined using quantitative pulse radiolytic and fast kinetic spectrophotometric methods[70-72]. The equilibrium constant, K, for the electron transfer process may be estimated either directly, by quantitation of the transient absorption spectra following pulse irradiation, or indirectly by monitoring the separate rate(s) of approach to equilibrium[71]. Since the one-electron reduction potentials at pH 7, $E_7^1(Q/Q^{\cdot -})$, of the redox indicators are known, it is possible to determine the required redox potential ($E_7^1(S/S^{\cdot -})$) of the sensitiser compound (at pH 7):

$$E_7^1(S/S^{\cdot -}) = E_7^1(Q/Q^{\cdot -}) - 59 \log K$$

The more positive the E_7^1 value for a compound, the greater the electron affinity and hence ease of reduction; the one-electron reduction potential for molecular oxygen, $E_7^1(O_2/O_2^{\cdot -})$ is $-155\,mV$. Reference redox indicators are available to enable E_7^1 values for compounds to be estimated in the range of

-100 to $-900\,\text{mV}$[71,72]. Polarographic half-wave potentials may also be used as a measure of one-electron reduction if the values are determined under thermodynamically reversible conditions. An inverse correlation exists between the unpaired spin density on the nitro-group of radical anions, as revealed by ESR, and $E_7^{\frac{1}{2}}$ values for the parent nitroarenes[70,73].

Wardman[63] has described an elegant chemical basis for hypoxic cell radiosensitisation by nitroaromatic compounds in which the separate biological and chemical properties are used to develop quantitative structure-activity relationships. The redox properties of nitroaromatic compounds were established as clearly important in determining both radiosensitising efficiency and cytotoxicity *in vitro*.

Multiple regression analysis for 44 nitroarenes with various aromatic nuclei, examples of which are listed in Table 13.2, reveals that the radiosensitising efficiency for mammalian cells *in vitro* $(C_{1.6})$ can be fitted to a structure-activity relationship[74] of the general form:

$$-\log C_{1.6} = b_0 + b_1 E_7^{\frac{1}{2}} + b_2 \log P + b_3 (\log P)^2$$

where P is the usual partition coefficient for the nitro-compound, normally measured at physiological pH (pH 7.4). Statistical tests show that the terms involving partition behaviour may be ignored, confirming earlier observations that lipophilicity is unimportant *in vitro*. The following empirical structure-activity relationship is thus observed:

$$-\log(C_{1.6}/\text{mol dm}^{-3}) = 6.54 + 8.21(E_7^{\frac{1}{2}}/\text{V})$$

Quantitatively similar behaviour is found using radiobiological end-points other than a 1.6-fold dose-modifying factor (ER value). Thus, a 100 mV alteration in $E_7^{\frac{1}{2}}$ leads to an approximately 10-fold change in the concentration of added sensitiser $(C_{1.6})$ required to effect a fixed level (ER = 1.6) of enhanced radiosensitivity, with the more electron-affinic agents showing the lowest $C_{1.6}$ values.

These conclusions provide quantitative support for the electron transfer mechanism proposed for radiosensitisation by electron-affinic compounds[55],

Table 13.2 One-electron redox potentials, sensitiser efficiencies and chronic aerobic cytotoxicities *in vitro* for typical nitroarenes (from Refs. 69, 74, 75, 140)

Compound	$E_7^{\frac{1}{2}}(\text{mV})$	$C_{1.6}(\text{mmol dm}^{-3})$	$C_{50\%}(\text{mmol dm}^{-3})$
Nitrofurazone	-257	0.05	0.018
(34)	-346	0.06	—
(19)	-355	—	0.13
Benznidazole	-380	0.24	0.75
(29)	-388	0.54	—
(27)	-389	0.3	1.3
(21)	-486	4.0	6.5

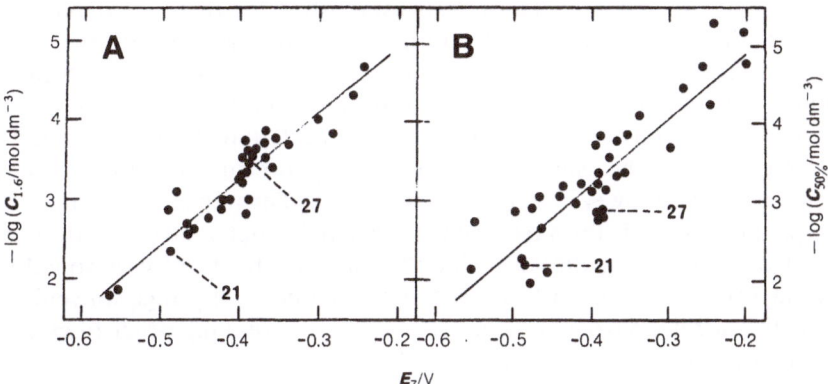

Figure 13.1 Dependence of (a) radiosensitisation efficiency ($C_{1.6}$), and (b) chronic aerobic cytotoxicity ($C_{50\%}$) on one-electron reduction potential (E_7^1) for a range of nitroarenes and -heteroarenes (after Adams *et al.*[(74,75)]).

whereby such agents behave as one-electron oxidants which 'fix' or render irreversible a level of radiation-induced chemical damage at sensitive cellular sites or biotargets (e.g. DNA). Such events necessarily involve fast radiation-induced free radical events rather than slow biochemical processes, and probably lead to the formation of unstable radical cation or onium-type DNA species. The dependence of hypoxic cell radiosensitisation efficiency on E_7^1 for a range of nitroarenes *in vitro* is shown in Figure 13.1a.

Simultaneous determination of the chronic aerobic *in vitro* cytotoxicities[75] for various nitroarenes reveals that the data can be fitted to an equation (Figure 13.1b) of the form:

$$-\log(C_{50\%}/\text{mol dm}^{-3}) = 6.59 + 8.40(E_7^1/\text{V})$$

where $C_{50\%}$ is the molar concentration of compound required to reduce cellular survival by 50%. Drug lipophilicity is found to have little influence upon *in vitro* cytotoxic activity, although this property would be likely to affect the *in vivo* response[75]. Thus, a 100 mV alteration in E_7^1 leads to an approximately 10-fold change in the concentration necessary to effect a fixed level of cytotoxic killing of aerobic cells. Qualitatively similar effects are seen if the acute aerobic and hypoxic cytotoxicities, rather than the chronic aerobic cytotoxicities, are used as biological end-points[1,75]. It is interesting to note that the slope and intercept values for the two biological responses are of a similar magnitude, suggesting that closely related chemical processes may be responsible.

The ratio $C_{50\%}/C_{1.6}$ thus provides a qualitative therapeutic index of the *in vitro* effectiveness of a radiosensitiser, with the better agents having the higher quotient values[75]. It is noteworthy that $C_{50\%}/C_{1.6}$ appears to be independent of either the complexion of the aromatic residue or the redox potential, E_7^1.

The structure-activity relationships established for these nitroarenes encompass the hypoxic radiosensitisation and aerobic cytotoxicity afforded by many electroreducible agents, including oxygen and oxidants such as the highly electron-affinic quinones (e.g. 1,4-benzoquinone, $E_7^1 = +99$ mV; menadione, $E_7^1 = -203$ mV[40,63]). Indeed, these correlations have been found to correctly predict the effectiveness of a wide spectrum of electron-affinic radiosensitisers. However, extensive *in vitro* studies indicate that an active compound should ideally have a one-electron redox potential (E_7^1) in the range -350 to -500 mV if radiosensitising efficiency is to be effective without dose-limiting cytotoxicity towards normal or oxic tissues. This range conveniently spans the nitroimidazole class of agents, with the ranking order of electron affinity given by: 2-nitro- > 5-nitro- > 4-nitro-[62,63].

The natural antibiotic and antiprotozoal agent (**22**) (azomycin) is one of the simplest known biologically active imidazoles and provided the early stimulus for the development of nitroimidazoles with trichomonacidal activity[65]. It is, however, the least accessible of the isomeric nitroimidazoles and cannot be prepared by usual electrophilic nitration of the imidazole nucleus, which always proceeds at the C4 and C5 ring positions[76]. Thus, nitration of the parent imidazole or 1-alkylimidazoles affords the tautomeric $1H$-4(5)-nitro-derivative or a mixture of isomeric 4- and 5-nitro-1-alkylimidazoles. The 2-nitro isomer (**22**) is instead prepared via diazotisation of 2-aminoimidazole at high dilution[77,78]. Kirk has recently shown[79] that 2-lithioimidazoles may be exploited to generate both the 2-nitro derivatives and other imidazoles where functionalisation is required at the C2 position.

Synthetic effort in this area has largely been directed towards the synthesis of N1-substituted nitroimidazoles. Indeed, most attention has focused upon the development of 1-alkyl-2-, -4- and -5-nitroimidazoles[65]. The alkyl moieties that have been introduced range in complexity from unsubstituted alkyl or alkenyl groups to glycosides and mono- or poly-substituted alkyl groups containing functional residues (e.g. —OH, —OR, —NR$_2$, —halogen, —CO$_2$H, —SO$_2$R, —SR, etc.). Several alternative strategies have been adopted for the synthesis of these compounds: (i) 1-alkylation of nitro-imidazoles; (ii) nitration of 1-alkylimidazoles, and (iii) aromatic ring or sidechain chemical modification of existing 1-alkylnitroimidazoles. The advantages and disadvantages of each synthetic strategy have been reviewed[65].

The strongly electron-withdrawing nitro-group in simple nitroimidazoles renders the π-system electron-deficient and hence facilitates ionisation of the

N1 proton. The basicities of these acids are influenced[65] by the position of the nitro-group ((22) pK_a 7.15; (23) pK_a 9.2), and 1-alklation is thus generally facile. Indeed, N1-substitution normally proceeds under relatively mild conditions in the presence of weak bases; this reactivity suggests an application for either organosilicon or sulphonamide N-blocking groups for the introduction of additional ring functionality[80]. Since $1H$-4(5)-nitroimidazole (23) can exist in two tautomeric forms, treatment with usual alkylating or N-blocking reagents affords a mixture of 4- and 5-nitroimidazole isomers, (24) and (25); the isomeric composition of the alkylation product can be influenced by appropriate choice of reagent and/or reaction conditions[65]. Thus, reflux treatment of the 2-methyl derivative of (23) with 2-bromoethanol or oxirane favours formation of the 5-nitro isomer (21) if strongly acidic media are used; basic or neutral conditions give the 4-nitro isomer as the major or sole product[65].

$$
\begin{array}{l}
22 \\
23
\end{array}
\quad
\xrightarrow[\text{20--70\%}]{\overset{\triangle O -(K_2CO_3)}{-X}}
$$

X = OH, OR, H
R, halogen, etc.

24: 2-NO$_2$; R=H
25: 4-NO$_2$; R= 2-Me
26: 5-NO$_2$; R= 2-Me
27: 2-NO$_2$; R=H; X=OMe

Thermal condensation of nitroimidazoles, either as the parent compound or the derived sodium salt, with substituted oxiranes (usually in the presence of a weak inorganic base catalyst if the sodium salt is not used) leads to N1-alkylation via nucleophilic ring-opening of the oxirane. Thus, treatment of azomycin (22) with 1,2-epoxypropyl compounds (substituted 2-methyloxiranes) gives the corresponding 2-propanol derivatives (24)[65,81]. Hoffer and Grunberg[82] demonstrated that this reaction can also be used to generate the analogous 4- and 5-nitroimidazoles (e.g. (25) and (26)), with acidic solvent media again favouring formation of the 5-nitro isomers. The yields of the thermodynamically preferred 4-nitro isomers are increased if neutral or weakly basic reaction conditions are used. The nitroimidazolyl-2-propanol derivatives (24–26) have marked chemotherapeutic activity as both antiprotozoal and antibacterial agents and good tolerance in vivo[81,82].

The 2-propanol derivative (27) (misonidazole, Ro.07-0582), prepared in 72% yield by alkylation of (22) with the corresponding oxirane[81], was found to be a superior radiosensitiser for both hypoxic bacteria and mammalian cells in vitro[83]. This result is in accord with the greater electron affinity (E_7^1) determined for (27) (see Table 13.2) and related 2-nitroimidazoles. In vivo radiosensitiser evaluation revealed that (27) shows a broad spectrum of activity in experimental tumours of very many different types, including tumours grown as xenografts in immune-deprived mice (for review, see Ref. 67).

In vivo sensitisation by (27) (misonidazole) thus appears to be a fairly general phenomenon in solid experimental rodent and human tumours, particularly when large single doses of radiation are employed. The influences of tumour type and size, together with fractionation of radiation (and, hence, inter-fraction tumour reoxygenation), upon the sensitising efficiency achievable with (27) have been discussed[1]. The fractionation regimen adopted has been shown[84] to be critical in influencing the local control of the MT tumour in WH mice[84] irradiated with either single or multiple fractions of X-rays over variable periods.

Clinical studies revealed that large single doses of (27) are tolerated and pharmacological evidence[85] showed that, in most cases, tumour penetration is generally good. It was soon found, however, that (27) induces acute peripheral neuropathy in patients, particularly when administered in multiple dose regimens[86,87]. The cumulative neurotoxic properties associated with (27) limit the total dose of drug that can be administered to about $12 \, g \, m^{-2}$ over 6 weeks. Attempts to ameliorate the unpleasant side effects of neuropathy using dexamethasone rescue[88] have not been entirely successful. The failure to observe significant radiosensitisation by this compound in most clinical trials is attributed to dosage limitations which prevent the agent from attaining optimal tumour concentrations[2,4,89].

Studies with (27) and structurally related compounds of similar electron affinity suggested that the neurotoxicity of these agents is partly related to their lipophilic properties[90–92]. Brown and Workman[91,92] showed that lipophilicity largely determines the pharmacokinetics of these compounds and related changes in the penetration of neural tissue and rapid renal clearance to toxicity and therapeutic ratio. This situation contrasts with conclusions from earlier *in vitro* studies that the lipophilicity factor is essentially unimportant[75]. Compounds with values of partition coefficient P from 0.014 to 30, determined from their partition behaviour between aqueous buffer and 1-octanol, were examined for their ability to penetrate into the brain tissue of experimental animals. The levels of drug attained in the brain and plasma were used to assess total drug exposure from measurements of AUC (area under the curve multiplied by time). Compounds with values of P similar to or greater than that for (27) (i.e. $P \geqslant 0.43$) show ratios of around 100%, indicating no tendency to concentrate in or be excluded from the brain. However, for more hydrophilic compounds where $P < 0.43$ the brain/plasma ratio decreases as a function of P due to the lipid nature of the blood-brain barrier, indicating that these agents are less able to penetrate neural tissue.

Tumour penetration by 2-nitroimidazoles is independent of P over the range 0.026–3.1[92]. Further, i.v. administration of more hydrophilic drugs often results in higher peak plasma levels than those achieved with (27) given either i.v. or orally. Thus, in terms of therapeutic ratio, compounds of lower P than (27) will give similar levels of radiosensitisation since similar con-centrations of drug can be achieved in tumours, but their toxicity should be considerably reduced as a result of exclusion from the brain. These conclusions promoted the quest for electron-affinic agents which are less lipophilic (i.e. more hydrophilic) to overcome the dose-limiting problems associated with neurotoxicity[93,94].

28: X = OMe
29: X = NH(CH$_2$)$_2$OH

30: n = 2–11

The ethanolamide compound (**29**) (etanidazole, SR-2508), prepared by stepwise alkylation of azomycin (**22**) and aminolysis of the formed ester (**28**) with ethanolamine[81], was selected on this basis as an efficient radiosensitiser[95]. This neutral 2-nitroimidazole is much less lipophilic ($P = 0.046$) than (**27**) and has a similar redox behaviour ($E_7^1 = -388$ mV); (**27**) and (**29**) have comparable *in vitro* and *in vivo* radiosensitising efficiencies. Further, (**29**) is only slowly accumulated in the brain and peripheral nerves[91], shows much faster rates of plasma and renal clearance and is more slowly metabolised *in situ*. Clinical studies[96,97] indicate that (**29**) causes no acute effects at individual doses of $\geqslant 3$ g m^{-2}, but does induce peripheral neuropathy at cumulative doses of 30 g m^{-2}. The total tolerated drug dose is greater than for (**27**) and neuropathy is induced at a considerably higher dosage. Encouraged by the positive results from preliminary studies, further clinical trials with (**29**) as a radiosensitiser for hypoxic tumours are currently in progress[4,98].

Adams and his colleagues examined the effects of chain length and terminal basic substituents upon the sensitising efficiency of nitroimidazoles by developing an homologous series of ω-aminoalkyl derivatives (**30**), where $n = 2$–11 and $-$NR$_2$ represents a dialkylamine or cyclic amine function[99]. Methiodide quaternary salt forms were also examined. Two alternative synthetic strategies were adopted: (i) catalysed condensation of (**22**) with a 1-bromo-ω-chloroalkane and subsequent treatment with a secondary amine in the presence of a catalytic quantity of sodium iodide, or (ii) alkylation of (**22**) with preformed N-(ω-chloroalkyl)amines. The first route is preferred since it is both higher-yielding and provides flexibility with regard to choice of amine[65,99].

The radiosensitising behaviour of all these compounds towards hypoxic Chinese hamster cells *in vitro* is qualitatively similar to that of nitroimidazoles previously reported[69,75]. It is clear, however, that sidechain length (n) markedly affects both the radiosensitising efficiency and chronic aerobic toxicity since, within an homologous series of analogues, as n increases, the value of $C_{1.6}$ decreases to a minimum at $n = 5$. In contrast, the compounds become progressively more toxic to aerobic cells as n increases, in that lower concentrations are required to effect inhibition of colony-forming ability. Optimal sensitiser 'effectiveness' is apparent for compounds where $n = 5$. Lipophilicity appears not to be important within this series of compounds and the reasons for this redox-independent phenomenon remain obscure[1].

It was noted that about an order of magnitude increase in hypoxic radiosensitising efficiency is observed for certain amines of general formula (30) *without* any increase in chronic aerobic cytotoxicity when compared to simple uncharged nitroimidazoles such as (27)[99]. Further, this increased efficiency is greater than expected on the basis of measured redox behaviour, using the earlier structure-activity relationship[75]. It is evident that prototropic equilibria influence the *in vitro* effectiveness of nitro-compounds with basic moieties, with weakly basic functions affording little or no improved efficiency. Enhancement of sensitising efficacy is observed for basic agents (30) with $pK_a > 7.5$ (e.g. piperidino- and pyrrolidino-derivatives) where the compounds are essentially fully protonated at physiological pH. Involvement in such processes probably explains the superior radiosensitising properties observed for the water-soluble nitrobenzene (20) when compared with the parent ketone derivative (19). The influence of pK_a upon the protonation status of nitroarenes and their derived radical anions, particularly with regard to redox behaviour, has been discussed[63,100].

Although *in vitro* studies with ω-morpholinoalkyl homologues (30) indicate that compounds with $n \geq 4$ are 5–10 times more efficient than the non-basic 2-nitroimidazole (27)[99], Chaplin *et al.*[101] showed that this activity is not achieved *in vivo*. Failure to achieve enhanced sensitising efficiency was ascribed to either poor bioavailability or limited drug permeation into the tumour at physiological pH[101]. Nevertheless, the incorporation of basic, and hence protonatable, amine functions into 2-nitroimidazoles was shown to influence both the redox and partition behaviours. Such effects indicated a potential means of manipulating the molecular properties of hypoxic radiosensitisers, particularly lessening the inherently high lipophilic character, for therapeutic gain[102].

33 : NR$_2$ = amino
34 : NR$_2$ = piperidine

Smithen *et al.*[103] used this strategy to develop a series of chiral 1-(2-nitro-1-imidazolyl)propanolamines (33) by aminolysis of the oxirane (32) with primary and secondary amines. The intermediate oxirane (32), prepared by base-catalysed ring closure of the chlorohydrin (31) formed by alkylation of (22) using epichlorohydrin[81], has proved to be a versatile synthon for the synthesis of many hydrophilic 2-nitroimidazoles[65]. Reflux condensation of (31) with 2 mole equivalents of the amine provides an alternative route to (33) which avoids the intermediate oxirane. Since no stereoselectivity is afforded, the amino-compounds are formed as racemic mixtures of the two chiral isomers.

Physicochemical and *in vitro* studies revealed that many basic agents of general formula (33) have hypoxic cell radiosensitising efficiencies which are equal or superior to that of the neutral 2-nitroimidazole (27) (misonidazole). One such propanolamine (34) (pimonidazole, Ro.03-8799), where $-NR_2$ is a piperidine moiety, was identified as clearly superior to (27)[103] solely on the basis of its therapeutic ratio *in vitro* ($C_{50\%}/C_{1.6}$). Preliminary *in vivo* studies revealed that the bioavailability of (34) is not problematic and that the compound is apparently well tolerated.

The radiosensitising efficiency of (34) to mammalian cells *in vitro* is highly dependent upon the extracellular pH of the culture medium[104,105] and this phenomenon has been attributed to the prototropic behaviour of the compound (pK_a 8.71). Since the intracellular pH of hypoxic cells is necessarily lower compared to the external medium, it is argued that diffusion of the free base form of the drug facilitates its accumulation in the protonated form[100]. Differential uptake into tumours relative to plasma would thus be expected to lead to increased therapeutic response.

Further evaluation revealed that the lipophilic piperidine compound (34) is slightly more efficient than (27) as an hypoxic cell radiosensitiser in experimental animal tumour systems[106] and that drug penetration into the central nervous system is rapid[97,107]. Pharmacokinetic studies showed that the rates of renal and metabolic clearance, chiefly as the metabolite amine-*N*-oxide, were sufficient to minimise total drug exposure. Brain drill data show that (34) is not excluded from normal brain tissue. Further, there appears to be a significant degree of selectivity in uptake of (34) in human tumour tissues[108,109].

Phase I and II clinical studies with (34) show no apparent chronic toxicity problems in man and that peripheral neuropathy is not induced. The compound is, however, dose-limited by an acute CNS syndrome to a maximum tolerable single dose of 0.75–1 g m^{-2}[106–108]. Further clinical trials with this radiosensitiser, both as a single adjunct for radiotherapy and in combination with the neutral compound (29), are currently in progress[4,98,110].

The mixed-function propanolamine compound (35) (RSU-1069), in which the N1 sidechain contains a monofunctional alkylating group in the form of a basic aziridine, was found to be a particularly efficient hypoxic cell radiosensitiser both *in vitro* and *in vivo*. Indeed, (35) proved to be substantially

35 : $R^1 = R^2 = R^3 = R^4 = H$
36 : $R^1 = Me$; $R^2 = R^3 = R^4 = H$ 38 : $R^1 = R^2 = R^3 = R^4 = H$
37 : $R^1 = R^2 = R^3 = R^4 = Me$

superior to (27), using either tumour cell survival or tumour cure as radiobiological end-points[111,112]. The compound is, however, considerably more cytotoxic towards aerobic cells, despite its similar one-electron reduction potential ($E_7^1 = -398$ mV). In addition, (35) is able to potentiate the cytotoxic action (chemosensitisation) of melphalan and other active toxic drugs towards the MT tumour[111].

The propanolamines (35)–(37) are prepared in high yield by reflux condensation of the oxirane (32) with the corresponding aziridine, preferably in the presence of an added alkali or inert organic base (e.g. pyridine or triethylamine) to inhibit polymerisation[111,112]. Drug stability is ultimately limited as a result of acid-induced hydrolysis and/or polymerisation of the reactive aziridine moiety[112].

Hill et al.[113] found that the enhanced radiosensitising properties of (35) in vitro are moderated if drug exposure and irradiation are given at 277 K, with efficiency then becoming that expected for a typical 2-nitroimidazole. The enhancement effect was attributed to the accumulation of sub-lethal damage, possibly as a consequence of cellular nitro-reduction (see section 13.3), during the period prior to irradiation. Quantitative support for this conclusion is provided by rapid-mix and other cellular studies[114–116]. Anomalous or abnormal radiosensitisation has been demonstrated[117–119] for ortho-substituted nitroimidazoles and CB1954 (5-aziridino-2,4-dinitrobenzamide), and this property is ascribed to depletion of intracellular non-protein thiols via nucleophilic processes[120]. Stratford et al.[116] demonstrated, however, that (35) has no significant effect upon the intracellular level of GSH (reduced glutathione) in either aerobic or hypoxic cells. The contact time-dependent enhancement of radiosensitisation by (35) thus stems from slow cellular biochemical events unrelated to thiol depletion (e.g. bioreduction of the nitro-group) which are clearly distinct from the fast free radical ('electron-affinic') processes. The accumulation of site-specific cell damage during such events is likely to be responsible for the chemosensitisation phenomenon whereby tumours pre-treated with (35) exhibit enhanced chemosensitivity or lowered chemoresistance when subsequently exposed to conventional alkylating or carbamoylating cytotoxic agents[3,5,111].

Pharmacokinetic studies of (35) and aziridine ring-substituted analogues (e.g. (36) and (37)) reveal that these basic compounds are preferentially taken up by tumours[121], whereas other tissues, including brain, achieve a tissue/plasma drug ratio which is close to 100%[121]. These observations are in accord with the measured lipophilic and acid-base prototropic properties of these compounds and reveal a clear pH-dependence for both cellular and tumour drug accumulation. The radiosensitisation afforded by this series of agents *in vitro* shows a dependence of sensitising efficiency upon pK_a, reflecting the alterations in cellular uptake and alkylating reactivity of the aziridine group[112]. The tetramethylaziridine (37) (RB-7040, pK_a 8.45) shows exemplary hypoxic cell accumulation, but the aziridine ring has considerably reduced reactivity towards nucleophilic reagents[112].

Preliminary clinical studies with the unsubstituted aziridine (35) as a hypoxic cell radiosensitiser reveal a dose-limiting gut toxicity[122]. Evaluation studies with (poly)methyl-substituted aziridine derivatives (e.g. (36) and (37)), where the aziridine reactivity and inherent toxicity is moderated, and a less toxic prodrug form (38) (RB-6145), prepared by HBr-induced ring opening of (35)[123], are currently in progress.

The identification of cellular DNA as the key biotarget for radiation damage has prompted the targetting of nitroimidazoles to DNA via complexation with DNA-binding metal ligands. Thus, active cis-$[PtCl_2(NH_3)(nitroimidazole)]$ complexes containing (21) and (27) as the radiosensitising component have been found to be effective agents towards DNA *in vitro*[124]. This elegant approach to drug targetting, particularly of hypoxic cells, lends itself to the sequence-specific recognition of sensitive sites or genes on the DNA biomolecule.

39

40

Brown *et al.*[125] have recently examined a series of non-nitro compounds (39) (1,2,4-benzotriazine-1,4-dioxides) with E_7^1 values in the range -300 to -450 mV which show activity as hypoxic cell radiosensitisers. Preliminary results, however, indicate that the lead compound from this series (39) ($R^1 = H$; $R^2 = -NH_2$, SR-4233) may be too cytotoxic for application *in vivo*.

In summary, four compounds have been identified as potential clinical replacements for the hypoxic cell radiosensitiser (27) (misonidazole). Two 2-nitroimidazole compounds, (29) (etanidazole) and (34) (pimonidazole), are currently being assessed in randomised clinical trials. The outcome of these trials is eagerly awaited. The remaining compounds, (35) (RSU-1069) and (39) (SR-4233), are both potent hypoxia-mediated agents which probably require tailoring in order to reduce toxicity prior to further clinical evaluation.

13.3   Hypoxia-mediated cytotoxins

The topic of drug activation via bioreduction is currently a major research area in the development of anticancer agents and largely stems from studies of antibacterial and antiparasitic chemotherapeutic agents[15]. Thus, bioreductive activation of an administered prodrug leads to the genesis of biologically active forms at or near the site of action.

Two major classes of compound are now known to be selectively toxic towards hypoxic rather than oxic cells through bioreductive activation. These include the mitomycin quinone antibiotics, of which mitomycin C (**40**) is a naturally occurring prototype[5,126], and the nitroimidazole radio-sensitisers[3,15,127] (for a review see Ref. 5).

In situ bioreduction of mitomycins generates a reactive and cytotoxic aziridine intermediate with the capacity to induce DNA interstrand cross-links which are considered to be the lesions responsible for the cytotoxic activity. Bioactivation requires 1- or 2-electron reduction to semiquinone or hydroquinone species and this process is negated by futile cycling in the presence of O_2; the agents are hence differentially toxic to hypoxic cells. The potential processes involved in the bioactivation of mitomycins have been reviewed[128]. Although (**40**) shows 5- to 10-fold increased toxicity to hypoxic cells *in vitro*, the compound fails to exhibit selective toxicity *in vivo*[126,128]. The quest for less toxic analogues and structurally related quinones continues[129].

The chemotherapeutic development of nitroimidazoles as hypoxia-mediated cytotoxins has been the major focus of attention. Nitroimidazoles such as (**21**) and (**27**) have long been known to be differentially toxic to hypoxic cells *in vitro* and the mechanism involved is probably similar to that whereby such agents behave as antibiotics[15,130]; both the aerobic and hypoxic cytotoxicities *in vitro* are clearly dependent upon redox processes[1,3,100]. The biochemistry of nitro-reduction and drug activation has been reviewed[131].

Prolonged exposure of hypoxic cells to ^{14}C-radiolabelled (**27**) leads to binding of the radioactivity to various intracellular constituents including DNA, RNA, protein and non-protein thiols[130]. This hypoxia-selective binding has been usefully exploited to develop probes to monitor (i) the tissue levels of oxygen at the cellular level, and (ii) the hypoxic fraction of tumours *in situ* by autoradiographic labelling[127,132]. The development of (poly)fluorinated sidechain analogues of (**27**) offers a prospect for non-invasive imaging of tumours *in situ* using either ^{18}F-positron emission tomography (^{18}F-PET) or ^{19}F-NMR.

The chemical, electrochemical or radiolytic reduction of (**27**) and related 2-nitroimidazoles in the presence of DNA leads to binding, irreversible base damage and subsequent induction of strand breaks[130,133–135]. The time-dependent accumulation of such damage is probably responsible for the ability of nitroimidazoles to act as efficient potentiators (chemosensitisers) of cytotoxic agents when used for pretreatment at sub-lethal doses[2,15,111].

The identity of the cytotoxic intermediate formed during bioreduction of 2-

nitroimidazoles is not yet established, but it is believed that the hydroxylamine 4-electron reduction product (41) plays a determining role[130]. The presence of oxygen results in futile redox cycling via the intermediate nitroarene radical anion and effective inhibition of reduction. Nitro-reduction to form DNA-reactive intermediates is thus effected primarily under conditions of hypoxia:

$$ArNO_2 \xrightarrow{e^-} [ArNO_2]^{\cdot-} \xrightarrow{e^-} [ArN{=}O] \xrightarrow{2e^-} ArNHOH \xrightarrow{2e^-} ArNH_2$$

$$O_2^- \qquad O_2 \qquad\qquad\qquad\qquad 41$$

futile cycling

glyoxal   [ArNH]⁺

42

DNA

DNA binding,
strand breaks, etc.

Aromatic ring fragmentation of reduced nitroimidazoles results in formation of glyoxal, which forms cyclic adducts with guanine, DNA and nucleic acid constituents[130]. This mechanism of action has, however, been shown to be relatively unimportant[134]. McClelland and his colleagues[136,137] have recently identified the possible formation of a reactive nitrenium species (42) as an electrophilic intermediate during the decomposition of unstable 2-hydroxylamino-imidazoles (41). This reactive ion, implicated in a Bamberger-type rearrangement, is suggested to have exquisite reactivity towards cellular nucleophiles (e.g. DNA bases, GSH, etc.), thus effecting electrophilic arylation. Stable nitroxyl free radicals have also been detected during the enzymatic reduction of nitroarenes[138].

The 2-nitroimidazole (27) (misonidazole) is about 10 times more toxic to hypoxic cells than aerobic cells[3,139] as a result of bioreductive activation. This differential factor appears to be similar for other 2-nitroimidazoles[140]. The incorporation of an alkylating moiety as in (35) improves this differential to about 100, but the improvement is dependent upon the reactivity of the aziridine residue. Methyl-substitution as in (36) and (37) has the effect of moderating the hypoxic/aerobic toxic behaviour[139].

O'Neill et al.[141] have demonstrated that, following radiolytic reduction in aqueous media, (35) and (36) induce single-strand breaks, alkali-labile base damage and DNA cross-links after incubation with DNA. The yields of DNA damage and cross-links induced by the reduced nitroimidazoles correspond with the hypoxic cytotoxicities determined in vitro. In contrast, the yields of DNA damage induced by the unreduced nitro-compounds appear to largely determine their aerobic cytotoxicities in vitro. These findings suggest that the induction of DNA cross-links by these agents plays a determining role in their effectiveness as hypoxia-selective cytotoxins[141]. Mechanistic support for this conclusion has come from studies with (35) and methyl-substituted aziridine derivatives, including (36) and (37)[142].

The mechanism(s) of hypoxia-mediated cytotoxic action of (35) is attributed to a switch from monofunctional alkylating behaviour to a reactive bifunctional character following metabolic reduction[139,141]. The 100-fold differential toxicity thus reflects the induction of DNA cross-links by a bifunctional agent, possibly the corresponding 2-hydroxylaminoimidazole, which is generated only in hypoxia. Qualitative support for this behaviour is provided by recent studies of the dinitrobenzene CB-1954, a monofunctional alkylating agent which induces DNA interstrand cross-links only following bioreductive activation[143].

The abnormal hypoxic cell radiosensitising efficiency of (35) and structurally related aziridine compounds, as detailed earlier, is attributed to pre-irradiation bioreductive activity[113,114]. Chaplin and Acker[144] have demonstrated that (35) is a highly efficient hypoxia-induced cytotoxin both *in vivo* and in spheroid systems. The antihypertensive vasoactive drug hydralazine was found to greatly potentiate the tumour toxicity of (35) *in vivo*, but to have no effect upon systemic toxicity. The chemotherapeutic gain afforded by hydralazine is attributed to a 'steal phenomenon' whereby induced vasodilation selectively reduces tumour blood flow and hence tumour oxygenation status. The improved tumour response to the bioreductive cytotoxic agent is hence a direct consequence of induced severe hypoxia[2,144]. Clinical trials with (35), as a sole agent and in conjunction with vasoactive drugs, are currently in progress.

Other mixed-function nitroarenes have recently been developed for evaluation as hypoxia-mediated cytotoxic agents, containing 2-chloroethylnitrosourea[145], hydroxamic acid[146] or simple ω-haloalkyl[147] functions. Denny and Wilson[6] have outlined the structural requirements for nitrophenyl mustard compounds to act as hypoxia-selective cytotoxins. The ultimate utility of each of these approaches remains to be established.

The benzotriazine-1,4-dioxide (39) ($R^1 = H$; $R^2 = -NH_2$, SR-4233) also shows potent hypoxia-selective cell killing *in vitro* and *in vivo*, with a hypoxic/aerobic toxicity factor of 15–50 in cells of human origin[125]. Mechanistic studies suggest that the radical anion formed by one-electron reduction undergoes homolytic cleavage to form OH radicals; DNA damage in the form of strand breaks is the most probable cytotoxic lesion. Recent investigations reveal that hypoxic cell killing by (39) in solid tumours is also potentiated by hydralazine; this activity may be clinically exploitable[148]. The evaluation of (39) as a hypoxic tumour cytotoxin is continuing.

13.4 Concluding remarks

It is clear that the relative radio- and chemoresistance of hypoxic tumour cells is likely to adversely affect their local control by scheduled radio- or chemotherapy. Agents which provide either (i) selective radiosensitisation, or (ii) hypoxia-selective cell killing will ultimately be essential if this therapeutic obstacle is to be overcome.

The limited metabolic stability of nitroxyl radiosensitisers *in vivo* has necessarily restricted their development for clinical utility. Primary attention has thus naturally focused upon the electron-affinic class of compounds, and nitroheteroarenes in particular. Developments in oxygen-mimetic radiosensitisers, and nitroimidazoles in particular, have resulted in chemical adjuncts with pronounced activity both *in vitro* and *in vivo*, but which are dose-limited for clinical application due to unscheduled toxicity. The current generation of hypoxic radiosensitisers is inherently less toxic by design, based upon the wealth of experience gained with earlier agents, and clinical evaluation is still in progress.

The results of clinical trials with the first-generation radiosensitiser (27) (misonidazole) have largely been disappointing. Many explanations have been proposed to account for this failure[89], but it is likely that tumour heterogeneity is ultimately a key factor[2]. There is an urgent requirement to identify those human tumours where hypoxia presents a particular problem before clinical effectiveness with a radiosensitiser is likely to be demonstrable.

Adams and Stratford[2] have outlined several new approaches to the modification of hypoxic tumour response, including adjunct radio- and chemotherapy with alternative agents. The suggestion of deliberate manipulation of tumour hypoxia for therapeutic gain provides an elegant strategy for further development and clinical evaluation. Thus, compounds which effect a temporary reduction in the level of tumour oxygenation by deliberate displacement of oxyhaemoglobin association or by altering the blood flow in tumours afford a valuable means of inducing severe tumour hypoxia[2,144].

Approaches of this type should help to maximise the clinical effectiveness of the radiosensitisers and bioreductively activated cytotoxins discussed in this chapter.

References

1. G. E. Adams, in *Radiosensitisers of Hypoxic Cells*, eds. A. Breccia, C. Rimondi and G. E. Adams, Elsevier, Amsterdam (1979) pp. 13, 245; *Cancer* **48** (1981) 696.
2. G. E. Adams and I. J. Stratford, in *Radiobiology in Radiotherapy*, ed. N. M. Bleehen, Springer-Verlag, London (1989) Chap. 14.
3. G. E. Adams and I. J. Stratford, *Biochem. Pharmacol.* **35** (1986) 71.
4. S. Dische, in *Radiobiology in Radiotherapy*, ed. N. M. Bleehen, Springer-Verlag, London (1988) Chap. 15.
5. K. A. Kennedy, *Anti-Cancer Drug Design* **2** (1987) 181.
6. W. A. Denny and W. R. Wilson, *J. Med. Chem.* **29** (1986) 879.
7. D. B. Carter and A. F. Phillips, *Nature (London)* **174** (1954) 121.
8. G. L. Rice, C. Hoy and R. T. Schimke, *Proc. Natl. Acad. Sci. USA* **83** (1986) 5978.
9. G. L. Rice, C. Hoy, V. Ling and R. T. Schimke, *Proc. Rad. Res. Soc.* **35** (1987) 105.
10. E. A. Bump, N. N. Yu and J. M. Brown, *Int. J. Radiat. Biol. Oncol. Biol. Phys.* **8** (1982) 439.
11. J. W. Harris, *Pharmacol. Ther.* **7** (1979) 375.
12. O. W. Griffith and A. Meister, *J. Biol. Chem.* **254** (1979) 7558.
13. J. E. Biaglow, M. E. Varnes, E. P. Clark and E. R. Epp, *Radiat. Res.* **95** (1983) 437.
14. E. P. Clark, *Int. J. Radiat. Oncol. Biol. Phys.* **12** (1986) 1121.
15. Collected papers, in *Bioreduction in the Activation of Drugs*, eds. P. Alexander, J. Gielen and A. C. Sartorelli, *Biochem. Pharmacol.* **35** (1986) 1–122.
16. E. G. Rozantsev, *Free Nitroxyl Radicals*, Plenum, New York (1970).

17. D. Griller and K. U. Ingold, *Acc. Chem. Res.* **9** (1976) 13.
18. A. K. Hoffmann and A. T. Henderson, *J. Am. Chem. Soc.* **83** (1961) 4671; A. Rassat, *Bull. Soc. Chim. Fr.* (1965) 3273.
19. A. R. Forrester, J. M. Hay and R. H. Thomson, *Organic Chemistry of Stable Free Radicals*, Academic Press, New York (1968).
20. J. F. W. Keana, *Chem. Rev.* **78** (1978) 37.
21. See : F.-P. Tsui, F. A. Robey, T. W. Engle, S. M. Ludeman and G. Zon, *J. Med. Chem.* **25** (1982) 1106.
22. P. T. Emmerson and P. Howard-Flanders, *Nature* **204** (1964) 1005; *Radiat. Res.* **26** (1965) 54.
23. E. G. Rozantsev and M. B. Neiman, *Tetrahedron* **20** (1964) 131.
24. P. T. Emmerson, *Radiat. Res.* **30** (1967) 841.
25. L. Parker, L. D. Skarsgard and P. T. Emmerson, *Radiat. Res.* **38** (1969) 493.
26. P. T. Emmerson, E. M. Fielden and I. Johansen, *Int. J. Radiat. Biol.* **19** (1971) 229.
27. R. I. Longley, Jr. and W. S. Emerson, *J. Am. Chem. Soc.* **72** (1950) 3079.
28. H. C. Beyerman, P. H. Enthoven and P. E. Verkade, *Recl. Trav. Chim.* **82** (1963) 1199.
29. R.-M. Dupeyre and A. Rassat, *J. Am. Chem. Soc.* **88** (1966) 3180; *Bull. Soc. Chim. Fr.* (1978) 612.
30. T. C. Jenkins, unpublished work.
31. G. D. Mendenhall and K. U. Ingold, *J. Am. Chem. Soc.* **95** (1973) 6395.
32. G. Scholes, in *Radiation Chemistry of Aqueous Systems*, ed. G. Stein, Interscience, London (1968), 259.
33. P. B. Roberts and E. M. Fielden, *Int. J. Radiat. Biol.* **20** (1971) 363.
34. P. O'Neill, T. C. Jenkins and E. M. Fielden, *Radiat. Res.* **82** (1980) 55.
35. S. Hornsey, *Int. J. Radiat. Biol.* **22** (1972) 91.
36. P. L. Olive, W. R. Inch and R. M. Sutherland, *Radiat. Res.* **52** (1972) 618.
37. N. M. Blackett, W. E. Wooliscroft, E. M. Fielden and S. C. Lillicrap, *Radiat. Res.* **58** (1974) 361.
38. R. P. Hill, E. M. Fielden, S. C. Lillicrap and J. A. Stanley, *Int. J. Radiat. Biol.* **27** (1975) 499.
39. B. C. Millar, E. M. Fielden and C. E. Smithen, *Radiat. Res.* **71** (1977) 516.
40. P. O'Neill and T. C. Jenkins, *J. Chem. Soc. Faraday I* **75** (1979) 1912.
41. B. C. Cooke, E. M. Fielden, M. Johnson and C. E. Smithen, *Radiat. Res.* **65** (1976) 152.
42. J. W. Harris, C. J. Koch, J. A. Power and J. E. Biaglow, *Radiat. Res.* **70** (1977) 585.
43. B. C. Millar, E. M. Fielden and C. E. Smithen, *Br. J. Cancer* **37** Suppl. III (1978) 73.
44. G. M. Rosen, *J. Med. Chem.* **17** (1974) 358.
45. B. C. Millar, T. C. Jenkins and E. M. Fielden, *Radiat. Res.* **90** (1982) 271.
46. B. C. Millar, T. C. Jenkins, E. M. Fielden and S. Jinks, *Radiat. Res.* **96** (1983) 160.
47. B. C. Millar, T. C. Jenkins, C. E. Smithen and S. Jinks, *Radiat. Res.* **101** (1985) 111.
48. P. Rey and H. M. McConnell, *J. Am. Chem. Soc.* **99** (1977) 1637.
49. I. Johansen, R. Gulbrandsen, E. M. Fielden and O. Sapora, *Radiat. Res.* **70** (1977) 597.
50. O. Sapora, E. M. Fielden and P. S. Loverock, *Radiat. Res.* **69** (1977) 293.
51. B. C. Millar, E. M. Fielden and T. C. Jenkins, *Radiat. Res.* **88** (1981) 369.
52. M. J. Tilby and P. S. Loverock, *Radiat. Res.* **96** (1983) 309.
53. P. O'Neill and S. E. Davies, *Int. J. Radiat. Biol.* **49** (1986) 937.
54. G. E. Adams, B. D. Michael, J. C. Asquith, M. A. Shenoy, M. E. Watts and D. W. Whillans, in *Radiation Research: Biomedical, Chemical and Physical Perspectives*, eds. O. F. Nygaard, H. I. Adler and W. K. Sinclair, Academic Press, New York (1975) p. 719.
55. G. E. Adams and M. S. Cooke, *Int. J. Radiat. Biol.* **15** (1969) 457.
56. J. D. Chapman, R. G. Webb and J. Borsa, *Int. J. Radiat. Biol.* **19** (1971) 561.
57. G. E. Adams, J. C. Asquith, D. L. Dewey, J. L. Foster, B. D. Michael and R. L. Willson, *Int. J. Radiat. Biol.* **19** (1971) 575.
58. G. E. Adams, J. C. Asquith, M. E. Watts and C. E. Smithen, *Nat., New Biol.* **239** (1972) 23.
59. G. F. Whitmore, S. Gulyas and A. J. Varghese, *Radiat. Res.* **61** (1975) 325.
60. J. D. Chapman, J. A. Raleigh, J. Borsa, R. G. Webb and R. Whitehouse, *Int. J. Radiat. Biol.* **21** (1972) 475.
61. G. A. Infante, P. Gonzalez, D. Cruz, J. Correa, J. A. Myers, M. F. Ahmad, W. L. Whitter, A. Santos and P. Neta, *Radiat. Res.* **92** (1982) 296.
62. P. Wardman, in *Radiosensitisers of Hypoxic Cells*, eds. A. Breccia, C. Rimondi and G. E. Adams, Elsevier, Amsterdam (1979) p. 91.
63. P. Wardman, *Curr. Top. Radiat. Res. Q.* **11** (1977) 347.

64. E. Grunberg and E. H. Titsworth, *Annu. Rev. Microbiol.* **27** (1973) 317.
65. C. E. Smithen and C. R. Hardy, in *Advanced Topics on Radiosensitisers of Hypoxic Cells*, eds. A. Breccia, C. Rimondi and G. E. Adams, Plenum, New York (1982) p.1.
66. J. L. Foster and R. L. Willson, *Br. J. Radiol.* **6** (1973) 234.
67. J. F. Fowler and J. D. Denekamp, *Pharmacol. Therapeut.* **7** (1980) 413.
68. R. C. Urtasun, P. Band, J. D. Chapman, M. C. Feldstein, B. Mielke and C. Fryer, *N. Engl. J. Med.* **294** (1976) 1364.
69. G. E. Adams, I. R. Flockhart, C. E. Smithen, I. J. Stratford, P. Wardman and M. E. Watts, *Radiat. Res.* **67** (1976) 9.
70. D. Meisel and P. Neta, *J. Am. Chem. Soc.* **97** (1975) 5198.
71. P. Wardman and E. D. Clarke, *J. Chem. Soc. Faraday I* **72** (1976) 1377.
72. R. F. Anderson and K. B. Patel, *J. Chem. Soc. Faraday I* **80** (1984) 2693.
73. P. B. Ayscough, A. J. Elliot and G. A. Salmon, *J. Chem. Soc. Faraday I* **74** (1978) 511.
74. G. E. Adams, E. D. Clarke, I. R. Flockhart, R. S. Jacobs, D. S. Sehmi, I. J. Stratford, P. Wardman, M. E. Watts, J. Parrick, R. G. Wallace and C. E. Smithen, *Int. J. Radiat. Biol.* **35** (1979) 133.
75. G. E. Adams, E. D. Clarke, P. Gray, R. S. Jacobs, I. J. Stratford, P. Wardman, M. E. Watts, J. Parrick, R. G. Wallace and C. E. Smithen, *Int. J. Radiat. Biol.* **35** (1979) 151.
76. K. Schofield, M. R. Grimmett and B. R. T. Keene, *Heteroaromatic Nitrogen Compounds: The Azoles*, Cambridge University Press, Cambridge (1976).
77. A. G. Beaman, W. Tautz, T. Gabriel and R. Duschinsky, *J. Am. Chem. Soc.* **87** (1965) 389.
78. G. C. Lancini and E. Lazzari, *Experientia* **21** (1965) 83.
79. K. L. Kirk, *J. Org. Chem.* **43** (1978) 4381.
80. A. J. Carpenter and D. J. Chadwick, *Tetrahedron* **42** (1986) 2351.
81. A. G. Beaman, W. Tautz and R. Duschinsky, *Antimicrob. Agents Chemother.* **1967** (1968) 520.
82. M. Hoffer and E. Grunberg, *J. Med. Chem.* **17** (1979) 1019.
83. J. C. Asquith, M. E. Watts, K. B. Patel, C. E. Smithen and G. E. Adams, *Radiat. Res.* **60** (1974) 108.
84. P. W. Sheldon and J. F. Fowler, *Br. J. Cancer* **37** Suppl. III (1978) 242.
85. D. V. Ash, M. R. Smith and R. D. Bugden, *Br. J. Cancer* **39** (1979) 503.
86. S. Dische, M. I. Saunders, M. E. Lee, G. E. Adams and I. R. Flockhart, *Br. J. Cancer* **35** (1977) 567.
87. Collected papers, in *Radiation Sensitisers: Their Use in the Clinical Management of Cancer*, Cancer Management, Vol. 5, ed. L. W. Brady, Masson, New York (1980).
88. R. C. Urtasun, H. Tanasichuk, D. Fulton, O. Agboola, A. R. Turner, D. Koziol and J. Raleigh, *Int. J. Radiat. Oncol. Biol. Phys.* **8** (1982) 365.
89. H. B. Stone, *Int. J. Radiat. Oncol. Biol. Phys.* **14** (1988) 957.
90. C. Clarke, K. B. Dawson, P. W. Sheldon and I. Ahmed, *Int. J. Radiat. Oncol. Biol. Phys.* **8** (1982) 787.
91. J. M. Brown and P. Workman, *Radiat. Res.* **82** (1980) 171.
92. P. Workman, in *Advanced Topics on Radiosensitisers of Hypoxic Cells*, eds. A. Breccia, C. Rimondi and G. E. Adams, Plenum, New York (1982) p. 143.
93. J. M. Brown, *Int. J. Radiat. Oncol. Biol. Phys.* **8** (1982) 1491; **10** (1984) 425.
94. V. L. Narayanan and W. W. Lee, *Adv. Pharmacol. Chemother.* **19** (1982) 155.
95. J. M. Brown, N. Y. Yu, D. M. Brown and W. W. Lee, *Int. J. Radiat. Oncol. Biol. Phys.* **7** (1981) 695.
96. C. N. Coleman, R. C. Urtasun, T. H. Wasserman, S. Hancock, J. W. Harris, J. Halsey and K. V. Hirst, *Int. J. Radiat. Oncol. Biol. Phys.* **10** (1984) 1749.
97. H. F. V. Newman, N. M. Bleehen and P. Workman, *Int. J. Radiat. Oncol. Biol. Phys.* **12** (1986) 1113.
98. H. Bartelink, N. M. Bleehen and T. L. Phillips, *Int. J. Radiat. Oncol. Biol. Phys.* **14** (1988) S39.
99. G. E. Adams, I. Ahmed, E. D. Clarke, P. O'Neill, J. Parrick, I. J. Stratford, R. G. Wallace, P. Wardman and M. E. Watts, *Int. J. Radiat. Biol.* **38** (1980) 613.
100. P. Wardman, in *Advanced Topics on Radiosensitisers of Hypoxic Cells*, eds. A. Breccia, C. Rimondi and G. E. Adams, Plenum, New York (1982) p. 49.
101. D. J. Chaplin, P. W. Sheldon, I. J. Stratford, I. Ahmed and G. E. Adams, *Int. J. Radiat. Biol.* **44** (1983) 387.
102. I. Ahmed, I. J. Stratford and T. C. Jenkins, *Arzneim. Forsch./Drug Res.* **35**(II) (1985) 1763.
103. C. E. Smithen, E. D. Clarke, J. A. Dale, R. S. Jacobs, P. Wardman, M. E. Watts and M.

Woodcock, in *Radiation Sensitisers: Their Use in the Clinical Management of Cancer*, Cancer Management, Vol. 5, ed. L. W. Brady, Masson, New York (1980) p.22; UK Pat. Appl. GB 2003154A (1979).

104. M. F. Dennis, M. R. L. Stratford, P. Wardman and M. E. Watts, *Int. J. Radiat. Biol.* **47** (1985) 629.

105. M. E. Watts and N. R. Jones, *Int. J. Radiat. Biol.* **47** (1985) 645.

106. M. V. Williams, J. Denekamp, A. I. Minchinton and M. R. L. Stratford, *Br. J. Cancer* **46** (1982) 127.

107. M. I. Saunders, P. J. Anderson, M. H. Bennett, S. Dische, A. Minchinton, M. R. L. Stratford and M. Tothill, *Int. J. Radiat. Oncol. Biol. Phys.* **10** (1984) 1759.

108. J. T. Roberts, N. M. Bleehen, M. I. Walton and P. Workman, *Br. J. Radiol.* **59** (1986) 107.

109. S. Dische, M. I. Saunders, M. H. Bennett, E. P. Dunphy, C. Des Rochers, M. R. L. Stratford, A. I. Minchinton and P. Wardman, *Br. J. Radiol.* **59** (1986) 911.

110. H. F. V. Newman, R. Ward, P. Workman and N. M. Bleehen, *Int. J. Radiat. Oncol. Biol. Phys.* **15** (1988) 1073.

111. G. E. Adams, I. Ahmed, P. W. Sheldon and I. J. Stratford, *Br. J. Cancer* **49** (1984) 571; UK Pat. Appl. GB 2123816A (1984).

112. I. Ahmed, T. C. Jenkins, J. M. Walling, I. J. Stratford, P. W. Sheldon, G. E. Adams and E. M. Fielden, *Int. J. Radiat. Oncol. Biol. Phys.* **12** (1986) 1079.

113. R. P. Hill, S. Gulyas and G. F. Whitmore, *Br. J. Cancer* **53** (1986) 743.

114. K. R. J. Austen, T. J. Jenner, P. O'Neill and E. M. Fielden, *Int. J. Radiat. Biol.* **52** (1987) 281.

115. J. M. Walling, I. J. Stratford, G. E. Adams, A. R. J. Silver, I. Ahmed and T. C. Jenkins, *Int. J. Radiat. Oncol. Biol. Phys.* **12** (1986) 1083.

116. I. J. Stratford, J. M. Walling and A. R. J. Silver, *Br. J. Cancer* **53** (1986) 339.

117. I. J. Stratford, C. Williamson, S. Hoe and G. E. Adams, *Radiat. Res.* **88** (1981) 502.

118. I. J. Stratford, S. Hoe, G. E. Adams, C. Hardy and C. Williamson, *Int. J. Radiat. Biol.* **43** (1983) 31.

119. S. Rockwell, C. G. Irvin and M. Nierenburg, *Cancer Treat. Rep.* **70** (1986) 411.

120. P. Wardman, *Int. J. Radiat. Biol.* **41** (1982) 231.

121. J. M. Deacon, S. B. Holliday, I. Ahmed and T. C. Jenkins, *Int. J. Radiat. Oncol. Biol. Phys.* **12** (1986) 1087.

122. A. Horwich, S. B. Holliday, J. M. Deacon and M. J. Peckham, *Br. J. Radiol.* **59** (1986) 1238.

123. G. E. Adams, E. M. Fielden, T. C. Jenkins and I. J. Stratford, UK Pat. Appl. GB 8818348A (1988).

124. N. Farrell and K. A. Skov, *J. Chem. Soc. Chem. Commun.* (1987) 1043.

125. E. M. Zeman, J. M. Brown, M. J. Lemmon, V. K. Hirst and W. W. Lee, *Int. J. Radiat. Oncol. Biol. Phys.* **12** (1986) 1239.

126. A. M. Rauth, J. K. Mohindra and I. F. Tannock, *Cancer Res.* **43** (1983) 4154.

127. Collected papers, in *Chemical Modifiers of Cancer Treatment*, ed. J. M. Brown, *Int. J. Radiat. Oncol. Biol. Phys.* **12** (1986) 1019–1551.

128. A. C. Sartorelli, *Biochem. Pharmacol.* **35** (1986) 67.

129. See: W. Verboom, D. N. Reinhoudt, B. H. N. Lammerink, E. O. M. Orlemans, F. C. J. M. van Veggel and P. Lelieveld, *Anti-Cancer Drug Design* **2** (1987) 271.

130. See: G. F. Whitmore and A. J. Varghese, *Biochem. Pharmacol.* **35** (1986) 97.

131. J. E. Biaglow, M. E. Varnes, L. Roizen-Towle, E. P. Clarke, E. R. Epp, M. B. Astor and E. J. Hall, *Biochem. Pharmacol.* **35** (1986) 77.

132. D. J. van Os-Corby, C. J. Koch and J. D. Chapman, *Biochem. Pharmacol.* **36** (1987) 3487.

133. A. R. J. Silver, P. O'Neill and T. C. Jenkins, *Biochem. Pharmacol.* **34** (1985) 3537.

134. A. R. J. Silver, S. S. McNeil, P. O'Neill, T. C. Jenkins and I. Ahmed, *Biochem. Pharmacol.* **35** (1986) 3923.

135. A. Zahoor, M. V. M. Lafleur, R. C. Knight, H. Loman and D. I. Edwards, *Biochem. Pharmacol.* **36** (1987) 3299.

136. R. A. McClelland, J. R. Fuller, N. E. Seaman, A. M. Rauth and R. Battistella, *Biochem. Pharmacol.* **33** (1984) 303.

137. R. A. McClelland, R. Panicucci and A. M. Rauth, *J. Am. Chem. Soc.* **107** (1985) 1762.

138. J. A. Raleigh, F. Y. Shum and S. F. Liu, *Biochem. Pharmacol.* **30** (1981) 2921.

139. P. O'Neill, T. C. Jenkins, I. J. Stratford, A. R. J. Silver, I. Ahmed, S. S. McNeil, E. M. Fielden and G. E. Adams, *Anti-Cancer Drug Design* **1** (1987) 271.

140. T. C. Jenkins, I. J. Stratford and M. A. Stephens, *Anti-Cancer Drug Design* **4** (1989) 145.
141. P. O'Neill, S. S. McNeil and T. C. Jenkins, *Biochem. Pharmacol.* **36** (1987) 1787.
142. L. D. Dale, J. H. Tocher and D. I. Edwards, *Anti-Cancer Drug Design* **3** (1988) 169.
143. R. J. Knox, F. Friedlos, M. Jarman and J. J. Roberts, *Biochem. Pharmacol.* **37** (1988) 4661; see also 4671.
144. D. J. Chaplin and B. Acker, *Int. J. Radiat. Oncol. Biol. Phys.* **13** (1987) 579; see also 1361.
145. A. Carminati, J.-L. Barascut, E. Chenut, C. Bourut, G. Mathé and J.-L. Imbach, *Anti-Cancer Drug Design* **3** (1988) 57.
146. R. T. Mulcahy, D. J. Wustrow, R. R. Hark and A. S. Kende, *Radiat. Res.* **105** (1986) 296.
147. D. C. Heimbrook, K. Shyam and A. C. Sartorelli, *Anti-Cancer Drug Design* **2** (1988) 339.
148. E. M. Zeman and J. M. Brown, *Int. J. Radiat. Oncol. Biol. Phys.* **15** Suppl. 1 (1988) 128.

14 The chemistry and structure-activity relationships of some new prodrugs of vinca alkaloids

K. S. P. BHUSHANA RAO, M.-P. COLLARD,
J. P. DEJONGHE AND A. TROUET

14.1 Introduction

The alkaloids are natural products grouped together by their common possession of a basic nitrogen. The bisindole alkaloids comprise a large and complex group of naturally occurring organic compounds possessing the indole or dihydroindole nucleus. Among these we deal precisely with the subgroup from *Aspidosperma-cleavamine*. An illustration of the complexity of the structures encountered in this field are the antitumour alkaloids vinblastine (VLB) and vincristine (VCR). They are derived from the Madagascan periwinkle plant *Catharanthus roseus* G. Don. The biological, pharmacological, toxicological and clinical aspects of these alkaloids have been reviewed extensively[1,2].

VLB is useful in the treatment of various malignancies such as Hodgkin's disease, and VCR has shown effectiveness in acute lymphoblastic leukaemia[3]. Use of the former is restricted by leucopenia and of the latter by neurotoxicity[3].

The oncolytic activity of these dimeric alkaloids has been related to spindle poisoning resulting in mitotic arrest[4]. These alkaloids also change the rate of axoplasmic transport[5] by producing an alteration in neurotubules[6,7]. Cytotoxicity and peripheral neuropathy may be the result of the primary mode of action of these alkaloids by binding to tubulin[8].

This chapter is limited to a presentation of results of the new chemical modification products carefully designed from the vindoline moiety of vinblastine[9-12] with improved oncolytic activity.

After a brief historical introduction of these alkaloids, published hypotheses on structure-activity relationships are discussed. The chemistry of these compounds is then dealt with in detail and a general discussion of structure-activity relationships forms the final part of this chapter.

14.2 History

Vinca alkaloids possess a dimeric structure consisting of vindoline and catharanthine moieties. Four major active dimeric alkaloids, i.e. vinblastine (vincoleukoblastine, VLB), vincristine (VCR), vinleurosine (VLR) and vinrosidine (VRD), occurring in the periwinkle plant inhibit the growth of tumours[13-16]. Of these only vinblastine and vincristine are of proven clinical value.

14.3 Structure-activity relationships

VLB differs in molecular structure from VCR (Figure 14.1) in that it contains a methyl group instead of a formyl group and this minor structural difference leads to a different antitumour spectrum, potency and toxicity[3]. Certain arbitrary requirements for conserving the activity of these alkaloids were suggested[17]. Removal of the acetyl group at C-4 of the vindoline portion of VLB destroys its antileukaemic activity, as does acetylation of the free hydroxyl groups. Either hydrogenation of the C-6,7 double bond or reductive formation of carbinols diminishes or destroys the oncolytic activity of these alkaloids.

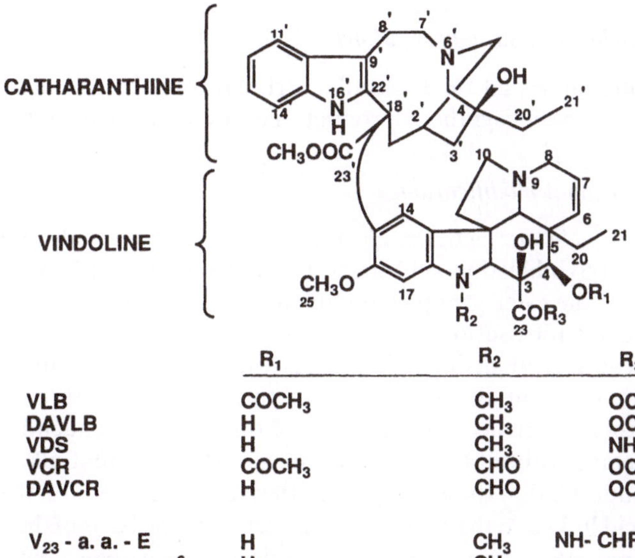

	R_1	R_2	R_3
VLB	$COCH_3$	CH_3	OCH_3
DAVLB	H	CH_3	OCH_3
VDS	H	CH_3	NH_2
VCR	$COCH_3$	CHO	OCH_3
DAVCR	H	CHO	OCH_3
V_{23} - a. a. - E	H	CH_3	$NH- CHR^5 - COR^4$
V_{23} - $(CH_2)_n$ - R^6	H	CH_3	$NH - (CH_2)_n - R^6$
V_4 - $(CH_2)_n$ - CH_3	$CO - (CH_2)_n - CH_3$	CH_3	OCH_3
V_4 - $(CH_2)_n$ - Mal	$CO - (CH_2)_n$ —Mal 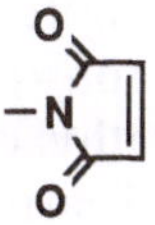	CH_3	OCH_3
V_4 - a. a. - Mal	$CO - CHR^5$ —Mal 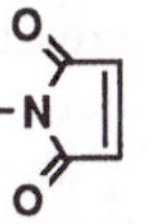	CH_3	OCH_3

Figure 14.1 Molecular structure of vinca alkaloids. $R^4 = $ -NH_2 or O-$(CH_2)_n$-CH_3; $R^5 = $ amino acid side chain; $R^6 = CH_3$ or NH_2; a.a. = amino acid; DAVLB = deacetyl vinblastine; DAVCR = deacetyl vincristine.

The deacetylvinblastine amide (VDS) is a semisynthetic derivative[3] of VLB and it differs slightly from VLB by having an amide group in place of the ester group at position C-23. The antineoplastic activity of vindesine (VDS) in rodents resembles that of VCR[18].

14.4   Chemical modifications

In the vindoline moiety of vinblastine we have modified the C-23 oyl and C-4 acyl moieties. As far as C-23 chemical modifications are concerned, two families of compounds have been synthesised: one includes condensation with amino acid derivatives and the other condensation with alkyl chains.

At the C-4 terminal, three types of chemical change have been accomplished: one comprises acylation with a long alkyl chain, the second involves condensation with a maleoyl-alkyl chain and the third one with maleoyl-amino acid.

14.4.1   Nomenclature and abbreviations

Figure 14.1 summarises all the molecular structures of the parent alkaloids (VLB, VCR) and the semisynthetic products, i.e. new congeners or 'prodrugs'.

14.4.2   C-23-terminal modification

14.4.2.1   Amino acid derivatives: chemistry and experimental antitumour activity.   Deacetylvinblastine acid azide is readily prepared from VLB by hydrazinolysis of the ester group of the dihydroindole moiety and deacetylation, followed by nitrosation of the resulting deacetylvinblastine monohydrazide[3]. Deacetylvinblastine azide reacts with amino esters in dichloromethane to give compounds of general structure V-23-a.a.-E (Figure 14.2).

We have selected seven derivatives[9,10] based on their antileukaemic activity in P388 leukaemia (Table 14.1) in comparison with the parent alkaloids.

The results indicate that the activity of the derivatives depends on the nature of the amino acid linked to the VLB-23-oyl moiety. On the basis of the polarity of the sidechains of the four main classes of amino acids, we can conclude that the activity is linked to non-polar or hydrophobic amino acids, i.e. derivatives (1), (4) and (6) (Table 14.1). The presence of an amino acid at the C-23-oyl moiety increases the therapeutic potency if there is a carboxylic ester present. Indeed, introducing an amide group (3) in place of the ester group (1) causes loss of the activity. Amino acid ethyl or methyl esters have similar potency in prolonging the lifespan of mice bearing the P388 leukaemia. When the alkyl chain length of the amino acid increases (7) the activity is maintained and the long-term survivors increase. As far as stereoisomerism of the amino acid is concerned (1,2) there seems to be an increase in activity with the D-isomer. From these results it is evident (in the case of (1) and (4)) that the regional increase in lipophilicity associated with the conversion of VLB to the compounds of type V-23-a.a.-E appears to be a desirable feature.

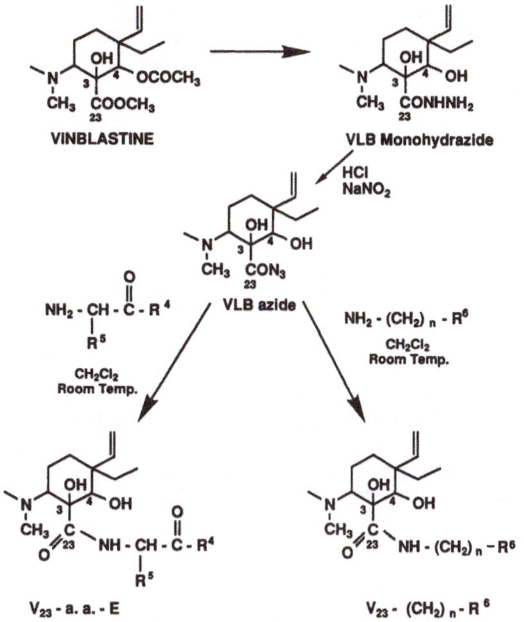

Figure 14.2 Chemical modifications at C_{23} of the vindoline moiety of vinblastine.

Table 14.1 Maximal chemotherapeutic activity of vinblastine, vindesine and amino acid derivatives against the P388 leukaemia[a] implanted intravenously in DBA_2 mice

			60th day survivors
No. Derivative[b]	Dose (mg/kg)	ILS[c] (%)	Total
Vinblastine	10	81	0/10
Vindesine	6	32	0/10
Vincristine	1.5	20.8	0/10
Deacetylvinblastine	4	81.3	0/10
(1) V_{23}-L-Leu-OEt	30	84.2	1/10
(2) V_{23}-D-Leu-OEt	10	>146	2/10
(3) V_{23}-L-Leu-Amide	40	19	0/10
(4) V_{23}-L-Ile-OEt	8	>183	0/10
(5) V_{23}-L-Trp-OMe	70	145.3	3/10
(6) V_{23}-L-Trp-OEt	60	>178	3/10
(7) V_{23}-L-Trp-O-n-Bu	80	>216	7/10

[a] Female DBA_2 mice were inoculated i.v. on day 0 with 10^4 P388 cells. Treatment was given i.v. on day 1.
[b] Compounds were tested as their sulphate salts.
[c] ILS: percent increase in lifespan by using all deaths and survivors following the equation: $[(T/C \times 100) - 100]$. Maximal median survival time (MST): 30 days. Control average day of death: 10.7 days.

Table 14.2 Maximal chemotherapeutic activity of vinblastine, vindesine and C_{23}-alkyl chain derivatives against the P388 leukaemia[a] implanted intravenously in DBA_2 mice

No. Derivative[b]	Dose (mg/kg)	ILS[c] (%)	60th day survivors Total
Vinblastine	8	78	0/9
Vindesine	1.75	35	0/10
Deacetylvinblastine	4.5	102	0/7
(8) V_{23}-$(CH_2)_5$-CH_3	3.13	74	0/7
(9) V_{23}-$(CH_2)_9$-CH_3	50	112	2/9
(10) V_{23}-$(CH_2)_{11}$-CH_3	150	167	1/7
(11) V_{23}-$(CH_2)_{12}$-NH_2	50	18	0/7
(12) V_{23}-$(CH_2)_{17}$-CH_3	175	3	0/5

[a] Female DBA_2 mice were inoculated i.v. on day 0 with 10^5 P388 cells. Treatment was given i.v. on day 1.
[b] Compounds were tested as their sulphate or methane sulphonate salts.
[c] See Table 14.1. Maximal median survival time (MST): 60 days. Control average day of death: 9 days.

14.4.2.2 Alkyl chain derivatives: chemistry and experimental antitumour activity. Deacetylvinblastine acid azide, obtained as described previously, reacts with alkylamines in dichloromethane to give compounds of general structure V3-$(CH_2)_n$-R[6]. A number of congeners were readily prepared from the azide and suitable alkylamines (Figure 14.2, Table 14.2).

The antitumour activity of these compounds has been assessed on the P388 leukaemia system and the maximal chemotherapeutic response of derivatives **(8)–(12)** is summarised in Table 14.2.

Derivative **(8)**, which possesses a 6-carbon alkyl chain, shows a similar activity to the reference derivatives and no long-term survivors. When the alkyl chain increases **(9, 10)**, superior activity is observed as well as long-term survivors. However, the maximal tolerated dose (MTD) also increases. P388 leukaemia becomes insensitive to the vinca alkaloid when the terminal alkyl group **(10)** of the carbonyl chain is functionalised by an amino group **(11)**. From these results we can conclude that the polarity of the C-23 substitutent determines the antitumour activity of the vinca alkaloid.

14.4.3 C-4 modification

14.4.3.1 Alkyl chain derivatives: chemistry and antitumour activity. First, deacetylvinblastine is obtained in good yield by treatment of VLB with sodium methoxide in methanol solution. This intermediate is then acylated at the C-4-terminal with an acyl chloride (in dichloromethane) at 0°C, to produce compounds of general structure V_4-$(CH_2)_n$-CH_3 (Figure 14.3).

Among the derivatives screened in the P388 leukaemia assay, derivative **(13)** emerges as the best candidate (Table 14.3). The optimal carbon chain length in

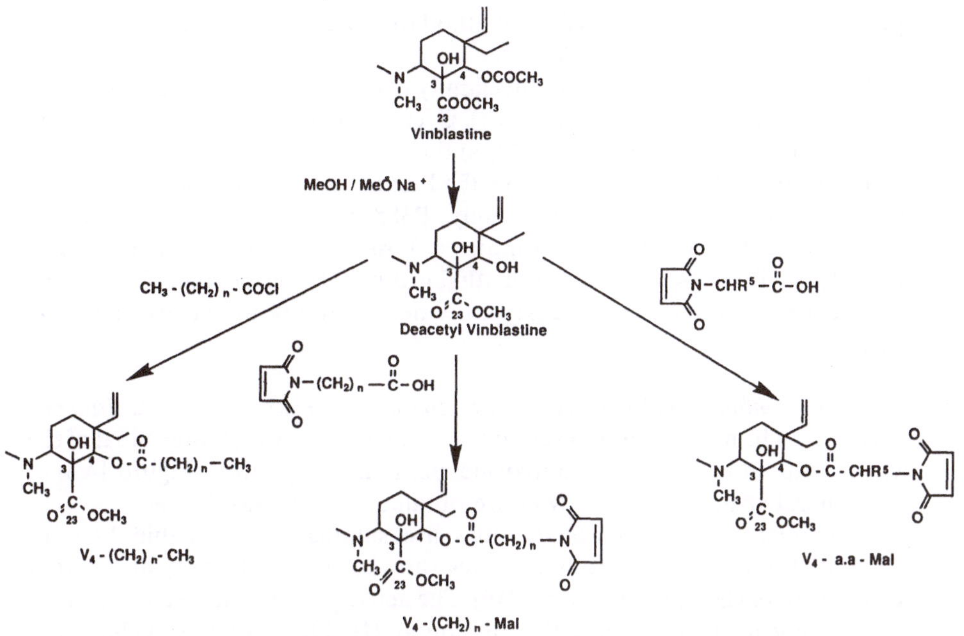

Figure 14.3 Chemical modifications at C_4 of the vindoline moiety of vinblastine.

this case also seems to be twelve carbons. The derivative is as active as DAVLB but at a much higher dose.

14.4.3.2 *Alkyl-maleoyl derivatives: chemistry and antitumour activity.* The condensation of the maleoylamino-alkylacid with deacetylvinblastine is

Table 14.3 Maximal chemotherapeutic activity of vinblastine, vincristine, a C_4-alkyl chain derivative and C_4-alkyl-maleoyl derivatives against the P388 leukaemia[a] implanted intraperitoneally in BDF_1 mice

No. Derivative[b]	Dose (mg/kg)	ILS[c] (%)	60th day survivors Total
Vinblastine	3	77	0/10
Vincristine	2.7	64	0/11
Deacetylvinblastine	4	>623	4/7
(13) V_4-$(CH_2)_{10}$-CH_3	100	>614	4/7
(14) V_4-$(CH_2)_5$-Mal	50	218	1/7
(15) V_4-$(CH_2)_{10}$-Mal	50	37	1/7
(16) V_4-$(CH_2)_{11}$-Mal	50	>659	5/8

[a] Female BDF_1 mice were inoculated i.p. on day 0 with 10^6 P388 cells. Treatment was given i.p. on day 1.
[b] Compounds were tested as their sulphate or methane sulphonate salts.
[c] See Table 14.1. Maximal median survival time (MST): 60 days. Control average day of death: 9.6 days.

performed by the mixed anhydride method to give compounds (Figure 14.3) of general structure V_4-$(CH_2)_n$-Mal[19].

The antitumour properties of this family of compounds have been assessed using the P388 leukaemia (Table 14.3). Great variations were observed in the antitumour activity: derivatives (14) and (16) which have 5 and 11 carbons, respectively, on the alkyl chain show the best activity and produce long-term survivors in this model. Surprisingly P388 leukaemia is insensitive to derivative (15). Derivative (16) contains a N-maleimide group instead of a H atom ((13), Table 14.3) at the last methyl group of the ester chain. This slight modification produces a decrease in the dose required to achieve the same activity.

14.4.3.3 *Amino acid-maleoyl derivatives: chemistry and antitumour activity.* The maleoyl amino acids were condensed on the C-4-terminal of the vinca alkaloid by the mixed anhydride method. Compounds (Figure 14.3) of the general structure V_4-a.a.-Mal are obtained in this way[19].

Experimental antitumoral activities are summarised in Table 14.4. In general all these amino acid derivatives show significant activity at different dose levels, particularly derivative (19). The activity is again linked to the non-polar amino acids as well as to the compounds (19,20) where the sidechains are protected. The non-biological amino acid analogues (18) and (22) also show biological activity.

Table 14.4 Maximal chemotherapeutic activity of vinblastine, vincristine and C_4-amino acid-maleoyl derivatives against the P388 leukaemia[a] implanted intraperitoneally in BDF$_1$ mice

No. Derivative[b]	Dose (mg/kg)	ILS[c] (%)	60th day survivors Total
Vinblastine	3	77	0/10
Vincristine	2.7	64	0/11
Deacetylvinblastine	4	> 623	4/7
(17) V_4-Ala-Mal	50	127	0/7
(18) V_4-β-Ala-Mal	75	307	3/7
(19) V_4-ε-Z-Lys-Mal	120	> 689	5/7
(20) V_4-Glu-γ-Methylester-Mal	100	111	0/10
(21) V_4-Ile-Mal	50	148	0/7
(22) V_4-Norleu-Mal	75	122	0/7

[a] Female BDF$_1$ mice were inoculated i.p. on day 0 with 10^6 P388 cells. Treatment was given i.p. on day 1.
[b] Compounds were tested as their sulphate or methane sulphonate salts.
[c] See Table 14.1. Maximal median survival time (MST): 60 days. Control average day of death: 9.6 days.

14.5 Discussion

It is well known that for the vinca alkaloids, the minor structural differences are responsible for differences in oncolytic spectrum, potency and toxicity. Knowing that minor changes in the vindoline moiety do not cause loss of clinical activity, this programme has been developed to produce chemical modifications at the C-23-oyl and C-4-acyl moieties. From the literature, removal of the acetyl group at C-4 of the vindoline portion of VLB destroys its antileukaemic activity. However, deacetylvinblastine (DAVLB) is one of the major metabolites of VLB in man[20,21,22]. Its biological activity in rodent model systems is also controversial. We found very good activity in P388 leukaemia (cf. Tables 14.1 and 14.2) when compared to VLB.

With the first family of the new compounds developed, i.e. V_{23}-a.a.-E, it is clear that the presence of a non-polar hydrophobic amino acid at the C-23 position increases the activity with long-term survivors at different dose levels. This activity seems to be linked to the presence of an ester group which might be interacting with the C-4 hydroxyl group via hydrogen bonds.

The presence of a lipophilic chain favours antitumour activity especially if it is 12 carbon atoms long in the V_{23}-$(CH_2)_n$-R^6 or V_4-$(CH_2)_n$-CH_3 families. Indeed in natural products we encounter molecules with paired rather than unpaired carbons.

The maleimide derivatives, i.e. V_4-$(CH_2)_n$-Mal or V_4-a.a.-Mal, also show an antitumour activity linked to alkyl chain length and the presence of non-polar hydrophobic amino acids. The addition of the maleimide group to the 12 carbon alkyl chain induces a decrease in the dose required. It is also possible that the maleimide group will interact with either proteins in the serum or receptors on the cell membrane. The antitumour activity found might also be due to a more favourable biodistribution in tissues or different metabolic alterations. Steric factors may also influence maximal binding and permeability across the cell membranes. In addition to the need for more physical investigations, feedback information from clinical experience will aid in molecule design.

It is evident from our study that these derivatives by themselves have a superior chemotherapeutic activity coupled with diminished toxicity compared to the parent alkaloids. Currently, (6) is undergoing phase I clinical trials[23] at Cliniques St Luc, UCL, Belgium and European Organization for Research on Treatment of Cancer (EORTC). The other derivatives will be available for clinical evaluation soon.

One approach to increase the selectivity of drugs is to couple the therapeutic agents to a carrier molecule that would carry the active molecule to the target cells, but has also to be endocytosed to gain access to the lysosomes, i.e. drug targeting. The efficiency of a drug carrier conjugate depends on the link between the drug and its carrier, which should be stable in the extracellular biological fluids and reversible in the presence of lysosomal enzymes (for review see Ref. 24). As a first step in the development of such a drug carrier conjugate we have developed the above mentioned families of 'prodrugs'. Indeed we hope the second family of derivatives (C_4-maleoyl), when conjugated to a specific carrier, will release DAVLB, which is an active metabolite with good chemotherapeutic activity.

References

1. P. Calabresi and R. E. Parks, Jr., in *The Pharmacological Basis of Therapeutics*, eds. L. S. Goodman and A. Gilman, 5th edn. MacMillan, New York (1975) p. 1284.
2. R. C. DeConti and W. A. Creasey, in *The Catharanthus Alkaloids*, eds. W. I. Taylor and N. R. Franworth, Marcel Dekker, New York (1975) p. 237.
3. C. J. Barnett, G. J. Cullinan, K. Gerzon, R. C. Hoying, W. E. Jones, W. M. Newlon, G. A. Poore, R. L. Robison, M. J. Sweeney, G. C. Todd, R. W. Dyke and R. L. Nelson, *J. Med. Chem.* **21** (1978) 88.
4. H. D. Weiss, M. D. Walker and P. H. Wiernik, *N. Engl. J. Med.* **291** (1974) 75; 127.
5. S. Ochs, *Ann. N. Y. Acad. Sci.* **228** (1974) 202.
6. S. S. Schochet, Jr., P. W. Lampert and K. M. Earle, *J. Neuropathol. Exp. Neurol.* **27** (1968) 645.
7. H. Wisniewski, M. L. Shelanski and R. D. Terry, *J. Cell Biol.* **38** (1968) 224.
8. R. J. Owellen, D. W. Donigian, C. A. Hartke, R. M. Dickerson and M. J. Kuhar, *Cancer Res.* **34** (1974) 3180.
9. K. S. P. Bhushana Rao, M. P. Collard, G. Atassi, J. P. DeJonghe, J. Hannart and A. Trouet, *J. Med. Chem.* **28** (1985) 1079.
10. K. S. P. Bhushana Rao, M. P. Collard and A. Trouet, 13th International Congress of Chemotherapy, Vienna, August 28–September 2 (1983) S.E.12.4.M/A (abstract).
11. K. S. P. Bhushana Rao, M. P. Collard and A. Trouet, *Anticancer Res.* **5** (1985) 379.
12. K. S. P. Bhushana Rao, M. P. Collard, A. Trouet, A. DeBruyn, M. Verzele, A. Noiret, J. P. DeJonghe and J. Hannart, *Bull. Soc. Chim. Belg.* **94** (1985) 133.
13. R. L. Noble, C. T. Beer and J. H. Cutts, *Biochem. Pharmacol.* **1** (1958) 347.
14. I. S. Johnson, H. F. Wright and G. H. Svoboda, *J. Lab. Clin. Med.* **54** (1959) 836.
15. J. H. Cutts, C. T. Beer and R. L. Noble, *Cancer Res.* **20** (1960) 1023.
16. G. H. Svoboda, *Lloydia* **24** (1961) 173.
17. W. A. Creasey, in *Handbook of Experimental Pharmacology*, Vol. 38. Part II: *Antineoplastic and Immunosuppressive Agents*, eds. A. C. Sartorelli and D. G. Johns, Springer-Verlag, Berlin (1975) p. 670.
18. M. J. Sweeney, G. J. Cullinan, G. A. Poor and K. Gerzon, *Proc. Am. Assoc. Cancer Res.* **15** (1974) 37.
19. K. S. P. Bhushana Rao, J. A. Hannart and A. Trouet, *Anticancer Res.* **9** (1989) 619.
20. R. J. Owellen, *Fed. Proc., Fed. Am. Soc. Exp. Biol.* **34** (1975) 808.
21. R. J. Owellen, D. W. Donigian, C. A. Hartke and F. O. Haines, *Biochem. Pharmacol.* **26** (1977) 1213.
22. R. J. Owellen, C. A. Hartke and F. O. Haines, *Cancer Res.* **37** (1977) 2597.
23. F. Ceulemans, Y. Humblet, A. Bosly, M. Symann and A. Trouet, *Cancer Chemother. Pharmacol.* **18** (1986) 44.
24. A. Trouet, *Eur. J. Cancer* **14** (1978) 105.

15 The chemistry of mitomycins

R. W. FRANCK AND M. TOMASZ

15.1 Introduction

The mitomycins are a group of antibiotics, discovered in Japan in the 1950s in fermentation cultures of certain strains of *Streptomyces*[1]. Observation of their significant antitumour activity soon followed and today, one of the drugs, mitomycin C, is widely used in clinical anticancer chemotherapy[2].

The structures of the various mitomycins are shown in Scheme 15.1. They reveal a diverse set of functional groups in an unusually compact arrangement within the molecule. These characteristics, as well as an early theory[3] relating the chemical properties of the mitomycins directly to their mode of action, have stimulated much activity in the area of their chemical transformations and this is still continuing today. The earlier work has been reviewed in several publications[4]. The present review focuses on newer acid, base and redox reactions and the bioreductive alkylation chemistry in model systems and with DNA. The usual trivial nomenclature of the mitomycins[5] will be used throughout (Scheme 15.2). It should be noted that the absolute stereochemistry of the mitomycins was revised in 1983[6]. Thus, configurations in the earlier literature appear reversed from those in use today.

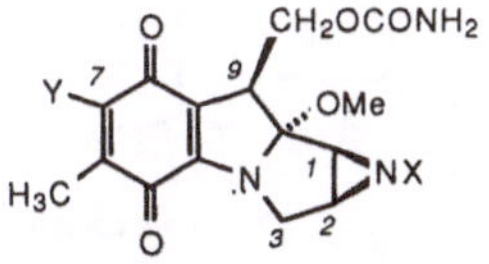

1 Y = OCH₃, X = H, Mitomycin A

2 Y = NH₂, X = H, Mitomycin C

3 Y = NH₂, X = CH₃, Porfiromycin

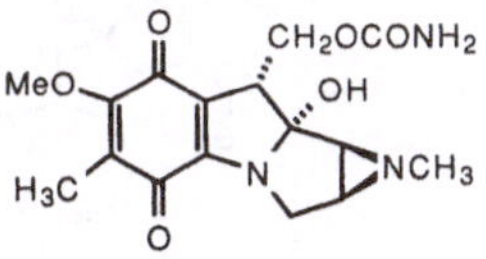

4 Mitomycin B

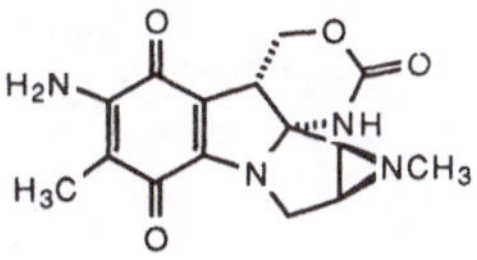

5 Mitiromycin

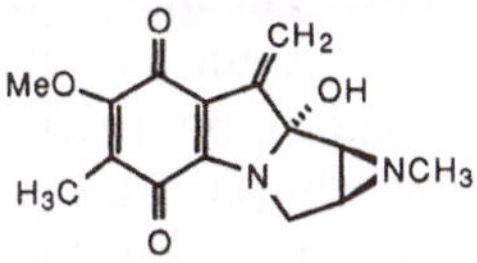

6 9,10- Dehydromitomycin B

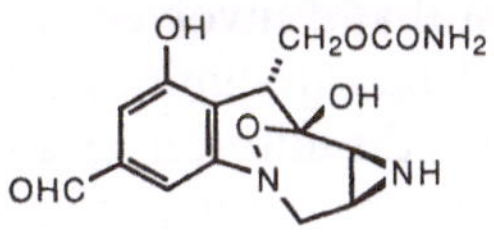

7 FR900482

Scheme 15.1

8 mitosane

9 mitosene

Scheme 15.2

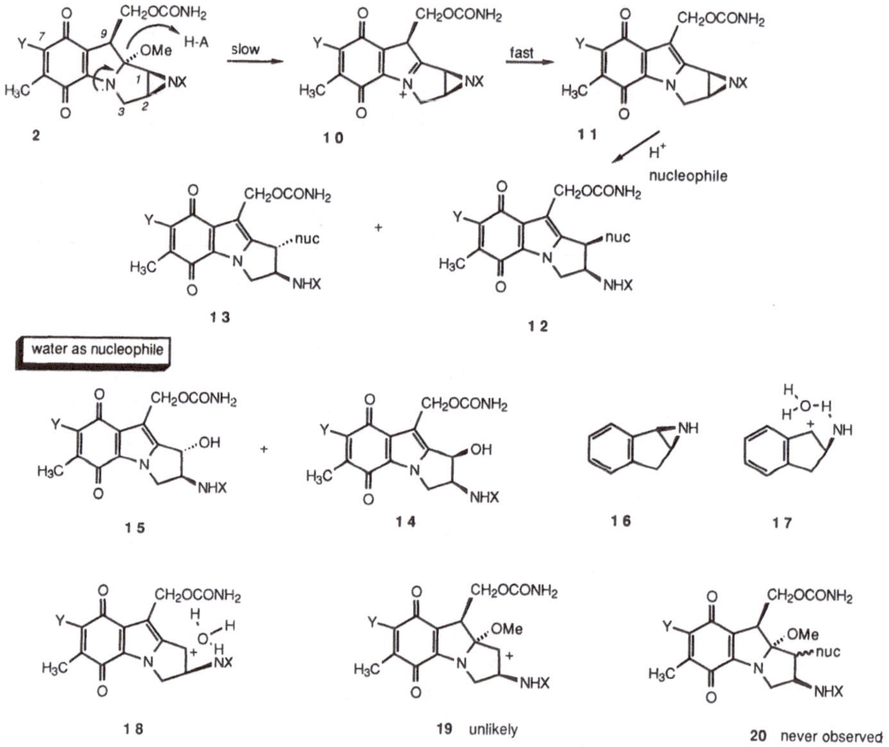

CH2OCONH2

H-A   slow

fast

2

10

11

H+
nucleophile

...nuc

nuc

NHX

NHX

13

12

water as nucleophile

CH2OCONH2

...OH

OH

NHX

NHX

15

14

16

17

18

19   unlikely

20   never observed

Scheme 15.3

15.2 Mitomycins and acids

The acid catalysed hydrolysis of mitomycin has been studied extensively; there is general agreement that it involves two separate steps[4]. The first step, which has been shown to be rate-limiting by careful kinetic studies, is the acid catalysed conversion of the mitosane framework to the aziridinomitosene (11) by elimination of water or methanol at C-9a, as shown in (2) and (10) (Scheme 15.3). Thus, the oxygen at C-9a must be protonated and the aziridine must remain unprotonated to avoid a coulombic repulsion[7]. This rationale accounts for linearity of the hydrolysis rate from pH 5.5 to 3, followed by a levelling off when the pH drops below 3; i.e. when the aziridine ($pK_a = 3$) is protonated, the C-9a oxygen is not. Furthermore, the inverse solvent isotope effect, the Brönsted coefficient and the buffer dependence are all consistent with oxygen protonation by a general acid followed by carbonium ion formation. In a fast step, the aziridinomitosene (11) is protonated and ring-opened to afford a mixture of isomers (12) and (13) where the trapping nucleophilic species have been (i) water, (ii) phosphate supplied by inorganic phosphate, 5'-UMP, 5'-UTP. UpU and 5'-FdUMP, (iii) N-2 and N-7 of a variety of guanine species, d(GpC), GpC, dG, G, and guanine residues of calf thymus DNA. The production of the C-1 hydroxy has been studied most thoroughly. A mixture of isomers, namely *cis* (14) and *trans* (15), is always obtained, with *cis* predominant. In simple aziridines, the ring is usually opened with clean inversion. However, Kohn[8] has shown in a study of aziridino-indan (16), which can be considered to be a simplified mitomycin model, that the *cis* hydroxy amino product also prevails. The simple rationale is that the protonated aziridine opens to the benzylic carbonium ion (17) and that the protonated amine function is solvated by water molecules which are properly situated to attack the carbonium ion to give the *cis* product[8]. This concept is easily extrapolated to the mitomycins via intermediate (18) and explains the unusual pH sensitivity of the hydrolytic stereochemistry reported by Underberg and Beijnen[9]. Thus, at low pH, the hydrolyses of mitomycins A(1), B(4), C(2) and porfiromycin (3) all gave *cis* products in ratios ranging from 6:1 to 4:1. As the pH was raised, the ratios changed over to nearly 1:1, the inflection point being about pH 3 for mitomycin C and porfiromycin and pH 2 for mitomycin A and B. Our speculation is that at the higher pH, a second ring-opening pathway, namely S_N2 inversion at C-1 of the aziridine, begins to compete with the open carbonium ion. Two other plausible mechanisms have been ruled out. First, the possibility of back-side participation by the C-10 carbamate blocking *trans* attack at the C-1 carbonium ion was dismissed when it was observed that the C-10 OH derivative, decarbamoyl-mitomycin, gave the identical product distribution upon acid treatment[10a,11]. Second, the possibility that the aziridine was cleaved in a fast step to (19) before the rate determining formation of the mitosene framework can be ruled out because compounds such as (20) have never been observed. If, on the other hand, cleavage to (19) were rate determining, all the kinetic data consistent with acid catalysed acetal hydrolysis would have to be disregarded. When mitomycin C was treated with trifluoroacetic acid in aprotic solvents, only *cis*-(14) was obtained, where X = TFA and Y = NH_2. Presumably, carbonium ion (18) was trapped as a C-1 trifluoroacetate which then underwent O to N acyl transfer[10b]. In their studies of acid catalysed cleavage of mitomycin C followed by trapping with inorganic phosphate and uridine phosphate derivatives, Tomasz and Lipman also observed predominant *cis* stereochemistry in product (21) (Scheme 15.4) where the phosphate oxygen became covalently linked to C-1[12]. More recently, Remers reported the same

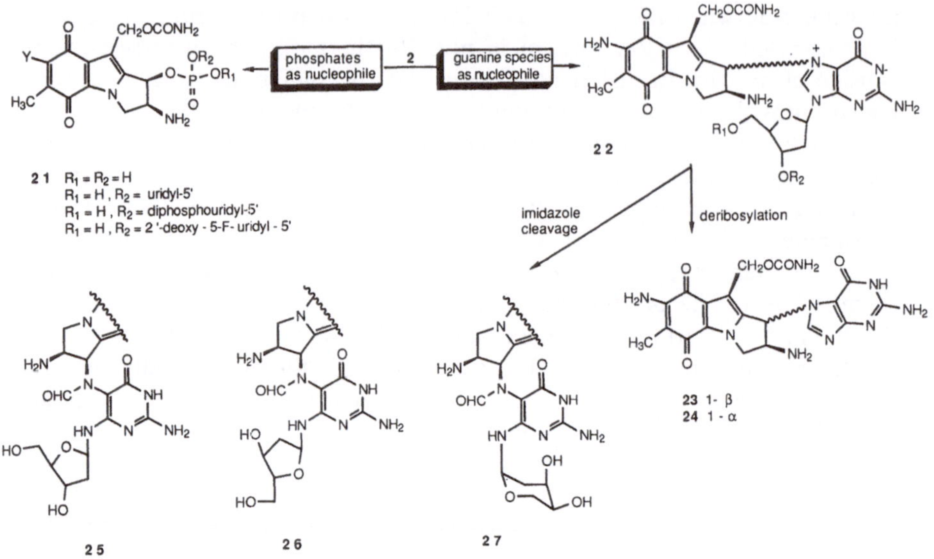

Scheme 15.4

outcome when 5-FdUMP was used as the trapping nucleophile[13]. In detailed studies of the acid catalysed hydrolysis of mitomycin C in the presence of calf thymus DNA and model guanines, the principal products were formed by alkylation at the N-7 of guanine. Since the initial alkylation product (**22**) is charged, further chemistry takes place to afford either deribosylated materials (**23**) and (**24**) or imidazolium ring-opened materials ((**25**)–(**27**)). The isolated yields in these experiments, based on starting mitomycin C, were less than 10%. Thus, the problem of assignment of alkylation stereochemistry of the mitomycin derivatives and the location of *N*-alkylation of the guanine required special analytical methods. The stereochemistry was assigned by the discovery that the circular dichroism spectra of the 1-α and 1-β diastereoisomers are mirror images in the 520–540 nm region, regardless of the functionality of the 1-substituent. Thus, the sign of the 520–540 extremum identifies the stereochemistry[10]. The major stereochemistry observed in these examples, like that from reductive activation methods, was *trans*. However, the guanine reactivity was different from the reductive activation case where N-2 alkylation was observed. This difference was rationalised by noting that N-7 is considered to be a 'harder' nucleophile than N-2 and thus it is reactive toward the C-1 carbonium ion formed via acid catalysis[14]. In the reductive activation case, C-1 is the neutral terminus of a quinone methide and reacts with the softer nucleophile at N-2 of guanine.

15.3   Mitomycins and base

The base catalysed chemistry of the mitomycins has not been actively pursued. Workers at Bristol-Myers, in an attempt to alkylate N-7 of mitomycin C, mitomycin C (2) $\xrightarrow[\text{RX}]{\text{NaH}}$

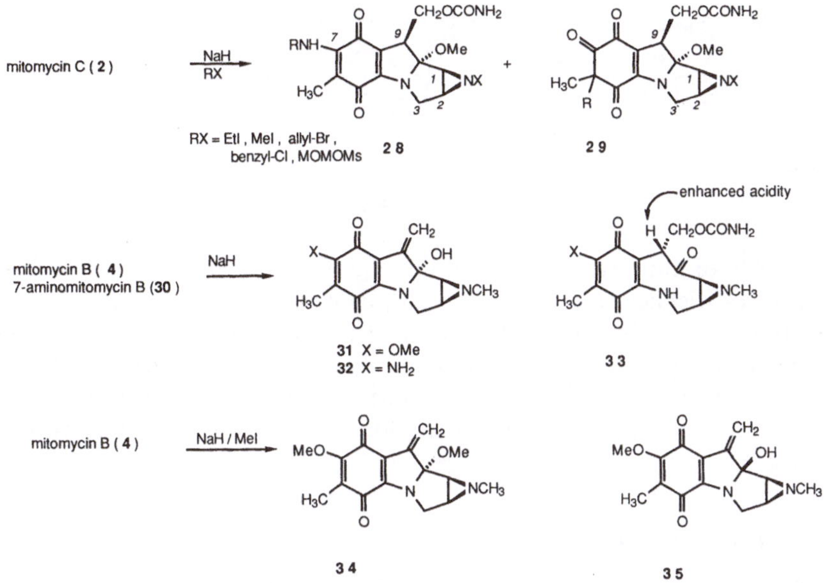

RX = EtI , MeI , allyl-Br , benzyl-Cl , MOMOMs   **2 8**

2 9

mitomycin B (**4**)
7-aminomitomycin B (**30**) $\xrightarrow{\text{NaH}}$

31 X = OMe
32 X = NH$_2$

enhanced acidity

3 3

mitomycin B (**4**) $\xrightarrow{\text{NaH / MeI}}$

3 4

3 5

Scheme 15.5

treated it with NaH in DMF to produce a deep blue solution. Reaction of this solution with a series of reactive halides afforded, in low yields, mixtures of the desired 7-alkyl mitomycin C (**28**) and the novel triketo derivative (**29**) (Scheme 15.5). It appears that (**29**) was obtained as a single stereoisomer, but the relative stereochemistry was not assigned[15]. In contrast to the ring A activation of mitomycin C, when mitomycin B and its 7-amino analogue (**30**), compounds with a free OH at C-9a, were treated with NaH and no alkylating agent, the dehydro compounds (**31**) and (**32**) were obtained in modest yield. The intermediate (**33**) is assumed to form whereupon the acidity of the proton at C-9 is enhanced and a facile elimination of the carbamate takes place, followed by recyclisation. When the reaction is repeated in the presence of MeI, the elimination still takes place, followed by alkylation of both the C-9a OH and the aziridine N to produce the N-methyldehydromitomycin A analogue (**34**)[16]. Danishefsky and Feigelson[17], in synthetic studies, have prepared an iso-dehydromitomycin (**35**). They were not able to equilibrate this material, via ring-opening, to a species like (**33**). It appears that the hybridization at C-9 of mitomycins with C-9a hydroxyls must be sp^5 for ring-opening to be facile.

15.4 Mitomycin rearrangements

Perhaps the most unusual chemistry of the mitomycins, discovered recently, is the series of Michael and retro-Michael reactions which interrelates mitomycin A, albomitomycin A (**36**) and isomitomycin A (**37**) (Scheme 15.6). Albomitomycin A, which is colourless, can be envisioned as the product of the

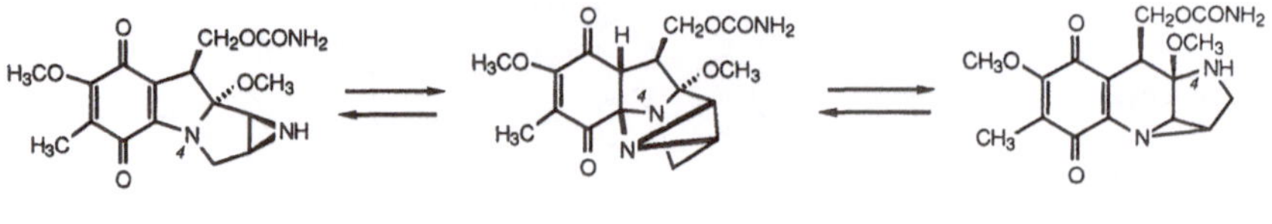

Scheme 15.6

aziridine N of mitomycin A having undergone a Michael reaction to C-4a. Then retro-Michael reaction of N-4 affords isomitomycin A. Each of these isomers, upon standing in MeOH, proceeds in about 3 days to an equilibrium mixture, with mitomycin A predominant. The equilibration can be accelerated by aluminium alkoxides[18]. This unusual equilibration served as the basis for a total synthesis of the mitomycins where isomitomycin was the initial synthetic target[19].

15.5 Chemistry at C-7

Functional group manipulation at the C-7 of mitomycin has long been a concern of the medicinal chemistry community. There have been thorough explorations of 7-amino and 7-alkoxy compounds, usually prepared by substitution of the OMe at C-7 in mitomycin A using conventional addition-elimination chemistry of alkoxyquinones[20]. Recently, workers at Bristol-Myers have developed alternative routes to C-7 analogues (Scheme 15.7). One method requires the hydroxide catalysed conversion of mitomycin C to 7-hydroxymitosane (38). Since hydroxyquinones are vinylogous acids, they are acidic enough to be alkylated by triazenes (39) which are prepared by condensing primary amines with aryl diazonium salts (Chapter 5). Thus, a variety of alkoxymitocins (40) were prepared from (38) in modest yields ranging from 27% to 41%[21]. In a novel method for achieving replacement of the amino function at C-7 of mitomycin C, the bisamidine (41) was first obtained from the antibiotic with a 50-fold excess of dimethyl formamide dimethyl acetal. Use of only an 8-molar excess of reagent amidinated the carbamate selectively. Aminolysis of (41) afforded 7-alkylaminomitomycins (42) along with minor amounts of mitomycin C. Hence, the amidine function at C-7 enhances quinone addition-elimination chemistry to permit the introduction of amines, a process which normally requires an alkoxy group at C-7[22].

15.6 Mitomycin-metal complexes

Although the natural mitomycins do not appear to form stable metal complexes, there has been some interest in preparing metal derivatives of N-7 modified mitomycins, for example (43). These complexes are not really part of mitomycin chemistry, but are of interest because of their potential in modifying the biological spectrum of the drug[23].

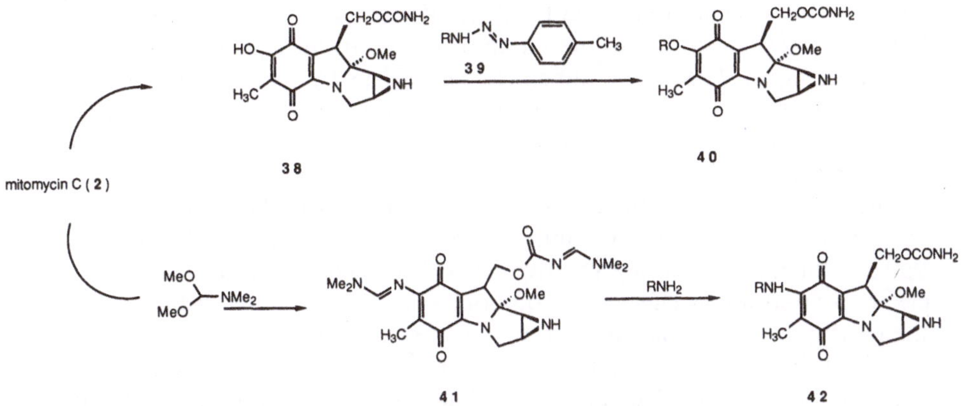

Scheme 15.7

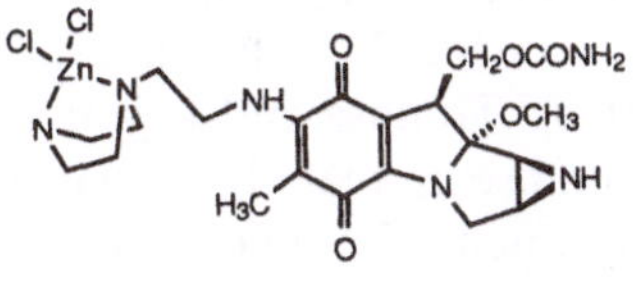

15.7 Chemistry in the reduced state

15.7.1 *Bioreductive alkylation as the mode of action of the mitomycins*

The reduction chemistry of the mitomycins has been the subject of great interest ever since the discovery by Iyer and Szybalski[3] that these substances cross-link the complementary strands of DNA irreversibly *in vivo* and *in vitro* and that the presence of a reducing agent is an absolute requirement for this phenomenon. The covalent DNA-binding and cross-linking activity of the mitomycins represents the molecular basis for their biological activity[3], as indicated by parallels with a number of known DNA-targeted agents: selective inhibition of DNA replication, induction of the SOS response and sister chromatid exchange, cross-resistance or cross-hypersensitivity of cell mutants to ultraviolet light and mitomycin[24].

Living cells possess enzymes which metabolise these drugs reductively[3,25]; in turn, this process activates the drug to bind and cross-link DNA[3]. This series of events can be mimicked in the test tube if chemical reducing agents or NADPH-dependent flavoreductases are added to a mixture of DNA and mitomycin in neutral aqueous buffer[3].

From their original observation, Iyer and Szybalski[3] postulated a mechanism for the reductive activation of mitomycins: the C-1 aziridine and C-10 carbamate groups are two masked alkylating functions, which become 'allylic' (therefore activated) upon reduction of the quinone system and consequent spontaneous elimination of methanol from the 9/9a position; then they may be displaced by two nucleophiles in DNA, resulting in a mitomycin-DNA cross-link. This mechanism led to designation of mitomycin C as a 'bioreductive alkylating agent'[26]. Moore[27] amended these ideas by speculating that both displacements are of S_N1 type, facilitated by resonance with the indolohydro-quinone system of reduced mitomycin C taking place sequentially as summarised in Scheme 15.8.

This theory addresses the peculiar lack of reactivity of the aziridine ring of the mitomycins in their native (quinone) form. In contrast to simple aziridines, those in the mitomycins are inactive towards nucleophilic attack, due to the electron-withdrawing effect of the quinone system, transmitted via overlap of the p orbitals of N-4 and the aziridine nitrogen[28]. This is manifested by the drastically lowered basicity (the pK_a of mitomycin C is 3.2 while that of ethyleneimine is 8.0)[29]. The unusual stability of the aminal or carbinolamine function at C-9 is also a consequence of the presence of the quinone. Reduction reactivates these 'masked' functional groups.

Solid experimental evidence for this scheme, which originated in 1964[3], has come forth only in the 1980s. Tomasz and Lipman[30] isolated and character-ised bioreductive (i.e. reductase activated) alkylation products and reductive metabolites of (2) formed in neutral aqueous phosphate buffer. The products (52), (53) and (54) (Scheme 15.9) were all quinones since isolation took place in the presence of air. The formation of the three sets of substances verified 'one half' of the postulated Scheme 15.8: The approximately 1:1 α/β stereochemis-try of (53) and (54) (both in the Y = OH and Y = phosphate series) at C-1 argues against alternative mechanisms *not* involving the quinone methide (46), since direct attacks on the aziridine should have given a single stereoisomer. More importantly, formulation of product (52) along with (53) and (54) is strong evidence for intermediacy of (46). Such an intermediate is expected to exhibit both nucleophilic and electrophilic reactivity at C-1, resulting in (53) and (54) in the former case and in (52), due to protonation, in the latter (Scheme 15.9). Consistently, the proportion of (52) to (53) and (54) increased with increased acidity of the medium[30]. Kohn and Zein[31] substantiated this

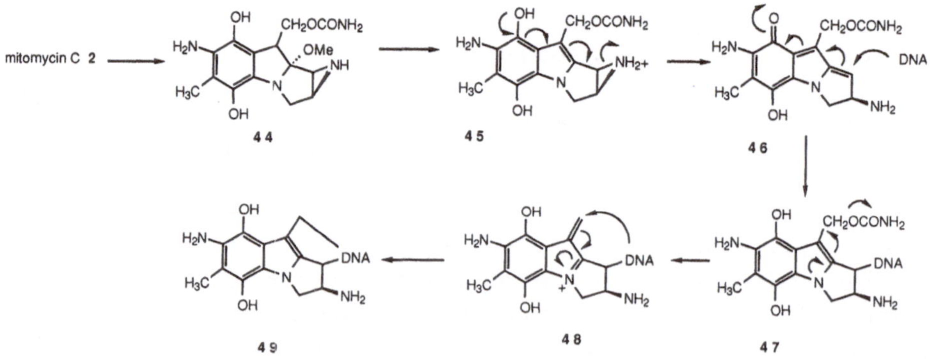

Scheme 15.8 Iyer–Szybalski–Moore hypothesis.

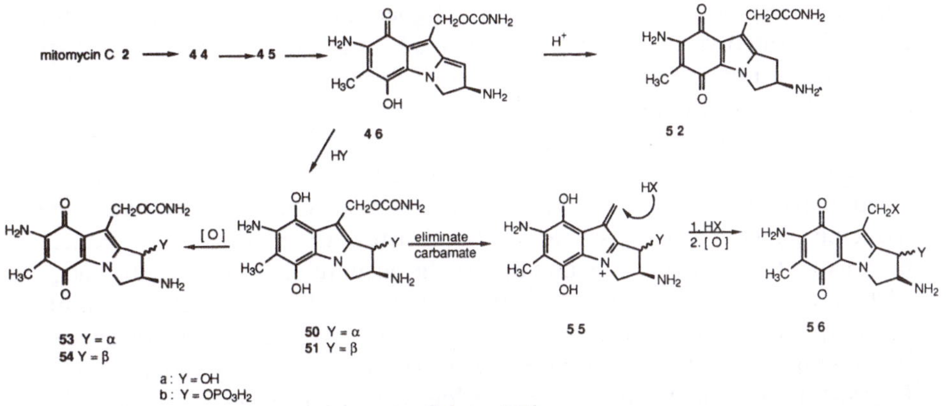

Scheme 15.9

mechanism, showing that the new hydrogen at C-1 of (52) comes from a proton. Use of H_2/PtO_2 rather than the microsomal enzyme preparation resulted in identical products[30]. Similarly, Pan et al.[32] observed the same set of products using the purified flavoenzymes NADPH-cytochrome P-450 reductase and xanthine oxidase.

These results indicated that the primary reductive alkylating function of the mitomycins is the C-1 aziridine position. The C-10 carbamate as the postulated other reactive function upon reduction (Scheme 15.8) remained inert under the bioreductive conditions. However, with the use of $Na_2S_2O_4$ as reducing agent and with the inclusion of some more powerful nucleophiles in the neutral aqueous reaction medium, such as potassium ethyl monothiocarbonate[33a], thiobenzoate[33b] or potassium ethyl xanthate[34], C-1, C-10-disubstituted products were isolated and characterised, providing the first evidence for the postulated nucleophilic reactivity at C-10.

A thorough and insightful study by Peterson and Fisher[35] has provided the answer as to why only the C-1 position was activated under the usual model bioreductive alkylation conditions. They studied the kinetics of formation of the products (52)–(54) using Old Yellow enzyme/NADPH for the reduction of (2) and discovered that the reduction and the formation of (53) and (54) was autocatalytic. The activation process was propagated by electron transfer from reduced mitosenes to mitomycin C quinone as demanded by the relative redox potentials of mitosanes and mitosenes (the latter are 100 mV more negative[41]). Thus, non-stoichiometric reduction of mitomycin resulted in stoichiometric activation in a chain reaction:

$$\text{mitomycin C}_{red} \longrightarrow \text{mitosene}_{red}$$

$$\text{mitosene}_{red} + \text{mitomycin C} \longrightarrow \text{mitosene} + \text{mitomycin C}_{red}$$

Net: mitomycin C $\longrightarrow$ mitosene

(2)   (53) + (54)

This chain reaction was also reported by Egbertson and Danishefsky[36] in the case of N-methyl mitomycin A. Peterson and Fisher[35] proposed that oxidation of mitosene$_{red}$ ((50) and (51) in Scheme 15.9) is so fast under the autocatalytic conditions that it competes effectively with ejection of the C-10 carbamate which requires the reduced state (formation of (55) in Scheme 15.9). This facile reoxidation could explain why no C-10 activation was observed when enzymatic or H_2/PtO_2 reduction was used. On the other hand, the reducing agent $Na_2S_2O_4$ led to C-10 substitution[33,34] because it reduces mitomycin C at a higher rate and allows the accumulation of the mitosene$_{red}$ ((50) and (51)) required for C-10 activation. These ideas were fully substantiated using DNA itself as the nucelophile[37], as discussed later. The existence of the iminium ion intermediates ((48) and (55)) was shown by Kohn et al.[40]: H_2/Pd-C reduction of (53) in ethanol yielded (58), interpreted as diagnostic for (55) as precursor since (55), also a quinone methide, is expected to be trapped by protonation to give (58) (Scheme 15.10).

Further evidence for the mechanism of reductive activation of mitomycins came from Danishefsky's laboratory[38]; these investigators isolated discrete intermediates of the 'mitomycin activation cascade' (cf. Scheme 15.8 (44)) and the corresponding mitomycin B and N-methyl mitomycin A derivatives (59) (leucomitomycins) as well as (60), a leucoaziridinomitosene. These intermediates were formed in pyridine, by H_2/Pd-C reduction of the corresponding quinones (Scheme 15.11). It was most surprising, however, that these compounds were relatively stable in the absence of air. Such systems were thought to eliminate methanol spontaneously. However, when (59) was mixed with an equal amount of N-methyl mitomycin A (1), methanol elimination took place rapidly. Since mixtures of quinones and hydroquinones yield semiquinones, elimination of the 9a-methoxyl function apparently occurred in the one-electron oxidation state only[39]. Somewhat less direct evidence indicated[36] that (60) is also more reactive in the one-electron than in the two-electron reduced state. Similar conclusions were reached by Kohn et al.[40] from electrolytic reduction of mitomycin C in absolute methanol: elimination of methanol was observed only at the one-electron stage. Andrews et al.[42] also reported electrochemical reduction of mitomycin C in dimethylformamide or dimethyl sulphoxide and observed the semiquinone radical anion (one-electron reduction product) of (2) by ESR. Addition of H_2O at this stage led to the usual 'trio' of (52)–(54) (see above and Scheme 15.9) and, in addition, to the 10-decarbamoylated form of each. Reduction to the mitomycin C dianion (two-electron) stage, however, yielded only two products: 10-decarbamoyl-(52) and a new over-reduced mitosene. The authors interpreted these results as indicating that (a) one-electron reduction of mitomycin C is sufficient to induce the 'activation cascade' and (b) one-electron reduction is also the major mode of bioreductive action, since only the one-electron reduction resulted in the usual[30,32] bioreductive (i.e. enzymatically reduced) products (52)–(54).

It must be kept in mind, however, that all of the above evidence for one-

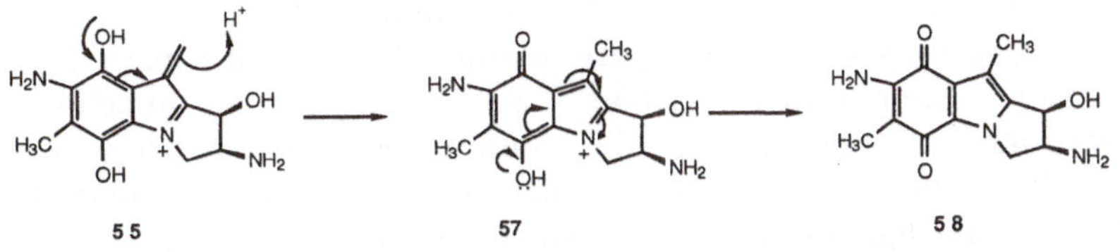

Scheme 15.10

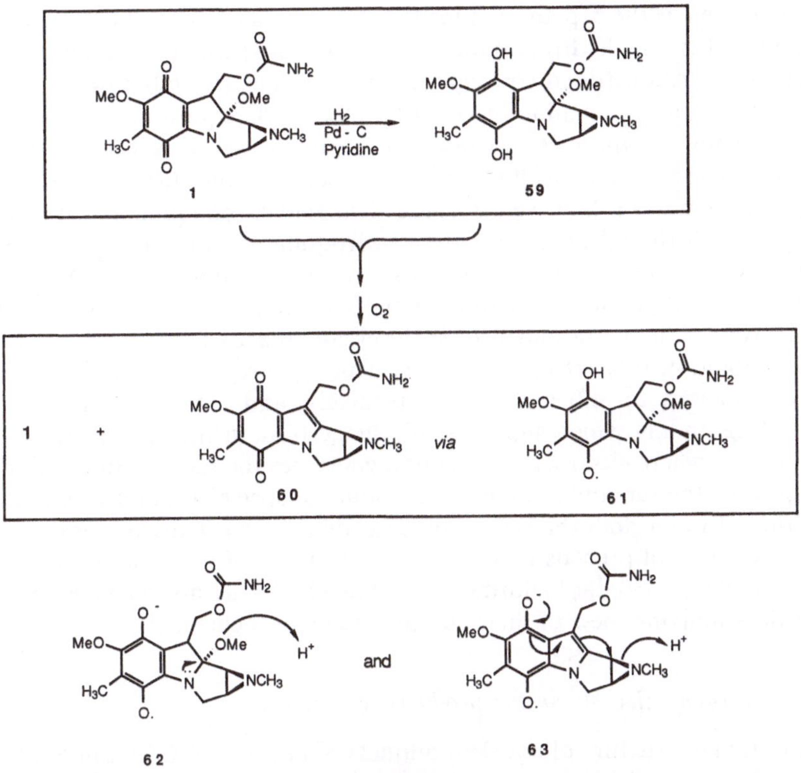

Scheme 15.11

electron activation of mitomycins involved organic solvents, in which semiquinones are much more stable than in water[43]. What is the reactive intermediate in water? This was addressed directly by Swallow's group[44] by generating the mitomycin C semiquinone cleanly and rapidly, using formate radicals produced radiolytically from sodium formate, in aqueous buffer. The products were the usual 'trio' again, (52)–(54). Pulse radiolysis was used to produce the semiquinone radical of mitomycin and ultraviolet spectral changes on the micro-second time scale were used to observe the nature and kinetics of the individual reactions that followed. The striking conclusion was that the semiquinone radical dismutated to mitomycin C and mitomycin C hydroquinone extremely fast and further reactions of the 'activation cascade' proceeded exclusively from the hydroquinone. Very similar experiments and conclusions were reported by others[45]. Swallow's group recommend[44] that previous indirect implications of the semiquinone of mitomycin in aqueous systems[32], including that by one of the present reviewers[46], should be re-evaluated in the light of their results, which show that only the hydroquinone is stable enough to act as the activated intermediate under physiological conditions.

What, then, is the explanation for the opposite behaviour of mitomycin in pyridine, where the hydroquinone form is unreactive and the semiquinone is reactive, as demonstrated so elegantly by Danishefsky's laboratory[38,39,36]? We venture that the elimination of the 9a-methoxyl, requiring protonation of the departing group, does not occur in pyridine *per se*. Hence, the leuco-mitomycin (**59**) is kinetically stable. The semiquinone (**61**) is capable of supplying a proton, however, because it is acidic: semiquinones are much stronger acids than the corresponding hydroquinones. For example, the pK_a of the hydroquinone and semiquinone of anthraquinone are 8 and 3, respectively[47]. This may account for the greater reactivity of (**61**) towards elimination of methanol, and also for the greater reactivity of the semiquinone form of (**60**) towards aziridine opening, which is also proton-catalysed: they both dissociate, to (**62**) and (**63**), respectively, and a proton. There is an interesting finding which supports this suggestion: in the aprotic system of Andrews[42], where electrolytic reduction generates the semiquinone anion as opposed to the semiquinone itself, this anionic species is stable. Only after addition of water does the activation cascade proceed. Thus we suggest that the availability of protons modulates the reactivity of both the one-electron and two-electron reduced forms of mitomycins. This notion may help to provide a unifying view of their complex reduction chemistry.

15.7.2   *Bioreductive alkylation products with DNA*

Isolation and structure of covalent adducts of mitomycin C and nucleotides from calf thymus DNA were first reported by Hashimoto *et al.*[48]. Catalytic hydrogenation was used as the activating agent in neutral aqueous buffer, the resulting drug-DNA complex was degraded by P_1 nuclease and the products analysed. Two mitomycin-5'-dGMP adducts and one mitomycin-5'-dAMP adduct were described. At the same time, the Tomasz and Nakanishi groups reported collaboratively, the isolation and structure of the major adduct formed between mitomycin C and d(GpC)[49], which was revised by them later from being a guanine-O^6-linked mitosene to a guanine-N^2-linked one (**64**). The minor 1″β-stereoisomer (**65**) was also characterised rigorously[50]. Structure assignments in this series were difficult due to scarcity of purified material and lack of non-exchangeable protons on the guanine ring for easy location of the mitomycin-guanine linkage. A battery of micro-scale spectroscopic methods were employed, including analysis of second-derivative FT-IR and UV difference spectra (adduct minus mitosene)[51] to arrive at these structures. The same adduct (64) was isolated from calf thymus DNA as the virtually exclusive product, under NADPH-cytochrome c reductase, xanthine oxidase, H_2/PtO_2 or H_2/Pd-C activating conditions[52a,b]. Isolation of the adduct required enzymatic degradation of the mitomycin-DNA complex and separation of the normal and modified nucleosides by HPLC, presenting a difficult analytical challenge, since the latter were formed only to the extent of 3–5%. Since Hashimoto et al.[48] reported that not one but three major adducts were formed in the same DNA, under the same conditions as used by Tomasz et al.[52], this discrepancy was re-investigated. There was only one difference in the methodologies of the two laboratories. The Tomasz group degraded DNA to the nucleoside level, using a DNase I, snake venom diesterase and alkaline phosphatase mixture, while Hashimoto used P_1 nuclease which degrades DNA to 5'-mononucleotides. Thus the characterised adducts were not directly comparable between the two laboratories. The Tomasz group found, however, that the 3'-phosphodiester bond of a mitomycin-substituted nucleotide in DNA is completely resistant to P_1 nuclease, in contrast to its susceptibility to snake venom diesterase. It was thus demonstrated that the adducts isolated from P_1 nuclease digestion are dinucleotide-mitomycin adducts, all of which gave the single adduct (64) upon further degradation by snake venom diesterase plus alkaline phosphatase. This led to the conclusion that the three adducts in Hashimoto's report[48] must have been dinucleotides, each consisting of (64) and another unmodified nucleotide, providing a source of error in their interpretation. For further confirmation, one of the adducts, the O^6-linked mitomycin-deoxyguanosine, claimed to be formed, was synthesised by an authentic route[53] and was shown to have properties different from those reported[48]. Another proposed guanine-O^6-linked mitomycin C adduct structure[54] was also discounted by this authentic synthesis.

A major puzzle remaining after completion of the above work[52] was that all of the identified adducts, accounting for 95% of the total, were monofunctional alkylation products of mitomycin C. What then is the origin of the observed 'cross-linked' behaviour of DNA exposed to this drug? A breakthrough occurred when $Na_2S_2O_4$ was used instead of flavoenzyme or H_2 as reducing agent in the in vitro reaction of mitomycin C with DNA: a formerly minor adduct was now major (70%), with another formerly minor one making up the other 30%. A sample of 4 mg of this purple solid substance was isolated from 300 mg of Micrococcus luteus DNA modified by mitomycin C under these conditions. Its structure was determined using ^{1}H-NMR, FT-IR, FAB mass spectroscopy and circular dichroism spectra[55]. In this substance (68), the drug is linked by its C-1″ and C-10″ positions to two N^2-atoms of deoxyguanosine residues. This adduct accounts fully for the 'cross-linked' behaviour of DNA[3], provided the two linked deoxyguanosines are in opposite strands of duplex DNA. The structure of the other (30%) adduct (66) was also determined[56]. In

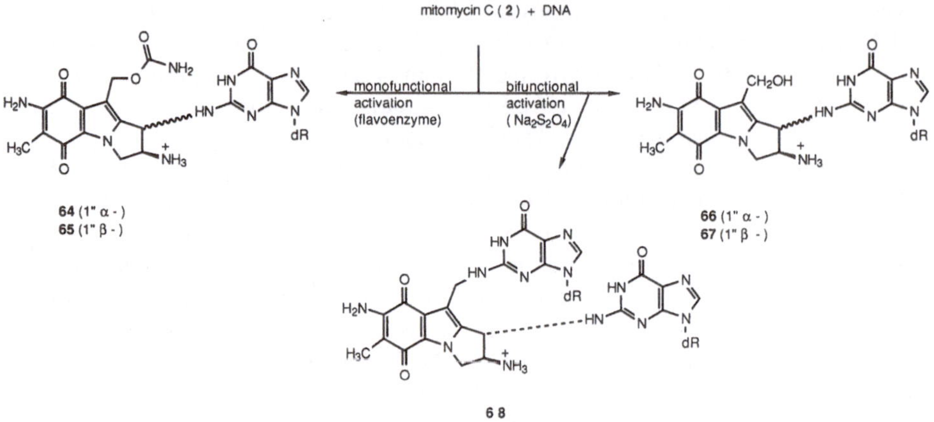

Scheme 15.12

vivo, the same cross-link adduct was isolated from DNA of rats injected with mitomycin C and the cross-link (68) and monofunctional adduct (66) were identified in CHO cells in mitomycin-treated cell culture[55,57].

A summary of the adducts formed with DNA is given in Scheme 15.12. It is apparent that their structures confirm fully the Iyer–Szybalski hypothesis (cf. Scheme 15.8)[3]. It is most interesting that two types of adducts can be obtained: flavoenzyme or H_2/PtO_2 activation of mitomycin C results in formation of adduct (64), reflecting monofunctional activation of the drug at C-1, while $Na_2S_2O_4$ activation gives products of bifunctional activation, at C-1 and C-10, i.e. (66) and (68) as a pair. Compound (66) originates from base sequences which have no other G in the vicinity for cross-link formation, while (68) comes from CpG and GpG sequences of duplex DNA, which are sterically favourable for cross-linking two guanines[55]. The types of reactions with DNA match fully the reactions in model studies in aqueous systems with respect to monofunctional versus bifunctional activating conditions (see above).

Since the cross-link is more lethal than monofunctional adducts of DNA, determination of the factors governing the cross-linking activity of the mitomycins is important. Tomasz et al.[37] designed a set of experiments, to test the idea, based on the original suggestion of Peterson and Fisher[35], that autocatalytic versus stoichiometric reducing conditions may be the crucial factor (see above). The experiments showed that all customary reducing agents were capable of inducing either monofunctional or bifunctional activation, depending only on the kinetic conditions (autocatalytic or stoichiometric reduction) chosen for the experiment. Oxygen was shown to be selectively inhibitory to bifunctional activation. This detailed understanding of the reductive alkylating action of the mitomycins *in vitro* should help in elucidating their mode of action as hypoxia-selective antitumour agents[58] *in vivo*.

References

1. T. Hata, Y. Sano, R. Sugawara, A. Matsumae, K. Kanamori, T. Shima and T. Hoshi, *J. Antibiot, (Japan) Ser. A* **9** (1956) 14.
2. S. T. Crooke, in *Cancer Chemotherapy, Vol. 3*, eds. S. T. Crooke and A. W. Prestayko, Academic Press, New York (1981) p. 49.
3. V. N. Iyer and W. Szybalski, *Science* **145** (1964) 55.
4. Earlier reviews: (a) W. A. Remers, in *The Chemistry of Antitumor Antibotics, Vol. 1*, Wiley, New York (1979) Chap. 5; (b) R. W. Franck, *Prog. Chem. Org. Natl. Prod.* **38** (1979) 1; (c) for a bibliography of recent synthetic work see S. Nakatsuka, H. Miyazaki and T. Goto, *Heterocycles* **24** (1986) 2109.
5. J. S. Webb, D. B. Cosulich, J. H. Mowat, J. B. Patrick, R. W. Broschard, W. E. Meyer, R. P. Williams, C. F. Wolf, W. Fulmore, C. Pidacks, and J. E. Lancaster, *J. Am. Chem. Soc.* **84** (1962) 3185.
6. N. Hirayama and K. Shirahata, *J. Am. Chem. Soc.* **105** (1983) 7199.
7. R. A. McClelland and K. Lam, *J. Am. Chem. Soc.* **107** (1985) 5182.
8. I-C. Chiu and H. Kohn, *J. Org. Chem.* **48** (1983) 2857.
9. W. J. M. Underberg and J. H. Beijnen, *Chem. Pharm. Bull.* **35** (1987) 4557.
10a. M. Tomasz, M. Jung, G. L. Verdine and K. Nakanishi, *J. Am. Chem. Soc.* **106** (1984) 7367.
10b. G. L. Verdine, B. F. McGuinness, K. Nakanishi and M. Tomasz, *Heterocycles* **25** (1987) 577.
11. U. Hornemann, P. J. Keller and K. Takeda, *J. Med. Chem.* **28** (1985) 31.
12. M. Tomasz and R. Lipman, *J. Am. Chem. Soc.* **101** (1979) 6063.
13. B. S. Iyengar, R. T. Dorr, W. A. Remers and C. D. Kowal, *J. Med. Chem.* **31** (1988) 1579.
14. M. Tomasz, R. Lipman, M. S. Lee, G. L. Verdine and K. Nakanishi, *Biochemistry* **26** (1987) 2010.
15. T. Kaneko, H. Wong and T. W. Doyle, *Tetrahedron Lett.* **26** (1985) 3923.
16. U. Urakawa, H. Tsuchiya, K-I. Nakano and N. Nakamura, *J. Antibiot.* **34** (1981) 1152.
17. G. B. Feigelson and S. J. Danishefsky, *J. Org. Chem.* **53** (1988) 3393.
18. M. Kono, Y. Saitoh, K. Shirahata, Y. Arai and S. Ishii, *J. Am. Chem. Soc.* **109** (1987) 7224.
19. T. Fukuyama and L. Yang, *J. Am. Chem. Soc.* **109** (1987) 7881.
20. See Ref. 4; see also S. M. Sami, B. S. Iyengar, S. E. Tarnow, W. A. Remers, W. T. Bradner and J. E. Schurig, *J. Med. Chem.* **27** (1984) 701.
21. D. M. Vyas, D. Benigni, R. A. Partyka and T. W. Doyle, *J. Org. Chem.* **51** (1986) 4307.
22. D. M. Vyas, Y. Chiang, D. Benigni and T. W. Doyle, *J. Org. Chem.* **52** (1987) 5601.
23. (a) B. S. Iyengar, T. Takahashi, W. A. Remers and W. T. Bradner, *J. Med. Chem.* **29** (1986) 144; (b) B. S. Iyengar, S. M. Sami, T. Takahashi, E. E. Sikorski, W. A. Remers and W. T. Bradner, *J. Med. Chem.* **29** (1986) 1760.
24. See Ref. 52 for specific citations.
25. S. R. Keyes, S. Rockwell and A. C. Sartorelli, *Cancer Res.* **45** (1985) 3642.
26. A. J. Lin, L. A. Cosby and A. C. Sartorelli, *ACS Symp. Ser.* **30** (1976) 71.
27. H. W. Moore, *Science* **197** (1977) 527.
28. A. Tulinsky and J. H. van den Hende, *J. Am. Chem. Soc.* **89** (1967) 2905.
29. C. L. Stevens, K. G. Taylor, M. E. Munk, W. S. Marshall, K. Noll, G. D. Shah, L. G. Shah and K. Uzu, *J. Med. Chem.* **8** (1964) 1.
30. M. Tomasz and R. Lipman, *Biochemistry* **20** (1981) 5056.
31. H. Kohn and N. Zein *J. Am. Chem. Soc.* **105** (1983) 4105.
32. S.-S. Pan, P. A. Andrews, C. J. Glover and N. R. Bachur, *J. Biol. Chem.* **259** (1984) 959.
33. (a) M. Bean and H. Kohn, *J. Org. Chem.* **48** (1983) 5033; (b) *J. Org. Chem.* **50** (1985) 293.
34. U. Hornemann, K. Iguchi, P. J. Keller, H. M. Vu, J. F. Kozlowski and H. Kohn, *J. Org. Chem.* **48** (1983) 5026.
35. D. M. Peterson and J. Fisher, *Biochemistry* **25** (1986) 4077.
36. M. Egbertson and S. J. Danishefsky, *J. Am. Chem. Soc.* **109** (1987) 2204.
37. M. Tomasz, A. K. Chawla and R. Lipman, *Biochemistry* **27** (1988) 3182.
38. S. J. Danishefsky and M. Ciufolini, *J. Am. Chem. Soc.* **106** (1984) 6424.
39. S. J. Danishefsky and M. Egbertson, *J. Am. Chem. Soc.* **108** (1986) 4648.
40. H. Kohn, N. Zein, X. Q. Lin, J.-Q. Ding and K. M. Kadish, *J. Am. Chem. Soc.* **109** (1987) 1833.

41. G. M. Rao, A. Begleiter, J. W. Lown and J. A. Plambeck, *J. Electrochem. Soc.* **124** (1977) 199.
42. P. A. Andrews, S.-S. Pan and N. R. Bachur, *J. Am. Chem. Soc.* **108** (1986) 4158.
43. J. A. Chambers, *Chemistry of the Quinoids*, Part 2, ed. S. Patai, Wiley, New York, (1974) Chap. 14.
44. B. M. Hoey, J. Butler and A. J. Swallow, *Biochemistry* **27** (1988) 2608.
45. G. Machtalere, Ch. Houee-Levin, M. Gardes-Albert, C. Ferradini and B. Hickel, *C.R. Acad. Paris Serie II* **307** (1988) 17.
46. M. Tomasz, C. M. Mercado, J. Olson and N. Chatterjie, *Biochemistry* **13** (1974) 4878.
47. D. O. Wipf, K. R. Wehmeyer and R. M. Wightman, *J. Org. Chem.* **51** (1986) 4760.
48. Y. Hashimoto, K. Shudo and T. Okamoto, *Tetrahedron Lett.* **23** (1982) 677; *Chem. Pharm. Bull.* **31** (1983) 861; *Acc. Chem. Res.* **17** (1984) 403.
49. M. Tomasz, R. Lipman, J. K. Snyder and K. Nakanishi, *J. Am. Chem. Soc.* **105** (1983) 2059.
50. M. Tomasz, R. Lipman, G. L. Verdine and K. Nakanishi, *Biochemistry* **25** (1986) 4337.
51. G. L. Verdine and K. Nakanishi, *J. Am. Chem. Soc.* **107** (1985) 6118.
52. (a) M. Tomasz, D. Chowdary, R. Lipman, S. Shimotakahara, D. Veiro, V. Walker and G. L. Verdine, *Proc. Natl. Acad. Sci. USA* **83** (1986) 6702; (b) apparently the same adduct (**64**) was also isolated by S.-S. Pan, T. Tracki and N. R. Bachur, *Mol. Pharmacol.* **29** (1986) 622, although these authors identified it as the O^6-guanine-linked structure, being unaware of the revision[50] of the structure of (**64**).
53. B. F. McGuinness, K. Nakanishi, R. Lipman and M. Tomasz, *Tetrahedron Lett.* **29** (1988) 4673.
54. N. Zein and H. Kohn, *J. Am. Chem. Soc.* **109** (1987) 1576.
55. M. Tomasz, R. Lipman, D. Chowdary, J. Pawlak, G. L. Verdine and K. Nakanishi, *Science* **235** (1987) 1204.
56. M. Tomasz, R. Lipman, B. F. McGuinness and K. Nakanishi, *J. Am. Chem. Soc.* **110** (1988) 5892.
57. D. Chowdary and M. Tomasz, *Fed. Proc., Fed. Am. Soc. Exp. Biol.* **46** (1987) 2037.
58. A. C. Sartorelli, *Biochem. Pharmacol.* **35** (1986) 67.

16 The chemistry of activated bleomycin

16.1 Introduction

The bleomycins are a structurally related class of antitumour antibiotics that are utilised extensively for the treatment of human malignancies[1,2]. In common with many antitumour agents, the cellular locus for the action of bleomycin is thought to be DNA. More specifically, bleomycin can be activated for DNA strand scission and this transformation is believed to constitute the basis for the therapeutic utility of this class of compounds[2-7].

Bleomycin A$_2$

16.2 Activation of metallobleomycins

DNA strand scission by bleomycin (BLM) is an oxidative process that is mediated by activated metal chelates of these compounds. The most extensively studied metallobleomycins have been Fe·BLM[3,8] and Cu·BLM[9,10], although activated complexes of Mn·BLM[11,12], Co·BLM[13], Ru·BLM[14] and VO·BLM[15] have also been reported to degrade DNA. DNA cleavage by Fe·BLM and Cu·BLM has been shown to be dependent on the availability of oxygen and greatly facilitated by reducing agents such as dithiothreitol[3,8 – 10,16,17]. In addition to reductive activation of dioxygen, which leads to the formation of activated metallobleomycins, there are a number of examples of bleomycin activation by the use of oxygen surrogates such as H_2O_2[12,15,18,19], cumene hydroperoxide,[19,20] oxone,[21] iodosobenzene,[19,20,22,23] and sodium periodate[19,20].

The utilisation of activated metallobleomycins in oxidation and oxygenation processes, such as DNA strand scission or olefin epoxidation, results in the regeneration of the original metallobleomycins. Indeed, both Fe and Cu·BLMs can be utilised catalytically for DNA degradation[10,24,25]. On the basis of the accumulated data, it seems likely that Fe·BLM activation can be represented as shown in Scheme 16.1. According to this scheme, the reducing agent would potentiate the action of Fe·BLM both by allowing the 'recycling' of spent Fe(III)·BLM and by providing a reducing equivalent required for Fe·BLM activation *per se*. In the absence of reducing agents, the requisite electron would presumably have to be obtained from within the experimental system, e.g. by disproportionation of two Fe(II)·BLMs in the presence of O_2 to give one 'activated Fe·BLM' and one Fe(III)·BLM. This scheme makes several predictions consistent with available experimental evidence. These include the production of one DNA lesion for every two Fe(II)·BLMs employed in the absence of a reducing agent[26], the diminished efficiency of Fe(II)·BLMs utilised at concentrations that disfavour the bimolecular activation process suggested above[26], the dramatic effect of reducing agents in facilitating DNA degradation by Fe·BLM[3,8,16,25] and the consumption of $1/2\,O_2$ for each Fe(II)·BLM activated in the absence of a reducing agent[27].

At present, no activated metallobleomycin has been characterised in detail as a chemical entity. However, there is considerable evidence that activated bleomycins are ternary complexes consisting of bleomycin, a metal ion and oxygen. This view is supported by physicochemical measurements[18] and by the chemical behaviour of activated bleomycins, which strongly resemble

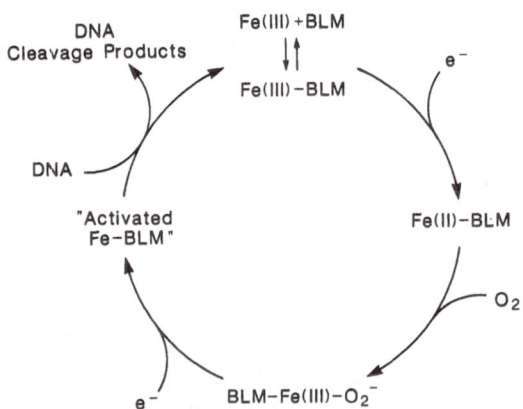

Scheme 16.1 Stoichiometry of Fe·BLM activation.

cytochrome P-450 and related metalloporphyrins[18,28]. The chemical behaviour of activated bleomycins differs in a number of respects from species that generate diffusible oxygen radicals. The distinctions include the indifference of activated BLM to radical scavengers[29], the production of DNA breaks at single, isolated sites[10,26] and the finding that Fe·BLM-mediated cleavage at a double strand (i.e. GC) cleavage site can be highly asymmetric as regards strand selectivity[25].

16.3 Chemistry of DNA strand scission

Bleomycin-mediated DNA degradation has been studied at the levels of chromation[30] and isolated DNA[5,31], as well as by the use of covalently closed circular DNAs[10,11,13] and ^{32}P-end labelled linear duplex DNAs[32-35] prepared by treatment of larger precursors with restriction endonucleases. In all systems studied, metallobleomycins have produced both single- and double-strand nicks; the latter, which may be of particular importance in a therapeutic context, occur at a level of abundance well in excess of what would be predicted from the random accumulation of single-strand breaks[36]. Bleomycin-mediated DNA strand scission is sequence selective, occurring with the greatest frequency at $^{5'}GC^{3'}$ and $^{5'}GT^{3'}$ sequences[32-35]. Although not studied exhaustively, the available evidence suggests that the sequence selectivity of the various metallobleomycins does not differ dramatically[10,13,15,32-35,37].

The actual chemistry of DNA strand scission has also been investigated. Such studies have been carried out using isolated DNAs from a number of species, including some that were specifically radiolabelled to faciliate chemical analysis[5]. Perhaps the most precise analysis of the chemistry of DNA strand scission has been achieved by the use of oligonucleotides capable of acting as efficient substrates for activated Fe·BLM[25,26,38]. For example, Sugiyama and co-workers[25,26] employed the self-complementary dodecanucleotide d($^{5'}$CGCTTTAAAGCG$^{3'}$) as a substrate for Fe(II)·BLM. In common with the other DNA substrates studied, two types of lesions were observed. One type involved the formation of strand breaks with the concomitant appearance of base propenals; the other involved the formation of alkali labile lesions and the release of free bases. However, the advantage of employing the dodecanucleotide d($^{5'}$CGCTTTAAAGCG$^{3'}$) as a substrate was that degradation occurred with exceptional efficiency and primarily at cytidine$_3$ and cytidine$_{11}$. The nature of the products resulting from DNA oligonucleotide degradation at cytidine$_3$ was established qualitatively and quantitatively by chemical synthesis of the putative products, or suitable derivatives thereof, and direct chromatographic comparison of those synthetic standards with the actual products of BLM-mediated dodecanucleotide degradation. As shown in Scheme 16.2, degradation of the dodecanucleotide at cytidine$_3$ was found to result in the formation of trans-3-(cytosin-1'-yl)propenal (cytosine propenal

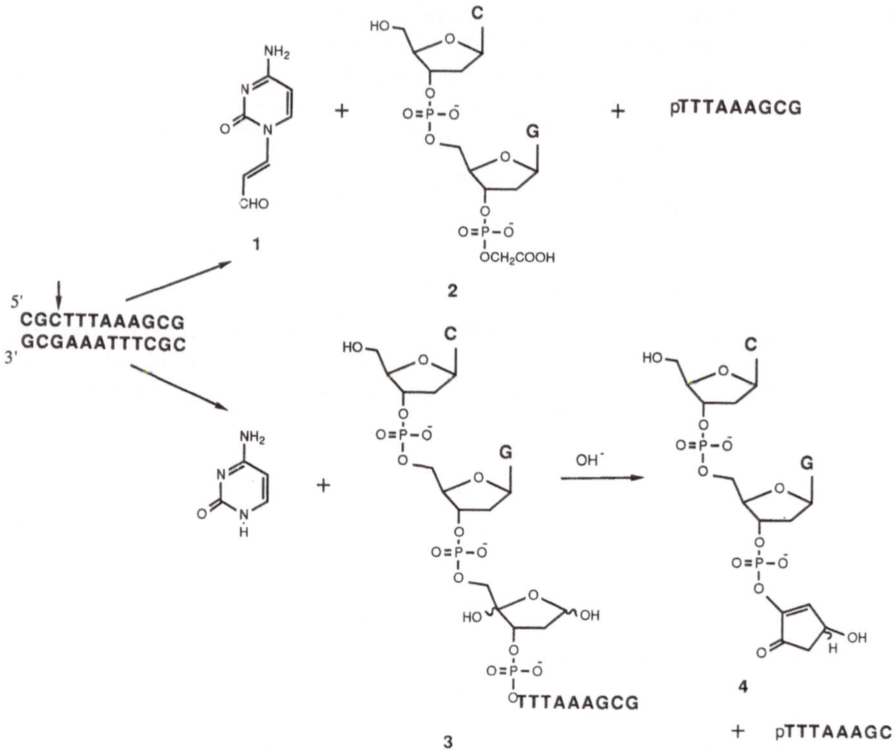

Scheme 16.2 Products resulting from the degradation of d(CGCT$_3$A$_3$GCG) by Fe·BLM.

(**1**)) and 2'-deoxycytidylyl(3' → 5')(2-deoxyguanosine-3'-(phosphoro-2"-*O*-glycolate (**2**))[26,31]. These products are believed to form via the intermediacy of a C-4' hydroperoxy derivative of cytidine$_3$; both reasonable chemical inference and a study that employed specifically radiolabelled DNAs suggested that the propenal moiety of (**1**) was derived from carbon atoms C-1'–C-3' of the cytidine$_3$ deoxyribose, with the remaining two atoms (C-4' and C-5') constituting the glycolate moiety of (**2**)[5,26,39].

Also produced from d(5'CGCT$_3$A$_3$GCG) via the agency of activated Fe·BLM was an alkali labile product (**3**), whose formation was accompanied by the appearance of cytosine. Base treatment of this species effected DNA strand scission and concomitant rearrangement of the deoxyribose of cytidine$_3$ to diastereomeric hydroxycyclopentenones (**4**) (Schemes 16.2 and 16.3) by a process that has substantial chemical precedent[40,41]. The alkali labile lesion was characterised further by treatment with hydrazine, which effected its efficient conversion to pyridazine derivative (**5**); the identity of (**5**) was confirmed by direct comparison of physiochemical properties with an authentic synthetic sample[41]. The alkali labile lesion has also been character-

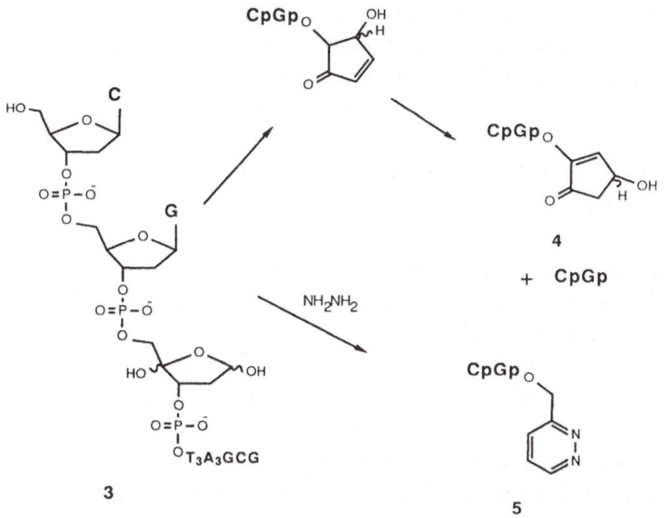

Scheme 16.3 Characterization of the alkali labile product resulting from degradation of d(CGCT$_3$A$_3$GCG) by Fe·BLM.

ised after treatment with sodium borohydride[42]. The alkali labile lesion is believed to form by the C-4′ hydroxylation of deoxyribose[39-42].

Much less has been reported concerning the chemical mechanisms by which other metallobleomycins mediate DNA degradation. However, DNA degradation by Cu·BLM and Co·BLM has been studied. For Cu·BLM, it was shown that the products resulting from degradation of d(CGCT$_3$A$_3$GCG) were qualitatively the same as those produced by Fe·BLM[10]. DNA degradation by Co·BLM requires light but not oxygen[13]; a recent analysis of the chemistry of DNA oligonucleotide degradation indicated that the alkali labile lesion was the exclusive product when this metallobleomycin was employed[43].

16.4 Bleomycin-mediated degradation of structurally altered DNA substrates

Early studies with bleomycin established convincingly that metallobleomycins could bind to DNA substrates having diverse primary structures[32-35]; in all cases DNA cleavage was observed to occur primarily at 5′-GT-3′ and 5′-GC-3′ sequences. Single-stranded DNA was not a good substrate for cleavage by BLM[35].

More recently, Hertzberg and co-workers[44,45] have studied DNA fragments containing sequences that were substrates for restriction methylases. By the use of DNA fragments lacking or containing one or more 5-methylated cytidines (or N^6-methylated adenosines) and four structural congeners of bleomycin, they were able to demonstrate that the extent of DNA degradation by Fe·BLM could be diminished significantly when certain nucleotides were methylated. On the basis of the known[46,47] alteration of B-form DNA structure that attends methylation, it was suggested that the diminution of methylated DNA degradation by bleomycin resulted from the diminished efficiency of cleavage of these conformationally altered B-form DNAs. The alternative possible explanation, i.e. that there was a direct steric interaction between the methyl groups on DNA and bleomycin, was excluded by the distance between the methylated cytidines and bleomycin cleavage sites in certain sequences studied. A second observation inconsistent with steric interaction of bleomycin and the introduced methyl groups, which are located in the major groove of DNA, was that T4 DNA, which contains bulky 5-(glucosyloxy)methyl groups in the major groove, was degraded as efficiently by bleomycin A$_2$ and bleomycin B$_2$ as T4 DNA lacking these glucosyloxymethyl groups. It was shown in the same study that Z-form DNA was not a substrate for cleavage by Fe·BLM[45].

Other substrates for bleomycin have included DNA oligonucleotides containing a bulge at the bleomycin cleavage site[48] and heteroduplexes of RNA and DNA that have been reported to undergo degradation only within the DNA oligonucleotide strand[49,50]. On the basis of a study by Mascharak et al.[51] which demonstrated alteration of the sites of BLM mediated cleavage of a DNA plasmid that had been pretreated with cis-platinum, Gold et al.[52] have recently investigated an oligonucleotide containing a preferred platination site at one end, and a preferred BLM cleavage site at the other. Remarkably, they found that the platinated oligonucleotide was cleaved preferentially at a new site having the sequence 5'-TT-3'. This represents the first example of alternation of bleomycin cleavage specificity in a DNA oligonucleotide of defined structure. In addition to the potential implications for the interaction of these two antitumour agents, which are often co-administered in the clinic, this observation implies that recognition of a preferred site for BLM binding and cleavage need not involve the exocyclic 2-amino group of guanosine, which has been suggested[53] as a recognition element for bleomycin. Given the known distortion of the DNA duplex via platination[51,52], as well as the results cited above for DNA altered conformationally with restriction methylases[44,45], it seems likely that sequence selectivity is actually a function of the overall shape of the DNA duplex at individual nucleotide sites. Since the chemical degradation of DNA by metallobleomycins is mediated in the minor groove via the action of the metal centre[45,53], the latter of which has also been shown to participate in BLM-DNA interaction and the choice of sites for degradation[25,54], it seems logical to think that sequence selectivity of DNA degradation by a given metallobleomycin may be a strong function of the facility with which the metal centre can be accommodated within the minor groove of DNA at individual positions.

Acknowledgement

The study of the mechanism of action of bleomycin has been supported in my laboratory by P. H. S. Research Grants CA27603 and CA38544, awarded by the National Cancer Institute, DHHS.

References

1. S. K. Carter, in *Bleomycin: Current Status and New Developments*, eds. S. K. Carter, S. T. Crooke and H. Umezawa, Academic Press, New York (1978) p. 9.
2. H. Umezawa, in *Anticancer Agents Based on Natural Product Models, Medicinal Chemistry Series*, Vol. XVI, eds. J. M. Cassady and J. D. Douros, Academic Press, New York (1980) pp. 148ff.
3. E. A. Sausville, J. Peisach and S. B. Horwitz, *Biochemistry* **17** (1978) 2740.
4. R. M. Burger, J. Peisach and S. B. Horwitz, *J. Biol. Chem.* **257** (1982) 3372.
5. L. Giloni, M. Takeshita, F. Johnson, C. Iden and A. P. Grollman, *J. Biol. Chem.* **256** (1981) 8608.
6. S. M. Hecht, *Acc. Chem. Res.* **19** (1986) 383, and Refs. therein.
7. J. Stubbe and J. M. Kozarich, *Chem. Rev.* **87** (1987) 1107, and Refs. therein.
8. R. Ishida and T. Takahashi, *Biochem. Biophys. Res. Commun.* **66** (1975) 1432.
9. G. M. Ehrenfeld, L. O. Rodriguez, S. M. Hecht, C. Chang, V. J. Basus and N. J. Oppenheimer, *Biochemistry* **24** (1985) 81.
10. G. M. Ehrenfeld, J. B. Shipley, D. C. Heimbrook, H. Sugiyama, E. C. Long, J. H. van Boom, G. A. van der Marel, N. J. Oppenheimer and S. M. Hecht, *Biochemistry* **26** (1987) 931.
11. G. M. Ehrenfeld, N. Murugesan and S. M. Hecht, *Inorg. Chem.* **23** (1984) 1496.
12. R. M. Burger, J. H. Freedman, S. B. Horwitz and J. Peisach, *Inorg. Chem.* **23** (1984) 2215.
13. C.-H. Chang and C. F. Meares, *Biochemistry* **23** (1984) 2268.
14. R. Subramanian and C. F. Meares, *Biochem. Biophys. Res. Commun.* **133** (1985) 1145.
15. J. Kuwahara, T. Suzuki and Y. Sugiura, *Biochem. Biophys. Res. Commun.* **129** (1985) 368.
16. E. A. Sausville, R. W. Stein, J. Peisach and S. B. Horwitz, *Biochemistry* **17** (1978) 2746.
17. N. J. Oppenheimer, L. O. Rodriguez and S. M. Hecht, *Proc. Natl. Acad. Sci. USA* **76** (1979) 5616.
18. R. M. Burger, J. Peisach and S. B. Horwitz, *J. Biol. Chem.* **256** (1981) 11636.
19. D. C. Heimbrook, R. L. Mulholland, Jr. and S. M. Hecht, *J. Am. Chem. Soc.* **108** (1986) 7839.
20. D. C. Heimbrook, S. A. Carr, M. A. Mentzer, E. C. Long and S. M. Hecht, *Inorg. Chem.* **26** (1987) 3835.
21. M. Girardet and B. Meunier, *Tetrahedron Lett.* **28** (1987) 2955.
22. N. Murugesan, G. M. Ehrenfeld and S. M. Hecht, *J. Biol. Chem.* **257** (1982) 8600.
23. N. Murugesan and S. M. Hecht, *J. Am. Chem. Soc.* **107** (1985) 493.
24. L. F. Povirk, *Biochemistry* **18** (1979) 3989.
25. H. Sugiyama, R. E. Kilkuskie, L.-H. Chang, L.-T. Ma, S. M. Hecht, G. A. van der Marel and J. H. van Boom, *J. Am. Chem. Soc.* **108** (1986) 3852.
26. H. Sugiyama, R. E. Kilkuskie and S. M. Hecht, *J. Am. Chem. Soc.* **107** (1985) 7765.
27. H. Kuramochi, K. Takahashi, T. Takita and H. Umezawa, *J. Antibiot.* (Tokyo) **34** (1981) 576.
28. S. M. Hecht, *Fed. Proc.* **45** (1986) 2784, and Refs. therein.
29. L. O. Rodriguez and S. M. Hecht, *Biochem. Biophys. Res. Commun.* **104** (1982) 1470.
30. D. E. Berry, L.-H. Chang and S. M. Hecht, *Biochemistry* **24** (1985) 3207.
31. N. Murugesan, C. Xu, G. M. Ehrenfeld, H. Sugiyama, R. E. Kilkuskie, L. O. Rodriguez, L.-H. Chang and S. M. Hecht, *Biochemistry* **24** (1985) 5735.
32. A. D. D'Andrea and W. A. Haseltine, *Proc. Natl. Acad. Sci. USA* **75** (1978) 3608.
33. M. Takeshita, A. P. Grollman, E. Ohtsubo and H. Ohtsubo, *Proc. Natl. Acad. Sci. USA* **75** (1978) 5983.

34. C. K. Mirabelli, A. Ting, C.-H. Huang, S. Mong and S. T. Crooke, *Cancer Res.* **42** (1982) 2779.
35. J. Kross, W. D. Henner, S. M. Hecht and W. A. Haseltine, *Biochemistry* **21** (1982) 4310.
36. C. W. Haidle, R. S. Lloyd and D. L. Robberson, in *Bleomycin: Chemical, Biochemical and Biological Aspects*, ed. S. M. Hecht, Springer-Verlag, New York (1979) pp. 222ff.
37. T. Suzuki, J. Kuwahara and Y. Sugiura, *Nucleic Acids Res., Symp. Ser.* **15** (1984) 161.
38. S. Uesugi, T. Shida, M. Ikehara, Y. Kobayashi and Y. Kyogoku, *Nucleic Acids Res.* **12** (1984) 1581.
39. J. C. Wu, J. W. Kozarich and J. Stubbe, *J. Biol. Chem.* **258** (1983) 4694.
40. H. Sugiyama, C. Xu, N. Murugesan and S. M. Hecht, *J. Am. Chem. Soc.* **107** (1985) 4104.
41. H. Sugiyama, C. Xu, N. Murugesan, S. M. Hecht, G. A. van der Marel and J. H. van Boom, *Biochemistry* **27** (1988) 58.
42. L. Rabow, J. Stubbe, J. W. Kozarich and J. A. Gerlt, *J. Am. Chem. Soc.* **108** (1986) 7130.
43. I. Saito, T. Morii, H. Sugiyama, T. Matsuura, C. F. Meares and S. M. Hecht, *J. Am. Chem. Soc.* **111** (1989) 2307.
44. R. P. Hertzberg, M. J. Caranfa and S. M. Hecht, *Biochemistry* **24** (1985) 5285.
45. R. P. Hertzberg, M. J. Caranfa and S. M. Hecht, *Biochemistry* **27** (1988) 3164.
46. C.-W. Chen, J. S. Cohen and M. J. Behe, *Biochemistry* **22** (1983) 2136.
47. H.-Y. Wu and M. J. Behe, *Biochemistry* **24** (1985) 5499.
48. L. D. Williams and I. H. Goldberg, *Biochemistry* **27** (1988) 3004.
49. C. W. Haidle and J. Bearden, Jr., *Biochem. Biophys. Res. Commun.* **65** (1975) 815.
50. C. R. Krishnamoorthy, D. E. Vanderwall and J. W. Kozarich, *J. Am. Chem. Soc.* **110** (1988) 2008.
51. P. K. Mascharak, Y. Sugiura, J. Kuwahara, T. Suzuki and S. J. Lippard, *Proc. Natl. Acad. Sci. USA* **80** (1983) 6795.
52. B. Gold, V. Dange, M. A. Moore, A. Eastman, G. A. van der Marel, J. H. van Boom and S. M. Hecht, *J. Am. Chem. Soc.* **110** (1988) 2347.
53. J. Kuwahara and Y. Sugiura, *Proc. Natl. Acad. Sci. USA* **85** (1988) 2459.
54. M. J. Levy and S. M. Hecht, *Biochemistry* **27** (1988) 2647.

17 Actinomycins

A. B. MAUGER

17.1 Introduction

The actinomycins are a family of chromopeptide antibiotics isolated from various *Streptomyces* strains. First isolated in 1940[1], they were found in 1952 to possess antitumour activity[2] and were introduced clinically[3]. They have been used mainly in the treatment of Wilm's tumour[4], choriocarcinoma[5] and embryonal testicular carcinoma[6]; the clinical applications have been reviewed[7,8]. All contain the same phenoxazinone chromophore (red, max. 445 nm) attached amide-wise to two variable pentapeptide lactone units[9]. The best known example is actinomycin D (AMD, Figure 17.1)[10] and analogues will be abbreviated by reference to this structure. The natural variants and artificial analogues and their comparative biological activities have been reviewed[11,12], covering the literature up to 1979. In what follows, references cited in those reviews are not repeated.

Comparisons of biological activities are dubious when different parameters have been measured for different compounds, and for some only antimicrobial activities have been reported. Among the compounds more thoroughly examined, roughly parallel rankings are usually found between antitumour potency, antimicrobial activity, inhibition of RNA synthesis and DNA-binding proclivity. Antitumour efficacy, however, may be unrelated to potency.

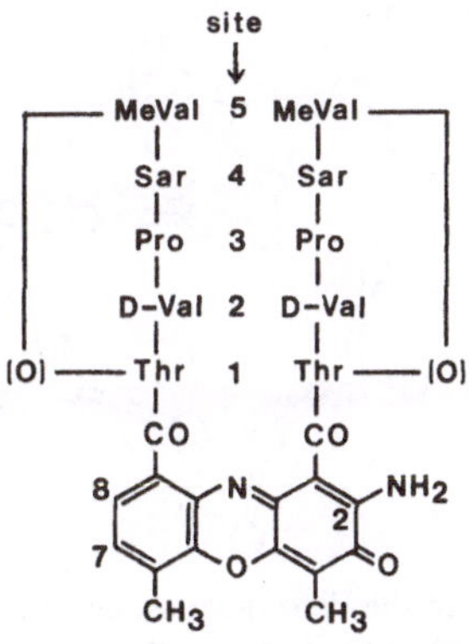

Figure 17.1 The structure of actinomycin D (AMD).

17.2 Mode of action

The primary mode of action of the actinomycins depends upon their inhibition of DNA-dependent RNA synthesis, a consequence of their complex formation with DNA. The intercalative model for this interaction received support from an X-ray crystallographic study of the 2:1 deoxyguanosine-AMD complex (Figure 17.2), and extrapolation of this structure provided a model for the AMD-DNA complex. X-ray crystal structures of AMD complexes with d(GC)[14] and d(ATGCAT)[15] have subsequently appeared. Various NMR studies support an AMD solution conformation approximating that present in these complexes and in the recently reported[16] crystal structure of AMD itself. In addition, NMR studies of the solution interaction of AMD with various nucleotides confirm the intercalative model and GC-related site specificity. The biological activity of actinomycins depends upon the much slower dissociation of their DNA complexes than those of simple phenoxazinones. The role of the peptide moieties in the binding, involving hydrogen

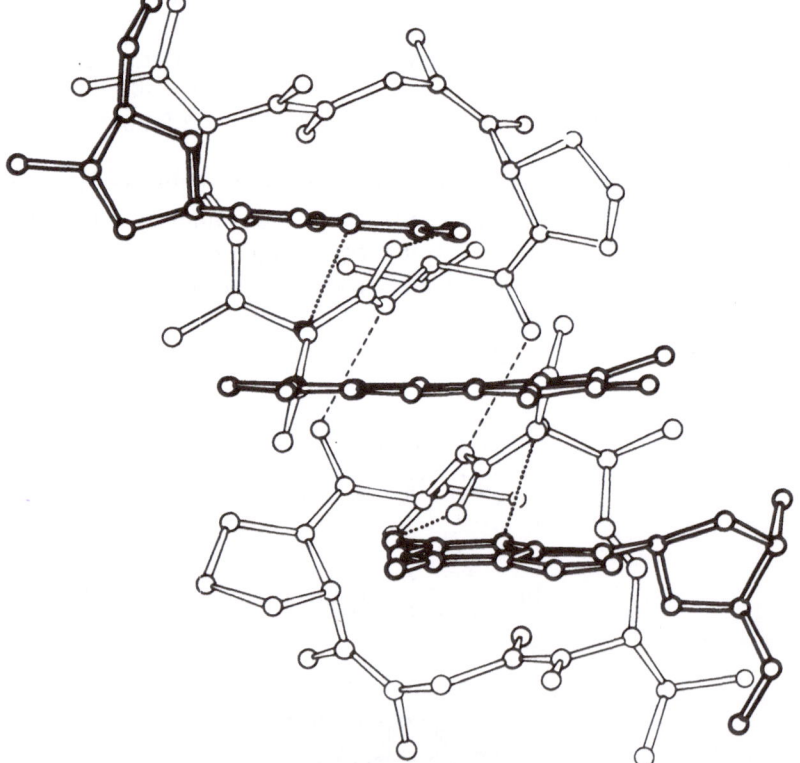

Figure 17.2 The crystal structure of the deoxyguanosine:actinomycin D complex. The AMD chromophore and deoxyguanosines are indicated with bold lines. The dotted lines represent hydrogen bonds. Original artwork used without alteration, from S.C. Jain and H.M. Sobell (1972)[13].

bonds and hydrophobic interactions, has been discussed[17]. One postulate envisions binding of AMD to β-DNA within the transcriptional complex, thereby blocking elongation of the RNA chains[18].

Another proposed mode of action, which explains the observed DNA damage, involves enzymatic reduction of AMD to a free radical, with transfer of an electron to oxygen to generate superoxide anions[19].

17.3 Natural structural variations

Whereas *Streptomyces parvulus* produces only AMD, other *Streptomyces* species generally yield mixtures ('complexes') of closely related compounds, which fall into three main types. Type 1, from *S. antibioticus* etc., contains actinomycin (**IV**) (identical with AMD), (**I**) (with one hydroxyproline at site 3) and (**V**) (with one 4-ketoproline at site 3). It also contains trace components (**III**) and (**II**) having, respectively, one or both prolines replaced by sarcosine. Type 2, from *S. chrysomallus*, contains C_1 (identical with AMD) in addition to C_2 and C_3 which have, respectively, one or both D-valines replaced by D-*allo*isoleucine. In type 3, from *S. fradiae*, all components have one threonine replaced by γ-hydroxythreonine and one N-methylvaline replaced by N-methylalanine. Z_1 has 3-hydroxy-5-methyl-proline and 4-keto-5-methyl-proline at the 3-sites, while Z_5 has 5-methylproline (*cis*) and 4-keto-5-methylproline at those sites. The other components of the Z complex are less well defined; none contain proline.

In comparisons with AMD, replacement of proline by 4-ketoproline reduces antitumour activity but not antimicrobial activity, while hydroxyproline reduces both. At site 2, D-valine and D-*allo*isoleucine are interchangeable with little effect. In the Z series, the deactivating effect of a site 3 hydroxyl group is again apparent, as Z_1 is the least active. These hydroxyl groups may weaken the hydrophobic interaction between the peptide surface and the narrow groove of DNA.

17.4 Biosynthetic analogues

Limited variations in the amino acids at sites 3 and 5 have been effected by 'directed biosynthesis', in which an actinomycin-producing *Streptomyces* strain is cultivated in the presence of an added amino acid which is incorporated in place of a natural precursor. Thus, addition of isoleucine results in the incorporation of N-methyl*allo*isoleucine into one or both 5-sites with formation of actinomycins E_1 and E_2, respectively. These compounds have slightly lower antimicrobial activities than AMD.

Proline analogues which have been incorporated into one or both 3-sites include azetidine-2-carboxylic acid, pipecolic acid, *cis*- and *trans*-4-chloroproline and *cis*- and *trans*-4-methylproline. In the case of pipecolic acid, analogues containing *trans*-4-hydroxypipecolic and 4-ketopipecolic acids were also isolated. In general, replacement of one proline residue reduces biological activity moderately, while replacement of both causes drastic reduction. The observed diminution in activity produced by 4-hydroxypipecolic acid parallels that seen for hydroxyproline. In terms of antitumour efficacy, only the analogues incorporating azetidine-2-carboxylic acid and those containing one pipecolic acid or *cis*-4-chloroproline residue were comparable with AMD.

17.5 Total synthesis of peptidic analogues

Actinomycins D and C_3 have been synthesised; about seven different routes have been employed. The most effective approach involves construction of the 3-benzyloxy-4-methyl-2-nitrobenzoyl peptide lactone via ring closure at the Pro-Sar or Sar-MeVal peptide bond, followed by hydrogenation to the 3-hydroxy-4-methylanthraniloyl peptide lactone. This intermediate, which is also the biosynthetic precursor[20], is oxidised to the actinomycin with ferricyanide. Some of the analogues which have been synthesised are shown in Table 17.1. None has biological activity superior to that of AMD. Loss of activity in such analogues is usually due to alterations in conformation.

17.6 Total synthesis of chromophoric analogues

Analogues of AMD in which chromophoric 4- and 6-methyl groups are replaced by H, OMe, Et and *t*-Bu have been synthesised. All are virtually inactive except the diethyl compound which has one third of the antimicrobial activity of AMD.

An AMD analogue in which the entire chromophore is replaced by 1,9-dicarboxy-4-methylphenazine was synthesised; its ability to inhibit RNA synthesis is about one tenth that of AMD.

Table 17.1 Amino acid variations in synthetic actinomycin analogues

Site 1	Site 2	Site 3	Site 4	Site 5	Activity[a]
Ser	D-Val	Pro	Sar	MeVal	2
Dpr	D-Val	Pro	Sar	MeVal	1
Dbu	D-Val	Pro	Sar	MeVal	1
Thr	D-Ala	Pro	Sar	MeVal	0
Thr	D-Leu	Pro	Sar	MeVal	1
Thr	D-Val	Pro	Gly	Val	0
Thr	D-Val	–	Sar	MeVal	0
Thr	D-Val	Pro	Sar	MeAla	2[b]

[a] Biological activity is on a scale of 0 to 3; AMD = 3. Dpr = α, β-diaminopropionic acid; Dbu = α, β-diaminobutyric acid.

[b] Reference 21 (others cited in Refs. 11 and 12).

17.7 Partial synthesis of 2-substituted chromophoric analogues

Mild acid hydrolysis of AMD affords 2-hydroxy-2-deamino-AMD which is converted to the 2-chloro compound with thionyl chloride; both are biologically inactive. Hydrogenation of 2-chloro-2-deamino-AMD gave 2-deamino-AMD. This compound does not intercalate in DNA but has antitumour activity ascribed to its propensity to form ion radicals and thereby generate superoxide anions[22].

N^2-substituted AMD analogues have been synthesised by reaction of 2-chloro-2-deamino-AMD with various primary and secondary amines. Direct alkylation of the AMD amino group with dimethylsulphonium methylide, followed by an alkyl halide, has also been described. While the simple N-alkyl derivatives are inactive, antitumour efficacy was noted for the N^2-(γ-hydroxypropyl), N^2-(3'-aminopropyl) and N^2-(10'-aminodecyl) derivatives at higher doses than for AMD[23]. N^2-substituted spin labelled AMD analogues containing the 4-(2, 2, 6, 6-tetramethyl-1-piperidinyloxy) moiety have been synthesised and reported to possess better antitumour efficacy than AMD at ten times the dose[24].

17.8 Partial synthesis of 7-substituted chromophoric analogues

2-Chloro-2-deamino-AMD can be chlorinated or brominated in the 7-position; subsequent ammonolysis then restores the 2-amino group. 7-Nitro-AMD can be prepared by nitration of AMD and hydrogenated to afford 7-amino-AMD. 7-Hydroxy-AMD has been synthesised via oxidation of an oxazine derivative ((2), Figure 17.3) and methylated to give 7-methoxy-AMD. 7-Chloro- and 7-nitro-AMD have half the antimicrobial activity of AMD, while 7-hydroxy- and 7-amino-AMD have very weak activity. In contrast, 7-amino- and 7-nitro-AMD display antitumour activity comparable with that of AMD, while 7-hydroxy-AMD is efficacious at ten times the dose. 8-Amino-7-hydroxy-AMD ((4), Figure 17.3) was reported to have superior antitumour efficacy due to free radical formation[25].

Bulkier substitution at the 7-position has been achieved by alkylation or acylation of 7-hydroxy- or 7-amino-AMD. Most of these compounds are only weakly active. However, 7-O-(1'-adamantoyloxy)-AMD exhibited DNA binding affinity[23]. 7-(2, 3-Epoxypropoxy)-AMD was claimed to have antitumour efficacy superior to that of AMD[26], attributed to formation of covalent adducts with DNA[27]. Analogous compounds with the epoxy ring replaced by aziridine[28] and cyclopropyl[29] rings interact with DNA similarly to AMD.

The depsipeptide derivatives 7-L-valylglycolylamino-AMD and 7-(L-valylglycolylamino)$_2$-AMD were synthesised from 7-amino-AMD; they bind to DNA less strongly than AMD[30]. Two bis-AMD analogues were prepared by oxaloylation of these compounds[31]. These doubly intercalating derivatives have higher sequence specificity than AMD since each molecule carries two GC-specific centres[32].

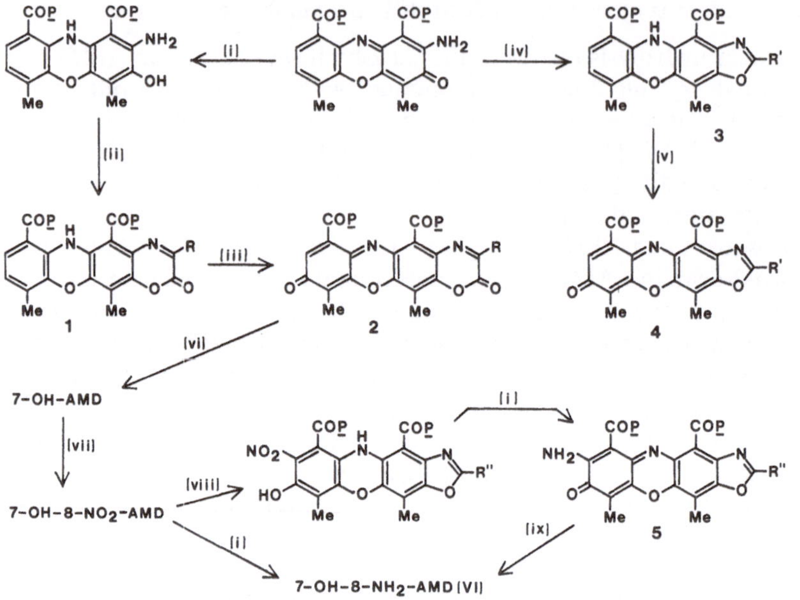

Figure 17.3 Transformation of AMD into tetracyclic chromophoric analogues. P represents a pentapeptide lactone moiety. $R = CH_3$, C_6H_{13}, C_6H_5, C_6F_5, etc. $R' = CH_3$, C_6H_5, C_6F_5. $R'' = CH_3$, C_6H_5, CH_2COOH, $CH_2CONH(CH_2)_4NH_2$. Reagents: (i) H_2/Pd; (ii) RCOCOOH; (iii) CrO_3 or $FeCl_3$; (iv) $R'CHO/Cl_3CHO$; (v) DDQ; (vi) pH 7.2; (vii) HNO_2; (viii) $R'CHO$; (ix) metabolism.

17.9 Tetracyclic chromophoric analogues

Reaction of dehydro-AMD with α-keto-acids produces a series of analogues having a fourth (oxazinone) ring on the chromophore ((**1**), Figure 17.3)[33]. These compounds are hydrolysed above pH 7 to AMD, and thus act as transport-modified AMD prodrugs. Their antitumour potency was tenfold lower than that of AMD, but their efficacy was reportedly superior, especially in the case of (**1**) ($R = C_6H_5$). Reaction of AMD with aldehydes produced another series of tetracyclic analogues containing an oxazole ring ((**3**), Figure 17.3)[33]. Antitumour activity is lost, but is restored upon oxidation to the iminoquinone (**4**) which metabolises to 7-hydroxy-AMD[25]. The oxazole analogues (**5**), which metabolise to the highly active 7-hydroxy-8-amino-AMD (**6**), were reportedly very effective antitumour agents[34], especially with $R'' = C_6H_5$ or $CH_2CONH(CH_2)_4NH_2$.

17.10 Summary

Actinomycins have a long history as antitumour agents and inhibitors of RNA synthesis. They belong to the DNA-intercalating class of drugs. Numerous modifications of the actinomycin structure have been effected by biosynthesis, partial synthesis and total synthesis. Analogues for which superior antitumour efficacy has been claimed either act by a different mechanism from actinomycin D or have different distribution profiles *in vivo*. These compounds generally involve modifications of the chromophore rather than the peptide moieties. Thus far no clinical trials of these analogues have been reported.

References

1. S. A. Waksman and H. B. Woodruff, *Proc. Soc. Exp. Biol. Med.* **45** (1940) 609.
2. C. Hackmann, *Z. Krebsforsch.* **58** (1952) 607.
3. G. Schulte, *Z. Krebsforsch.* **58** (1952) 500.
4. S. Farber and A. Mitus, in *Chemotherapy of Cancer*, ed. W. H. Cole, Lea and Febiger, Philadelphia (1970) p. 277.
5. G. T. Ross, L. L. Stohlbach and R. Herz, *Cancer Res.* **22** (1962) 1015.
6. A. R. MacKenzie, *Cancer* **19** (1966) 1369.
7. E. Frei, III, *Cancer Chemother. Rep.* Part 1 **58** (1974) 49.
8. T. G. Terent'eva, L. V. Egorov and N. I. Perevodchikova, *Antibiot. Med. Biotekhnol.* **30** (1985) 619.
9. H. Brockmann, G. Bohnsack, B. Franck, H. Gröne, H. Muxfeldt and C. Süling, *Angew. Chem.* **68** (1956) 70.
10. E. Bullock and A. W. Johnson, *J. Chem. Soc.* (1957) 3280.
11. J. Meienhofer and E. Atherton, in *Structure-Activity Relationships among the Semisynthetic Antibiotics*, ed. D. Perlman, Academic Press, New York (1977) p. 427.
12. A. B. Mauger, *Top. Antibiot. Chem.* **5** (1980) 223.
13. S. C. Jain and H. M. Sobell, *J. Mol. Biol.* **68** (1972) 1.
14. F. Takusagawa, M. Dabrow, S. Neidle and H. M. Berman, *Nature (London)* **296** (1982) 466.
15. F. Takusagawa, B. M. Goldstein, S. Youngster, R. A. Jones and H. M. Berman, *J. Biol. Chem.* **259** (1984) 4714.
16. S. Ginell, L. Lessinger and H. M. Berman, *Biopolymers* **27** (1988) 843.
17. F. Takusagawa, *J. Antibiot.* **38** (1985) 1596.
18. H. M. Sobell, *Proc. Natl. Acad. Sci. USA* **82** (1985) 5328.
19. H. Nakazawa, F. E. Chou, P. A. Andrews and N. R. Bachur, *J. Org. Chem.* **46** (1981) 1493.
20. U. Keller, *J. Biol. Chem.* **259** (1984) 8226.
21. A. B. Mauger, O. A. Stuart, J. A. Ferretti and J. V. Silverton, *J. Am. Chem. Soc.* **107** (1985) 7154.
22. R. K. Sehgal, S. K. Sengupta, D. J. Waxman and A. I. Tauber, *Anticancer Drug Design* **1** (1985) 13.
23. S. K. Sengupta, J. E. Anderson, Y. Kogan, D. H. Trites, W. R. Beltz and M. S. Madhavarao, *J. Med. Chem.* **24** (1981) 1052.
24. B. K. Sinha, M. G. Cox and C. F. Chignell, *J. Med. Chem.* **22** (1979) 1051.
25. S. K. Sengupta, C. Kelly and R. K. Sehgal, *J. Med. Chem.* **28** (1985) 620.
26. S. K. Sengupta, J. E. Anderson and C. Kelly, *J. Med. Chem.* **25** (1982) 1214.
27. S. K. Sengupta, J. Blondin and J. Szabo, *J. Med. Chem.* **27** (1984) 1465.
28. R. K. Sehgal, B. Almassian, D. P. Rosenbaum, R. Zadrozny and S. K. Sengupta, *J. Med. Chem.* **30** (1987) 1626.
29. R. K. Sehgal, B. Almassian, D. P. Rosenbaum, R. Zadrozny and S. K. Sengupta, *J. Med. Chem.* **31** (1988) 790.
30. M. V. Mikhailov, S. M. Nikitin, A. S. Zasedatelev, A. Z. Zhuze, G. V. Gurskii and B. P. Gottikh, *Mol. Biol. (Moscow)* **13** (1979) 1322.
31. S. M. Nikitin, S. L. Grokhovskii, A. L. Zhuze, M. V. Mikhailov, A. S. Zasedatelev, G. V. Gurskii and B. P. Gottikh, *Bioorg. Khim.* **7** (1981) 542.
32. M. V. Mikhailov, S. M. Nikitin, A. L. Zhuze, A. S. Zasedatelev, G. V. Gurskii and B. P. Gottikh, *FEBS Lett.* **136** (1981) 53.
33. S. K. Sengupta, M. S. Madhavarao, C. Kelly and J. Blondin, *J. Med. Chem.* **26** (1983) 1631.
34. S. K. Sengupta, Y. Kogan, C. Kelly and J. Szabo, *J. Med. Chem.* **31** (1988) 768.

18 Ellipticines

M. SAINSBURY

18.1 Isolation and structure of the pyrido[4,3-b]carbazole alkaloids

The alkaloid ellipticine, 5,11-dimethyl-6H-pyrido[4,3-b]carbazole, ((1) R = H) was first isolated in 1959[1] from the leaves of the evergreen tree *Ochrosia elliptica* Labill (Apocynaceae) which grows wild in Oceania. At the time its biological properties were not recognised, but in 1967 during an examination of Australian plants for anticancer activity the same alkaloid and its 9-methoxylated derivative ((1) R = OMe) were obtained from the extracts of two other shrubs, *O. moorei* F. Muell. and *Excavatia coccinea* (Tejs. and Bin.) Mgf[2]. Together the two ellipticines accounted for the activity of the crude plant extracts towards the experimental tumours sarcoma 180, adenocarcinoma and L 1210 lymphoid leukaemia implanted in mice, and human carcinoma of the nasopharynx carried in cell culture[3].

Other alkaloids related to ellipticine are now known; these include 5-hydroxymethyl-11-methyl-6H-pyrido[4,3-b]carbazole (2, R^1 = H; R^2 = CH_2OH), and 5-formyl-11-methyl-6H-pyrido[4,3-b]carbazole (so-called, '5-formylellipticine'), (2, R^1 = CHO; R^2 = H)[4], ellipticine-N-oxide (3)[5], 3,4-dihydroellipticine (4)[6], 1,2,3,4-tetrahydro-2-methylellipticine (5)[7], and the 'dimer' strellidimine (6)[8] which has been isolated from *Strychnos dinklagei*

Gilg. The isomer of ellipticine, olivacine (7), and its 1,2,3,4-tetrahydro-2-methyl derivative, guatambuine (8), also occur in nature[9].

The best source of 9-methoxyellipticine is perhaps the Fijian tree *Bleekeria vitiensis* (Markgraf) A. C. Smith syn. *Ochrosia vitiensis*[10]. Here the dry stem bark contains 0.17% of the alkaloid, with only minor amounts in the wood, leaves or roots of the plant. Smaller amounts of ellipticine are found in this tree, but mainly in the roots. The principal extractive of the leaves is isoreserpiline (9), a typical representative of a group of indole alkaloids widely distributed through plants of the Apocynaceae. Although there is still some debate[11] as to the precise mechanism, the biosyntheses of the two types of alkaloids are probably related and share common precursors in the form of a tryptamine unit (from tryptophan or tryptamine, (11) $R = CO_2H$, or $R = H$, respectively) and a C9, or C10 fragment which is terpenoid in origin and is utilised by the plant in the form of the glucoside secologanin (10)[12].

(9)   (10)   (11)

18.2 Synthesis

Ellipticine is a planar molecule, but 'arc shaped' due to the inclusion of a 5-membered pyrrole ring B in the tetracyclic system. For an indole alkaloid, ellipticine has a very simple structure and its interesting biological properties have motivated many organic chemists to devise syntheses of it, yet despite this activity few of these routes are totally free from difficulties when they are adapted towards the preparation of analogues. Much of the work in this area has already been reviewed[13-15] and the following account focuses on those syntheses which seem to offer the greatest flexibility in terms of access of derivatives for the optimisation of therapeutic activity.

In order to classify the approaches to the ellipticines it is useful to consider them on the basis of which ring B, C, or D is formed last in the synthetic sequence (in every case, described so far, the benzenoid ring A is present in one of the starting materials).

18.2.1 *Syntheses involving the formation of ring B*

The formation of the B ring of the ellipticines and their analogues normally involves recourse to standard procedures which are employed for the preparation of simple indoles. One such ellipticine synthesis has been described by Miller and Stowell[16] in which the indolic ring is formed through

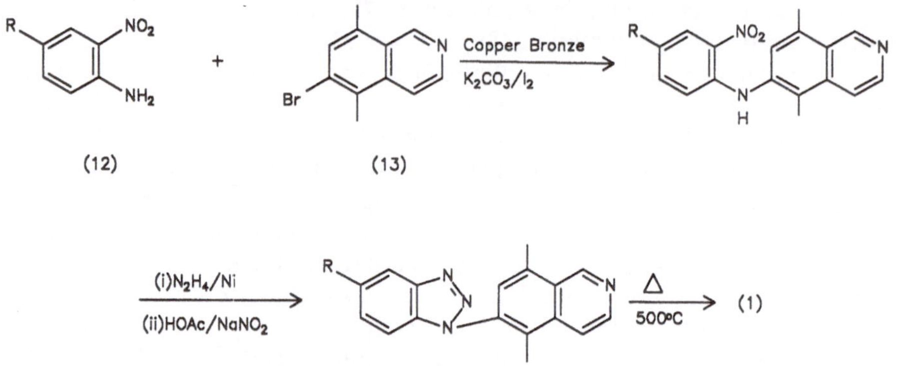

Scheme 18.1 Miller and Stowell's synthesis of ellipticine and 9-methoxyellipticine.

the thermolysis of the benzotriazole (**14**). The starting material is itself prepared in two steps from 2-nitroaniline ((**12**) R = H) and the bromoiso-quinoline (**13**).

Although the temperature of the thermolysis step is approx. 500°C, the reaction is carried out in a flow system thereby minimising the risk of product decomposition. This short synthesis may be adapted to the production of 9-methoxyellipticine in an overall yield of 31% from 4-methoxy-2-nitroaniline ((**12**) R = OMe) (Scheme 18.1).

18.2.2 Syntheses requiring the formation of ring C as the last step in the formation of the tetracycle

18.2.2.1 Cycloadditions. In principle since indoles and pyridines are very familiar compounds, syntheses of ellipticines involving these two starting materials are very appealing, and recently May and Moody[17] have completed a brief construction of ellipticine which requires the Diels Alder cycloaddition of pyrano[4,3-*b*]indole-3-one (**15**) and 3,4-pyridyne (**16**). Unfortunately the cycloaddition step is not regioselective and both ellipticine and isoellipticine (**17**) are formed in equal amounts (Scheme 18.2).

18.2.2.2 Tandem metallations. Clearly this lack of selectivity may be overcome if the two addends are already linked through either the indole-2, or

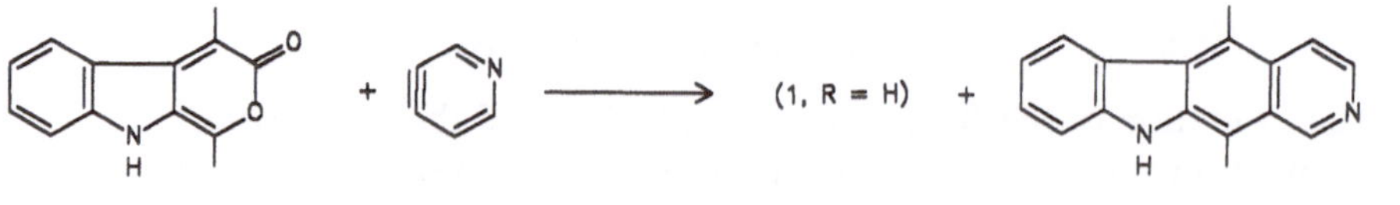

Scheme 18.2 May and Moody's ellipticine synthesis.

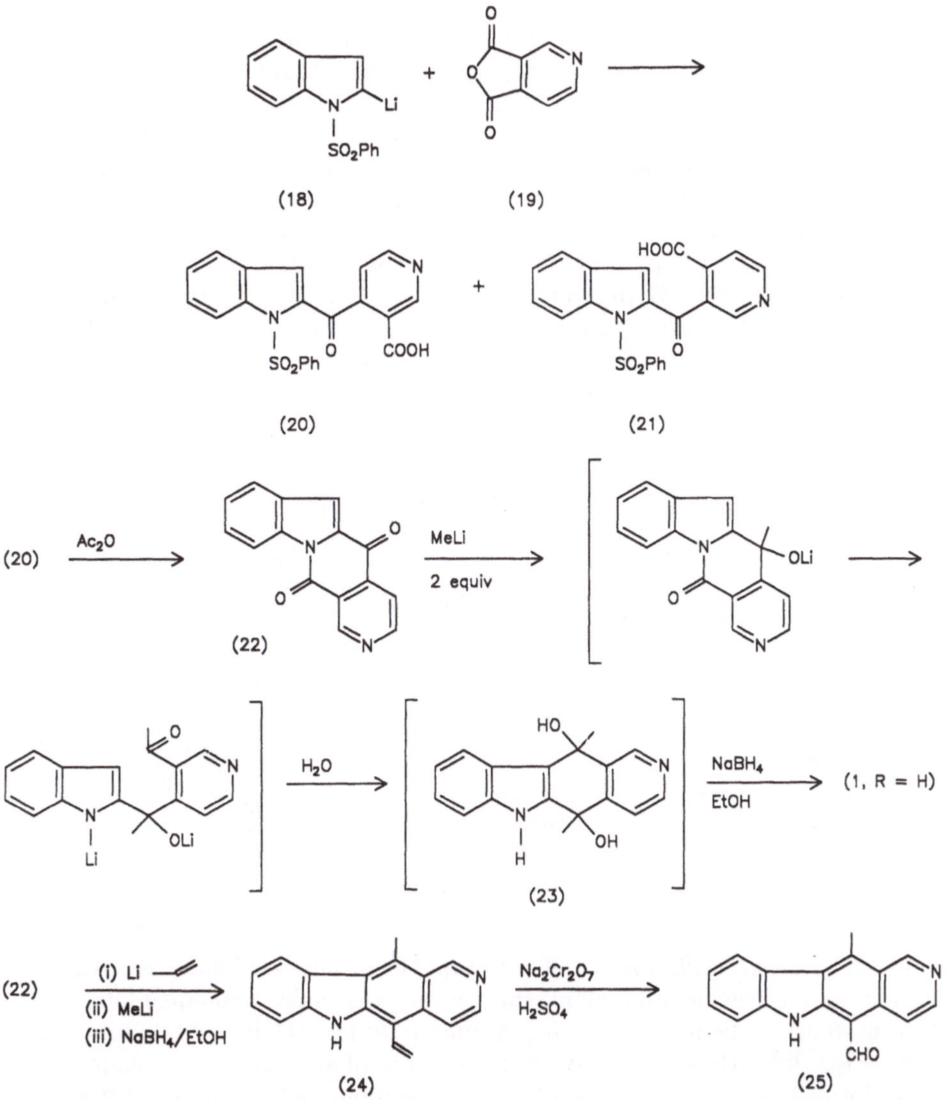

Scheme 18.3 Gribble and Saulnier's synthesis of ellipticine and 5-formyl-11-methyl-6*H*-pyrido[4, 3-*b*]pyridocarbazole.

-3 positions, and Saulnier and Gribble[18] have reacted *N*-benzenesulphonyl-2-lithioindole (**18**) and cinchonomeric anhydride (**19**) to obtain the 2-acylindole (**20**) as the major product and only a minor amount of the alternative isomer (**21**) (92 : 8). Cyclisation of the acylindole (**20**) through reaction with acetic anhydride gives the ketolactam (**22**) and this compound when alkylated with methyl lithium affords the dihydroxydihydroellipticine (**23**), or its equivalent, which without isolation can be dehydrated and reduced to ellipticine in an overall yield of 54% (Scheme 18.3). 9-Methoxyellipticine is similarly obtained in an overall yield of 47%. A very important feature of this route is its applicability to the construction of pyrido[4,3-b]carbazoles unsymmetrically substituted at C-5 and C-11. This is possible since the first molecular equivalent of the alkyl lithium reagent reacts at the ketone carbonyl group of the ketolactam (22) rather than at the less active amide unit[19]. In this way two different alkyl lithiums may be utilised and this has led to the preparation of, for example, the vinyl compound (24), and thence to the natural product 5-formyl-11-methyl-6H-pyrido[4,3-b]pyridocarbazole (25).

A similar feature is present in syntheses developed independently by Snieckus[20] and by Joule[21], in which the linear quinones (26), can be used in place of the angular ketolactam (22) (Scheme 18.4).

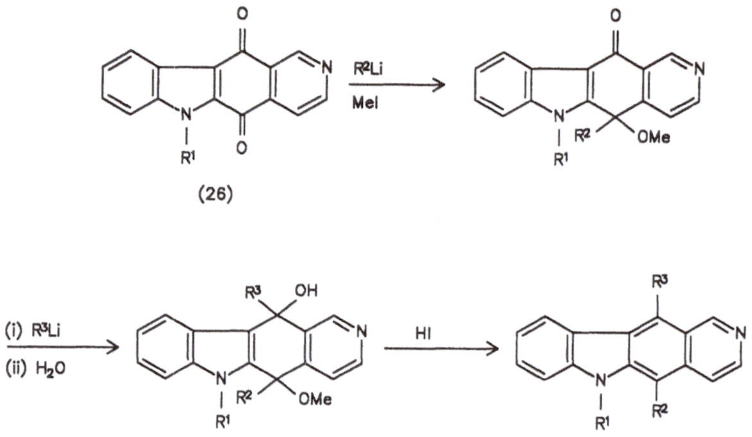

Scheme 18.4 Ellipticine synthesis due to Snieckus and to Joule.

18.2.2.3 *Electrocyclisations.* Kano *et al.*[22] have illustrated that it is possible to obtain olivacine and ellipticine by the thermally promoted electro-cyclisations of the indol-2-ylvinylpyridines ((28) $R^1 = H, R^2 = Me$) and ((28) $R^1 = Me, R^2 = H$), respectively, in the presence of air. The starting materials are formed by the alkylation of N-benzenesulphonyllithioindole (18) with the appropriate 4-acetylpyridines (27) (Scheme 18.5). Unfortunately the acetylpyridines needed for this approach are not commercially available.

Once more the inspiration for this type of approach stems from earlier work and the route is a variation of Bergman's ellipticine synthesis[23] (Scheme 18.6), wherein the 2-ethylindol-3-ylvinylpyridine (31), prepared from the reaction between 2-ethylindole (29) and 3-acetylpyridine (30), is converted into its 2-benzyl salt and heated at approx. 400°C. Although 72% of ellipticine is produced in the final pyrolysis step, 10% of isoellipticine is also formed and this presents an extra problem of separation.

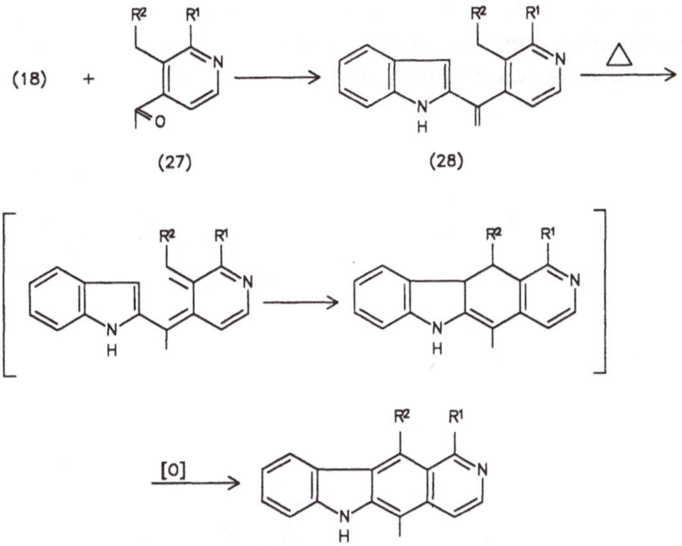

Scheme 18.5 Kano and co-workers' synthesis of ellipticine and olivacine.

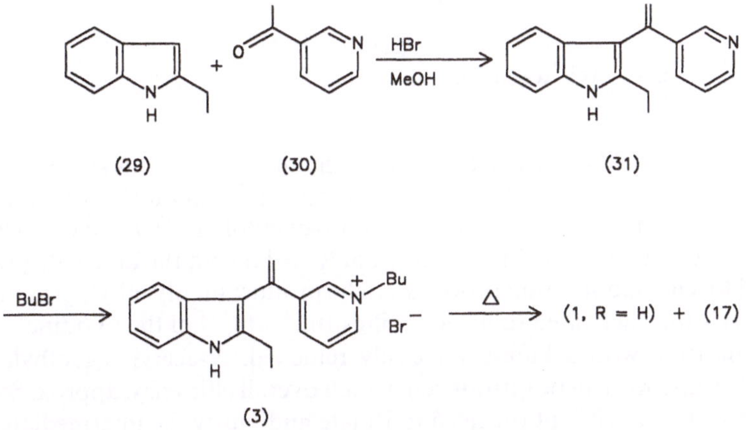

Scheme 18.6 Bergman and Carlsson's synthesis of ellipticine and isoellipticine.

18.2.2.4 *Electrophilic substitution at C-2 of indolyl precursors.* Pyrolysis methods are acceptable when the products are as thermally stable as ellipticine and olivacine, but for more complex and less stable targets, less drastic conditions must be employed. In the first ellipticine synthesis announced by Woodward as long ago as 1959[24] indole (**32**) was reacted with 3-acetylpyridine (**33**) in the anticipation of forming a 1:1 adduct; unfortunately such a product cannot be isolated and the reaction continues to yield the bis-indolylethylpyridine (**34**). When this product is treated with zinc and acetic anhydride, the diacetyldihydropyridine (**35**) derivative is obtained which on pyrolysis in high vacuum cyclises to ellipticine (Scheme 18.7).

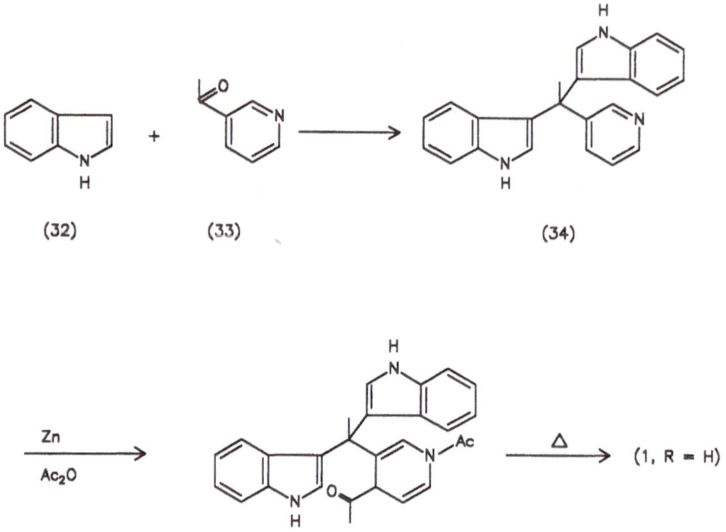

Scheme 18.7 Woodward and co-workers' synthesis of ellipticine.

The overall yield is extremely poor, yet the simplicity of this strategy is very attractive and it acted as the template for one of the author's own ellipticine syntheses[25]. In this work we restrict the ratio of indole and pyridine moieties to 1:1 by reacting the phenylhydrazine ((**36**) $R^1 = H$) with the aldehyde (**37**) in a typical Fischer indolisation procedure. The product indolylethylpyridine ((**38**) $R^1 = H$) is then activated to nucleophilic attack at C-4 of the pyridine ring by quaternisation with a bulky, yet easily removed, *N*-acetyl-*N*-methylamino unit. The function can be introduced in high overall efficiency, approx. 85%, in three operations without the need to isolate and purify the intermediates (see Scheme 18.8).

When the salt ((**39**) $R^1 = H$) is reacted with sodium cyanide in the presence of ammonium chloride and the intermediate dihydropyridine, without separation, is exposed to sunlight, the nitrile ((**40**) $R^1 = H$) is formed. Reaction of the nitrile with methyl lithium affords the corresponding imine which again is not isolated, but hydrolysed directly by reaction with acetic acid to give ellipticine in 65% overall yield from the indolylethylpyridine. A variety of other alkyl and aryl lithium reagents can be used in the penultimate step thus giving rise to a range of C-5 substituted 6*H*-pyrido[4,3-*b*]carbazoles[26,27]. From 2- and 4-substituted phenylhydrazines the corresponding 7- and 9-

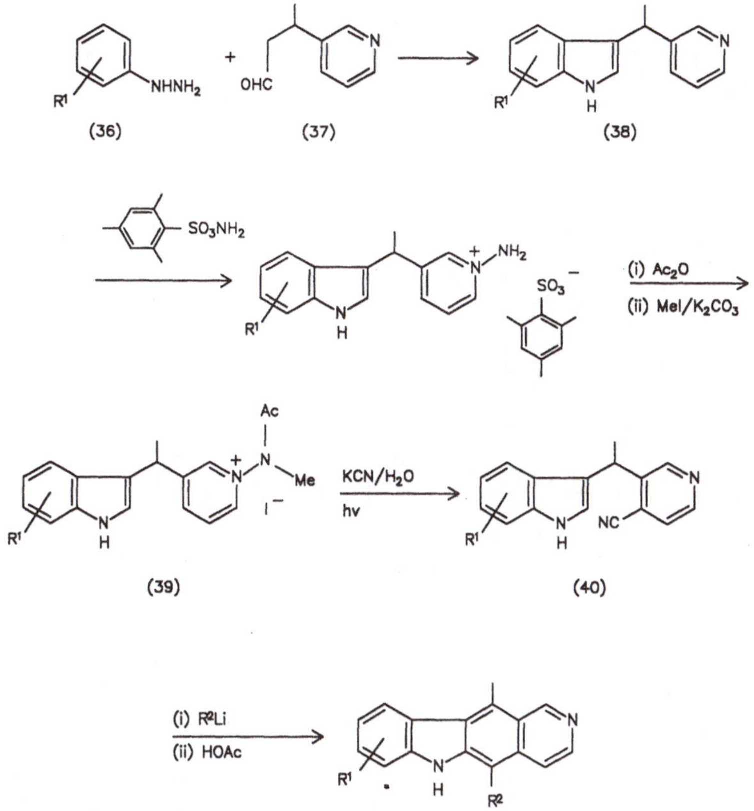

Scheme 18.8 Sainsbury and co-workers' synthesis of ring A and C-5 substituted ellipticines.

substituted ellipticines, respectively, can be synthesised, although, of course, if 3-substituted phenylhydrazines are used mixtures of 4- and 6-substituted indolylethylpyridines are producd.

Recently we have modified this route to obtain the natural product 5-formyl-11-methyl-6H-pyrido[4,3-b]pyridocarbazole ((4) R = H) and its methoxylated analogue ((41) R = OMe)[28] *en route* to dehydrodimers of the type (42) where the linking unit is an adipic acid residue joined to N-methylaminomethyl groups appended to C-5 (Scheme 18.9).

18.2.2.5 *Nucleophilic attack at C-4 of 2-ethylindolylmethylpyridines.* The order of reactions can obviously be reversed in approaches of the above type and Pandit[29] has developed a strategy which embodies the construction of 2-alkylated indolylpyridylmethanones (45) through the acylation of the 2-ethylindolyl ester (43) with the nicotinoyl chlorides (44), these compounds can then be activated to nucleophilic attack by N-benzylation and cyclised in the presence of base to tetracyclic 1,4-dihydropyridines (46). Aromatisation of the

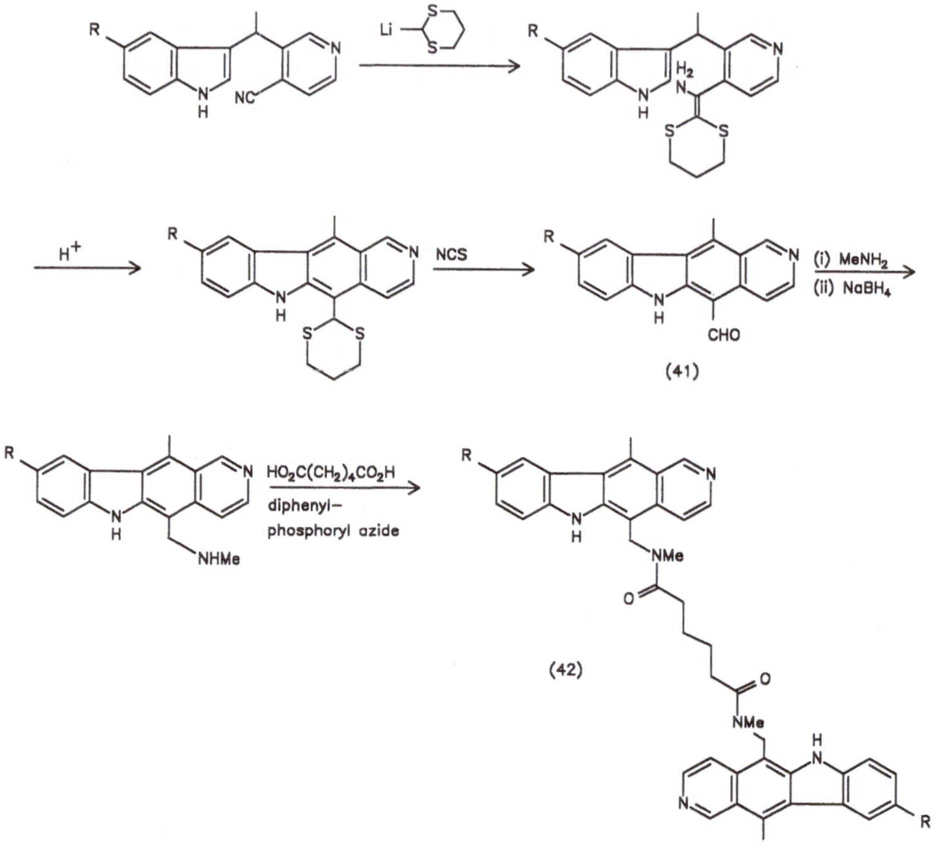

Scheme 18.9 The synthesis of dehydrodimers of ellipticines linked through C-5.

pyridine ring can then be achieved by oxidation with N-benzylacridinium bromide (47), and the N-benzyl group is removed by hydrogenation over palladium on carbon and treatment with sodium ethoxide to yield the tetracyclic ketone ((48) R = H). 6-Methylellipticine (49) is obtained from the ketone by methylation, either by a Wittig reaction with triphenylmethylene phosphorane, or through a reaction with methylmagnesium bromide. In either event de-esterification and decarboxylation may then be carried out very simply to afford the final product. 6-Methylolivacine (50) is similarly formed from the ketone ((48) R = Me) by reduction with Red-Al (Scheme 18.10).

Obviously the presence of the keto group in intermediates of the type (48) allows the introduction of other groups into this site and in further work Pandit[30] has recorded the synthesis of a range of 11-alkyl and aminoalkyl analogues of ellipticine. Weller and Ford[31] have further extended this approach by reacting 2-methoxycarbonylmethylindole (51) with 3-acetylpy-

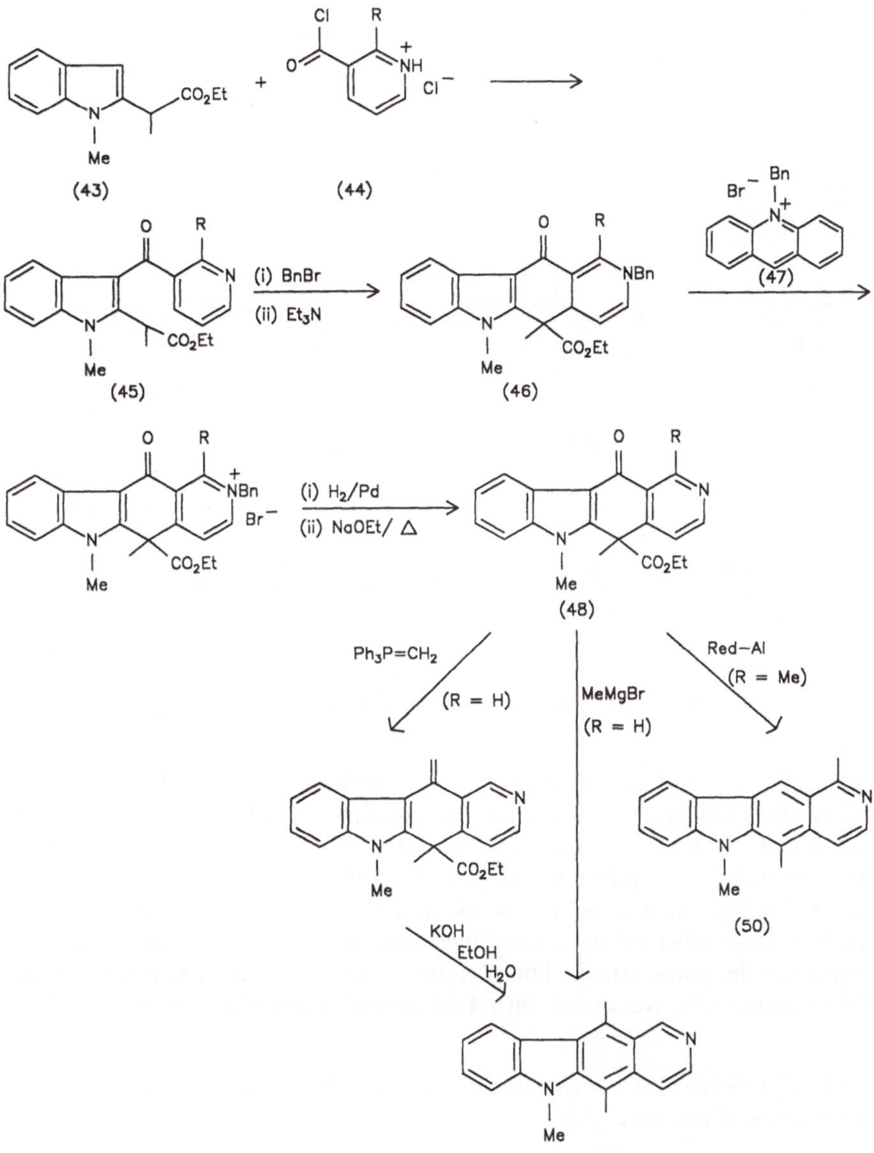

Scheme 18.10 Pandit and co-workers' synthesis of 2-methylellipticine and 2-methylolivacine.

ridine **(52)**. This affords the ester **(53)** which, when N-methylated and the product treated with base, gives the dihydropyridocarbazole **(54)**. Oxidation then yields the corresponding pyridinium salt **(55)** which on reduction and re-oxidation gives an ellipticium metho salt **(56)**. This product is N-demethylated by heating it with sodium thiophenate (Scheme 18.11).

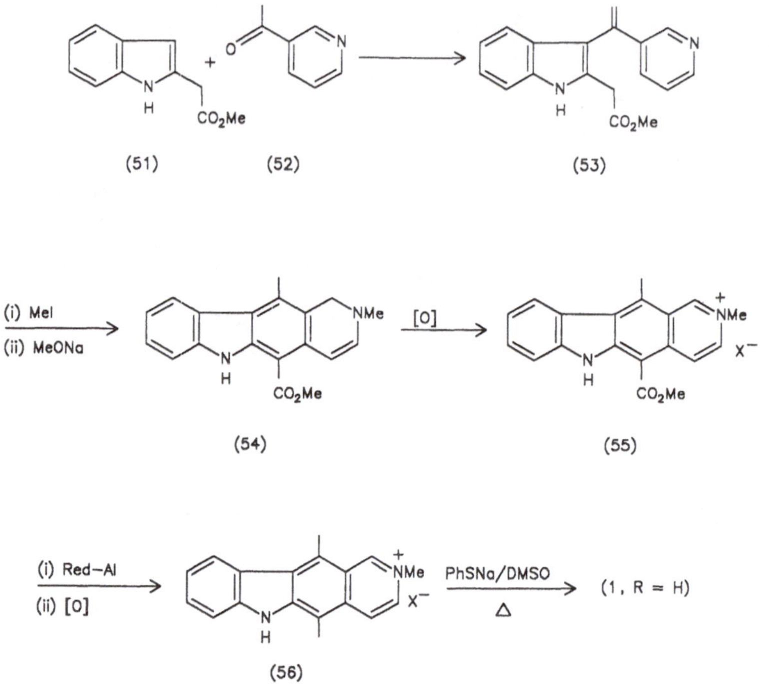

(51)   (52)   (53)

(54)   (55)

(56)

Scheme 18.11 Weller and Ford's ellipticine synthesis.

Demethylation of the pyridine nitrogen atom of ellipticine derivatives by catalytic methods is always troublesome, and this is why Weller and Ford used a chemical method of *N*-alkylation in the final step of their synthesis. Unfortunately, the optimum reaction conditions are equally severe, and Archer[32] reports in subsequent work that better results are obtained if 4-nitrobenzylation is used in place of methylation to activate the pyridine ring towards nucleophilic attack. The *N*-substituent may then be removed under mild conditions by treatment with 4-nitroso-*N*,*N*-dimethylaniline.

18.2.3   Syntheses utilising ring D formation as the last step in the construction of the tetracycle

Cranwell and Saxton's ellipticine synthesis was announced as long ago as 1962[33] and, with certain modifications, has withstood the test of time to remain one of the most successful and widely used pyrido[4,3-*b*]carbazole syntheses. 1,5-Dimethylcarbazole (**58**), from indole and hexan-2,5-dione (**57**), is subjected to a Vilsmeir–Haack type formylation and the product aldehyde (**59**) is reacted with 2,2-diethoxyethylamine (**60**) to produce the imine (**61**). In the original synthesis, the imine is reduced to the corresponding amine and cyclised to 1,2-dihydroellipticine by the action of hydrogen chloride. Conversion of the dihydro compound to the fully aromatic base is achieved by heating it with palladium on carbon at relatively high temperature. Clearly it would be preferable to cyclise the imine since it is at the correct oxidation level to give ellipticine directly, and Dalton *et al.*[2] showed that this reaction does occur in the presence of phosphoric acid. Despite this, however, the yield is not good unless the indolic system is activated by methoxylation at C-7[34]. This difficulty has been cleverly overcome by converting the amine into its *N*-tosyl derivative (62) prior to reaction with 6 M hydrogen chloride in dioxane. Now ring closure and elimination of 4-toluenesulphonic acid occur concomitantly[35].

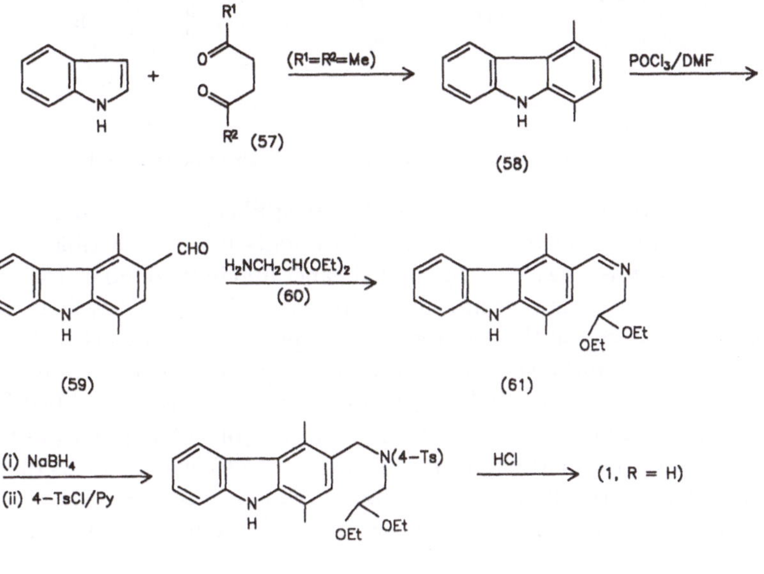

Scheme 18.12 Cranwell and Saxton's modified ellipticine synthesis.

One remaining problem with this route, however, is that unsymmetrically substituted 1,4-diketones when reacted with indoles give mixtures of carbazoles. We[36] have solved this by reacting indoles with α-cyanoacrylate esters (63) to affords adducts (64), which can then be alkylated with α-bromoketones. The products (65) can then be cyclised, hydrolysed and decarboxylated to variously substituted carbazoles (66) bearing a nitrile function at C-2. These may be alkylated or reduced to give the corresponding ketones or aldehydes, respectively (Scheme 18.3).

18.3   Structure-activity studies

Following the discovery of the action of ellipticine and its 9-methoxy derivative towards a range of experimental tumours, the effects of altering the

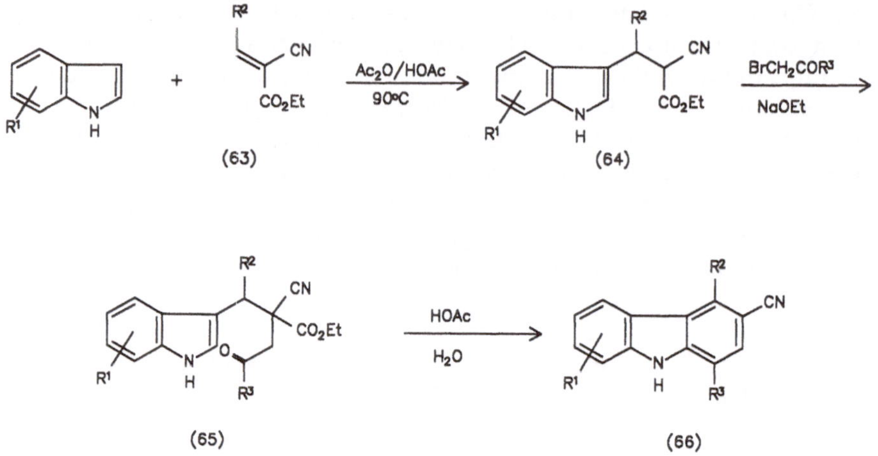

Scheme 18.13 Regioselective synthesis of 1,4-disubstitutedcarbazoles.

pattern of methylation on the tetracycle were studied[37]. Later an analysis of ellipticine analogues bearing various substituents became possible through the efforts of chemists based in Australia, France and in England[38]. The results of some of these experiments are summarised in Table 18.1; they show that a loss of activity occurs if the two methyl groups at C-5 and C-11 are absent. Although there is some dispute as to whether the *nor*-C-5-methyl compound is active or not, a shift of methylation from C-11 to C-1, as exemplified in the structure of olivacine, leads only to a small loss of potency. However, the presence of an extra methyl group at C-9, or at C-10, causes a loss of effect. Various isosteres of ellipticine bearing oxygen or sulphur atoms[39] or a methylene group[40] at position 6 have also been synthesised, but were found to be inactive.

One of the most potent of the compounds substituted in ring A is 9-

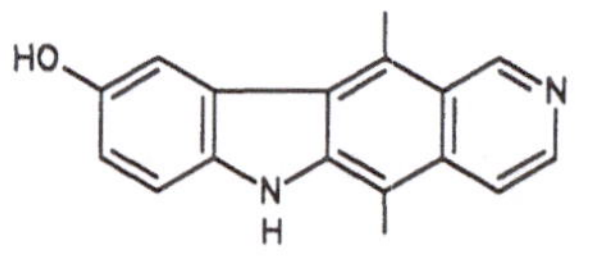

(67)

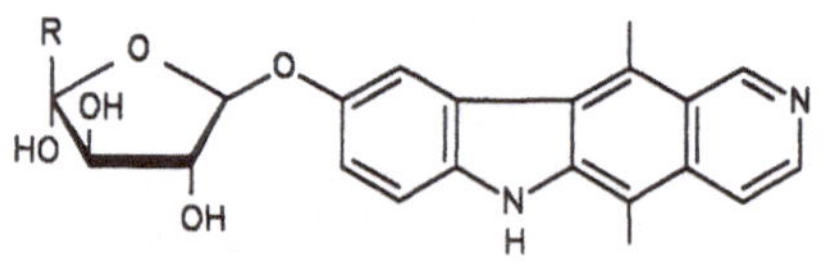

(68, R = CH$_2$OH or CO$_2$H)

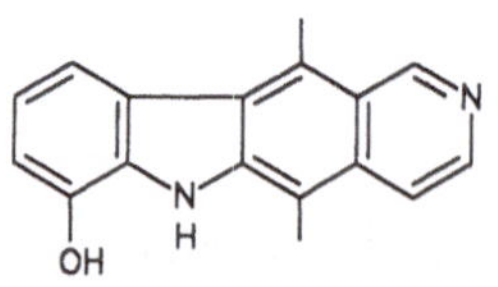

(69)

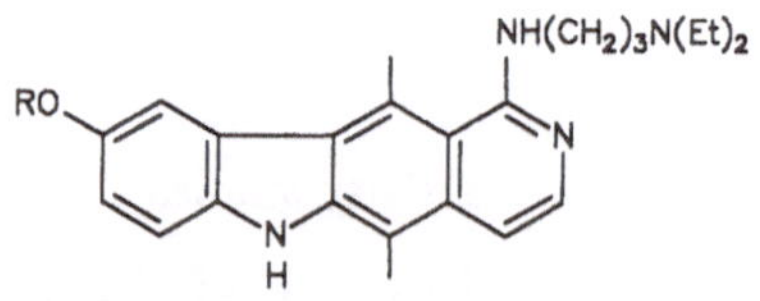

(70)

hydroxyellipticine (**67**), first obtained by *O*-demethylation of 9-methoxy-ellipticine[41], but now known as a natural product[4] in its own right. Indeed it was considered that those 9-alkoxyellipticines[42] which show anticancer properties do so because they suffer de-*O*-alkylation *in vivo*. *O*-Glucosides of 9-hydroxyellipticine (**68**), for example, are much less cytotoxic towards L 1210 cells in culture than is the parent 9-hydroxyellipticine[43]. Certainly 9-methoxyellipticine is metabolised to its hydroxyl derivative in rats[44], and ellipticine itself is oxidised to a mixture of 7- (**69**) and 9-hydroxyellipticines in mammals, the former being inactive[45].

Table 18.1 Activity of some pyrido[4,3-*b*]carbazoles

| Substitution pattern | | | | | | | | Activity[a] | |
1	3	5	6	7	8	9	11	*in vitro*	*in vivo*
			H					−[b]	−
	Me		H					−	?
			H				Me		−
Me		Me	H						+ +
			Bu	H			Me	+	
		Me	H				Me	+ +	+ +
		Me	Me				Me	+ +	?
		Me	Et				Me		?
		Me	Me			OMe	Me		+
		Me	Ac			OMe	Me	+	
		Me	H			Me	Me		−
		Me	H			Ph	Me		−
		Me	H			F	Me		−
		Me	H			Cl	Me		−
		Me	H			Br	Me		−
		Me	H			NH$_2$	Me		−
		Me	H			OAc	Me		+ +(a, c)
		Me	H			OBn	Me		−(a, c)
		Me	H			OMe	Me		+ +
		Me	H			OH	Me	+ +	+ +
Me		Me	H			OH			+(b, d)
		Me	H	OH			Me	−	
		Me	H	F			Me	−	
		Me	H	Me			Me		
		Me	H	(CH$_2$)$_3$NEt$_2$			Me	+	
		Me	H		OH				+(b)

[a] Data from the author's own files, and from M. Stuffness and G. A. Cordell, Antitumour Alkaloids Chap. 1, pp. 3–345 in *The Alkaloids*. Vol. XXV, ed. A Brossi, Academic Press, London (1985) (with permission) and from Refs. cited therein.
[b] Symbols: + +, highly active; +, active; −, inactive. Unless specified *in vitro* and *in vivo* test systems used L 1210 leukaemia; other systems were (a) sarcoma 180; (b) P 388 leukaemia; (c) Ehrlich ascites (d) B16 melanoma.

A number of ellipticines substituted at C-1 by aminoalkylamino groups have been synthesised and their cytotoxicity assessed[46]. The best compound in the series is the methoxylated derivative ((70) R = Me). The hydroxy analogue ((70) R = H) is less useful and it is possible that the methoxy compound is acting at a different biological target. This also appears to be the case for the azaellipticine (71)[47] which is a candidate for clinical evaluation. One important fact is that whereas all ellipticines which show activity *in vivo* exhibit DNA binding, DNA binding alone, as noted for 9-aminoellipticine (72)[48], does not guarantee anticancer activity. Similarly, the diquaternised salt (73) which is inactive *in vivo*, also shows a strong propensity to bind to DNA[49].

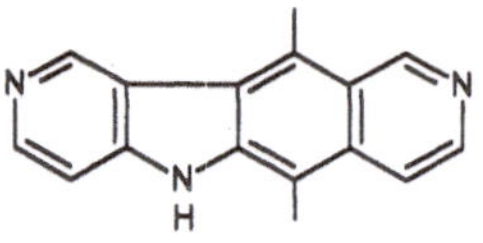

(71) 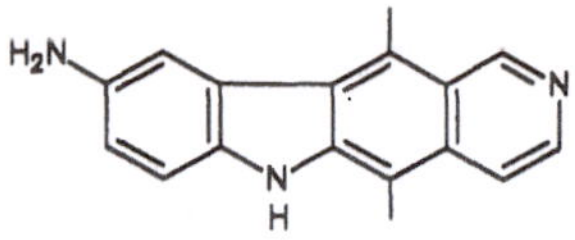   (72)

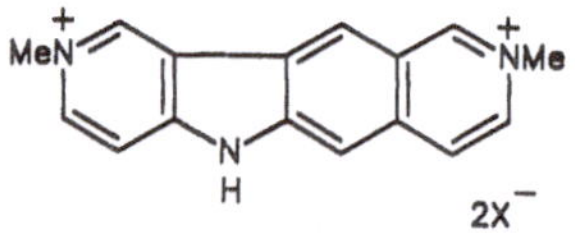

(73) 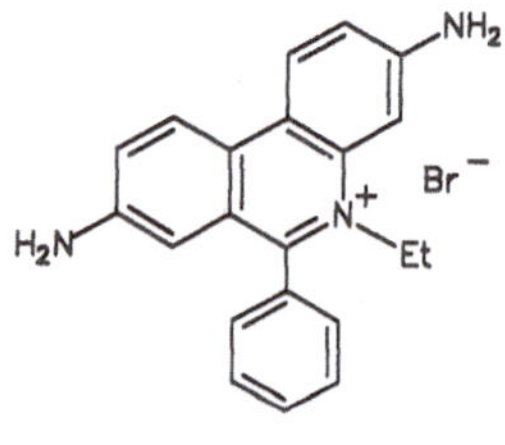   (74)

Ethidium bromide (74) is known to bind strongly to DNA[50] and this has led to the synthesis and testing of a series of N-2 alkylated ellipticinium salts in the expectation of useful anticancer activity and also increased water solubility[41,51] (see Table 18.2). Of these the N-methyl and N-ethyl salts ((75) R = Me, or Et) bind well to the nucleic acid and also show anticancer activity comparable to the free base. In general N-alkylation at the indolic nitrogen atom usually leads to inactive or weakly active compounds[52] and it is possible that those compounds of this type which are active are N-dealkylated *in vivo*. No biological information is available for the glucosides (76), but N-acetylellipticine (77) is a highly active molecule[53].

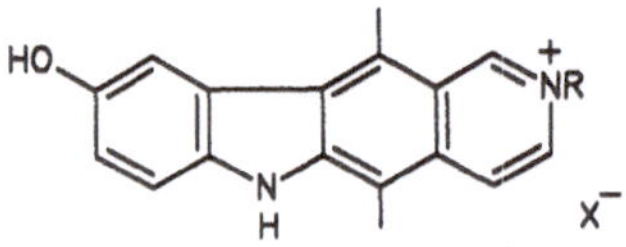

(75)

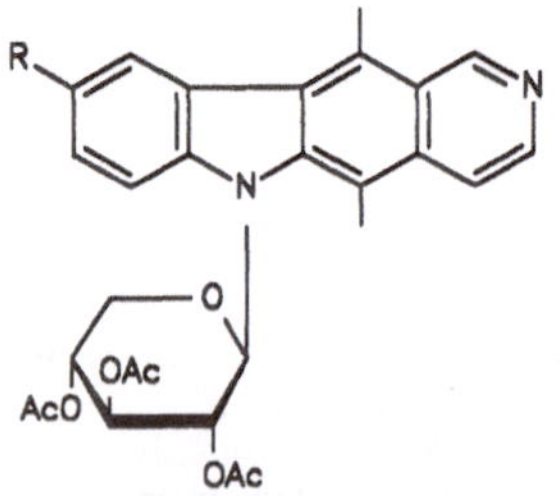

(76, R = H or OH)

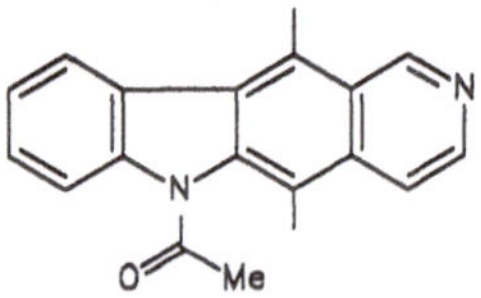

(77)

Table 18.2 Activity of some pyrido[4,3-b]carbazolium salts

Substitution pattern					Activity[a]	
1	2	6	9	11	*In vitro*	*In vivo*
	Me	H		Me	+[b]	+
	Me	H	OMe	Me	+ +	+ +
	Me	H	OH	Me	+ +	+ +
	Et	H	OH	Me	+ +	+ +
	Me	Me	OH	Me	+ +	+ +
	Et	H	OAc	Me		+
	Me	H	OMe	Me		+
Me	Me	H	OMe	Me		−
Me	Me	H	OH	Me		+

[a] Data from the M. Stuffness and G. A. Cordell, Antitumour alkaloids, Chap. 1, in *The Alkaloids*, Vol. XXV, ed. A Brossi, Academic Press, London (1985) (with permission) and from Refs. cited therein.
[b] Symbols: + +, highly active; +, active; −, inactive. *In vitro* test system P 388, *in vivo* system murine L 1210 leukaemia.

X-ray crystallographic analyses of several ellipticine derivatives have been carried out[54] in order to obtain data for detailed computer simulation studies of the docking of the bases with DNA. A striking result of this work is the 'snug fit' of the tetracyclic system of certain ellipticines, such as 5-*n*-butyl-11-methyl-6H-pyrido[4,3-b]carbazole, within the base pair cavity of the nucleic acid[55], however, substitution at any of the 2, 5, 6, 8, 9, 10, or 11-positions would not be expected to produce adverse non-bonded interactions with the sugar phosphate backbone *once the ellipticine is in place*. Sidechains linked to positions 5-, 6- and 8-, appear to cause the least degree of disturbance.

Bis-intercalation of two ellipticine units into the same strand of DNA affords the chance of greater cytotoxic effect and a benefit of the above calculations is the identification of sites in the ellipticine monomer where linking chains might be anchored. Interestingly, however, Roques *et al.*[56] note that dimeric salts of the type (**78**) are totally inactive in the L 1210 screen. Similarly dehydrodimers (**79**)[57] or (**80**)[52] based on unions through the C-11 methyl group, or through position 6, fail to exhibit significant activity. On the other hand, the 9-methoxylated dehydrodimer ((**81**) R = OMe) structured through C-5 and recently synthesised in the author's laboratory is active against L 1210 *in vitro*[28]. The desmethoxy analogue ((**81**) R = H) is less effective.

Preliminary investigations of the properties of methoxylated dimers (**82**) based on inactive 7H-pyrido[4,4-c]carbazole monomers show them to inhibit

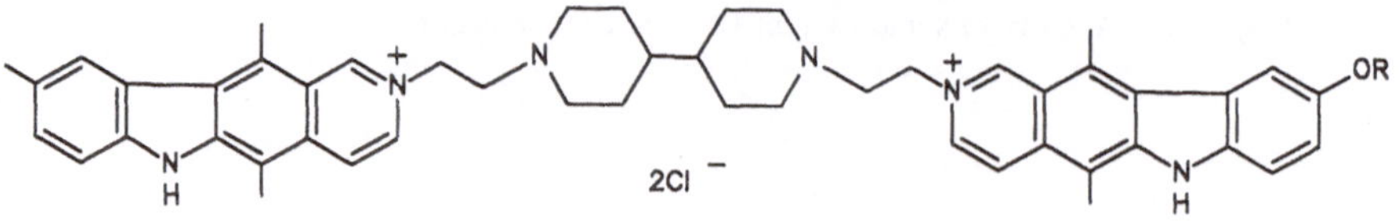

(78, R = H or Me)

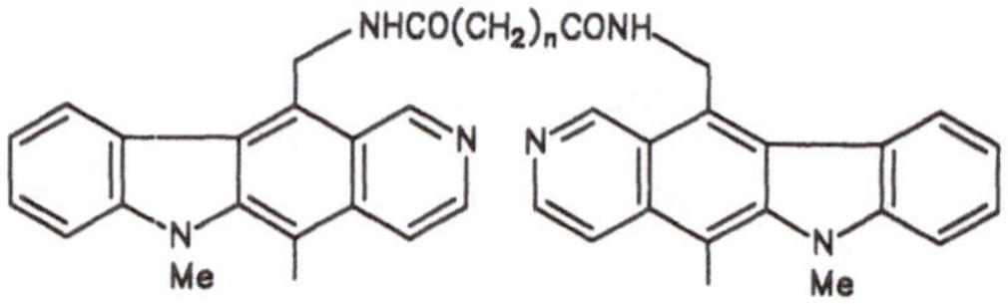

(79, n = 2 or 3)

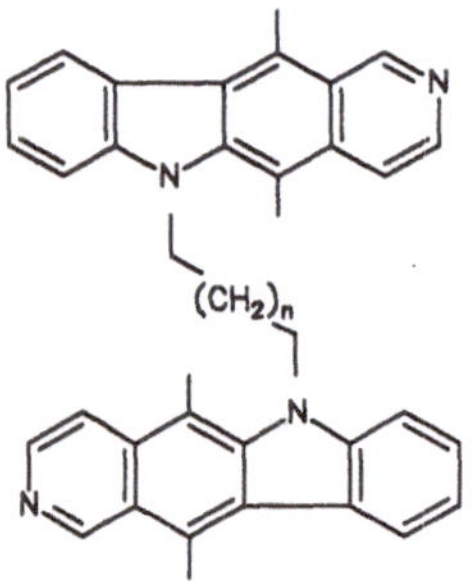

(80, n = 2 or 3)

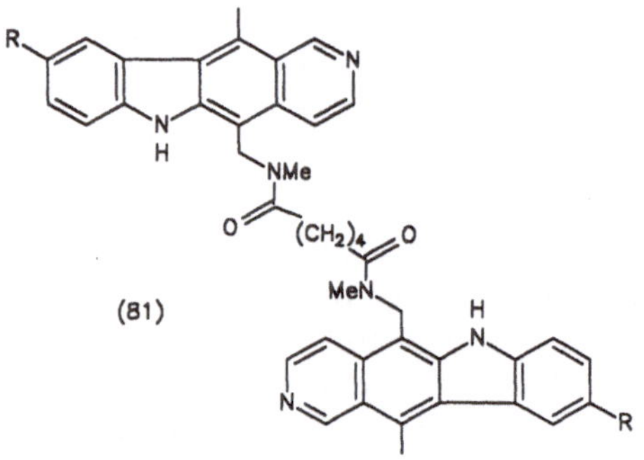

(81)

the growth of L 1210 leukaemia and Chinese hamster lung cells[58]. To account for this discrepancy it may be that the mechanisms of action of the dimers and the monomers are different, and if this is so then these dimers could represent a new class of cytotoxic agents. It is noteworthy that whereas the introduction of an alkyl group into C-6 of these compounds has no appreciable effect on their biology, insertion of a methyl group elsewhere on the tetracycle leads to a reduction of activity. As stated above in the ellipticine series, alkylation at C-6 normally causes a drop in activity.

One problem with the ellipticines as candidates for human clinical trials is their relative insolubility in aqueous media and in many organic solvents. Further elaborations of the chromophore must take this problem into consideration and it is interesting to note that the amine (83), synthesised recently at Bath[59], is a good deal more soluble, at least in organic solvents, than any of the other ellipticines handled by the author. It is also moderately active against some tumour systems *in vitro*.

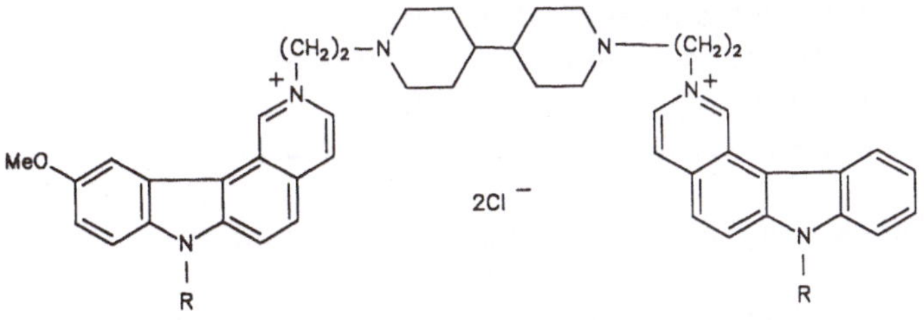

(82)

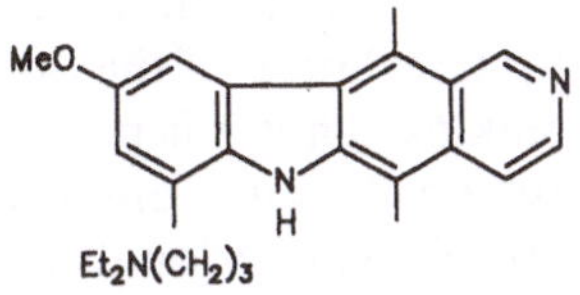

(83)

18.4   Theories of the mode of action

At present, the precise role, or roles, of the ellipticines as cytotoxic agents is uncertain because there is evidence for at least three types of biological effect[60].

18.4.1   *Intercalation and the effect on topoisomerases*

A large number of anticancer drugs are believed to act either by binding to the surface of the double helix of DNA, or by intercalating between the base pairs which link the sugar-phosphate chains. Ellipticine is a planar molecule well suited to the latter type of interaction and its potential to combine with DNA through intercalation has been known for some years. It is possible, for example, to measure the ^{1}H-NMR spectrum of a DNA-ellipticine complex[61], and to co-crystallise the molecule with 5-iodocytidylyl(3',5') guanosine and to determine the X-ray structure of the product[62]. The effect of intercalation is to destroy the kinoplastic and base pairing activity of DNA and this could be the cause of the drug's cytotoxicity. Additionally ellipticines act to modify the activity of topoisomerase II[63], an enzyme which has the ability to break and reseal both strands of doubly stranded DNA. Single-strand breaks controlled by the action of the sister enzyme topoisomerase I are also noted when mammalian cells are exposed to ellipticine and 9-hydroxy-2-ellipticinium salts[64], but only at low concentrations. For the salts there is a linear increase in the number of breaks in the DNA up until a drug concentration of 35 μM, when the effect levels off. In the case of ellipticine, invalid results are obtained above a concentration of 20 μM because at this point the breaks in the nucleic acid are no longer protein-associated.

Several established anticancer drugs are also known to be inhibitors of topoisomerase II, and to promote the formation of protein associated strand breaks. Most of these drugs are able to intercalate into DNA, and it has been proposed that the intercalated species imposes local pertubations which lead to weakness and/or vulnerability. These defects are exploited by the nuclease which, after cleaving the nucleic acid, remains bound to one terminus of the break site[65].

Liu and his co-workers using highly purified topoisomerase from calf thymus have shown[66] that site specific DNA scission occurs in the presence of 9-hydroxy-2-methylellipticinium acetate, and that the enzyme remains bonded to the 5'-end of the cleavage point via a phosphotyrosine residue. However, it was noted that intercalation of the drug may not be an essential for protein linked strand breaking. Thus it is probable that intercalation may enable the drug to react specifically with topoisomerase II and to interfere with the resealing action of the enzyme through stabilisation of the cleavable complex. Within the cell a vast amount of DNA is packed into the confines of the nucleus, where it is associated with histones and acidic proteins in the form of a complex termed chromatin. 9-Hydroxy-2-methylellipticinium acetate has been shown to unwind chromatin leading to greater access of the linker DNA to nucleases and to the unravelling of the core DNA from the histone assembly. Furthermore the drug binds preferentially to sections of the chromatin which correspond to D-nase-1 sensitised regions of the genome[67].

Unfortunately the relationship between these processes and cell death is difficult to substantiate. For example, ellipticine has been observed[68] to generate a greater frequency of DNA protein associated breaks than adriamycin, even though ellipticine is a less potent compound in cytotoxicity tests. On the other hand, this may be explained simply if the ellipticine breaks are more easily repaired than those caused by adriamycin. Indeed it could be argued that in many of the early studies direct comparisons between the activities of drugs of different structural types are unfair in view of the possibility of competition between factors other than DNA cleavage.

When drugs of a similar type are studied a clear relationship is unmasked, and for instance ellipticine, 9-hydroxyellipticine, 9-aminoellipticine, and 9-hydroxy-2-methylellipticinium acetate cause protein associated DNA breaks at relative concentrations that correspond to their order of cytotoxicity[69].

18.4.2 *Alkylation of macro-biomolecules*

It is known that if 9-hydroxy-2-methylellipticinium acetate is treated either with horse radish peroxidase, or with human myeloperoxide in the presence of hydrogen peroxide[70], a phenoxy radical is generated which decays by a dismutation process to yield the iminoquinone (**84**). This species acts to trap potential nucleophiles such as pyridine, cysteine and glutathione added to the medium regiospecifically at the C-10 position (Scheme 18.14)[71].

The corresponding adducts (**85**), ((**86**), R = H) and (**87**) can be isolated, and it is interesting that the urinary metabolites of the drug in mammals are the cysteinyl adduct, its *N*-acetylcysteinyl derivative ((**86**), R = Ac), the *O*-glucuronide (**88**) and the *S*-glutathionide (**88**)[72]. The formation of both C- and O-bonded compounds indicates that two competing metabolic pathways are operating in parallel.

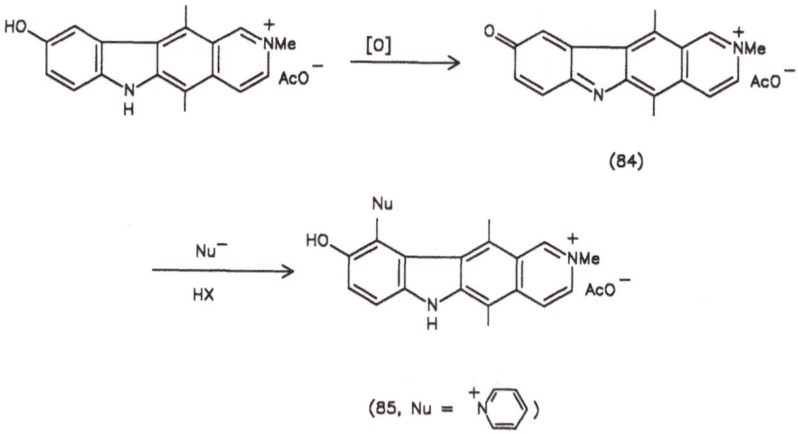

(84)

(85, Nu = $\overset{+}{N}$)

[86, Nu = $-SCH_2(CO_2H)NHR$]

[87, Nu = $-SCH_2CH(CONHCH_2CO_2H)NHCOCH_2CH_2CH(CO_2H)NH_2$]

Scheme 18.14 Oxidation of 9-hydroxy-2-methylellipticium salts and the reactions of nucleophiles with the product iminoquinone (**84**).

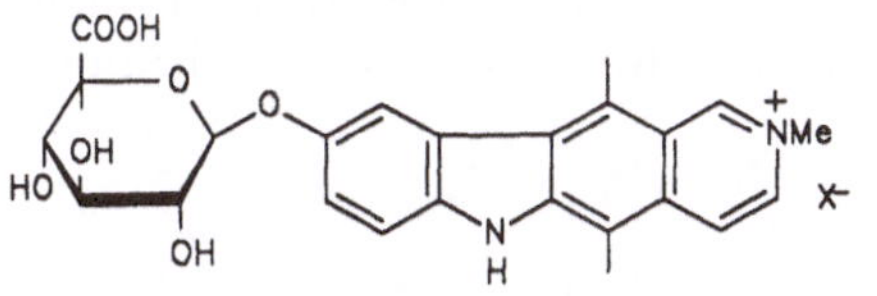

(88)

None of the adducts described above exhibits greater cytotoxicity than the parent salt and it has been argued that the iminoquinone (**84**), generated within the cell, is the activated form of the drug and this species behaves as an electrophile reacting with important macromolecules in the manner proposed for other drugs which are activated as quinonemethides. Further studies[73-76] have in fact established that under oxidative conditions the quaternised iminoquinone reacts with ribonucleosides and ribonucleotides to produce adducts of the type (**89**) and (**90**), respectively, where B, B$_1$ and B$_2$ represent various bases found in nucleic acids.

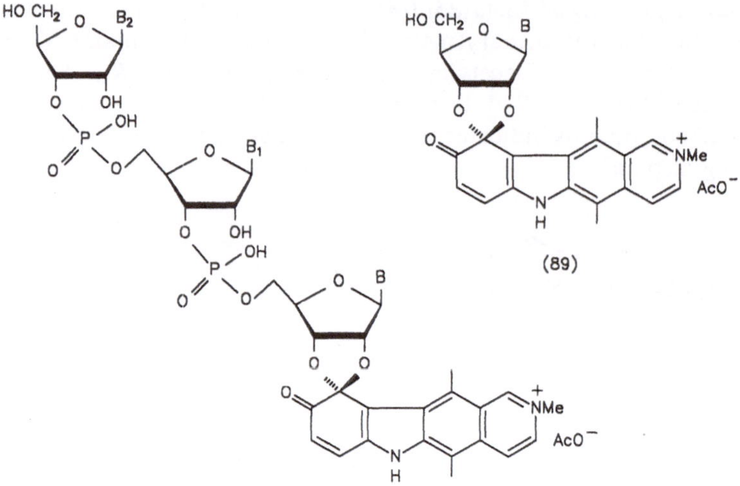

(89)

(90)

In view of these results it has been suggested that the hydroxylated ellipticines may exert some of their antitumour activity through alkylation of either t- or m-RNA, leading to an inhibition of protein synthesis. However, since Potier's group have shown[77] that adducts of the type (91) are formed between the iminoquinone from 9-hydroxyellipticine and alkylamines (RCH_2NH_2) under oxidative conditions, or amino acids $[RCH(NH_2)CO_2H]$, it is also possible that the oxidised forms of the 9-hydroxyellipticines might act to trap amines, amino acids, peptides, or proteins *in vivo*, and thus inhibit protein biosynthesis[78]. Support for this suggestion comes from the demonstration that bovine serum albumin and other proteins[79] bind irreversibly to the oxidised form of 9-hydroxy-2-methylellipticinium acetate, but against

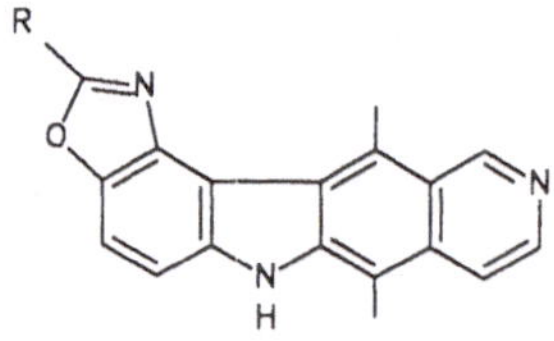

(91)

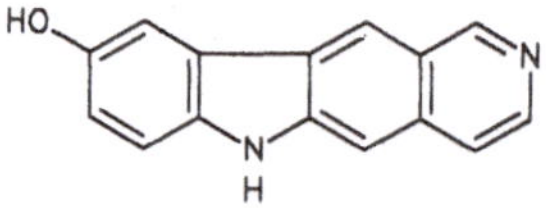

(92)

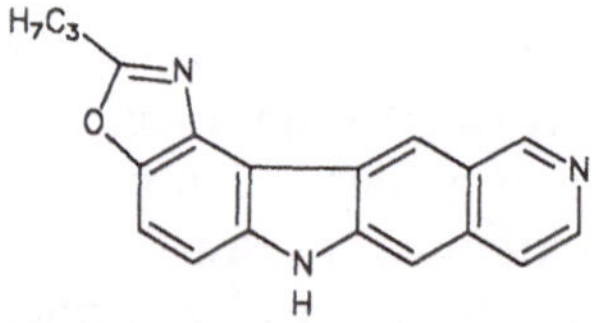

(93)

this is the discovery[80] that if 9-hydroxy-2-methylellipticinium acetate is incubated with L 1210 murine leukaemia cells, covalent binding occurs to both DNA and RNA, but not to proteins.

Archer[32] suggests that iminoquinones are not responsible for the biological effect of the ellipticines since 9-hydroxy-5,11-desmethylellipticine (92) is inactive against tumour systems, but may still be oxidised and coupled with *n*-butylamine to form a Potier type adduct (93). For anticancer action Archer argues that a 5-methyl group is necessary, which *in vivo* is oxidised and phosphorylated to furnish a leaving group at this position. The product (94) could then behave as an alkylating agent for marcomolecules such as nucleic acids or nucleases (Scheme 18.15).

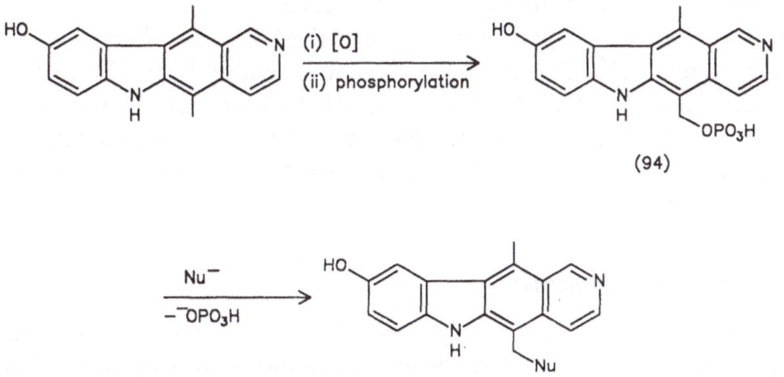

(94)

Scheme 18.15 Nu = DNA, topoisomerase II, Archer's suggested mode of action of 9-hydroxyellipticine.

18.4.3 Radical and superoxide formation

Another possible mode of action has been proposed[81] wherein the iminoquinone (84), or its equivalent, undergoes reduction to a semiquinone radical. This species might then be re-oxidised by molecular oxygen with the concomitant formation of superoxide ions. Superoxide ions are known to combine with water and produce highly reactive hydroxide radicals which could damage DNA or initiate cell destruction through lipid peroxidation[82].

18.5 Clinical applications

Human clinical evaluation of ellipticine was considered some years ago, but there are problems of toxicology, thus the alkaloid exerts a rapid haemolytic effect[83,84] and also decreases the heart rate in mammals. At toxic doses in monkeys, irreversible respiratory depression occurs, however, haemolysis in these animals is inhibited by the use of a citrate buffer[85]. It is now known[86] that ellipticine strongly interacts with cell membranes through binding to acidic phospholipids, while only minor amounts are associated with the haemoglobin. A phase I evaluation of 9-methoxyellipticine in 1970 was abandoned because of toxicological problems[87], even though some remissions in patients suffering from acute myeloblastic leukaemia were reported[88]. A major problem with the ellipticines is their insolubility in aqueous media and a study of 9-hydroxyellipticine in France was discontinued[89] because insufficient of the drug appeared to reach the desired site of action. The result of this investigation led to the use of the quaternary salt 9-hydroxy-2-methylellipticinium acetate which proved to be more soluble and is effective against thyroid and renal tumours, and bone metastases associated with advanced breast cancer[90,91]. A number of toxicity problems have emerged from the use of this drug[92,93] and, although it has been marketed by the SANOFI company, so far, there is little published work relating to the evaluation of the compound outside of France.

Recently there has been renewed interest the USA in ellipticine solubilised as its acetate salt[94]. Results show that in a phase II trial[95] two partial remissions were observed from a group of thirteen women with advanced breast cancer. Rather better results were obtained in a study with patients with metastatic renal cancer[96] and ten responses from a group of twenty-two adults suffering from the disease are claimed.

18.6   Subjective comments

The ellipticines and their allies have a long history, much is known about their biology, and their syntheses have been worked out. Various possible modes of action are proposed yet, so far, despite some limited clinical success with 9-hydroxy-2-methylellipticinium acetate, their potential as anticancer drugs has not been fulfilled. However, the remarks of the author in 1977[97] that the compounds tested tend to reflect ease of access rather than a determined effort to optimise activity seems, with one or two notable exceptions, largely true today.

There is no evidence that the ellipticines currently available show selectivity of action, but this could change as biological targets are recognised and become more clearly defined, and particularly as our understanding of the role of the topoisomerases in regulating the behaviour of nucleic acids develops. Armed with this knowledge and backed by the mass of synthetic methodology which has already accumulated, tailor made drugs could be designed and constructed. After all since the ellipticines represent some of the simplest anticancer agents in terms of structure and pose no problems of stereochemical control during manufacture, they should offer a good investment in terms of the development of a clinically useful drug.

Some ellipticinium dimers linked through N-2-peptide units have recently been shown to intercalate into DNA[98]. A series of very active 2-glycosides of 9-hydroxyellipticine have been prepared[99], and some glycosides of 5,6-dimethylpyrido[4,3-b]carbazoles linked through a hydroxyethyl group at C-

11 have been described[100]. These last compounds are quite soluble in aqueous media, a very necessary feature for administration to patients. The ellipticine story is a long one but worth the reader's attention, and it still might become a best seller!

References

1. S. Goodwin, A. F. Smith and E. C. Horning, *J. Am. Chem. Soc.* **81** (1959) 1903.
2. L. K. Dalton, S. Demerac, B. C. Elmes, J. W. Loder, J. M. Swan and T. Teitei, *Austr. J. Chem.* **20** (1967) 2715.
3. G. H. Svoboda, G. A. Poore and M. L. Montfort, *J. Pharm. Sci.* **57** (1968) 1720.
4. S. Michel, F. Tillequin, M. Koch and L. Ake Assi, *J. Natl. Prod.* **43** (1980) 294; S. Michell, F. Tillequin and F. Koch, *Tetrahedron Lett.* **21** (1980) 4027.
5. J. Bruneton and A. Cavé, *Phytochemistry* **11** (1972) 846; A. Ahond, H. Fernandez, M. J.-Moore, C. Poupat, V. Sanchez, P. Potier, S. K. Kan and T. Sevenet, *J. Natl. Prod.* **44** (1981) 193.
6. J. Bruneton and A. Cavé, *Ann. Pharm. Fr.* **30** (1972) 629; G. Büchi, B. P. Mayo and F. A. Hochstein, *Tetrahedron* **15** (1961) 167; H. Lehner and J. Schmutz, *Helv. Chim. Acta* **44** (1961) 444.
7. J. Schmutz and F. Hunzicker, *Helv. Chim. Acta* **41** (1958) 288; R. H. Burnel and D. D. Casa, *Can. J. Chem.* **45** (1967) 89.
8. S. Michel, F. Tillequin and M. Koch, *J. Chem. Soc., Chem. Comun.* (1987) 229.
9. H. G. Boit, *Ergebnisse der Alkaloid Chemie bis 1960*, Akademie-Verlag, Berlin (1961).
10. M. Sainsbury and B. Webb, *Phytochemistry* **11** (1972) 2337.
11. M. Rueffer, N. Nagakura and M. H. Zenk, *Tetrahedron Lett.* (1978) 1593; A. R. Battersby, N. G. Lewis and J. M. Tippett, *Tetrahedron Lett.* (1978) 4849.
12. P. Potier and M. M. Janot, *C.R. Hebd. Seances Acad. Sci. Ser. C* **276** (1973) 1727.
13. M. J. E. Hewlins, A.-M. Oliviera Campos and P. V. R. Shannon, *Synthesis* (1948) 289.
14. G. W. Gribble and M. G. Saulnier, *Heterocycles* **23** (1985) 1277.
15. M. Stuffness and G. A. Cordell, in *The Alkaloids*, Vol. XXV, ed. A. Brossi, Academic Press, London (1985) pp. 3–345.
16. R. B. Miller and J. G. Stowell, *J. Org. Chem.* **48** (1983) 886.
17. C. May and C. J. Moody, *J. Chem. Soc., Chem. Commun.* (1984) 926.
18. M. G. Saulnier and G. W. Gribble, *J. Org. Chem.* **47** (1982) 2810.
19. M. G. Saulnier and G. W. Gribble, *Tetrahedron Lett.* **24** (1983) 3831.
20. M. Watanabe and V. Snieckus, *J. Am. Chem. Soc.* **102** (1980) 1457; V. Snieckus, *Heterocycles* **14** (1980) 1649.
21. D. A. Taylor, M. M. Baradarani, S. J. Martinez and J. A. Joule, *J. Chem. Res. Synop.* (1979) 387.
22. S. Kano, E. Sugino and S. Hibino, *Heterocycles* **19** (1982) 1673; S. Kano, N. Mochizuki, S. Hibino and S. Shibuya, *J. Org. Chem.* **47** (1982) 3566.
23. J. Bergman and R. Carlsson, *Tetrahedron Lett.* (1978) 4051.
24. R. B. Woodward, G. A. Iacobucchi and F. A. Hochstein, *J. Am. Chem. Soc.* **81** (1959) 4434.
25. M. Sainsbury and R. F. Schinazi, *J. Chem. Soc., Chem. Commun.* (1975) 540.
26. M. Driver, I. T. Matthews and M. Sainsbury, *J. Chem. Soc. Perkin Trans. 1* (1979) 2506.
27. M. Sainsbury, D. K. Weerasinghe and D. Dolman, *J. Chem. Soc. Perkin Trans. 1* (1982) 587.
28. A. J. Ratcliffe, M. Sainsbury, A. Smith and D. I. C. Scopes, *J. Chem. Soc. Perkin Trans. 1* (1988) 2933.
29. M. J. Wanner, G.-J. Koomen and U. K. Pandit, *Heterocycles* **17** (1982) 59, **19** (1982) 2295.
30. M. J. Wanner, G.-J. Koomen and U. K. Pandit, *Tetrahedron* **39** (1983) 3673.
31. D. D. Weller and D. W. Ford, *Tetrahedron Lett.* **25** (1984) 2105.
32. S. Archer and B. S. Ross, *Tetrahedron Lett.* **27** (1986) 5343.
33. P. A. Cranwell and J. E. Saxton, *J. Chem. Soc.* (1962) 3482.
34. J. Gilbert, D. Rousselle, C. Gasser and C. Viel, *J. Heterocycl. Chem.* **16** (1979) 7.
35. A. J. Birch, A. H. Jackson and P. V. R. Shannon, *J. Chem. Soc. Perkin Trans. 1* (1974) 2185; A. H. Jackson, P. R. Jenkins and P. V. R. Shannon, *J. Chem. Soc. Perkin Trans. 1* (1977) 1698.

36. I. Hogan, P. Jenkins and M. Sainsbury, *Tetrahedron Lett.* **29** (1988) 6505.
37. L. K. Dalton, S. Demerac and T. Teitei, *Austr. J. Chem.* **22** (1969) 185; J.-B. LePecq, N. Dat-Xuong, C. Gosse and C. Paoletti, *Proc. Natl. Acad. Sci. USA* **71** (1974) 5078; C. Paoletti, S. Cros, N. Dat-Xuong, P. Leconite and A. Moisand, *Chem.-Biol. Interact.* **25** (1979) 45.
38. M. Hayat, G. Maté, M. M. Janot, P. Potier, N. Dat-Xuong, A. Cavé, T. Sevenet, C. Kan-Fan, J. Poisson, J. Miet, J. LeMen, F. LeGoffic, A. Gouyette, A. Ahond, L. K. Dalton and T. A. Connors, *Biomedicine* **21** (1974) 101.
39. A. N. Fugiwara, E. M. Acton and L. Goodman, *J. Heterocycl. Chem.* **5** (1968) 853; **6** (1969) 379.
40. V. M. Dixit, J. M. Khanna and N. Anand, *Indian J. Chem.* **16B** (1978) 124.
41. C. Paoletti, J.-B. LePecq, N. Dat-Xuong, P. Juret, H. Garnier, J.-L. Amiel and J. Rousse, *Recent Results Cancer Res.* **74** (1980) 107.
42. R. W. Guthrie, A. Brossi, F. A. Mennona, J. G. Mullin, R. W. Kierstad and E. Grunberg, *J. Med. Chem.* **18** (1975) 755.
43. B. Dugué, B. Meunier and C. Paoletti, *Eur. J. Med. Chem.-Chim. Ther.* **19** (1983) 551.
44. P. Lecointe and A. Puget, *C.R. Hebd. Seances Acad. Sci. Ser. D* **296** (1983) 279.
45. J. Y. Lallemand, P. Lemaitre, L. Beeley, P. Lesca and D. Mansuy, *Tetrahedron Lett.* (1978) 1261.
46. E. Bisangi, C. Ducrocq, J.-M. Lhoste, C. Rivalle and A. Civier, *J. Chem. Soc. Perkin Trans. 1* (1979) 1706.
47. M. Marty, C. Jasmin, P. Pouillart, C. Gisselbrecht, G. Gouveia and H. Magdalainat, *Proc. Am. Soc. Clin. Oncol.* **22** (1981) 360.
48. R. Lideraeau, J. C. Chermann, J. Gruest, L. Montagier, C. Ducrocq, C. Rivalle and E. Bisagni, *Bull. Cancer* **67** (1980) 1.
49. M. T.-Perrin, F. Pochon, C. Ducrocq, C. Rivalle and E. Bisagni, *Bull. Cancer* **67** (1980) 9.
50. G. Pak and G. Loew, *Biochim. Biophys. Acta* **519** (1978) 163; N E. Nuss, F. J. Marsh and P. A. Kollman, *J. Am. Chem. Soc.* **101** (1979) 825.
51. C. Paoletti, J.-B. LePecq, N. Dat-Xuong, P. Lesca and P. Lecointe, *Curr. Chemother. Proc. Int. Congr. Chemother. 10th.* **2** (1978) 1195; C. Paoletti, S. Cros, N. Dat-Xuong, P. Lecointe and A. Moisand, *Chem.-Biol. Interact.* **25** (1979) 45.
52. J.-B. LePecq, C. Gosse, N. Dat-Xuong and C. Paoletti, *C.R. Hebd. Seances Acad. Sci. Ser. D* **277** (1973) 2289.
53. M. Bessodes, N. Dat-Xuong and K. Antonakis, *C.R. Hebd. Seances Acad. Sci. Ser. C.* **282** (1976) 1001; *J. Carbohydr. Nucleosides Nucleotides* **4** (1977) 215.
54. A. Aggrawal, S. Neidle and M. Sainsbury, *Acta Crystallogr.* **C39** (1983) 631; R. Kuroda and M. Sainsbury, *J. Chem. Soc. Perkin Trans. 2* (1984) 1751.
55. P. D.-Osguthorpe and M. Sainsbury, Unpublished.
56. B. P. Roques, D. Pelaprat, I. LeGuen, G. Porcher, C. Gosse and J.-B. LePecq. *Biochem. Pharmacol.* **28** (1979) 1811; D. Pelaprat, A. Delbarre, I. LeGuen and B. P. Roques, *J. Med. Chem.* **23** (1980) 1336.
57. M. J. Wanner, G.-J. Koomen and U. K. Pandit, *Tetrahedron* **39** (1983) 3673.
58. D. Pelaprat, R. Oberlin, I. LeGuen and and B. P. Roques, *J. Med. Chem.* **23** (1980) 1330; J. P. Bendirdjian, C. Delaporte, B. P. Roques and A. J. Sablon, *Biochem. Pharmacol.* **33** (1984) 3681.
59. M. Sainsbury, A. Smith, A. Vong and D. I. C. Scopes, *J. Chem. Soc. Perkin Trans. 1* (1988) 2945.
60. C. Auclair, *Arch. Biochem. Biophys.* **259** (1987).
61. J. Feigon, W. A. Denny, W. Leupin and D. R. Kearns, *J. Med. Chem.* **27** (1984) 450.
62. S. C. Jain, K. K. Bhandary and K. K. Sobell, *J. Mol. Biol.* **135** (1979) 813.
63. W. E. Ross, D. L. Daubiger and K. W. Kohn, *Biochim. Biophys. Acta* **519** (1978) 23.
64. L. A. Zwelling, S. Milchaels, D. Kerrigan, Y. Prommier and K. W. Kohn, *Biochem. Pharmacol.* **31** (1982) 3261.
65. E. M. Nelson, K. M. T. Tewey and L. F. Liu, *Proc. Natl. Acad. USA* **81** (1984) 1361; K. M. Tewey, G. L. Chen, E. M. Nelson and L. F. Liu, *J. Biol. Chem.* **259** (1984) 9182; K. M. Tewey, T. C. Rowe, L. Yang, B. D. Halligan and L. F. Liu, *Science* **226** (1984) 466.
66. L. F. Liu, *CRC Crit. Rev. Biochem.* **15** (1984) 1.
67. L. Larue, M. Quesne and J. Paoletti, *Biochem. Pharmacol.* **36** (1987) 3563.
68. W. E. Ross, D. L. Glaubiger and K. W. Kohn, *Biochim. Biophys. Acta* **562** (1978) 41.
69. C. Paoletti, C. L. Lesca, S. Cros, C. Malvy and C. Auclair, *Biochem. Pharmacol.* **28** (1979) 345.

70. C. Auclair and C. Paoletti, *J. Med. Chem.* **24** (1981) 289.
71. G. Meunier, B. Meunier, C. Auclair, J. Bernadou and C. Paoletti, *Tetrahedron Lett.* **24** (1983) 365.
72. M. Maftouh, B. Monsarrat, R. C. Rao, B. Meunier and C. Paoletti, *Drug. Dispos.* **12** (1984) 111; M. Maftouh, B. Monsarrat, G. Meunier, B. Dugué, J. Bernadou, J. P. Armand, C. P.-Fraire, B. Meunier and C. Paoletti, *Biochem. Pharmacol.* **32** (1983) 3887; B. Monsarrat, M. Maftouh, G. Meunier, J. Bernadou, J. P. Arrmand, C. Paoletti and B. Meunier, *J. Pharm. Biomed. Anal.* **5** (1987) 341; G. Meunier, D. DeMontauzon, J. Bernadou, G. Grassy, M. Bonnafous, S. Cros and B. Meunier, *Mol. Pharmacol.* **33** (1988) 93.
73. J. Bernadou, B. Menuier, G. Menuier, C. Auclair and C. Paoletti, *Proc. Natl. Acad. Sci. USA* **81** (1984) 1297.
74. V. K. Kansal, P. Potier, B. Gillet, E. Guittet, J. Y. Lallemand, T. H.-Dinh and J. Igolen, *Tetrahedron Lett.* **26** (1985) 2891.
75. V. K. Kansal, S. Funakoshi, P. Mangeney, B. Gillet, E. Guittet, J. Y. Lallemand and P. Potier, *Tetrahedron* **41** (1985) 5107.
76. G. Pratviel, J. Bernadou, T. B. T. Ha, G. Meunier, S. Cros, B. Meunier, B. Gillet and E. Guittet, *J. Med. Chem.* **29** (1986) 1350; T. B. T. Ha. J, Bernadou and B. Meunier, *Nucleosides Nucleotides* **6** (1987) 691.
77. V. K. Kansal, R. Sundaramoorthi, B. C. Das and P. Potier, *J. Chem. Soc., Chem. Commun.* (1986) 371.
78. C. Auclair, B. Meunier and C. Paoletti, *Biochem. Pharmacol.* **32** (1983) 3883.
79. B. Dugué, C. Auclair and B. Meunier, *Cancer Res.* **46** (1986) 3828.
80. M. Roy, N. Fernandez and P. Lesca, *Eur. J. Biochem.* **172** (1988) 593.
81. C. Paoletti, C. Auclair, J.-F. Tocanne, C. Malvy and M. Pinto, *Cancer Treat. Rep.* **65** (1981) 107; J. Bernadou, B. Meunier, C. P Paoletti and G. Meunier, *J. Med. Chem.* **26** (1983) 574.
82. S. D. Nelson, *J. Med. Chem.* **25** (1982) 753.
83. E. H. Herman, D. P. Chadwick and R. M. Mhartre, *Cancer Chemother Rep.* **58** (1974) 637.
84. E. H. Herman, J. Vick and B. Burka, *Toxicol. Appl. Pharmacol.* **18** (1971) 743.
85. E. H. Herman, I. P. Lee, R. M. Mhatre and D. P. Chadwick, *Cancer Chemother. Rep.* **58** (1974) 171.
86. E.-S. M. ElMashak, C. Paoletti and J.-F. Tocanne, *FEBS Lett.* **107** (1979) 155; E.-S. M. ElMashak and J.-F. Tocanne, *Eur. J. Biochem.* **105** (1980) 593.
87. G. Mathé, M. Hayat, F. DeVassal, L. Schwartzenburg, M. Schneider, J. R. Schlumberger, C. Jasmin and C. Rosenfeld, *Rev. Eur. Etud. Clin. Biol.* **15** (1970) 541.
88. B. M. Ansari and E. N. Thompson, *Postgrad. Med.* **51** (1975) 103.
89. C. Paoletti, J.-B. LePecq, N. Dat-Xuong, P. Juret, H. Garnier, J.-L. Amiel and J. Rouesse, *Recent Res. Cancer Res.* **74** (1980) 107.
90. J. Rouesse, T. Tursz, T. LeChavalier, D. Huertas, J.-L. Amiel, D. Brule, B. Callet, J. P. Droz, P. M. Voisin, H. S.-Garnier, J.-B. LePecq and C. Paoletti, *Nouv. Presse Med.* **10** (1981) 1997.
91. P. Juret, J. F. Heron, J. E. Couette, T. Delozier and J. P, LeTalaer, *Cancer Treat. Rep.* **66** (1982) 1909.
92. A. L. Harris and D. Spence, *Ann. Intern. Med.* **95** (1981) 125.
93. A. Buzdar, H. Y. Yap, A Goodman, C. Chou, G. Hortobagyi, G. Blumenschein, R. Benjamin and G. Bodey, *Proc. Am. Assoc. Cancer Res.* **25** (1984) 182.
94. A. Clarysse, A. Bruarolas, P. Siegenthaler, R. Abeler, F. Cavalli, R. DeJager, G. Renard, M. Rozencweig and H. H. Hansen, *Eur. J. Clin. Oncol.* **20** (1984) 243.
95. P. Byrne, K. Grady, F. Smith, R. Goldberg, A. Goodman and P. S. Schein, *Proc. Am. Soc. Clin. Oncol.* **3** (1984) 127.
96. J.-L. Amiel, J. Rouesse, J. P. Droz, P. Caille, J. P. Travagi, C. Theodore, T. LeChavalier, J. P. Ducret, M. Bidart and H. S. Garnier, *Nouv. Presse Med.* **13** (1984) 1555.
97. M. Sainsbury, *Synthesis* (1977) 437.
98. P. Rigaudy, C. G.-Jaureguiberry, A. J.-Sablon, J.-B. LePecq and B. P. Roques, *Int. J. Peptide Protein Res.* **30** (1987) 347.
99. T. Honda, M. Inoue, M. Kato, K. Shima and T. Shimamoto, *Chem. Pharm. Bull.* **35** (1987) 3975.
100. A. Langendoen, G.-Koomen and U. K. Pandit, *Tetrahedron* **44** (1988) 3627.

Index